AF355997

Novel Antiviral Agents: Synthesis, Molecular Modelling Studies and Biological Investigation

Novel Antiviral Agents: Synthesis, Molecular Modelling Studies and Biological Investigation

Editor

Simone Brogi

Basel • Beijing • Wuhan • Barcelona • Belgrade • Novi Sad • Cluj • Manchester

Editor
Simone Brogi
Department of Pharmacy
University of Pisa
Pisa
Italy

Editorial Office
MDPI
St. Alban-Anlage 66
4052 Basel, Switzerland

This is a reprint of articles from the Special Issue published online in the open access journal *Viruses* (ISSN 1999-4915) (available at: www.mdpi.com/journal/viruses/special_issues/novelantiviral_agents).

For citation purposes, cite each article independently as indicated on the article page online and as indicated below:

Lastname, A.A.; Lastname, B.B. Article Title. *Journal Name* **Year**, *Volume Number*, Page Range.

ISBN 978-3-7258-0050-6 (Hbk)
ISBN 978-3-7258-0049-0 (PDF)
doi.org/10.3390/books978-3-7258-0049-0

Contents

Editorial

Novel Antiviral Agents: Synthesis, Molecular Modelling Studies and Biological Investigation

Simone Brogi [1,2]

1 Department of Pharmacy, University of Pisa, Via Bonanno 6, 56126 Pisa, Italy; simone.brogi@unipi.it
2 Bioinformatics Research Center, School of Pharmacy and Pharmaceutical Sciences, Isfahan University of Medical Sciences, Isfahan 81746-73461, Iran

Citation: Brogi, S. Novel Antiviral Agents: Synthesis, Molecular Modelling Studies and Biological Investigation. *Viruses* **2023**, *15*, 2042. https://doi.org/10.3390/v15102042

Received: 16 September 2023
Accepted: 24 September 2023
Published: 2 October 2023

Representing more than 20% of all deaths occurring worldwide, infectious diseases remain one of the main factors in both human and animal morbidity and mortality. Approximately one-third of these deaths are attributable to viruses. Accordingly, over the past two decades, emerging and re-emerging viral agents, including the most recent SARS-CoV-2 and related coronaviruses (SARS and MERS), enteroviruses, Zika virus (ZIKV), and avian influenza A (H5N1, H1N1, and H7N9) viruses, have posed serious dangers to public health. Therefore, it is critical to identify novel antiviral drugs, vaccines, therapeutic strategies primarily based on the repurposing of marketed drugs, and early diagnostic and prevention strategies, considering the likelihood of future outbreaks. Recently, the World Health Organization (WHO) released a list of pathogens that could cause future pandemics and outbreaks. This list includes several viruses such as SARS-CoV-2 (COVID-19), Lassa fever, Ebola virus disease, Crimean-Congo hemorrhagic fever, Marburg virus disease, Middle East respiratory syndrome (MERS) and severe acute respiratory syndrome (SARS), ZIKV, Rift Valley fever, Nipah and henipaviral diseases, and Disease X. Furthermore, the WHO recommended that the scientific community invest in research to develop tests, treatments, and vaccines for the next outbreak [1].

In this case, computational techniques, such as cutting-edge machine learning methods, may speed up the development of efficient antivirals and therapeutic approaches. For this purpose, in July 2022, a Special Issue entitled *"Novel Antiviral Agents: Synthesis, Molecular Modelling Studies and Biological Investigation"* was launched, attracting the attention of researchers in the field. In this Special Issue, 15 papers on the development of antiviral agents against different types of clinically relevant viruses are published and in particular, 12 original research articles, 2 communication articles, and 1 review article are included. Accordingly, in this editorial article, we analyze and discuss the published papers, grouping them according to the research topic.

Starting from the studies conducted to identify possible antiviral agents against SARS-CoV-2 and related coronavirus species, Varbanov and coworkers synthesized and evaluated a series of antimicrobial diterenoids as possible anti-coronavirus agents. In particular, the authors synthesized a series of (+)-ferruginol derivatives using the abietane diterpene as a starting point because of its pharmacological profile, including its antimicrobial and, most importantly, antiviral activities. Compound C18-functionalized semisynthetic abietanes were synthesized from readily available methyl dehydroabietate and (+)-dehydroabietylamine. The resulting compounds were then computationally evaluated to assess the ADMET profile and select the most interesting derivatives for in vitro tests. Among the synthesized compounds, following the in silico indications, the authors selected eight compounds to evaluate their antiviral profile against human coronavirus 229E (HCoV-229E, PHE/NCPV-0310051v). The virus was established in the MRC-5 cell line, and its titers were determined as 50% infectious doses in the cell culture ($CCID_{50}$/mL). Interestingly, two novel ferruginol analogs inhibited a cytopathic effect and significantly decreased viral titers. Notably, the bioactive compounds showed an interesting ADMET profile that encourages further research to evaluate possible activities in closely related virus families because they could have a broad range of antiviral effects [2]. Possible anti-SARS-CoV-2 agents were identified by Zhang and collaborators using artificial intelligence. In particular, to improve the performance of the transfer learning model for identifying SARS-CoV-2 Mpro inhibitors, the researchers adopted a data augmentation technique. On an external test set, this approach, based on a deep learning technique, appeared to perform better than Chemprop, random forest, and graph convolutional neural network. Using the developed computer-based model, naturally and de novo produced compound libraries were screened. As screening results, by combining this method with other in silico analyses (e.g., molecular docking and PAINS evaluation), twenty-seven molecules were submitted to biological evaluation to confirm anti-Mpro activity. Gossypol acetic acid and hyperoside, two of the selected computational hit compounds, both exhibited inhibitory activity against SARS-CoV-2 Mpro (gossypol acetic acid IC_{50} = 67.6 μM; hyperoside IC_{50} = 235.8 μM). Despite having micromolar potency, these two compounds provide useful scaffolds for future drug optimization in the fight against COVID-19. Accordingly, the findings of this study may offer a useful method for identifying lead compounds to develop anti-SARS-CoV-2 and other coronavirus agents. Remarkably, artificial intelligence in drug discovery is increasingly being used, considering the significant advantages in lowering the costs and time associated with drug research and the development trajectory, particularly for emerging diseases [3]. Hassan and colleagues used a combined approach that coupled computer-based and experimental methods to identify potential TMPRSS2 inhibitors. This approach is very interesting because, considering that host proteases are necessary for coronaviruses to facilitate the entry of the virus into host cells, targeting the conserved host-based entry process, as opposed to the viral proteins that are constantly changing, may have advantages. Camostat and nafamostat have been identified as TMPRSS2 inhibitors, showing irreversible binding to the target protein. Considering the chemical structure of nafamostat, the researchers rationally designed a small series of related rigid analogs that were evaluated in silico using a molecular docking study employing the experimentally solved structure of TMPRSS2 (PDB: 7MEQ). The computer-based evaluation provided the compounds to be prioritized for synthesis. Accordingly, six of the best predicted molecules were synthesized using pentamidine as a starting point and experimentally validated in vitro. Four of them showed some bioactivities. Two compounds effectively inhibited TMPRSS2 in the micromolar range, but they showed modest activity in cell-based assays. Interestingly, although one compound did not show appreciable inhibitory activity against TMPRSS2, it exhibited possible cellular activity in a low micromolar range (IC_{50} = 10.87 μM) in inhibiting the fusion of membranes, indicating that another molecular target may be responsible for its effect. Overall, the results of this investigation provided hit compounds that might be used as a basis for the discovery of novel viral entry inhibitors with potential usage against coronaviruses [4]. Mohammad and colleagues developed a virtual screening protocol for identifying possible antiviral agents targeting SARS-CoV-2 NSP3 macrodomain-1, which plays a crucial role in viruses' attack on the innate immune system. The authors virtually screened, against the mentioned target, a library of natural compounds using different computer-based techniques such as molecular dynamic (MD) simulations (100 ns) coupled to the MM/GBSA calculation for estimating binding free energies. Among the screened compounds, the authors identified two promising hit compounds (3,5,7,4'-tetrahydroxyflavanone 3'-(4-hydroxybenzoic acid) and 2-hydroxy-3-O-beta-glucopyranosyl-benzoic acid), showing a stable binding mode within the SARS-CoV-2 NSP3 macrodomain-1 binding site, with significant binding affinity (molecule A ΔG_{bind} = −61.98 ± 0.9 kcal/mol; molecule B ΔG_{bind} = −45.125 ± 2.8 kcal/mol). To confirm the predicted activity of the selected drugs, computational bioactivity evaluation and dissociation constant (K_D) measurements were performed for both complexes. The findings highlighted that these two potential antivirals could block SARS-CoV-2 NSP3 macrodomain-1 functions, thereby directly boosting the host's immune response [5]. Finally, in an article featuring communications, Nydegger and coworkers described the application of microscale thermophoresis (MST) as a reliable binding assay to be employed in coronavirus drug discovery. This article is important because considering that new SARS-CoV-2 mutations keep surfacing as the COVID-19 epidemic spreads, the necessity is emerging of creating efficient methods to research such variants, as well as the emergence of future potential coronaviruses. This approach could be helpful in the development of novel antivirals. The researchers applied MST in three different types of tests: (a) binding of the SARS-CoV-2 spike receptor binding domain (RBD) to the host target ACE2; (b) binding of RBD to ACE2 in complex with the amino acid transporter SLC6A20/SIT1 and the mutated rs61731475 (I529V); and (c) binding of peptide-based agents to RBD as an approach to preclude virus entry. The findings showed that MST is an extremely accurate method for investigating protein–ligand and/or protein–protein interactions in anti-coronavirus drug discovery, indicating that it is a perfect technique for examining viral variations to develop effective antivirals. Furthermore, the MST approach can be used to examine whether or not membrane proteins are expressed in intracellular membranes. Notably, the strategy outlined in this study could be applied to different coronaviruses, including novel viral variants [6].

Dengue virus (DENV), which has a high global incidence of infections, causes dengue, an acute febrile illness. No medication is currently available for the treatment of this disease. Accordingly, Garcia-Ariza and coworkers reported a virtual screening protocol for identifying indole-containing molecules from natural compounds able to inhibit the DENV NS5 protein, evaluating its antiviral potential in vitro. To this end, using AutoDock Vina software, molecular docking on NS5 was performed, and chemical agents with interesting ADMET properties and pharmacological profile, assessed using the SwissADME web server, were chosen. NS1 protein expression was measured to assess its preliminary antiviral activity. Furthermore, NS5 production, utilizing the DENV-2 Huh-7 replicon using ELISA, and the quantification of viral RNA, using RT-qPCR, were used to assess the impact on viral genome replication and/or translation. The results of virtual screening identified 15 possible DENV NS5 inhibitors. Among them, the computational hit M78 exhibited a strong effect on the lifecycle of the DENV-2 virus. Compound M78 tested at a concentration of 50 µM reduced viral RNA (1.7 times), whereas NS5 protein expression was reduced by 70%, indicating that the replication and/or translation of the viral genome is strongly influenced by M78. Accordingly, studies of M78 function in NS5, pre-treatment activity, and virucidal effects on additional Dengue serotypes and flaviviruses are highly desired [7].

Castro-Amarante and colleagues explored the potential anti-ZIKV effects of the C-terminal region of the E protein (stem) from DENV. Notably, among flaviviruses, this region is highly conserved; consequently, it is attractive for developing peptide-based antiviral agents. Considering that the stem region of the ZIKV and DENV viruses share this sequence, the purpose of the research was to assess the cross-inhibition of ZIKV by the stem-based DV2 peptide (419–447), which has previously been demonstrated to inhibit DENV serotypes. Because the stem regions of the DENV and ZIKV viruses are similar, the researchers investigated whether the stem-based DV2 peptide (419–447), which has been demonstrated to inhibit all DENV serotypes, may also inhibit ZIKV. To determine the antiviral effect of the DV2 peptide against ZIKV, the researchers conducted in vitro and in vivo experiments. Computer-based methods were used to explore the binding mode of the DV2 peptide. The findings clearly indicated that the DV2 peptide was able to target amino acid residues exposed on the surface of the ZIKV envelope (E) protein. The DV2 peptide effectively reduced ZIKV infectivity in the Vero cell line, but had no discernible harmful effects on eukaryotic cells. Notably, the DV2 peptide also decreased mortality and morbidity in mice severely infected with a ZIKV strain discovered in Brazil. Overall, the findings of this work indicated that the DV2 peptide could be a promising agent for treating ZIKV infections, possibly preventing or reducing neurological impairment in neonates linked to infection in pregnant women. Moreover, this work provided indications for developing anti-flavivirus drugs from synthetic stem-based peptides [8].

Scior and collaborators focused their work on the discovery of possible antiviral agents against the human influenza A virus. The researchers developed a computer-based protocol to virtually screen a commercial chemical library containing 660,961 chemical entities against the drug target of human influenza A virus, neuroaminidase (viral strain A/Vietnam/1203/2004; H5N1). By applying a series of sequential filters, such as molecular fingerprints, pharmacophore modeling, molecular docking, and ADMET profile prediction, virtual screening was performed, taking into account several reference compounds for neuroaminidase (natural substrate sialic acid, substrate-like DANA, known antivirals such as peramivir, laninamivir, zanamivir, and oseltamivir). Notably, prior to use, all computational filtering tools underwent validation. Two of the top compounds currently have successful patent filings. Accordingly, the identified lead compounds could be used for further optimization to develop more focused ligands that act as antivirals against the human influenza A virus [9].

Next, two works published within this Special Issue involved the human immunodeficiency virus (HIV-1) to identify possible anti-HIV agents. In particular, Alvarez-Rivera and coworkers described the characterization of a bioactive extract polysaccharide peptide (PSP) from *Coriolus versicolor* by quantitative proteomic analysis. In THP1 monocytic cells, PSP showed anti-HIV replicative capabilities, although the antiviral mechanism were not elucidated. Accordingly, the researchers explored the possible mechanisms by which PSP exerts an anti-HIV profile. In this particular article, the authors investigated the involvement of PKR and IRE1α in the phosphorylation of cofilin-1 and their HIV-1-restriction functions in THP1. The infected supernatant was used to evaluate the HIV-1 p24 antigen to assess PSP's potential for restriction. Immunoblots were used to test the biomarkers PKR, IRE1α, and cofilin-1, and proteome biomarkers were validated using RT-qPCR. Furthermore, Western blot analysis was performed to confirm viral entrance and cofilin-1 phosphorylation using PKR/IRE1 inhibitors. The results demonstrated that treatment with the bioactive peptide PSP reduced viral infectivity, indicating that cofilin-1 phosphorylation and viral limitation were controlled by PKR and IRE1, which is reflected in the early entrance phase. Accordingly, new molecular insights into how UPR, IFN-IP, and cytoskeletal events interact are provided by the current study. Interestingly, the information presented in the article suggested that PSP may be utilized as a natural alternative to stop HIV-1 entrance [10]. In a communication article authored by Zhu and collaborators, the mechanism of the molecular recognition of the HIV-1 gp120 V3 loop by the coreceptor CXC chemokine receptor 4 (CXCR4) was investigated using a construct containing the V3 loop (full-length). It has been established that when HIV-1 enters a cell, one of its main coreceptors, CXCR4, is recognized by the virus through the third variable loop (V3 loop) of the HIV-1 envelope glycoprotein, gp120. In this study, a disulfide bond was used to covalently connect two ends of the V3 loop, creating a constrained cyclic peptide. Furthermore, an all-D-amino acid analog of the L-V3 loop peptide was created to test the impact of the altered conformation of the side-chains of the peptide on CXCR4 recognition. Similar binding recognition was shown for both of these cyclic L- and D-V3 loop peptides to the CXCR4 receptor, but not to the CCR5 receptor, indicating their selective interactions with CXCR4. Computational studies have highlighted that numerous negatively charged Asp and Glu residues on CXCR4 play crucial roles, likely through electrostatic interactions with the positively charged Arg residues found in these peptides. These findings confirm the flexibility of the HIV-1 gp120 V3 loop-CXCR4 interface for ligands of various chiralities, which may be important for the virus's capacity to maintain coreceptor recognition despite mutations at the V3 loop. The discovery of D-amino acid accommodation by the CXCR4 ligand-binding surface paves the way for the creation of more stable D-peptide analogs for antiviral use [11].

The foot-and-mouth disease virus (FMDV), a (+)-ssRNA virus that belongs to the Picornaviridae family, represents a commercially significant pathogen that affects livestock with cloven hooves. Theerawatanasirikul and coworkers utilized a computer-based approach to identify possible inhibitors of FMDV RNA-dependent RNA polymerase (RdRp) (3Dpol). The authors evaluated 5596 compounds in silico using a blind docking approach against

3Dpol, and selected 21 compounds with satisfactory predicted binding affinity. The selected computational hits were tested in vitro, and four compounds (NSC65850, NSC670283 (spiro compound), NSC217697 (quinoline), and NSC292567 (nigericin)) exhibited dose-dependent antiviral effects in vitro (BHK-21 cell-based test), showing an EC_{50} ranging from 0.78 to 3.49 μM. Notably, without affecting FMDV's primary protease, 3Cpro, these substances were able to strongly inhibit FMDV 3Dpol activity in the cell-based 3Dpol inhibition experiment with significant IC_{50} values ranging from 0.8 nM to 0.22 μM. The drugs' 3Dpol inhibitory activities were dose dependently commensurate with the reduced viral load and generation of $(-)$-ssRNA. Finally, the authors have discovered possible FMDV 3Dpol inhibitors that bind to the active areas of the enzyme and could prevent viral multiplication. These substances may be helpful in the treatment of FMDV and other picornaviruses [12].

The lack of therapeutic options for clinical isolates of carbapenem-resistant *Pseudomonas aeruginosa* (CRPA) necessitates the development of novel therapeutic approaches. Accordingly, Mabrouk and colleagues characterized and evaluated two locally isolated phages in a hydrogel formulation in vitro and in vivo and directed them against CRPA clinical isolates. Both phages were examined using phenotypic, genomic, in vitro, and rat models of cutaneous thermal damage caused by *Pseudomonas aeruginosa*. The two siphoviruses, vB_Pae_SMP1 and vB_Pae_SMP5, are members of the Caudovirectes class. Each phage showed an icosahedral head measuring 60 ± 5 nm and a flexible, non-contractile tail measuring 170 ± 5 nm, with the exception of vB_Pae_SMP5, which had an extra base plate with a 35 nm fiber visible at the tail's end. The phage lysate from CRPA propagation was combined with 5% *w/v* carboxymethylcellulose (CMC), which has a spreadability coefficient of 25 and a titer of 108 PFU/mL, to prepare the hydrogel. In the groups treated with Phage vB_Pae_SMP1, vB_Pae_SMP5, or a two-phage cocktail, hydrogel cellular subepidermal granulation tissues with remarkable records of fibroblastic activity and mixed inflammatory cell infiltrates were present. These tissues also showed records of 17.2%, 25.8%, and 22.2% dermal mature collagen fibers. Overall, phage vB_Pae_SMP1 or vB_Pae_SMP5, or the two-phage cocktails made into hydrogels, were effective at controlling CRPA infection in burn wounds and promoting healing at the injury site, as shown by histological inspection and a reduction in animal death rate. Thus, each phage lysate or the two-phage cocktail created as a hydrogel is a useful topical recipe to utilize in burn wounds after severe CRPA infections. To ensure that the corresponding phage hydrogels are appropriate for clinical use in humans, more clinical investigations should be conducted [13].

Fabrizi and colleagues described the possible use of pan-genotypic antivirals to treat hepatitis C virus (HCV) in chronic kidney disease. It is well-established that patients with chronic renal disease (CKD) often contract HCV. Recent research has shown that adult populations with chronic HCV are at increased risk of developing CKD. A relationship between positive anti-HCV serologic status and an elevated incidence of CKD was shown by combining the findings of longitudinal research (n = 2,299,134 distinct individuals) in a systematic review with a meta-analysis of clinical investigations (overall adjusted HR estimated was 1.54 (95% CI, 1.26–1.87), $p = 0.0001$). Furthermore, according to new guidelines, pan-genotypic medications, which are effective against all HCV genotypes, should be used as the initial HCV therapy because they have the potential to be both efficient and secure even in the presence of CKD. A paradigm shift in the treatment of HCV infection has been brought about by the development of direct-acting antiviral medications. The objective of the selected report was to present the most pertinent information regarding pan-genotypic direct-acting antiviral drugs in advanced CKD (CKD stages 4 or 5). Accordingly, through computerized databases and the gray literature, the authors gathered studies using several keywords (e.g., 'Hepatitis C' AND 'Chronic kidney disease' AND 'Pan-genotypic agents'). The results of this investigation showed that glecaprevir/pibrentasvir (GLE/PIB) and sofosbuvir/velpatasvir (SOF/VEL) are the two most significant pan-genotypic drug combinations for treating HCV in severe CKD. GLE/PIB combined therapy in CKD stage 4/5 provided SVR12 rates ranging from 86% to 99%, according to two clinical trials (EXPEDITION-4 and EXPEDITION-5) and other "real-world" studies (n = 6). Clinical trials (n = 1) and "real life" studies (n = 6) demonstrating the effectiveness of SOF/VEL were retrieved. The results showed that the SVR rate in experiments using the SOF/VEL antiviral combination was 100%. In patients on SOF/VEL, the dropout rate (caused by AEs) ranged from 0% to 4.8%. Although there are currently few trials on this topic with small sample sizes, pan-genotypic drug combinations, such as SOF/VEL and GLE/PIB, seem efficient and secure for treating HCV in progressed CKD. Studies are being conducted to determine if effective antiviral treatment with direct-acting antiviral drugs will result in improved survival in individuals with advanced CKD [14].

The WHO global eradication initiatives are aimed at eliminating poliovirus (PV), the cause of poliomyelitis. Following the elimination of type 2 and 3 wild-type PVs, vaccine-derived PV and type 1 wild-type PV, remain as an important threat to eradication. Therefore, antivirals could be used to effectively control a possible outbreak; nonetheless, no anti-PV medicines are available on the market. In this context, Arita and collaborators developed a virtual screening of a chemical library of compounds present in edible plant extracts (6032 extracts) to identify anti-PV agents. The results showed that anti-PV action was discovered in extracts from seven distinct plant species. Chrysophanol and vanicoside B (VCB) extracted from *Rheum rhaponticum* and *Fallopia sachalinensis* were identified as promising anti-PV agents. VCB inhibited the activity of PI4KB in vitro (IC$_{50}$ = 5.0 μM) while targeted the host PI4KB/OSBP pathway for anti-PV action (EC$_{50}$ = 9.2 μM). Furthermore, additional studies are expected to confirm the antiviral potential of VCB. Although the anti-PV activity was clearly specific to the PI4KB/OSBP pathway, we could not rule out the possibility that the observed anti-PV activity was influenced by the off-target effect of VCB (CC$_{50}$ = 27 μM) that is reflected in a limited therapeutic window (total protection from infection caused by PV1[Sabin 1] at 20 μM) [15].

Finally, in the last studies to be included in this Editorial article, Manna and collaborators presented a review article that explored the antiviral profile of various oxa- and aza-heterocycles. The main goal of this review article was to explore the biological importance, synthetic pathways, and antiviral activity of these ring structures. Furthermore, the structure–activity relationship (SAR) of the selected molecules is presented, together with a list of their key features, to create a useful framework for scientists in the field. The synergistic findings are crucial for introducing innovative instruments for the next antiviral drug discovery programs [16].

The *Viruses* Editorial staff, all the authors and co-authors for their significant contributions to this Special Issue, and all the reviewers for their work in evaluating the submissions deserve my sincere gratitude as Guest Editor. The success of the research topic as a whole was a result of all of these combined efforts. We hope that this subject will assist scientists in the development of efficient antivirals and serve as a substantial source of knowledge and inspiration for researchers and students. You may obtain this Special Issue for free by visiting the following link: https://www.mdpi.com/journal/viruses/special_issues/novelantiviral_agents.

Conflicts of Interest: The author declares no conflict of interest.

References

1. Available online: https://www.who.int/news/item/21-11-2022-who-to-identify-pathogens-that-could-cause-future-outbreaks-and-pandemics (accessed on 13 September 2023).
2. Varbanov, M.; Philippot, S.; Gonzalez-Cardenete, M.A. Anticoronavirus Evaluation of Antimicrobial Diterpenoids: Application of New Ferruginol Analogues. *Viruses* **2023**, *15*, 1342. [CrossRef] [PubMed]
3. Zhang, H.; Liang, B.; Sang, X.; An, J.; Huang, Z. Discovery of Potential Inhibitors of SARS-CoV-2 Main Protease by a Transfer Learning Method. *Viruses* **2023**, *15*, 891. [CrossRef] [PubMed]
4. Hassan, A.H.E.; El-Sayed, S.M.; Yamamoto, M.; Gohda, J.; Matsumoto, T.; Shirouzu, M.; Inoue, J.I.; Kawaguchi, Y.; Mansour, R.M.A.; Anvari, A.; et al. In Silico and In Vitro Evaluation of Some Amidine Derivatives as Hit Compounds towards Development of Inhibitors against Coronavirus Diseases. *Viruses* **2023**, *15*, 1171. [CrossRef]

5. Mohammad, A.; Alshawaf, E.; Arefanian, H.; Marafie, S.K.; Khan, A.; Wei, D.-Q.; Al-Mulla, F.; Abubaker, J. Targeting SARS-CoV-2 Macrodomain-1 to Restore the Innate Immune Response Using In Silico Screening of Medicinal Compounds and Free Energy Calculation Approaches. *Viruses* **2023**, *15*, 1907. [CrossRef]
6. Nydegger, D.T.; Pujol-Gimenez, J.; Kandasamy, P.; Vogt, B.; Hediger, M.A. Applications of the Microscale Thermophoresis Binding Assay in COVID-19 Research. *Viruses* **2023**, *15*, 1432. [CrossRef]
7. Garcia-Ariza, L.L.; Gonzalez-Rivillas, N.; Diaz-Aguirre, C.J.; Rocha-Roa, C.; Padilla-Sanabria, L.; Castano-Osorio, J.C. Antiviral Activity of an Indole-Type Compound Derived from Natural Products, Identified by Virtual Screening by Interaction on Dengue Virus NS5 Protein. *Viruses* **2023**, *15*, 1563. [CrossRef]
8. Castro-Amarante, M.F.; Pereira, S.S.; Pereira, L.R.; Santos, L.S.; Venceslau-Carvalho, A.A.; Martins, E.G.; Balan, A.; Souza Ferreira, L.C. The Anti-Dengue Virus Peptide DV2 Inhibits Zika Virus Both In Vitro and In Vivo. *Viruses* **2023**, *15*, 839. [CrossRef]
9. Scior, T.; Cuanalo-Contreras, K.; Islas, A.A.; Martinez-Laguna, Y. Targeting the Human Influenza a Virus: The Methods, Limitations, and Pitfalls of Virtual Screening for Drug-like Candidates Including Scaffold Hopping and Compound Profiling. *Viruses* **2023**, *15*, 1056. [CrossRef]
10. Alvarez-Rivera, E.; Rodriguez-Valentin, M.; Boukli, N.M. The Antiviral Compound PSP Inhibits HIV-1 Entry via PKR-Dependent Activation in Monocytic Cells. *Viruses* **2023**, *15*, 804. [CrossRef]
11. Zhu, R.; Sang, X.; Zhou, J.; Meng, Q.; Huang, L.S.M.; Xu, Y.; An, J.; Huang, Z. CXCR4 Recognition by L- and D-Peptides Containing the Full-Length V3 Loop of HIV-1 gp120. *Viruses* **2023**, *15*, 1084. [CrossRef] [PubMed]
12. Theerawatanasirikul, S.; Semkum, P.; Lueangaramkul, V.; Chankeeree, P.; Thangthamniyom, N.; Lekcharoensuk, P. Non-Nucleoside Inhibitors Decrease Foot-and-Mouth Disease Virus Replication by Blocking the Viral 3D(pol). *Viruses* **2022**, *15*, 124. [CrossRef] [PubMed]
13. Mabrouk, S.S.; Abdellatif, G.R.; Abu Zaid, A.S.; Aziz, R.K.; Aboshanab, K.M. In Vitro and Pre-Clinical Evaluation of Locally Isolated Phages, vB_Pae_SMP1 and vB_Pae_SMP5, Formulated as Hydrogels against Carbapenem-Resistant Pseudomonas aeruginosa. *Viruses* **2022**, *14*, 2760. [CrossRef] [PubMed]
14. Fabrizi, F.; Tripodi, F.; Cerutti, R.; Nardelli, L.; Alfieri, C.M.; Donato, M.F.; Castellano, G. Recent Information on Pan-Genotypic Direct-Acting Antiviral Agents for HCV in Chronic Kidney Disease. *Viruses* **2022**, *14*, 2570. [CrossRef] [PubMed]
15. Arita, M.; Fuchino, H. Characterization of Anti-Poliovirus Compounds Isolated from Edible Plants. *Viruses* **2023**, *15*, 903. [CrossRef]
16. Manna, S.; Das, K.; Santra, S.; Nosova, E.V.; Zyryanov, G.V.; Halder, S. Structural and Synthetic Aspects of Small Ring Oxa- and Aza-Heterocyclic Ring Systems as Antiviral Activities. *Viruses* **2023**, *15*, 1826. [CrossRef]

viruses

Article

Anticoronavirus Evaluation of Antimicrobial Diterpenoids: Application of New Ferruginol Analogues

Mihayl Varbanov [1,2], Stéphanie Philippot [1] and Miguel A. González-Cardenete [3,*]

[1] Université de Lorraine, CNRS, L2CM, F-54000 Nancy, France
[2] Laboratoire de Virologie, CHRU de Nancy Brabois, 54500 Vandoeuvre-lès-Nancy, France
[3] Instituto de Tecnologia Química (UPV-CSIC), Universitat Politècnica de Valencia-Consejo Superior de Investigaciones Científicas, Avenida de los Naranjos s/n, 46022 Valencia, Spain
* Correspondence: migoncar@itq.upv.es

Abstract: The abietane diterpene (+)-ferruginol (**1**), like other natural and semisynthetic abietanes, is distinguished for its interesting pharmacological properties such as antimicrobial activity, including antiviral. In this study, selected C18-functionalized semisynthetic abietanes prepared from the commercially available (+)-dehydroabietylamine or methyl dehydroabietate were tested in vitro against human coronavirus 229E (HCoV-229E). As a result, a new ferruginol analogue caused a relevant reduction in virus titer as well as the inhibition of a cytopathic effect. A toxicity prediction based on in silico analysis was also performed as well as an estimation of bioavailability. This work demonstrates the antimicrobial and specifically antiviral activity of two tested compounds, making these molecules interesting for the development of new antivirals.

Keywords: antimicrobial; coronavirus; abietane; dehydroabietylamine; phenol; phthalimide

Citation: Varbanov, M.; Philippot, S.; González-Cardenete, M.A. Anticoronavirus Evaluation of Antimicrobial Diterpenoids: Application of New Ferruginol Analogues. *Viruses* **2023**, *15*, 1342. https://doi.org/10.3390/v15061342

Academic Editor: Simone Brogi

Received: 18 May 2023
Revised: 2 June 2023
Accepted: 6 June 2023
Published: 9 June 2023

1. Introduction

Viral diseases are a significant global health problem with millions of infections annually and insufficient antiviral drugs available. For viruses such as Zika (ZIKV), Ebola (EBOV), and many others, there is no available explicit antiviral treatment currently [1]. These health problems have led to the (re)evaluating of traditional medical approaches in order to focus on new molecules or new mechanisms of action. Therefore, naturally based medicines may be appropriate alternatives for treating viral infections [2]. In fact, about fifty percent of clinical anti-infective medications, of which one third are antiviral drugs, include either natural products (NPs), molecules derived from NPs or inspired by NP structures [3].

Viral respiratory diseases cause life-threatening infections in millions of patients worldwide every year. Human coronaviruses (HCoVs) are accountable for steady epidemic outbreaks, thus representing a global public health concern. HCoVs are enveloped, positive single-stranded RNA viruses which provoke a large portion of upper and middle respiratory tract infections, such as cold, bronchitis, and pneumonia. Coronavirus infection can be especially severe and fatal for newborns, young infants, the elderly, or immunosuppressed people [4].

The COVID-19 pandemic caused by SARS-CoV-2 has infected more than 700 million people and caused nearly seven million deaths [5]. At present, few antiviral drugs against SARS-CoV-2 are in or close to clinical use. These include molnupiravir (Merck), and Paxlovid (Pfizer). Paxlovid is a combination of the main viral protease Mpro (also known as 3-chymotrypsin-like protease, 3C-like protease or 3CLpro) inhibitor nirmatrelvir (PF-07321332) and ritonavir (a CYP3A4 inhibitor that slows down the clearance of nirmatrelvir) [6]. The potential of drug–drug interactions from ritonavir, however, may limit its use by many patients [7]. Additionally, it seems that the efficacy of the treatment with molnupiravir is low since it requires a high intake of pills (40 in 5 days). Given the spread and severity of the disease, it is crucial to develop efficient treatments and rapidly available solutions that can supplement active immunization efforts, which are still challenged by high viral transmissibility, re-infection, and immune escape variants.

In 2021, Skaltsounis et al. [8] reviewed natural and nature-derived products targeting human coronaviruses since their discovery in the 1960s, and realized the scarce research on this subject. However, it is known that a number of diterpenes are potential antiviral agents, including towards coronaviruses [9]. Very recently, Jantan et al. [10] have reviewed the antiviral effects of a number of phytochemicals against SARS-CoV-2, including virtual screening carried out over the last three years.

The abietane-type diterpenoids are naturally occurring compounds isolated from plants of different families such as *Araucariaceae*, *Cupressaceae*, *Phyllocladaceae*, *Pinaceae*, *Podocarpaceae*, *Asteraceae*, *Celastraceae*, *Hydrocharitaceae*, and *Lamiaceae* [11]. These metabolites show an extensive range of promising pharmacological activities, including antimicrobial properties [11]. Several semisynthetic derivatives have also displayed interesting pharmacological properties [12]. For example, natural (+)-ferruginol (**1**) and diacetate (**2**) (3β,12-diacetoxyabieta-6,8,11,13-tetraene) have shown important anti-SARS-CoV activities with a high cytopathogenic effect (CPE) reduction and effective replication inhibition (SARS-CoV, EC_{50} = 1.39 and 1.57 µM, respectively, in Vero E6 cells) [13]. Wen et al. (2007) [13] found that ferruginol (**1**) did not show inhibition for SARS-CoV 3CL protease (3CLpro), a key target for antiviral drug design, below 100 µM; later, other researchers reported significantly better inhibitory effects on 3CLpro (IC_{50} = 49.6 µM) [14]. Additionally, the semisynthetic phthalimide ferruginol analogue **3** (Figure 1) has shown anti-dengue and anti-herpes properties [15], and anti-Zika activity against Brazilian strains [16] as well as anti-Zika and anti-chikungunya activities against Colombian strains [17]. Carnosic acid (**4**) has displayed inhibitory effects on human immunodeficiency virus 1 (HIV-1) protease and HIV-1 virus [18], and on human respiratory syncytial virus replication [19]. Its semisynthetic analogue **5**, however, demonstrated a better anti-herpes profile than carnosic acid itself [20]. Semisynthetic dehydroabietinol acetate (**6**) showed mild anti-herpes activity [21]. In recent years, several C19-functionalized abietane-type acids, i.e., jiadifenoic acid C (**7**), and its derivatives isolated from *Illiciaceae* plants (jiadifenoic acids from the roots of *Illicium jiadifengpi*) have exhibited important anti-Coxsackie virus B activities [22].

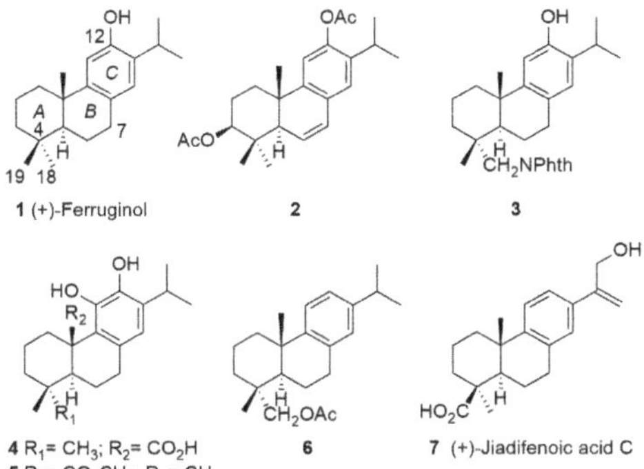

Figure 1. Natural and semisynthetic antiviral abietanes **1**–**7**.

Based on this background of antimicrobial activities of abietane-type diterpenoids, in this work, we have extended the knowledge of antiviral activities of several C18-functionalized ferruginol analogues. The current research aimed to find potential molecules to inhibit the old coronavirus associated infection as a model for broad-spectrum anticoronavirus agents. Thus, we report on the in vitro antiviral study of ferruginol (**1**), ferruginol analogues **3** and **12**–**16** (Scheme 1) against *human coronavirus 229E*.

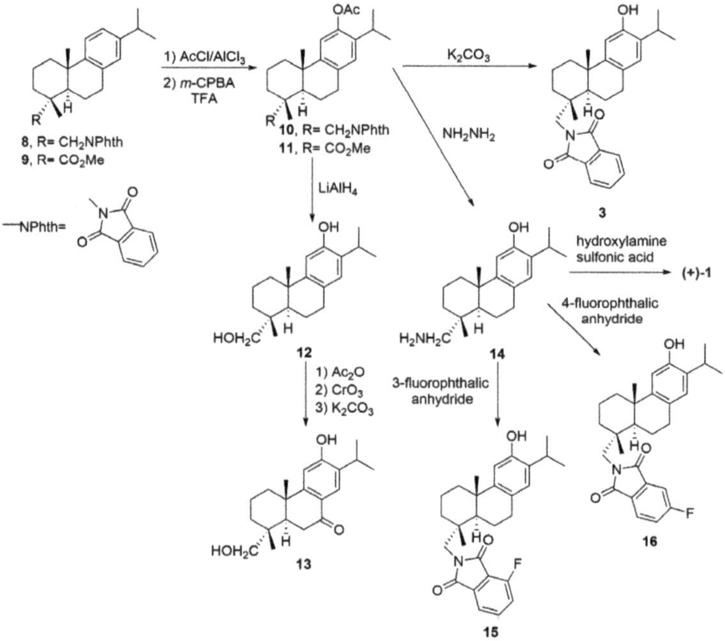

Scheme 1. Synthetic sequence for the preparation of tested molecules **1**, **3**, and **12–16**.

2. Materials and Methods

2.1. Chemistry: General Experimental Procedures

Specific rotation was measured using a 10 cm cell in a Jasco P-2000 polarimeter in DCM. NMR data were collected on a 300 MHz spectrometer. All spectra were recorded in $CDCl_3$ as solvent. Reactions were monitored by TLC using Merck silica gel 60 F254 (0.25 mm-thick) plates. Compounds on TLC plates were visualized under UV light at 254 nm and by immersion in a 10% sulfuric acid solution and heating with a heat gun. Purifications were performed by flash chromatography on Merck silica gel (230–400 mesh). Commercial reagent grade solvents and chemicals were used as purchased. Combined organic extracts were washed with brine, dried over anhydrous MgSO4, filtered, and concentrated under reduced pressure.

(+)-Dehydroabietylamine (ca. 60%) was purchased from Aldrich (Saint Louis, MO, USA) and (−)-abietic acid >70% from TCI Europe (Zwijndrecht, Belgium). Carbon numbering of compounds matched to that of natural products. All known compounds prepared in this work displayed spectroscopic data in agreement with the reported data [17,23–25]. Purity of final compound was 95% or higher. Copies of NMR spectra for new compounds **15** and **16** are available in the supplementary information.

2.2. Synthesis

18-Aminoferruginol or 12-hydroxydehydroabietylamine (**14**).

To a stirred yellow suspension of phthalimide **10** (3.5 g, 7.4 mmol) [17] in ethanol (65 mL) excess hydrazine monohydrate (50%, 2.8 mL, 44.4 mmol) was added, turning it to pink. Then, it was refluxed for 5 h. Then, without cooling, a white solid was filtered off and washed with fresh ethanol (3 × 20 mL). The filtrate was concentrated to give crude amino-phenol **14** (6.7 g). This crude solid was treated with 80 mL of aqueous 2 M NaOH and stirred for 2 h 30 min at 700 rpm. The resulting basic solution was neutralized with approx. 57.5 mL 10% HCl, and then extracted with dichloromethane (5 × 60 mL). The extract was dried over $MgSO_4$ and concentrated to give 2.10 g (94%) of amino-phenol **14** as a pale solid which was used without further purification with spectral data in agreement with the reported data [23].

12-Hydroxy-N,N-(3-fluorophthaloyl)dehydroabietylamine (**15**).

A mixture of 3-fluorophthalic anhydride (120 mg, 0.73 mmol) and 12-hydroxydehydroa bietylamine **14** (200 mg, 0.66 mmol) in AcOH (2.5 mL) was refluxed for 2.5 h, then cooled to rt and 30 mL of water was added. The resulting solid was filtered off, washed with water, and dried under vacuum to give compound **15** (150 mg, 50%) as a beige solid: $[\alpha]^{20}_D$–20.5 (c 1.0, CH_2Cl_2); ^{1}H NMR (300 MHz) δ 7.73–7.62 (2H, m), 7.35 (1H, t, J = 9.0), 6.86 (1H, s), 6.60 (1H, s), 3.67 (1H, d, J = 13.8), 3.49 (1H, d, J = 13.8), 3.10 (1H, sept., J = 6.9), 2.95–2.89 (2H, m), 2.28–2.10 (2H, m), 1.85–1.62 (3H, m), 1.52–1.30 (4H, m), 1.22 (3H, d, J = 6.9), 1.22 (3H, s), 1.20 (3H, d, J = 6.9), 1.04 (3H, s); ^{13}C NMR (75 MHz) δ_C 168.1 (d, $J_{C,F}$ = 3), 165.9 (s), 157.4 (d, $J_{C,F}$ = 264), 150.6 (s), 148.2 (s), 136.4 (d, $J_{C,F}$ = 7.5), 134.2 (d, $J_{C,F}$ = 1.5), 131.6 (s), 127.2 (s), 126.8 (d), 122.3 (d, $J_{C,F}$ = 20), 119.4 (d, $J_{C,F}$ = 4), 117.7 (d, $J_{C,F}$ = 12), 110.5 (d), 49.0 (t), 45.1 (d), 39.4 (s), 38.1 (t), 37.5 (s), 36.9 (t), 29.3 (t), 26.8 (d), 25.7 (q), 22.7 (q), 22.6 (q), 19.5 (t), 19.1 (q), 18.5 (t); ^{19}F NMR (282 MHz) δ_F − 113.0; HRMS (ESI) m/z 450.2441 [M+1]+, calcd for $C_{28}H_{33}FNO_3$: 450.2444; Anal. calcd. for $C_{28}H_{32}FNO_3$: C, 74.8; H, 7.2; N, 3.1; Found: C, 74.5; H, 7.3; N, 3.1.

12-Hydroxy-N,N-(4-fluorophthaloyl)dehydroabietylamine (**16**).

A mixture of 4-fluorophthalic anhydride (332 mg, 1.99 mmol) and 12-hydroxydehydroa bietylamine **14** (305 mg, 1.01 mmol) in pyridine (2.5 mL) was heated at reflux for 4 h, then cooled to rt and 15 mL of water was added, followed by extraction with diethyl ether (5 × 6 mL). The combined extract was washed with 10% HCl (2 × 5 mL), water (2 × 5 mL) and brine (5 mL), dried over $MgSO_4$, and concentrated under vacuum. The resulting beige foamy residue (360 mg) was purified by column chromatography eluting with *n*-hexane-ethyl acetate 7:3 to give compound **16** (256 mg, 56%) as a pale oil which became a light beige solid after trituration with *n*-hexane: $[\alpha]^{20}_D$–29.6 (c 1.0, CH_2Cl_2); ^{1}H NMR (300 MHz) δ 7.81 (1H, dd, J = 8.1, 4.5), 7.48 (1H, dd, J = 7.2, 2.1), 7.35 (1H, ddd, J = 8.4, 8.4, 2.2), 6.86 (1H, s), 6.59 (1H, s), 4.58 (1H, s, OH), 3.67 (1H, d, J = 14.0), 3.50 (1H, d, J = 14.0), 3.10 (1H, sept., J = 6.9), 2.92–2.89 (2H, m), 2.24–2.06 (2H, m), 1.87–1.59 (4H, m), 1.52–1.30 (4H, m), 1.23 (3H, d, J = 6.9), 1.21 (3H, d, J = 6.9), 1.21 (3H, s), 1.04 (3H, s); ^{13}C NMR (75 MHz) δ_C 168.2 (s), 167.9 (d, $J_{C,F}$ = 3), 166.3 (d, $J_{C,F}$ = 255), 150.6 (s), 148.2 (s), 134.8 (d, $J_{C,F}$ = 9), 131.6 (s), 127.7 (d, $J_{C,F}$ = 3), 127.2 (s), 126.8 (d), 125.6 (d, $J_{C,F}$ = 10), 120.9 (d, $J_{C,F}$ = 24), 111.0 (d, $J_{C,F}$ = 25), 110.5 (d), 49.1 (t), 45.1 (d), 39.4 (s), 38.1 (t), 37.5 (s), 36.9 (t), 29.3 (t), 26.8 (d), 25.7 (q), 22.7 (q), 22.6 (q), 19.5 (t), 19.1 (q), 18.5 (t); ^{19}F NMR (282 MHz) δ_F − 102.0; HRMS (ESI) m/z 450.2431 [M+1]$^+$, calcd for $C_{28}H_{33}FNO_3$: 450.2444; Anal. calcd. for $C_{28}H_{32}FNO_3$: C, 74.8; H, 7.2; N, 3.1; Found: C, 74.4; H, 6.8; N, 2.9.

2.3. Anti-HCoV 229E Assay

2.3.1. Cell Culture and Virus

For virus production and for the antiviral assay MRC-5 human lung fibroblast were used (ECACC, ref 05090501, Sigma-Merck, Saint Quentin, France). The cells were grown in antibiotic-free Minimum Essential Medium (MEM, M4655, Sigma-Merck, Saint Quentin, France) supplemented with 10% fetal calf serum (FCS, CVFSVF00-01, Eurobio, Les Ulis, France), 2 mM glutamine (G7513, Sigma-Merck, France), 1 mM sodium pyruvate (SH30239.01, Hyclone GE Healthcare, Logan, UT, USA), and 1 × MEM non-essential amino acids (11140-035, Gibco Life Technologies) (i.e., growth medium) at 37 °C in 5% CO_2. For the antiviral assays the medium was supplemented only with 2% FCS (i.e., maintenance medium). The cultures were maintained at 33 °C in humidified 5% CO_2 atmosphere.

The human coronavirus HCoV 229E (PHE/NCPV 0310051v) was produced and quantified in MRC-5 cells. Initial virus titers were calculated as 50% cell-culture infectious doses ($CCID_{50}$/mL), defined as the dilution of the virus required to infect 50% of the cell culture, according to Reed and Muench (1938) [26]. All virus stocks were stored at −70 °C until further use. The infectivity of the viral samples was titrated on 96-well microtiter plates containing about 80% confluent MRC-5 cells. A volume of 100 μL of serial 10-fold dilutions of the virus from 10^{-1} to 10^{-3} in DMEM medium with 2% FCS was added to the MRC-5 cells, in 8 replicas. The infected cells were incubated at 33 °C in 5% CO_2 for 6 days. The appearance of cytopathogenic effect (CPE) was recorded daily.

The lyophilized compounds were dissolved at 50 mg/mL in DMSO (D8418, Sigma-Merck, France) at room temperature. Stock solutions of compounds were kept at 4 °C until used. The concentration of DMSO in biological assays was of 0.1%, including in cell controls.

2.3.2. Cytotoxicity

For cytotoxicity evaluation, MRC-5 cells were seeded at 15×10^3 cells per well in 96-well tissue-culture plates. The DMSO-dissolved compounds were diluted in MEM complete medium with 2% FCS. At 24 h after the seeding, a limited volume of 100 μL of each compound were added to the cell monolayers at increasing concentrations (from 3.6–82.7 μM). The plates were then incubated for 72 h at 37 °C in a 5% CO_2 atmosphere. For assessment of the cell viability the MTT assay was used [27]. Percentages of survival compared to control were calculated and the maximal concentration with no cytotoxic effect was determined. The cytotoxicity assays were performed in triplicates. Values are presented as means of these independent experiments.

2.3.3. Antiviral Assay

The antiviral activity of the compounds was evaluated using the reduction in the viral titer and the reduction in the virus-induced cytopathogenic effect (CPE), characterized mostly by extensive host cell death and degradation of the cell monolayer.

The cells were seeded in 96-well plates at 15,000 cells/well. After 24 h of incubation, two different multiplicities of infection (MOI) of the virus (viral batch of 22 February 2021, 2.4×10^4 IP/mL) were evaluated, in cascade, with serial dilutions at 1/10 in the cell culture medium (2% fetal calf serum) containing or not the compound to be tested (at 3 μg/mL final concentration). Cell plates were emptied, then the cell monolayers were treated with 100 μL of the dilutions (one column per condition, n = 8). An untreated control and a 0.1% DMSO control were added to each plate. The plates were incubated at 33 °C for 72 h. The cytopathic effect of the virus was evaluated by reading the plate under a microscope (Zeiss, Rueil-Malmaison, France) and the viral titers were determined according to the Reed and Muench method [26]. The plates were then stained with crystal violet and the cytopathogenic effect of each well was calculated using a plate reader, measuring the optical density of each well at 540 nm.

Wells were tested the same way for each assay, including control or blank (n = 8). The test wells were assessed for the determination of viral CPE post-infection when cell destruction of infected untreated cultures was at its maximum. The median tissue culture infectious dose ($TCID_{50}$) was determined, and titrations were performed in triplicate. Titres were calculated as ($TCID_{50}$)/mL. The antiviral assays were performed in triplicates. Classic tissue culture infectious dose 50% ($TCID_{50}$/mL) were performed in triplicates of a single experiment.

2.4. In Silico Simulations

2.4.1. Calculation of Molecular Properties (Drug-Likeness)

The structures of tested compounds were manually drawn in ChemDraw Professional 15.0 software (PerkinElmer Informatics, Inc., Waltham, MA, USA), and the SMILES notation was obtained for each molecule. To calculate the molecular polar surface area [28] and the parameters of Lipinski's rule of five (see Table 1) [29,30], the SMILES notation in Molinspiration online 2023 software (available from https://www.molinspiration.com/cgi-bin/properties, accessed on 27 January 2023, Molinspiration Cheminformatics, Slovensky Grob, Slovak Republic, 2023, https://www.molinspiration.com) was employed.

Table 1. Molecular properties (drug-likeness) by Molinspiration online software for compounds **1, 3** and **12–16** [a,b].

Compound	miLog P	MW	n-HBA	n-HBD	TPSA	Lipinski's Violation
1	6.41	286.46	1	1	20.23	1
3	7.08	431.58	4	1	59.30	1
12	5.24	302.46	2	2	40.46	1
13	4.33	316.44	3	2	57.53	0
14	4.67	301.47	2	3	46.25	0
15	6.88	449.57	4	1	59.30	1
16	6.25	449.57	4	1	59.30	1
Rule of five	not >5	<500	not >10	not >5		1 violation allowed

[a] Values were calculated using Molinspiration Cheminformatics Online 2023 software (Molinspiration, accessed on 27 January 2023, Slovensky Grob, Slovak Republic, 2023, http://www.molinspiration.com). [b] P = partition coefficient; MW = molecular weight; n-HBA = number of hydrogen bond accepting groups; n-HBD = number of hydrogen bond donating groups; TPSA = total polar surface area.

2.4.2. Toxicity Prediction (GUSAR)

Compounds **1**, **3**, and **12–16** were investigated for acute rat toxicity properties in silico (see Table 2). The acute toxicity in rodent models of the tested compounds were predicted by GUSAR [31]. The analysis was based on the quantitative neighborhoods of atom descriptors and the prediction of activity spectra for substances algorithm. It correlates the results with the SYMYX MDL toxicity database and further classifies the tested compounds according to the Organization for Economic Co-operation and Development (OECD) manual.

Table 2. Acute rat toxicity prediction by GUSAR (on the Basis of PASS Prediction).

Acute Rat Toxicity Parameters	Compounds						
	1	**3**	**12**	**13**	**14**	**15**	**16**
	Rat LD$_{50}$ Values						
Rat IP [a] LD$_{50}$ (mg/kg)	0.00	1820.00	847.50	807.50	607.20	567.60	657.60
Rat IV [b] LD$_{50}$ (mg/kg)	0.00	30.90	33.52	34.50	17.84	24.57	32.71
Rat Oral [c] LD$_{50}$ (mg/kg)	0.00	1015.00	2175.00	2639.00	2167.00	1385.00	1429.00
Rat SC [d] LD$_{50}$ (mg/kg)	0.00	400.80	602.90	295.00	882.00	227.30	420.30
Acute rodent toxicity classification of compounds by OECD project							
Rat IP LD$_{50}$ Classification	Class 1	Non-toxic	Class 5	Class 5	Class 5	Class 5	Class 5
Rat IV LD$_{50}$ Classification	Class 1	Class 1	Class 3	Class 3	Class 3	Class 3	Class 3
Rat Oral LD$_{50}$ Classification	Class 1	Class 4	Class 5	Class 5	Class 5	Class 4	Class 4
Rat SC LD$_{50}$ Classification	Class 1	Class 4	Class 4	Class 4	Class 4	Class 4	Class 4

[a] IP—Intraperitoneal route of administration [b] IV—Intravenous route of administration [c] Oral—Oral route of administration [d] SC—Subcutaneous route of administration.

Toxic doses are expressed as median lethal dose, representing the dose of the compound that is lethal to 50% of the experimental animals exposed to it. The LD$_{50}$ values are expressed as the weight of the chemical per unit of body weight (mg/kg). The compounds were classified based on the toxicity scale used in the Gosselin, Smith and Hodge scale: class 1 (LD$_{50} \leq 5$ mg/kg) being super toxic, class 2 ($5 <$ LD$_{50} \leq 50$ mg/kg) being extremely toxic, class 3 ($50 <$ LD$_{50} \leq 500$ mg/kg) being very toxic, class 4 ($500 <$ LD$_{50} \leq 5000$ mg/kg) being moderately toxic, class 5 ($5000 <$ LD$_{50} \leq 15{,}000$ mg/kg) being slightly toxic, and practically non-toxic compounds with LD$_{50} > 15{,}000$ mg/kg [32].

3. Results

The tested compounds **1**, **3** and **12–14** are known and were obtained following synthetic routes previously reported, as follows (see above Scheme 1): compounds **1** and **3** following our recently optimized conditions [17] through intermediate **10**; meanwhile compounds **12** and **13** where obtained from intermediate **11** [24]; for the synthesis of compound **14** (12-hydroxydehydroabietylamine) we optimized our previous work [23] starting from **10**, changing slightly the protocol to achieve a better yield. Compound **14** was the intermediate for novel compounds **15** and **16**. In particular, compounds **1**, **3** and **14–16** were obtained from commercially available (+)-dehydroabietylamine, which by condensation with phthalic anhydride gives phthalimide **8** [17,25]. To make the C-12 phenolic moiety Friedel–Crafts and Baeyer–Villiger reactions were used (Scheme 1). Thus, Friedel–Crafts acylation of **8** followed by Baeyer–Villiger oxidation gave acetate **10** [25], which by mild basic methanolysis afforded ferruginol analogue **3** named 18-(phthalimid-2-yl) ferruginol [17]. By contrast, the treatment with hydrazine of compound **10** furnished phenol-amine **14** which served as an intermediate for the synthesis of (+)-ferruginol (**1**), by deamination [23], and fluorinated phthalimides **15** and **16** by condensation with either 3-fluorophthalic anhydride or 4-fluorophthalic anhydride, respectively. Similarly, starting from methyl dehydroabietate (**9**) [21], prepared from (−)-abietic acid by esterification and aromatization with Pd/C catalyst, followed by Friedel–Crafts reaction and Baeyer–Villiger oxidation, gave acetate **11** which was first converted into alcohol **12** by reduction with LiAlH$_4$, and then into keto-alcohol **13** in a three-step sequence [24].

The antiviral activities of the compounds were evaluated against a respiratory virus, human coronavirus 229E (*HCoV 229E*). These compounds were investigated at different concentrations depending on cytotoxicity (Figure 2). The antiviral of the compounds against *HCoV 229E* were determined according to the Reed and Muench end-point dilution method [26] and are shown in Figures 3 and 4 (see below). Three of the compounds reduced the viral titer, with low to no cytotoxicity at 3 µg/mL (Figure 2). The compound **12** moderately reduced the viral titer with close to 0.5-log, while the compounds **14**, **15** and **16** showed significant reduction in the viral titer with approximately 2-logs (Figure 3). The analysis of the reduction in the viral-induced cytopathogenic effect (CPE) confirmed the antiviral activity of compound **16** (at 3.0 µg/mL or 6.7 µM), where the cytopathogenic effect was totally abolished at all the applied MOIs (0.01–0.001). Compounds **3** and **15** also showed important reductions in the viral titer, similar to the one induced by **16**, but they also induced modification of the cell morphology in photonic microscope observation. The rest of the compounds also produced important reductions in the viral CPE, but mostly at lower MOIs (0.001), except jiadifenoic acid C (**7**) which showed no activity (Figure 4).

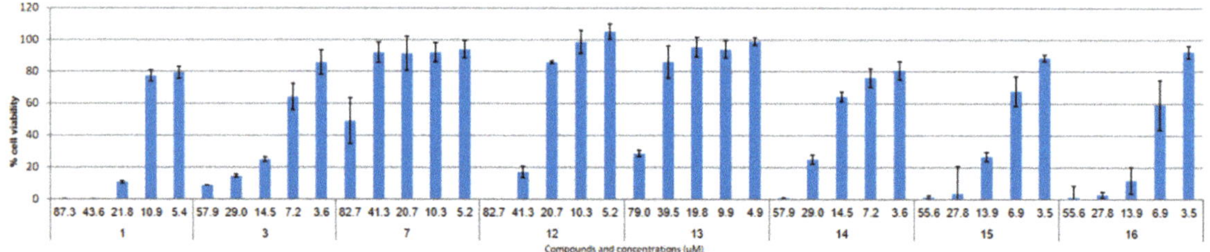

Figure 2. Cell viability after 72 h incubation with the compounds at 3.6–82.7 µM (n = 3).

In silico computation is, nowadays, an important part of the drug design development process. Using the Molinspiration Cheminformatics software (Molinspiration, accessed on 27 January 2023, Slovensky Grob, Slovak Republic, 2023, http://www.molinspiration.com), through an easily obtained SMILES notation for each molecule, we can calculate certain very useful molecular properties such as the Lipinski's rule of five [29]. The compounds **1**, **3**, **12**, **13**, **14**, **15** and **16** were subjected to this software and we obtained very promising results of drug-likeness (Table 1).

The computational prediction of compound toxicities in drug design allows the quick estimation of the lethal doses in animals, and avoids the use of animal experiments. The rat acute toxicity of the compounds **1**, **3**, **12**, **13**, **14**, **15** and **16** was predicted by the in silico software tool GUSAR (General Unrestricted Structure—Activity Relationships, version 16.0) [31] on the basis of PASS (Prediction of Activity Spectra for Substances) technology [33,34]. The quantitative in silico prediction was applied to four different types of administration: oral, intravenous (IV), intraperitoneal (IP), and subcutaneous (SC). As is shown in Table 2, the differences in the predicted LD_{50} values obtained for the four different routes, based on the structural formulae of the tested compounds, are reported along with the acute rodent toxicity classification of the compounds. Extreme/super toxicities (class 1) are documented for compound **1** for all routes of administration, and for compound **3** in the case of intravenous administration, where the value of 0.00 indicates that the compound falls out of the domain of applicability of the proposed route of administration. Compounds **13**, **14**, **15** and **16** are to be considered as very toxic (class 3) in the event of intravenous administration. Depending to the route of administration (mainly IP, oral and SC) compounds **3–16** were estimated to have moderate (class 4), slight (class 5), and practically non–toxic behavior based on the calculated LD_{50} values.

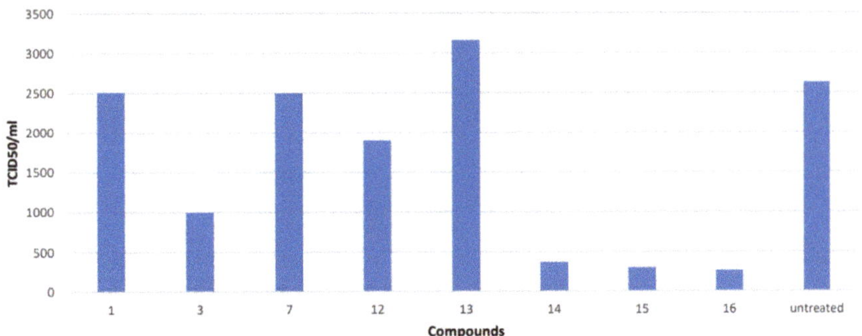

Figure 3. Effect of the compounds on the viral titer of HCoV 229E (n = 1).

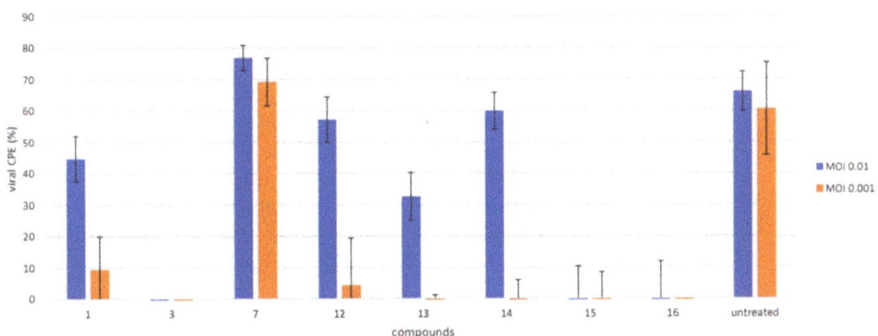

Figure 4. Antiviral effect based on the variation of the cell monolayer density and on the HCoV 229E-induced CPE (n = 3).

4. Discussion

Nature provides a huge diversity of chemicals to explore and develop drugs for the treatment of various ailments including viral diseases; however, it is difficult to estimate adverse effects since often the available studies are in vitro models [35]. Natural products and their derivatives against coronavirus have been reviewed recently, showing promising results [8,36]. To date, most of the studies have focused on SARS-CoV models with proposed mechanisms of action associated with 3CL protease (Mpro) [37]. We reported, in 2016, that several abietane-type diterpenoids exhibited antiviral activity in viral models of dengue with $EC_{50} < 5.0$ µM and herpes with $EC_{50} < 20$ µM [15]. The phthalimide-ferruginol analogue, compound **3**, was the most promising and potent. In recent years, we have explored the antiviral properties of other semisynthetic abietane diterpenoids, including compound **3**, in models of Brazilian Zika virus [16] and Colombian Zika and Chikungunya viruses [17]. Ferruginol (**1**), compound **3** as well as some of the ferruginol analogues herein studied are characterized by a 2-isopropylphenol moiety which is typical of the monoterpenoid thymol (2-isopropyl-5-methylphenol). Thymol and its source essential oil have been also investigated as antiviral agents [38]. The precedent that ferruginol (**1**) had demonstrated a high cytopathogenic effect (CPE) reduction and effective replication inhibition for SARS-CoV (EC_{50} = 1.39 µM) [13] encouraged us to study this molecule as well as other readily synthesized analogues in other coronavirus models. Recently, some in silico studies of abietanes (for example, sugiol or 7-oxoferruginol) have demonstrated the potential inhibition of 3CLpro (Main protease, Mpro) of SARS-CoV-2 [39] and, more importantly, some extracts of medicinal plants containing abietanes (Rosemary, i.e., (+)-carnosic acid) have demonstrated anti-SARS-CoV-2 activity in cell assays [40]. Tanshinones are related abietane diterpenoids first isolated from the roots of *Salvia miltiorrhiza* "tanshen", a well-known Traditional Chinese Medicine [41]. Park et al. [41] investigated, in 2012, several isolated tanshinones on SARS-CoV 3CLpro with very promising results. Recently, tanshinone IIA has shown anti-SARS-CoV-2 activity with IC_{50} of 4.08 ng/µL [42]. The studies in this field are rising as other semisynthetic abietane derivatives have also shown, very recently, anti-SARS-CoV-2 pseudovirus activity [43]. Our results with 12-hydroxydehydroabietylamine (**14**) and 12-hydroxy-N,N-(4-fluorophthaloyl)dehydroabietylamine (**16**) point out that these molecules do reduce the cytopathogenic effect of human CoV 229E and possess molecular characteristics useful for further studies of drug development such as drug-likeness and safety. On the one hand, compound **16** is a fluorinated analogue which at 6.7 µM removed all CPE with a reduction in the viral titer of 2-logs approximately. Its molecular structure draws only one violation of the well-known Lipinski's rule of five, in particular, its calculated partition coefficient which may lead to the design of pro-drugs for enhanced bioavailability. On the other hand, compound **14**, which also reduces the viral titer in 2-logs, follows the Lipinski's rules and has a reasonably good predicted safety profile of administration in rats. However, further validation through toxicological reporting will be needed as well as studies of lead optimization, product formulation and efficacy.

5. Conclusions

In conclusion, our results underline the potential of the ferruginol derivatives and analogues as antivirals, and particularly in this case of in the treatment of coronavirus infections. They are accessed in short synthetic sequences in good overall yield from either the easily available methyl dehydroabietate, via commercial (−)-abietic acid, or commercial (+)-dehydroabietylamine. They do have druggable features encouraging additional studies of the mechanism as well other activities in related viruses, as they have demonstrated broad spectrum antiviral properties.

Supplementary Materials: The following supporting information can be downloaded at: https://www.mdpi.com/article/10.3390/v15061342/s1, Figures S1–S8: ^1H, ^{13}C, DEPT and ^{19}F NMR spectra for compounds **15** and **16**.

Author Contributions: Conceptualization, methodology, supervision, and funding acquisition M.A.G.-C. and M.V.; M.A.G.-C. synthesized the compounds; S.P. performed the antiviral experiments and the cellular experiments; M.A.G.-C. and M.V. analyzed the results; M.A.G.-C. and M.V. wrote or contributed to the writing of the manuscript; M.A.G.-C. and M.V., funding acquisition and project management. All authors have read and agreed to the published version of the manuscript.

Funding: The funds were granted by the Universitat Politècnica de Valencia (grant ADSIDEO AD 1902 to M.A.G.-C) and the Université de Lorraine and CNRS (M.V.). The project was partially supported by the Erasmus+ Programme of the European Union, Key Action 2: Strategic Partnerships, Project No. 2020-1-CZ01-KA203-078218.

Institutional Review Board Statement: Not applicable.

Informed Consent Statement: Not applicable.

Data Availability Statement: The data is available in the manuscript and the Supporting Information of this article. Raw data are available upon request.

Acknowledgments: The authors greatly acknowledge the PhotoNS platform of the L2CM Laboratory, University of Lorraine.

Conflicts of Interest: The authors declare no conflict of interest.

References

1. Ji, X.; Li, Z. Medicinal chemistry strategies toward host targeting antiviral agents. *Med. Res. Rev.* **2020**, *40*, 1519–1557. [CrossRef] [PubMed]
2. Ben-Shabat, S.; Yarmolinsky, L.; Porat, D.; Dahan, A. Antiviral effect of phytochemicals from medicinal plants: Applications and drug delivery strategies. *Drug Deliv. Trans. Res.* **2020**, *10*, 354–367. [CrossRef] [PubMed]
3. Newman, D.J.; Cragg, G.M. Natural Products as Sources of New Drugs over the Nearly Four Decades from 01/1981 to 09/2019. *J. Nat. Prod.* **2020**, *83*, 770–803. [CrossRef] [PubMed]
4. Thabti, I.; Albert, Q.; Philippot, S.; Dupire, F.; Westerhuis, B.; Fontanay, S.; Risler, A.; Kassab, T.; Elfalleh, W.; Aferchichi, A.; et al. Advances on antiviral activity of *Morus* spp. plant extracts: Human coronavirus and virus-related respiratory tract infections in the spotlight. *Molecules* **2020**, *25*, 1876. [CrossRef]
5. WHO Coronavirus (COVID-19) Dashboard. Available online: https://covid19.who.int (accessed on 25 May 2023).
6. Dong, J.; Varbanov, M.; Philippot, S.; Vreken, F.; Zeng, W.-B.; Blay, V. Ligand-based discovery of coronavirus main protease inhibitors using MACAW molecular embeddings. *J. Enzym. Inhib. Med. Chem.* **2023**, *38*, 24–35. [CrossRef]
7. Wanounou, M.; Caraco, Y.; Levy, R.H.; Bialer, M.; Perucca, E. Clinically relevant interactions between ritonavir-boosted nirmatrelvir and concomitant antiseizure medications: Implications for the management of COVID-19 in patients with epilepsy. *Clin. Pharmacokinet.* **2022**, *61*, 1219–1236. [CrossRef]
8. Vougogiannopoulou, K.; Corona, A.; Tramontano, E.; Alexis, N.M.; Skaltsounis, A.-L. Natural and nature-derived products targeting human coronaviruses. *Molecules* **2021**, *26*, 448. [CrossRef]
9. Wardana, A.P.; Aminah, N.S.; Rosyda, M.; Abdjan, M.I.; Kristanti, A.N.; Tun, K.N.W.; Choudhary, M.I.; Takaya, Y. Potential of diterpene compounds as antivirals, a review. *Heliyon* **2021**, *7*, e07777. [CrossRef]
10. Jantan, I.; Arshad, L.; Septama, A.W.; Haque, M.A.; Mohamed-Hussein, Z.-A.; Govender, N.T. Antiviral effects of phytochemicals against severe acute respiratory syndrome coronavirus and their mechanisms of action: A review. *Phytother. Res.* **2023**, *37*, 1036–1056. [CrossRef]
11. González, M.A. Aromatic abietane diterpenoids: Their biological activity and synthesis. *Nat. Prod. Rep.* **2015**, *32*, 684–704. [CrossRef]
12. González, M.A. Synthetic derivatives of aromatic abietane diterpenoids and their biological activities. *Eur. J. Med. Chem.* **2014**, *87*, 834–842. [CrossRef] [PubMed]
13. Wen, C.-C.; Kuo, Y.-H.; Jan, J.-T.; Liang, P.-H.; Wang, S.-Y.; Liu, H.-G.; Lee, C.-K.; Chang, S.-T.; Kuo, C.-J.; Lee, S.-S.; et al. Specific plant terpenoids and lignoids possess potent antiviral activities against severe acute respiratory syndrome Coronavirus. *J. Med. Chem.* **2007**, *50*, 4087–4095. [CrossRef] [PubMed]
14. Ryu, Y.B.; Jeong, H.J.; Kim, J.H.; Kim, Y.M.; Park, J.-Y.; Kim, D.; Naguyen, T.T.H.; Park, S.-J.; Chang, J.S.; Park, K.H.; et al. Biflavonoids from Torreya nucifera displaying SARS-CoV 3CLpro inhibition. *Bioorg. Med. Chem.* **2010**, *18*, 7940–7947. [CrossRef] [PubMed]
15. Roa-Linares, V.C.; Brand, Y.M.; Agudelo-Gomez, L.S.; Tangarife-Castaño, V.; Betancur-Galvis, L.A.; Gallego-Gomez, J.C.; González, M.A. Anti-herpetic and anti-dengue activity of abietane ferruginol analogues synthesized from (+)-dehydroabietylamine. *Eur. J. Med. Chem.* **2016**, *108*, 79–88. [CrossRef]
16. Sousa, F.T.G.; Nunes, C.; Romano, C.M.; Sabino, E.C.; González-Cardenete, M.A. Anti-Zika virus activity of several abietane-type ferruginol analogues. *Rev. Inst. Med. Trop. São Paulo* **2020**, *62*, e97. [CrossRef] [PubMed]
17. González-Cardenete, M.A.; Hamulić, D.; Miquel-Leal, F.J.; González-Zapata, N.; Jimenez-Jarava, O.J.; Brand, Y.M.; Restrepo-Mendez, L.C.; Martinez-Gutierrez, M.; Betancur-Galvis, L.A.; Marín, M.L. Antiviral profiling of C18- or C19-functionalized semisynthetic abietane diterpenoids. *J. Nat. Prod.* **2022**, *85*, 2044–2051. [CrossRef]
18. Pariš, A.; Štrukelj, B.; Renko, M.; Turk, V.; Pukl, M.; Umek, A.; Korant, B.D. Inhibitory effect of carnosolic acid on HIV-1 protease in cell-free assays. *J. Nat. Prod.* **1993**, *56*, 1426–1430. [CrossRef]
19. Shin, H.-B.; Choi, M.-S.; Ryu, B.; Lee, N.-R.; Kim, H.-I.; Choi, H.-E.; Chang, J.; Lee, K.-T.; Jang, D.S.; Inn, K.-S. Antiviral activity of carnosic acid against respiratory syncytial virus. *Virol. J.* **2013**, *10*, 303. [CrossRef]
20. Gigante, B.; Santos, C.; Silva, A.; Curto, M.; Nascimento, M.; Pinto, E.; Pedro, M.; Cerqueira, F.; Pinto, M.; Duarte, M.; et al. Catechols from Abietic Acid: Synthesis and Evaluation as Bioactive Compounds. *Bioorg. Med. Chem.* **2003**, *11*, 1631–1638. [CrossRef]
21. González, M.A.; Pérez-Guaita, D.; Correa-Royero, J.; Zapata, B.; Agudelo, L.; Mesa-Arango, A.; Betancur-Galvis, L. Synthesis and biological evaluation of dehydroabietic acid derivatives. *Eur. J. Med. Chem.* **2010**, *45*, 811–816. [CrossRef]
22. Zhang, G.-J.; Li, Y.-H.; Jiang, J.-D.; Yu, S.-S.; Qu, J.; Ma, S.-G.; Liu, Y.-B.; Yu, D.-Q. Anti-Coxsackie virus B diterpenes from the roots of Illicium jiadifengpi. *Tetrahedron* **2013**, *69*, 1017–1023. [CrossRef]
23. González, M.A.; Pérez-Guaita, D. Short syntheses of (+)-ferruginol from (+)-dehydroabietylamine. *Tetrahedron* **2012**, *68*, 9612–9615. [CrossRef]

24. Hamulić, D.; Stadler, M.; Hering, S.; Padrón, J.M.; Bassett, R.; Rivas, F.; Loza-Mejía, M.A.; Dea-Ayuela, M.A.; González-Cardenete, M.A. Synthesis and biological studies of (+)-Liquiditerpenoic acid A (abietopinoic Acid) and representative analogues: SAR studies. *J. Nat. Prod.* **2019**, *82*, 823–831. [CrossRef] [PubMed]

25. Malkowsky, I.M.; Nieger, M.; Kataeva, O.; Waldvogel, S.R. Synthesis and properties of optically pure phenols derived from (+)-dehydroabietylamine. *Synthesis* **2007**, *2007*, 773–778. [CrossRef]

26. Reed, L.J.; Muench, H. A simple method of estimating fifty percent endpoints. *Am. J. Epidemiol.* **1938**, *27*, 493–497. [CrossRef]

27. Mosmann, T. Rapid colorimetric assay for cellular growth and survival: Application to proliferation and cytotoxicity assays. *J. Immunol. Methods* **1983**, *65*, 55–63. [CrossRef]

28. Ertl, P.; Rohde, B.; Selzer, P. Fast calculation of molecular polar surface area as a sum of fragment-based contributions and its application to the prediction of drug transport properties. *J. Med. Chem.* **2000**, *43*, 3714–3717. [CrossRef]

29. Lipinski, C.A.; Lombardo, F.; Dominy, B.W.; Feeney, P.J. Experimental and computational approaches to estimate solubility and permeability in drug discovery and development settings. *Adv. Drug Deliv. Rev.* **1997**, *23*, 3–25. [CrossRef]

30. Veber, D.F.; Johnson, S.R.; Cheng, H.-Y.; Smith, B.R.; Ward, K.W.; Kopple, K.D. Molecular properties that influence the oral bioavailability of drug candidates. *J. Med. Chem.* **2002**, *45*, 2615–2623. [CrossRef]

31. Lagunin, A.; Zakharov, A.; Filimonov, D.; Poroikov, V. QSAR modeling of rat acute toxicity on the basis of PASS prediction. *Mol. Inform.* **2011**, *30*, 241–250. [CrossRef]

32. Canadian Center for Occupational Health and Safety. What Is an LD50 and LC50? Available online: http://www.ccohs.ca/oshanswers/chemicals/LD50.html#_1_6 (accessed on 28 January 2023).

33. Sadym, A.; Lagunin, A.; Filimonov, D.; Poroikov, V. Prediction of biological activity spectra via the Internet. *SAR QSAR Environ. Res.* **2003**, *14*, 339–347. [CrossRef] [PubMed]

34. Poroikov, V.V.; Filimonov, D.A.; Ihlenfeld, W.-D.; Gloriozova, T.A.; Lagunin, A.A.; Borodina, Y.B.; Stepanchikova, A.V.; Nicklaus, M.C. PASS biological activity spectrum predictions in the enhanced open NCI database browser. *J. Chem. Inf. Comput. Sci.* **2003**, *43*, 228–236. [CrossRef] [PubMed]

35. Denaro, M.; Smeriglio, A.; Barreca, D.; De Francesco, C.; Occhiuto, C.; Milano, G.; Trombetta, D. Antiviral activity of plants and their isolated bioactive compounds: An update. *Phytother. Res.* **2020**, *34*, 742–768. [CrossRef]

36. Islam, M.T.; Sarkar, C.; El-Kersh, D.M.; Jamaddar, S.; Uddin, S.J.; Shilpi, J.A.; Mubarak, M.S. Natural products and their derivatives against coronavirus: A review of the non-clinical and pre-clinical data. *Phytother. Res.* **2020**, *34*, 2471–2492. [CrossRef]

37. Verma, S.; Twilley, D.; Esmear, T.; Oosthuizen, C.B.; Reid, A.-M.; Nel, M.; Lall, N. Anti-SARS-CoV natural products with the potential to inhibit SARS-CoV-2 (COVID-19). *Front. Pharmacol.* **2020**, *11*, 561334. [CrossRef] [PubMed]

38. Kowalczyk, A.; Przychodna, M.; Sopata, S.; Bodalska, A.; Fecka, I. Thymol and thyme essential oil –new insights into selected therapeutic applications. *Molecules* **2020**, *25*, 4125. [CrossRef]

39. Diniz, L.R.L.; Perez-Castillo, Y.; Elshabrawy, H.A.; Bezerra-Filho, C.S.M.; Pergentino de Sousa, D. Bioactive terpenes and their derivatives as potential SARS-CoV-2 proteases inhibitors from molecular modeling studies. *Biomolecules* **2021**, *11*, 74. [CrossRef] [PubMed]

40. Jan, J.-T.; Cheng, T.-J.R.; Juang, Y.-P.; Ma, H.-H.; Wu, Y.-T.; Yang, W.-B.; Cheng, C.-W.; Chen, X.; Chou, T.-H.; Shie, J.-J.; et al. Identification of existing pharmaceuticals and herbal medicines as inhibitors of SARS-CoV-2 infection. *Proc. Natl. Acad. Sci. USA* **2021**, *118*, e20211579118. [CrossRef]

41. Park, J.-Y.; Kim, J.H.; Kim, Y.M.; Jeong, H.J.; Kim, D.W.; Park, K.H.; Kwon, H.-J.; Park, S.-J.; Lee, W.S.; Ryu, Y.B. Tanshinones as selective and slow-binding inhibitors for SARS-CoV cysteine proteases. *Bioorg. Med. Chem.* **2012**, *20*, 5928–5935. [CrossRef]

42. Elebeedy, D.; Elkhatib, W.F.; Kandeil, A.; Ghanem, A.; Kutkat, O.; Alnajjar, R.; Saleh, M.A.; El Maksoud, A.I.A.; Badawy, I.; Al-Karmalawy, A.A. Anti-SARS-CoV-2 activities of tanshinone IIA, carnosic acid, rosmarinic acid, salvianolic acid, baicalein, and glycyrrhetinic acid between computational and in vitro insidghts. *RSC Adv.* **2021**, *11*, 29267–29286. [CrossRef]

43. Tret'yakova, E.V.; Ma, X.; Kazakova, O.B.; Shtro, A.A.; Petukhova, G.D.; Klabukov, A.M.; Dyatlov, D.S.; Smirnova, A.A.; Xu, H.; Xiao, S. Synthesis and evaluation of diterpenic mannich bases as antiviral agents against influenza A and SARS-CoV-2. *Phytochem. Lett.* **2022**, *51*, 91–96. [CrossRef] [PubMed]

Article

Discovery of Potential Inhibitors of SARS-CoV-2 Main Protease by a Transfer Learning Method

Huijun Zhang [1,2], Boqiang Liang [3], Xiaohong Sang [1], Jing An [4] and Ziwei Huang [1,4,5,*]

1 Cechanover Institute of Precision and Regenerative Medicine, School of Medicine, The Chinese University of Hong Kong (Shenzhen), Shenzhen 518172, China
2 School of Life Sciences, University of Science and Technology of China, Hefei 230026, China
3 Nobel Institute of Biomedicine, Zhuhai 519080, China
4 Division of Infectious Diseases and Global Public Health, Department of Medicine, School of Medicine, University of California San Diego, La Jolla, CA 92093, USA
5 School of Life Sciences, Tsinghua University, Beijing 100084, China
* Correspondence: zwhny@yahoo.com

Abstract: The COVID-19 pandemic caused by SARS-CoV-2 remains a global public health threat and has prompted the development of antiviral therapies. Artificial intelligence may be one of the strategies to facilitate drug development for emerging and re-emerging diseases. The main protease (M^{pro}) of SARS-CoV-2 is an attractive drug target due to its essential role in the virus life cycle and high conservation among SARS-CoVs. In this study, we used a data augmentation method to boost transfer learning model performance in screening for potential inhibitors of SARS-CoV-2 M^{pro}. This method appeared to outperform graph convolution neural network, random forest and Chemprop on an external test set. The fine-tuned model was used to screen for a natural compound library and a *de novo* generated compound library. By combination with other in silico analysis methods, a total of 27 compounds were selected for experimental validation of anti-M^{pro} activities. Among all the selected hits, two compounds (gyssypol acetic acid and hyperoside) displayed inhibitory effects against M^{pro} with IC50 values of 67.6 μM and 235.8 μM, respectively. The results obtained in this study may suggest an effective strategy of discovering potential therapeutic leads for SARS-CoV-2 and other coronaviruses.

Keywords: deep learning; SARS-CoV-2 M^{pro}; transfer learning; drug development; natural compound

check for **updates**

Citation: Zhang, H.; Liang, B.; Sang, X.; An, J.; Huang, Z. Discovery of Potential Inhibitors of SARS-CoV-2 Main Protease by a Transfer Learning Method. *Viruses* **2023**, *15*, 891. https://doi.org/10.3390/v15040891

Academic Editor: Simone Brogi

Received: 13 February 2023
Revised: 26 March 2023
Accepted: 27 March 2023
Published: 30 March 2023

1. Introduction

SARS-CoV-2, first reported in the beginning of 2020 [1], has caused over 759 million confirmed infection cases including 6.8 million deaths as of March of 2023 as reported to the World Health Organization (WHO). SARS-CoV-2 is a novel coronavirus which shares 79.5% sequence similarity with SARS-CoV [2], both of which belong to the Coronaviridae family, which contains positive single-stranded encapsulated viruses [3]. The virus genome contains several open-reading frames (ORFs) that encode four structure proteins (sps), 16 non-structure proteins (nsps) and several accessory proteins [4,5]. Nsp5 is the main protease (M^{pro}), which is also known as 3-Chymotrypsin like protease ($3CL^{pro}$). It has been characterized as one of the potential druggable targets of SARS-CoV-2 owing to its essential role in viral replication and transcription [6]. Active M^{pro} consists of a homodimer while each protomer has three domains (I–III) [7]. The active site of M^{pro} locates in the cleft between domains I and II and features the catalytic Cys-His dyad (Cys145-His41) [8–10]. After ORF1a/b translates into two polyproteins pp1a and pp1ab, M^{pro} cleavages at 11 distinct sites to release functional polypeptides [6,11,12]. The core recognition sequence is Leu-Gln↓ (Ser/Ala/Gly) [7,13]. Moreover, the high conservatism of M^{pro} among coronaviruses and the absence of homologues with similar cleavage specificity in humans make it an attractive target for antiviral drug discovery [14,15].

Many clinical trials have been initiated in the search for the prevention and treatment of coronavirus disease 2019 (COVID-19). At the time of writing, several vaccines have been approved by the U.S. Food and Drug Administration (FDA), including ones by Pfizer/BioNTech, Moderna and Johnson and Johnson/Jassen (JnJ) [16]. There have also been attempts in preclinical development of multiple formulations of vaccine candidates [17]. However, the continuing mutations in the viral genome may affect the protective effects of current vaccines. Notably, the emergence of the Omicron (B.1.1.529) VoC which contains a high number of mutations in the viral spike protein has an increased reinfection risk [18]. As the pandemic threat continues and vaccines cannot provide complete and lasting protection [19], the need for antiviral agents to treat infected patients remains. Drug repurposing, for the advantage of already confirmed clinical profiles data, is considered to be a fast and low-cost approach to find potential effective therapeutic agents against COVID-19 [20–22]. At present, there are only three drugs approved by the FDA for the treatment of COVID-19, including Actemra (Tocilizumab), Veklury (Remdesivir) and Olumiant (baricitinib) [23]. There are several authorized products under an EUA for the clinical treatment of COVID-19 as well, including two anti-viral drugs which are Paxlovid (nirmatrelvir and ritonavir) and Lagevrio (molnupiravir), three immune modulators, five SARS-CoV-2-targeting monoclonal antibodies, sedatives and renal replacement therapies. Hundreds of drugs are undergoing clinical trials for COVID-19, such as favipiravir, lopinavir, ribavirin, ritonavir, and tocilizumab, which have shown positive effects in vitro [17,24]. Dexamethasone and hydroxychloroquine have been withdrawn from treatment options because of the insignificant protection benefits and serious side effects [24–26].

Drug discovery and development is a time-consuming process in which computational methods can help speed up the identification and application of drug candidates. Deep learning techniques have recently received wide attention and been applied to drug discovery [27]. To facilitate efforts in exploring the chemical space against various therapeutic targets for SARS-CoV-2, deep learning combined with computer-aided drug design (CADD) methodologies such as docking and molecular dynamics simulation have been extensively used [20,28–35]. However, labeled data scarcity remains a challenge for supervised learning due to time-consuming and laborious benchwork testing. To better solve this problem, transformer pre-training by making use of large amounts of unlabeled data plus downstream task-specific finetuning has become a powerful architecture for learning representation of texts, i.e., natural language processing (NLP) [36–40]. Compared with many previous approaches such as graph neural networks (GNNs), modern transformers display substantial gain of efficiency and throughput [41,42]. Given the availability of millions of Simplified Molecular-Input Line-Entry system (SMILES) strings, different molecular property prediction tasks can be tackled by using learned representations of functional groups and atoms learned by the model [43–45].

In the present study, we used pre-trained ChemBERTa [39] which is based on RoBERTa [37] transformer implementation from HuggingFace and fine-tuned it on a dataset which contains over 280,000 molecules screened against SARS-CoV-1 M^{pro} [29]. Considering the fact that natural compounds have been sources of pharmacologically active molecules for a long history and that the *de novo* design of novel scaffolds might expand the chemical space of active drug candidates, we made predictions of two libraries, a natural compound library (TargetMol) and a *de novo* generated compound library from the literature by Santana et al. [29], to seek molecules against SARS-CoV-2 M^{pro}. The predicted active molecules were evaluated using molecular docking and PAINS filtering. In vitro enzyme activity inhibition experiments were performed to validate the selected hits.

2. Materials and Methods

2.1. Dataset Preparation

Due to the high sequence similarity (~76%) shared between SARS-CoV-2 M^{pro} and SARS-CoV-1 M^{pro}, we selected a dataset which contains over 280,000 molecules against SARS-CoV-1 M^{pro} as the fine-tuning dataset. Obtained from the publication of Santana and Silva-Jr [29], it consisted of 629 active molecules and 288,940 inactive molecules. Based on the fact that one molecule can be represented by more than one SMILES strings, and that the augmented dataset with enumerated SMILES could help improve model performance [46], we used the same approach to augment the dataset. Different ratios of SMILES enumeration were calculated with a python script, which is available at https://github.com/Ebjerrum/SMILES-enumeration (accessed on 1 July 2020).

2.2. Chemical Space Analysis

Morgan fingerprints for each molecule using radius 2 and 2048 bits fingerprint vectors were determined after obtaining the canonical SMILES by rdkit in Python. Then, t-Distributed Stochastic Neighbor Embedding (t-SNE) clustering analysis was performed by the scikit-learn package in Python. Data points were reduced from 2048 dimensions to 2 dimensions by t-SNE. All t-SNE parameters were Scikit-learn's default values.

2.3. Model Performance Evaluation

The fine-tuned model performance was evaluated with five-fold cross-validation. Scaffold splitting was used to ensure that the training/validation set is more structurally different, which, as a result, is more challenging for the model. Additionally, an external independent test dataset which was collected from results of a screening assay against SARS-CoV-2 M^{pro} using X-ray crystallography (at Diamond Light Source, Oxfordshire, United Kingdom) [47] was used. It consisted of 880 molecules with 78 hits. The performance of Chemprop [48], which is a freely available message passing neural network (MPNN) (http://chemprop.csail.mit.edu/predict (accessed on 27 October 2021)) on the same dataset, was also determined for comparison. Various evaluation metrics including area under the receiver–operator characteristic curve (au_roc), area under the precision–recall curve (au_prc), recall score, accuracy score, precision score and f1 score were calculated. Recall = TP/(TP + FN). Accuracy = (TP + TN)/(TP + FN + TN + FP). Precision = TP/(TP + FN). F1 = 2 × recision × Recall/(Precision + Recall). TP, TN, FP and FN stand for true positive, true negative, false positive and false negative, respectively. Figures were plotted by matplotlib in Python.

2.4. Compound Libraries and Compounds

The Natural Compound Library obtained from Targetmol (L6000) contains 2364 compounds after 228 compounds with large molecular weight were removed. The *de novo* generated compound library of Santana and Silva-Jr contains 66,392 generated molecules. PF-07321332 and Boceprevir were purchased from MedChemExpress. Compounds T2983, T3872, T2765, T2950, T2730, T2755, T2957, T3012, T2133, T3227, T1016, T2844, T2775, T1648, T1400, T1160, T2570, T3232, T5429, T2727, T5497, T1035, T1609, T6S1529, T3149, T3S1612 and TL0006 were purchased from TargetMol.

2.5. PAINS Filtering

All predicted active compounds were submitted to FAF-DRUGS4 server (available at http://fafdrugs4.mti.univ-paris-diderot.fr (accessed on 25 November 2021)) by evaluating their physicochemical properties [49]. Molecules with suspicious substructure features were flagged out by Pan Assay Interference Compounds (PAINS) filter.

2.6. Molecular Docking Protocol

Crystal structures of SARS-CoV-2 M^{pro} bound with inhibitor PF-07321332 (PDB ID: 7VH8) and inhibitor N3 (PDB ID: 6LU7) were accessed from the RCSB Protein Data Bank. The M^{pro} protein and inhibitor ligands were prepared using AutoDockTools by removing water atoms and adding polar hydrogen atoms and charges. Prepared protein and ligand files were converted to PDBQT format. Molecular docking was carried out using AutoDock Vina-1.2.0 software while M^{pro} in the structure of 7VH8 was used as the docking protein due to its higher resolution. The redocking of PF-07321332 and N3 was performed in order to validate the performance of the docking model; then, the docking model was determined for the virtual screening process. The grid box center was set at X: -18.217, Y: 17.605, Z: -25.603 and box dimension was set to X: 20, Y: 26, Z: 24. The binding affinities of the compounds with M^{pro} protein were calculated and ranked.

2.7. Protein Expression and Purification of SARS-CoV-2 M^{pro}

The plasmid pET-28b-SARS-CoV-2-M^{pro} was a kind gift from Professor George Fu Gao from the Institute of Microbiology, Chinese Academy of Sciences. The expression plasmid was transformed into *E. coli* strain BL21 cells and then cultured in LB medium containing 50 µg/mL kanamycin in a shaking incubator at 37 °C. When the cells were grown to an OD_{600} of 0.6–0.8, 0.6 mM IPTG was added to the cell culture to induce the protein expression at 16 °C. After 18 h, the cells were harvested by centrifugation at 4000 rpm for 20 min at 4 °C. The cell pellets were washed twice by PBS, resuspended in lysis buffer (50 mM HEPES, 300 mM NaCl, 10 mM imidazole, pH 7.5), lysed by sonication on ice for 3 s ON time 5 s OFF time for 30 min of total time and then clarified by ultracentrifugation at 18,000 rpm at 4 °C for 40 min to remove debris. The supernatants were then purified by TALON metal affinity resin and washed with washing buffer (25 mM HEPES, 500 mM NaCl, pH 7.5) to remove unspecific binding proteins. The His-tagged M^{pro} was eluted by elution buffer (25 mM HEPES, 500 mM NaCl, 300 mM imidazole, pH 7.5). His-tagged SUMO protease (home-made) was added to remove the His-tag, His-tagged SUMO protease and uncleaved His-tag protein overnight at 4 °C. The M^{pro} was further purified by His60 Ni superflow resin. The quality of M^{pro} was checked by SDS-PAGE, and the concentration of M^{pro} was determined via a BCA Protein Assay Kit. The purified M^{pro} was stored in (10 mM Tris-HCl, 1 mM DTT, 1 mM EDTA, 10% glycerol, pH 7.5).

2.8. FRET-Based M^{pro} Enzyme Activity Inhibition Assay

Fluorescence resonance energy transfer (FRET)-based M^{pro} enzyme activity inhibition assay was conducted as follows. First, 5 µL serially diluted concentrations of candidate compounds were incubated with 35 µL 150 nM M^{pro} in Assay Buffer (10 mM Tris-Hcl, pH 7.5; 1 mM DTT; 1 mM EDTA; 0.01% Triton X-100) in a 96-well plate at room temperature for 30 min. This was followed with the adding of 10 µL 20 µM fluorogenic substrate (Dabcyl-KTSAVLQSGFRKME-Edans, P9733-5 mg, purchased from Beyotime) in Assay Buffer on ice, after which the plate was shaken for 1 min and then transferred to a 37 °C incubator for 30 min of incubation. Fluorescence signals (excitation wavelength at 340 nm and emission wavelength at 490 nm) were measured using a PerkinElmer Envision multimode plate reader. Experiments were performed in triplicate. Experimental data were plotted by GraphPad Prism 8.0.

3. Results

3.1. Dataset Preprocessing and Chemical Space Analysis

Because of the highly conserved sequence and the similar substrate binding site of M^{pro} between SARS-CoV-1 and SARS-CoV-2, the previously described inhibitors targeting SARS-CoV-1 M^{pro} could be used as templates for the design of novel inhibitors against SARS-CoV-2. Thus, the dataset used for fine-tuning was collected from PubChem (AID:1706) and from the literature, which contains 629 active molecules and 288,940 inactive molecules [29,50]. Structural relationships between active compounds and inactive compounds using t-SNE (t-distributed stochastic neighbor embedding) were calculated (Figure 1A). Analysis details were provided in the Supplementary Information. Data obtained from PubChem were the result of a QFRET-based biochemical high-throughput screening assay. Two inactive molecules were dropped due to long SMILES length, which is over 150. Scaffold-based 5-fold split was used to split the data. Due to the high imbalance of the lab dataset, data augmentation via a SMILES enumeration script was used to create more copies of active molecules. As shown in Table 1, different ratios of augmentation were conducted for later comparison to seek the optimum dataset size. To confirm the scaffold differences among the five-fold compounds, we also analyzed the structural relationships among the five-fold molecules using t-SNE (Figure 1B).

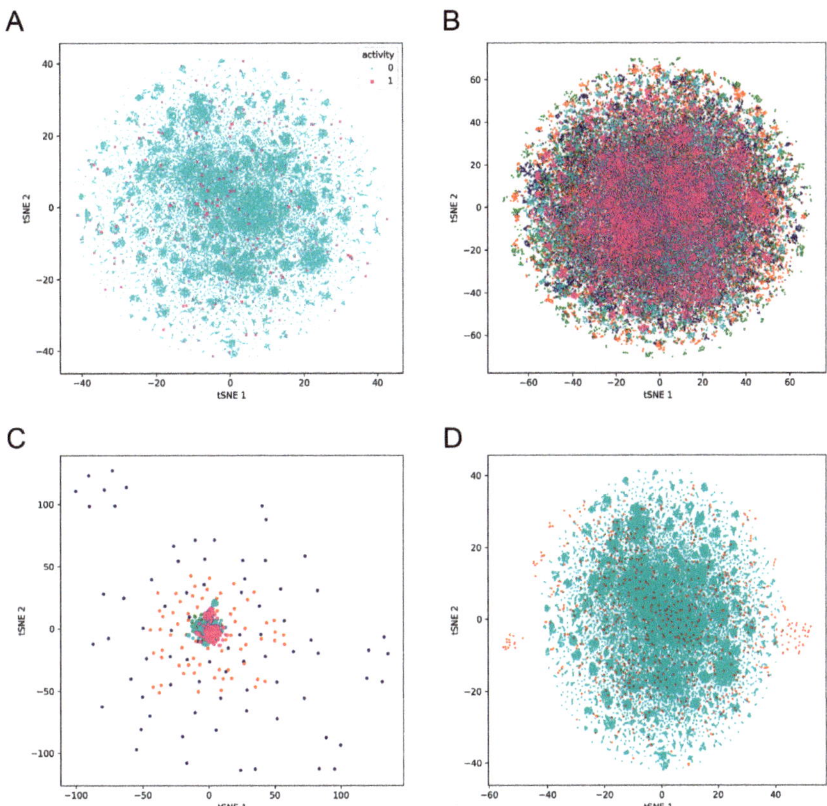

Figure 1. t-Distributed stochastic neighbor embedding (t-SNE) analysis of (**A**) active molecules (magenta) and inactive molecules (cyan) of the original dataset; (**B**) molecules in five subsets; (**C**) active molecules in five subsets; (**D**) molecules in original dataset (cyan) and the independent test dataset (red).

Table 1. Augmented dataset.

	Fold 1		Fold 2		Fold 3		Fold 4		Fold 5	
Label	1	0	1	0	1	0	1	0	1	0
Original	142	57,771	75	57,838	168	57,754	164	57,749	80	57,833
Augmentation_10	1420	57,771	750	57,838	1680	57,754	1640	57,749	800	57,833
Augmentation_20	2840	57,771	1500	57,838	3260	57,754	3280	57,749	1600	57,833
Augmentation_80	11,360	57,771	6000	57,838	13,440	57,754	13,120	57,749	6400	57,833

3.2. Performance of the Fine-Tuned Model

We used transfer learning to fine-tune a classification model for M^{pro} target bioactivity prediction. A pre-trained ChemBERTa model was downloaded from huggingface. To compare the performance of the classifier on different datasets, we calculated various evaluation scores using five-fold cross-validation. As shown in Table 2, the pre-trained model using augmented training data displayed better predictive ability on the validation dataset than no augmented data. An obvious improvement of evaluation scores was observed in all augmented datasets, especially in datasets with augmented active molecules 20 and

80 times. In addition, with augmented datasets, the pre-trained model for downstream task learning outperformed Graph Convolution Neural Network (GCNN) and baseline model Random Forest (RF).

Table 2. Performance of fine-tuned model, GCNN model and RF on validation dataset.

Model		Transfer Learning			GCNN		Random Forest	
Dataset	Original	Active× 10	Active× 20	Active× 80	Active× 20	Active× 80	Active× 20	Active× 80
mcc	0.06931	0.88577	0.77580	0.97618	0.17995	0.26405	0.47148	0.46608
tp	1.4	1021.4	2294.6	9712	192	2160	948	3792
tn	57787.4	57757.8	57761.6	57755	57489.2	54236.6	57784.2	57783.8
fp	0	29.8	26	32.6	298.4	3551	3.4	3.8
fn	124.4	236.6	219.6	352	232.4	7904	1568	6272
auroc	0.52137	0.96239	0.98226	0.99226	0.76543	0.76389	0.77730	0.77897
auprc	0.03054	0.88366	0.95221	0.98621	0.3119	0.50636	0.52265	0.65047
recall	0.01532	0.80871	0.90753	0.96350	0.08885	0.19786	0.31199	0.31759
accuracy	0.99785	0.99550	0.98794	0.99436	0.95663	0.83353	0.97394	0.90750
precision	0.36667	0.97514	0.98818	0.99568	0.59492	0.84163	0.97671	0.99222
f1	0.02881	0.88391	0.94605	0.97931	0.11763	0.20417	0.38709	0.39592

To assess the model performance more realistically, we also evaluated on an external test dataset [47]. The external test dataset contains 880 fragments including 78 hits, which were screened through a combined mass spectrometry and X-ray approach against SARS-CoV-2 M^{pro}. The structural diversity between the training and external datasets was also analyzed using t-SNE, as shown in Figure 1D. As shown in Table 3, a drop in performance on the external dataset was observed compared with the performance on the validation dataset, which was expected because no molecules in the test dataset were learned by the model before. The F1 score is one of the most meaningful metrics because it represents the harmonic mean of recall and precision. Datasets with 20 times more active molecules exhibited the highest f1 score of 0.34793, while GCNN and RF using the same training dataset only scored 0.0788 and 0.02025, respectively. Au_prc and au_roc were two other evaluation metrics for imbalanced data, while the former is more sensitive to the improvements of the positive class, which is a better indicator. In fine-tuned models, training datasets with 10 and 20 times more active molecules achieved similar au_prc scores, of 0.28671 and 0.28472, respectively, while the 80 times augmented datasets achieved a lower au_prc of 0.23152.

Table 3. Performance of fine-tuned model, GCNN model and RF on an external dataset.

Model		Transfer Learning			GCNN		Random Forest	
Dataset	Original	Active× 10	Active× 20	Active× 80	Active× 20	Active× 80	Active× 20	Active× 80
mcc	0	0.30798	0.30973	0.26022	0.05526	0.08691	0.08652	0.08652
tp	0	16.6	22.6	26.2	4.8	11.8	0.8	0.8
tn	802	789.8	774	746.2	787.8	754.2	802	802
fp	0	12.2	28	55.8	14.2	47.8	0	0
fn	78	61.4	55.4	51.8	73.2	66.2	77.2	77.2
auroc	0.50025	0.66905	0.67788	0.68109	0.66972	0.68249	0.68298	0.71836
auprc	0.08868	0.28671	0.28427	0.23152	0.20616	0.23784	0.35956	0.40239
recall	0	0.21282	0.28974	0.39990	0.06154	0.15128	0.01026	0.01026
accuracy	0.72909	0.91636	0.90523	0.87773	0.90068	0.87045	0.91227	0.91227
precision	0	0.58623	0.44416	0.31989	0.13992	0.23694	0.8	0.8
f1	0	0.29778	0.34973	0.32647	0.07880	0.10446	0.02025	0.02025

Having evaluated performances of various models and confirmed the advantages of data augmentation, we used the whole dataset as training input to compare the prediction abilities of transfer learning and a freely available classifier chemprop (http://chemprop.

csail.mit.edu/ (accessed on 27 October 2021)) on this external test dataset. Chemprop could be used for molecular property prediction through a Message Passing Neural Network (MPNN), which works directly on a molecular graph [48]. Transfer learning with a 20 times augmented dataset achieved the highest au_prc of 0.34433, while the AUC-PR of chemprop was 0.19321. The f1 score of transfer learning using 20 times augmentation data was 0.41321, while that of chemprop was 0.19048 (Table 4).

Table 4. Performance of fine-tuned model and Chemprop on an external dataset.

Model	Input	Mcc	Auroc	Auprc	Recall	Accuracy	Precision	f1
Transfer	Active × 20	0.37804	0.68186	0.34433	0.34359	0.91341	0.51978	0.41321
Learning	Active × 80	0.29978	0.68306	0.26118	0.34359	0.89091	0.37632	0.35833
Chemprop	original	0.17636	0.68152	0.19321	0.12821	0.90341	0.37037	0.19048

3.3. Prediction of Bioactivities of Natural Compound and De Novo Generated Molecule Libraries

The fine-tuned model using a 20 times augmented dataset was then used for making predictions of the Targetmol natural compound library and a *de novo* generated molecule library. Scoring ranks were the average results of five independent predictions. A total of 385 natural compounds and 66 *de novo* generated molecules were predicted as bioactive. The lists of predicted active compounds are provided in Tables S1 and S2. The top ranked 20 compounds from the natural compound library and 20 from the *de novo* generated molecule library are shown in Figures 2 and 3, respectively.

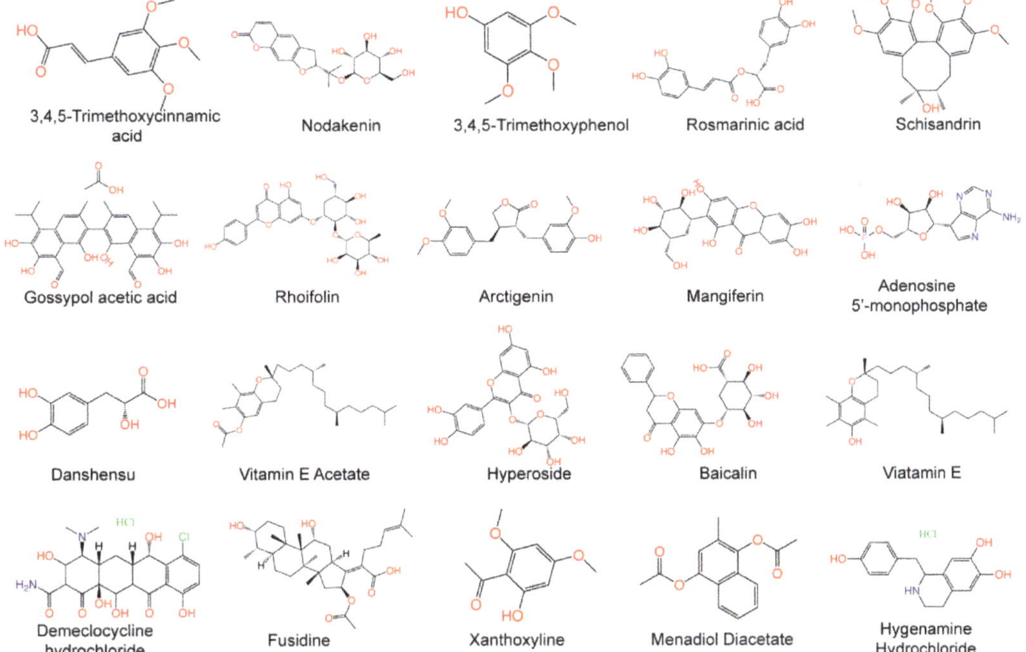

Figure 2. Top ranked 20 natural compounds screened by fine-tuned model. Rankings were from averages of five independent model predictions.

3.4. Molecular Docking Screening

We next submitted all the predicted active compounds to docking simulation using AutoDock Vina (version1.2.0). Crystal structures of SARS-CoV-2 M^{pro} in complex with inhibitor PF-07321332 (PDB:7VH8) and N3 (PDB:6LU7) were both downloaded from the Protein Data Bank. PF-07321332 (Paxlovid) is an oral SARS-CoV-2 M^{pro} inhibitor developed by Pfizer and has shown positive responses in Phase III trials in combination with Ritonavir [51]. N3 is a covalent inhibitor of SARS-CoV-2 M^{pro} derived from the inhibitor targeting SARS-CoV-1 M^{pro} [15]. After calculating the binding affinities of the compounds with M^{pro}, 46 compounds were selected for further binding pose analysis according to a cutoff score of -8.5 kcal/mol. After analysis of residue interactions in crystal structures of M^{pro} with PF-07321332 and N3, ligand interactions with F140 and E166 were considered critical for binding with M^{pro}. Twelve molecules were finally confirmed as hits due to more than two H-bonds formed with residues F140 and E166. These hits include 10 natural compounds (T5429, T2727, T5497, T1035, T1609, T6S1529, T3149, T3S1612, TL0006) and two *de novo* generated molecules (58353 and 52917). The binding poses of these hit compounds with M^{pro} are displayed in Table 5.

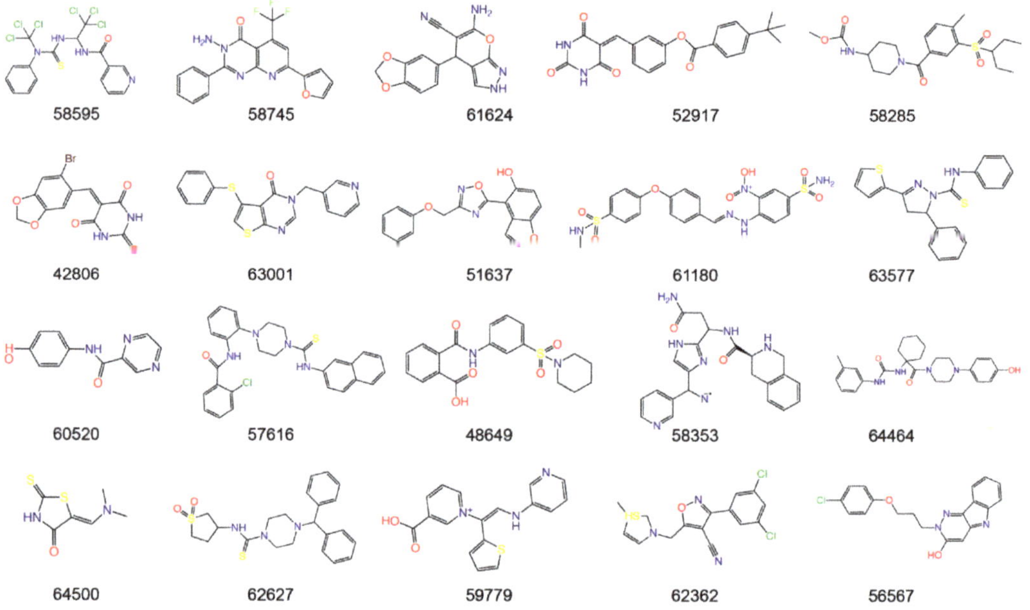

Figure 3. Top ranked 20 *de novo* generated molecules screened by fine-tuned model. Rankings were from averages of five independent model predictions.

3.5. PAINS Filtering

In the final round of the in silico analysis, we performed PAINS (pan assay interference compounds) filtering through a freely available web server FAF-Drugs4 to estimate potential molecules that may interfere with biological assays [49]. These compounds may display false positives in screening assays via a number of means and therefore represent poor choices for drug development [52]. We submitted all predicted active molecules to the server; 78 natural compounds and 5 *de novo* generated molecules were flagged as PAINS. For those natural compounds, among the top 20 predicted hits and 10 high-dock-scoring hits, T2765 (rosmarinic acid), T2730 (gossypol acetic acid), T3012 (mangiferin), T3227 (danshensu), T2844 (hyperoside), T2775 (baicalin), T3232 (higenamine hydrochloride), T5429 (theaflavin 3,3'-digallate), T2727 (salvianolic acid B), T6S1529 (1,5-Dicaffeoylquinic acid), T3149 (salvianolic acid C), TL0006 (chicoric acid) and T3242 (breviscapin) were flagged as PAINS. For those *de novo* generated molecules, among the top 20 predicted hits and two high dock-scoring hits, compound 52917, compound 42806, compound 64500 and compound 58353 were flagged as PAINS. However, virtual filters may not be perfect in identifying molecules that interfere with biological assays. Therefore, the judgement of PAINS should be taken with caution, and experimental confirmation is always necessary before any 'problematic' molecules are discarded.

Table 5. Summary of selected molecules screened against SARS-CoV-2 M^{pro} with their structures, binding affinities and interactive residues.

IDs	Name	Source	Structure	Docking Score (kcal/mol)	M^{pro} Residues Interacting with Molecules through H-Bond and Other Types
PF-07321332	-	-		−9.2	
T5429	Theaflavin 3,3′-digallate	Black tea		−10.4	
T2727	Salvianolic acid B	Slvia miltiorrhiza		−9.2	
T5497	AMAROGENTIN	Gentiana scabra		−8.9	
T1035	Hesperidin	Citrus sinensis		−8.8	
T1609	NAD+	Punica granatum		−8.8	
T6S1529	1,5-Dicaffeoylquinic acid	Lonicera japonica		−8.8	
T3149	Salvianolic Acid C	Slvia miltiorrhiza		−8.7	

Table 5. *Cont.*

IDs	Name	Source	Structure	Docking Score (kcal/mol)	M^{Pro} Residues Interacting with Molecules through H-Bond and Other Types
T3S1612	Kuwanon G	Morus alba		−8.7	
TL0006	Chicoric Acid	Cichorium intybus		−8.6	
T3242	Breviscapin	Erigeron		−8.5	
58353	-	-		−8.7	
52917	-	-		−8.6	

3.6. In Vitro Binding Assay Validation

In order to validate the in vitro binding activities of selected hits, we purchased 18 natural compounds from the top 20 scored active compounds predicted by deep learning and 9 selected natural compounds screened by molecular docking from Targetmol. PF-07321332 and Boceprevir were used as positive controls. These 27 compounds were tested by SARS-CoV-2 M^{Pro} inhibition assay at concentrations of 200 µM and 40 µM. As shown in Figure 4A, except for PF-07321332, only compound T2730 (Gossypol acetic acid) and T2844 (Hyperoside) had over 50% inhibitory effects against M^{Pro} catalytic activity at 200 µM, while all tested compounds exhibited less than 50% inhibitory effects at 40 µM. The IC50 values of compounds T2730 and T2844 were further determined in dose-dependent studies, which are 67.6 µM and 235.8 µM, respectively. Noteworthily, as many researchers have reported that some molecules self-associating into colloidal aggregates is one of the most common cause of non-specific inhibition [53,54], we added detergent triton X-100 in the experimental solvent; thus, the false positives caused by aggregate-based inhibition could be avoided. When treated with and without triton X-100, the inhibitory efficacies of the positive control Boceprevir and compound T2730 displayed no obvious differences within the experimental error, although a slight decrease in the inhibitory effects of T2844 when added with triton-X100 was observed. Gossypol acetic acid, a polyphenolic compound isolated form cottonseeds, has been reported to inhibit Bcl-2, Bcl-xL and Mcl-1 function and have antiproliferative effects on some cancer cells in vitro [55]. Hyperoside, a naturally occurring flavonoid compound isolated from Artemisia capillaris, shows myocardial protective, hepatoprotective, anti-redox and anti-inflammatory activities [56]. It is also a derivative of quercetin, which was predicted to potentially inhibit SARS-CoV-2 M^{pro} [57]. Recently, Dr. Souza's group has demonstrated a biflavonoid (agathisflavone) and two flavonols (myricetin and fisetin) as non-competitive inhibitors of SARS-CoV-2 M^{pro}, which indicated an interesting potential mode of action of these classes of compounds [58,59]. Further studies to deeper understand the mechanism of actions of these compounds are essential for chemical design to improve the activity profiles. Taken together, we have found that two natural compounds showed biological activity against M^{pro} in vitro.

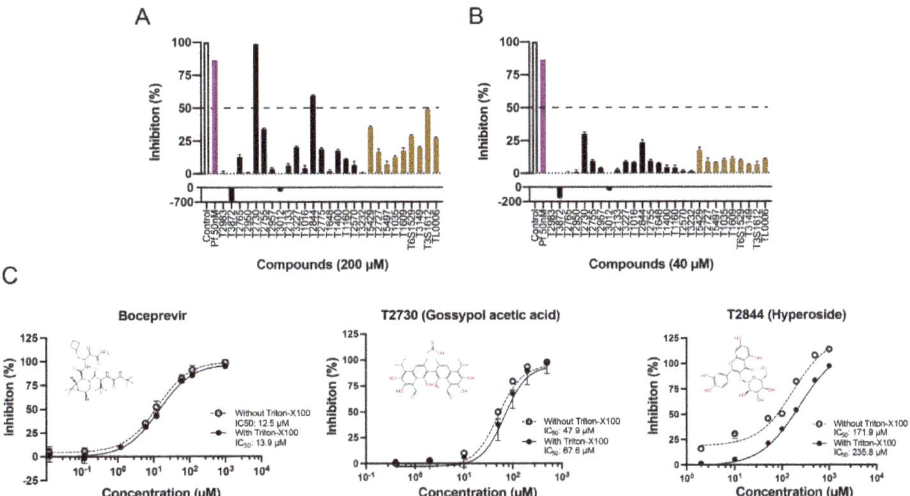

Figure 4. Inhibition of SARS-CoV-2 M^{pro}. (**A**) Inhibition percentage of selected compounds at concentrations of 200 μM. (**B**) Inhibition percentage of selected compounds at concentrations of 40 μM. (**C**) Representative curves of Boceprevir, compound T2730 and T2844. All data are from at least three independent experiments and shown as mean ± SD.

4. Discussion

Artificial intelligence-aided drug design is becoming extensively used especially for emerging diseases because of its potential advantage in saving the time and cost of the drug discovery and development process. Here, we used a data augmentation method to boost transfer learning model performance in the fine-tuned bioactivity prediction task. The model outperformed GCNN, RF and chemprop. A natural compound library and a *de novo* generated molecule library were screened by this fast and efficient model. In combination with frequently used CADD techniques, such as molecular docking and PAINS-filtering, this method allowed us to select a group of 27 commercial available compounds for further experimental validation. Among these experimentally tested compounds, gossypol acetic acid and hyperoside displayed inhibitory effects against M^{pro} with IC50 values of 67.6 μM and 235.8 μM, respectively. Even though these two compounds displayed only micromolar potency, they still provided valuable scaffolds for further drug design in searching for treatment of COVID-19. Follow-up cellular assays and in vivo experiments are also essentially necessary to ensure the efficacy and safety of these compounds and more deeply understand the mechanism of actions. Overall, our results demonstrated the feasibility of finding potential candidate compounds using a deep learning method, and the experimental outcome suggested that these natural products may merit further biological studies of their potential ability in blocking SARS-CoV-2 infection.

Supplementary Materials: The following supporting information can be downloaded at: https://www.mdpi.com/article/10.3390/v15040891/s1, the Fine-tuned model prediction results are provided in Tables S1 and S2 (.xls). Table S1: Predicted_Active_Natural_Compounds; Table S2: Predicted_Active_De_novo_Generated_Compounds; Table S3: Docking_scores.

Author Contributions: Conceptualization, H.Z.; methodology, H.Z., B.L. and X.S.; software, H.Z. and B.L.; validation, H.Z., B.L. and X.S.; formal analysis, H.Z.; investigation, H.Z.; resources, H.Z.; data curation, H.Z.; writing—original draft preparation, H.Z.; writing—review and editing, Z.H. and J.A.; visualization, H.Z.; supervision, Z.H.; project administration, Z.H. All authors have read and agreed to the published version of the manuscript.

Funding: This work was supported in part by the Ganghong Young Scholar Development Fund and fund from Shenzhen-Hong Kong Cooperation Zone for Technology and Innovation (HZQB0KCZYB-2020056).

Institutional Review Board Statement: Not applicable.

Informed Consent Statement: Not applicable.

Data Availability Statement: The pre-trained model used for transfer learning can be freely downloaded from huggingface (https://huggingface.co/seyonec/ChemBERTa-zinc-base-v1 (accessed on 15 June 2021)). AutoDock Vina (version 1.2.0) used for molecular docking can be downloaded from GitHub repository (https://github.com/ccsb-scripps/AutoDock-Vina (accessed on 11 November 2021)). The web server FAF-Drugs4 used for PAINS filtering is publicly available at https://fafdrugs4.rpbs.univ-paris-diderot.fr/. All relevant data are shown in figures, tables and Supporting Materials. The ChemBERTa model predicted active natural compound and *de novo* generated compound information are listed in Table S1 and Table S2, respectively. Molecular docking scores are provided in Table S3.

Conflicts of Interest: The authors declare no conflict of interest.

References

1. Wu, F.; Zhao, S.; Yu, B.; Chen, Y.M.; Wang, W.; Song, Z.G.; Hu, Y.; Tao, Z.W.; Tian, J.H.; Pei, Y.Y.; et al. A new coronavirus associated with human respiratory disease in China. *Nature* **2020**, *579*, 265–269. [CrossRef] [PubMed]
2. Lu, R.; Zhao, X.; Li, J.; Niu, P.; Yang, B.; Wu, H.; Wang, W.; Song, H.; Huang, B.; Zhu, N.; et al. Genomic characterisation and epidemiology of 2019 novel coronavirus: Implications for virus origins and receptor binding. *Lancet* **2020**, *395*, 565–574. [CrossRef] [PubMed]
3. Anand, K.; Ziebuhr, J.; Wadhwani, P.; Mesters, J.R.; Hilgenfeld, R. Coronavirus main proteinase (3CLpro) structure: Basis for design of anti-SARS drugs. *Science* **2003**, *300*, 1763–1767. [CrossRef]
4. Kim, D.; Lee, J.Y.; Yang, J.S.; Kim, J.W.; Kim, V.N.; Chang, H. The Architecture of SARS-CoV-2 Transcriptome. *Cell* **2020**, *181*, 914–921.e10. [CrossRef]
5. Chen, Y.; Liu, Q.; Guo, D. Emerging coronaviruses: Genome structure, replication, and pathogenesis. *J. Med. Virol.* **2020**, *92*, 418–423. [CrossRef] [PubMed]
6. Lee, J.; Worrall, L.J.; Vuckovic, M.; Rosell, F.I.; Gentile, F.; Ton, A.T.; Caveney, N.A.; Ban, F.; Cherkasov, A.; Paetzel, M.; et al. Crystallographic structure of wild-type SARS-CoV-2 main protease acyl-enzyme intermediate with physiological C-terminal autoprocessing site. *Nat. Commun.* **2020**, *11*, 5877. [CrossRef] [PubMed]
7. Wu, C.; Liu, Y.; Yang, Y.; Zhang, P.; Zhong, W.; Wang, Y.; Wang, Q.; Xu, Y.; Li, M.; Li, X.; et al. Analysis of therapeutic targets for SARS-CoV-2 and discovery of potential drugs by computational methods. *Acta Pharm. Sin. B* **2020**, *10*, 766–788. [CrossRef]
8. Dai, W.A.-O.; Zhang, B.A.-O.; Jiang, X.M.; Su, H.A.-O.; Li, J.; Zhao, Y.A.-O.; Xie, X.; Jin, Z.A.-O.X.; Peng, J.; Liu, F.A.-O.X.; et al. Structure-based design of antiviral drug candidates targeting the SARS-CoV-2 main protease. *Science* **2020**, *368*, 1331–1335. [CrossRef]
9. Yang, H.; Yang, M.; Ding, Y.; Liu, Y.; Lou, Z.; Zhou, Z.; Sun, L.; Mo, L.; Ye, S.; Pang, H.; et al. The crystal structures of severe acute respiratory syndrome virus main protease and its complex with an inhibitor. *Proc. Natl. Acad. Sci. USA* **2003**, *100*, 13190–13195. [CrossRef]
10. Anand, K.; Palm, G.J.; Mesters, J.R.; Siddell, S.G.; Ziebuhr, J.; Hilgenfeld, R. Structure of coronavirus main proteinase reveals combination of a chymotrypsin fold with an extra α-helical domain. *EMBO J.* **2002**, *21*, 3213–3224. [CrossRef]
11. Muramatsu, T.; Kim, Y.T.; Nishii, W.; Terada, T.; Shirouzu, M.; Yokoyama, S. Autoprocessing mechanism of severe acute respiratory syndrome coronavirus 3C-like protease (SARS-CoV 3CLpro) from its polyproteins. *FEBS J.* **2013**, *280*, 2002–2013. [CrossRef] [PubMed]
12. Zhang, L.; Lin, D.; Sun, X.; Curth, U.; Drosten, C.; Sauerhering, L.; Becker, S.; Rox, K.; Hilgenfeld, R. Crystal structure of SARS-CoV-2 main protease provides a basis for design of improved α-ketoamide inhibitors. *Science* **2020**, *368*, 6489. [CrossRef] [PubMed]

13. Günther, S.A.-O.; Reinke, P.A.-O.; Fernández-García, Y.A.-O.; Lieske, J.A.-O.; Lane, T.A.-O.; Ginn, H.A.-O.; Koua, F.A.-O.; Ehrt, C.A.-O.; Ewert, W.A.-O.; Oberthuer, D.A.-O.; et al. X-ray screening identifies active site and allosteric inhibitors of SARS-CoV-2 main protease. *Science* **2021**, *372*, 642–646. [CrossRef] [PubMed]

14. Pillaiyar, T.; Manickam, M.; Namasivayam, V.; Hayashi, Y.; Jung, S.H. An Overview of Severe Acute Respiratory Syndrome-Coronavirus (SARS-CoV) 3CL Protease Inhibitors: Peptidomimetics and Small Molecule Chemotherapy. *J. Med. Chem.* **2016**, *59*, 6595–6628. [CrossRef]

15. Jin, Z.; Du, X.; Xu, Y.; Deng, Y.; Liu, M.; Zhao, Y.; Zhang, B.; Li, X.; Zhang, L.; Peng, C.; et al. Structure of Mpro from SARS-CoV-2 and discovery of its inhibitors. *Nature* **2020**, *582*, 289–293. [CrossRef]

16. Shiravi, A.A.; Ardekani, A.; Sheikhbahaei, E.; Heshmat-Ghahdarijani, K. Cardiovascular Complications of SARS-CoV-2 Vaccines: An Overview. *Cardiol. Ther.* **2022**, *11*, 13–21. [CrossRef]

17. Venkadapathi, J.; Govindarajan, V.K.; Sekaran, S.; Venkatapathy, S. A Minireview of the Promising Drugs and Vaccines in Pipeline for the Treatment of COVID-19 and Current Update on Clinical Trials. *Front. Mol. Biosci.* **2021**, *8*, 637378. [CrossRef]

18. Amoutzias, G.D.; Nikolaidis, M.; Tryfonopoulou, E.; Chlichlia, K.; Markoulatos, P.; Oliver, S.G. The Remarkable Evolutionary Plasticity of Coronaviruses by Mutation and Recombination: Insights for the COVID-19 Pandemic and the Future Evolutionary Paths of SARS-CoV-2. *Viruses* **2022**, *14*, 78. [CrossRef]

19. Malik, J.A.; Ahmed, S.; Mir, A.; Shinde, M.; Bender, O.; Alshammari, F.; Ansari, M.; Anwar, S. The SARS-CoV-2 mutations versus vaccine effectiveness: New opportunities to new challenges. *J. Infect. Public Health* **2022**, *15*, 228–240. [CrossRef]

20. Jang, W.D.; Jeon, S.; Kim, S.; Lee, S.Y. Drugs repurposed for COVID-19 by virtual screening of 6218 drugs and cell-based assay. *Proc. Natl. Acad. Sci. USA* **2021**, *118*, e2024302118. [CrossRef]

21. Riva, L.; Yuan, S.; Yin, X.; Martin-Sancho, L.; Matsunaga, N.; Pache, L.; Burgstaller-Muehlbacher, S.; De Jesus, P.D.; Teriete, P.; Hull, M.V.; et al. Discovery of SARS-CoV-2 antiviral drugs through large-scale compound repurposing. *Nature* **2020**, *586*, 113–119. [CrossRef]

22. Kumar, Y.; Singh, H.; Patel, C.N. In silico prediction of potential inhibitors for the Main protease of SARS-CoV-2 using molecular docking and dynamics simulation based drug-repurposing. *J. Infect. Public Health* **2020**, *13*, 1210–1223. [CrossRef]

23. Yin, W.; Mao, C.; Luan, X.; Shen, D.-D.; Shen, Q.; Su, H.; Wang, X.; Zhou, F.; Zhao, W.; Gao, M.; et al. Structure basis for inhibition of the RNA-dependent RNA polymerase from SARS-CoV-2 by remdesvir. *Science* **2020**, *368*, 1499–1504. [CrossRef] [PubMed]

24. Srivastava, K.; Singh, M.K. Drug repurposing in COVID-19: A review with past, present and future. *Metab. Open* **2021**, *12*, 100121. [CrossRef] [PubMed]

25. Hall, K.; Mfone, F.; Shallcross, M.; Pathak, V. Review of Pharmacotherapy Trialed for Management of the Coronavirus Disease-19. *Eurasian J. Med.* **2021**, *53*, 137–143. [CrossRef] [PubMed]

26. Molina, J.M.; Delaugerre, C.; Le Goff, J.; Mela-Lima, B.; Ponscarme, D.; Goldwirt, L.; de Castro, N. No evidence of rapid antiviral clearance or clinical benefit with the combination of hydroxychloroquine and azithromycin in patients with severe COVID-19 infection. *Med. Mal. Infect.* **2020**, *50*, 384. [CrossRef] [PubMed]

27. Zhu, H. Big Data and Artificial Intelligence Modeling for Drug Discovery. *Annu. Rev. Pharmacol. Toxicol.* **2020**, *60*, 573–589. [CrossRef]

28. Zhang, H.; Yang, Y.; Li, J.; Wang, M.; Saravanan, K.M.; Wei, J.; Tze-Yang Ng, J.; Tofazzal Hossain, M.; Liu, M.; Zhang, H.; et al. A novel virtual screening procedure identifies Pralatrexate as inhibitor of SARS-CoV-2 RdRp and it reduces viral replication in vitro. *PLoS Comput. Biol.* **2020**, *16*, e1008489. [CrossRef]

29. Santana, M.V.S.; Silva-Jr, F.P. De novo design and bioactivity prediction of SARS-CoV-2 main protease inhibitors using recurrent neural network-based transfer learning. *BMC Chem.* **2021**, *15*, 8. [CrossRef]

30. Zhang, H.; Saravanan, K.M.; Yang, Y.; Hossain, M.T.; Li, J.; Ren, X.; Pan, Y.; Wei, Y. Deep Learning Based Drug Screening for Novel Coronavirus 2019-nCov. *Interdiscip. Sci.* **2020**, *12*, 368–376. [CrossRef]

31. Ton, A.T.; Gentile, F.; Hsing, M.; Ban, F.; Cherkasov, A. Rapid Identification of Potential Inhibitors of SARS-CoV-2 Main Protease by Deep Docking of 1.3 Billion Compounds. *Mol. Inform.* **2020**, *39*, 2000028. [CrossRef] [PubMed]

32. Tahir Ul Qamar, M.; Alqahtani, S.M.; Alamri, M.A.; Chen, L.L. Structural basis of SARS-CoV-2 3CL(pro) and anti-COVID-19 drug discovery from medicinal plants. *J. Pharm. Anal.* **2020**, *10*, 313–319. [CrossRef] [PubMed]

33. Nand, M.; Maiti, P.; Joshi, T.; Chandra, S.; Pande, V.; Kuniyal, J.C.; Ramakrishnan, M.A. Virtual screening of anti-HIV1 compounds against SARS-CoV-2: Machine learning modeling, chemoinformatics and molecular dynamics simulation based analysis. *Sci. Rep.* **2020**, *10*, 20397. [CrossRef] [PubMed]

34. Joshi, T.; Joshi, T.; Pundir, H.; Sharma, P.; Mathpal, S.; Chandra, S. Predictive modeling by deep learning, virtual screening and molecular dynamics study of natural compounds against SARS-CoV-2 main protease. *J. Biomol. Struct. Dyn.* **2021**, *39*, 6728–6746. [CrossRef]

35. Beck, B.R.; Shin, B.; Choi, Y.; Park, S.; Kang, K. Predicting commercially available antiviral drugs that may act on the novel coronavirus (SARS-CoV-2) through a drug-target interaction deep learning model. *Comput. Struct. Biotechnol. J.* **2020**, *18*, 784–790. [CrossRef]

36. Devlin, J.; Chang, M.-W.; Lee, K.; Toutanova, K. BERT: Pre-training of deep bidirectional transformers for language understanding. In Proceedings of the NAACL-HLT, Minneapolis, MN, USA, 2–7 June 2019; pp. 4171–4186.

37. Liu, Y.; Ott, M.; Goyal, N.; Du, J.; Joshi, M.; Chen, D.; Levy, O.; Lewis, M.; Zettlemoyer, L.; Stoyanov, V. RoBERTa: A Robustly Optimized BERT Pretraining Approach. *arXiv* **2019**, arXiv:1907.11692.

38. Rogers, A.; Kovaleva, O.; Rumshisky, A. A Primer in BERTology: What We Know About How BERT Works. *Trans. Assoc. Comput. Linguist.* **2020**, *8*, 842–866. [CrossRef]
39. Chithrananda, S.; Grand, G.; Bharath, R. ChemBERTa: Large-Scale Self-Supervised Pretraining for Molecular Property Prediction. *arXiv* **2020**, arXiv:2010.09885.
40. Wolf, T.; Debut, L.; Sanh, V.; Chaumond, J.; Delangue, C.; Moi, A.; Cistac, P.; Rault, T.; Louf, R.; Funtowicz, M.; et al. Transformers: State-of-the-Art Natural Language Processing. In Proceedings of the 2020 EMNLP, Online, 16–20 November 2020; pp. 38–45.
41. Raffel, C.; Shazeer, N.; Roberts, A.; Lee, K.; Narang, S.; Matena, M.; Zhou, Y.; Li, W.; Liu, P.J. Exploring the limits of transfer learning with a unified text-to-text transformer. *J. Mach. Learn. Res.* **2020**, *21*, 5485–5551.
42. Wang, S.; Guo, Y.; Wang, Y.; Sun, H.; Huang, J. Smiles-Bert: Large Scale Unsupervised Pre-Training for Molecular Property Prediction. In Proceedings of the 10th ACM International Conference on Bioinformatics, Computational Biology and Health Informatics, Niagara Falls, NY, USA, 7–10 September 2019; pp. 429–436.
43. Honda, S.; Shi, S.; Ueda, H.R. SMILES transformer: Pre-trained molecular fingerprint for low data drug discovery. *arXiv* **2019**, arXiv:1911.04738.
44. Schwaller, P.; Laino, T.; Gaudin, T.; Bolgar, P.; Hunter, C.A.; Bekas, C.; Lee, A.A. Molecular Transformer: A Model for Uncertainty-Calibrated Chemical Reaction Prediction. *ACS Cent. Sci.* **2019**, *5*, 1572–1583. [CrossRef] [PubMed]
45. Maziarka, Ł.D.; Tomasz Mucha, S.; Rataj, K.; Tabor, J.; Jastrzębski, S. Molecule attention transformer. *arXiv* **2020**, arXiv:2002.08264.
46. Bjerrum, E.J. SMILES Enumeration as Data Augmentation for Neural Network Modeling of Molecules. *arxiv* **2017**, arXiv:1703.07076.
47. Douangamath, A.; Fearon, D.; Gehrtz, P.; Krojer, T.; Lukacik, P.; Owen, C.D.; Resnick, E.; Strain-Damerell, C.; Aimon, A.; Abranyi-Balogh, P.; et al. Crystallographic and electrophilic fragment screening of the SARS-CoV-2 main protease. *Nat. Commun.* **2020**, *11*, 5047. [CrossRef]
48. Yang, K.; Swanson, K.; Jin, W.; Coley, C.; Eiden, P.; Gao, H.; Guzman-Perez, A.; Hopper, T.; Kelley, B.; Mathea, M.; et al. Analyzing Learned Molecular Representations for Property Prediction. *J. Chem. Inf. Model.* **2019**, *59*, 3370–3388. [CrossRef]
49. Lagorce, D.; Bouslama, L.; Becot, J.; Miteva, M.A.; Villoutreix, B.O. FAF-Drugs4: Free ADME-tox filtering computations for chemical biology and early stages drug discovery. *Bioinformatics* **2017**, *33*, 3658–3660. [CrossRef]
50. Tang, B.; He, F.; Liu, D.; Fang, M.; Wu, Z.; Xu, D. AI-aided design of novel targeted covalent inhibitors against SARS-CoV-2. *bioRxiv* **2020**. [CrossRef]
51. Owen, D.R.; Allerton, C.M.N.; Anderson, A.S.; Aschenbrenner, L.; Avery, M.; Berritt, S.; Boras, B.; Cardin, R.D.; Carlo, A.; Coffman, K.J.; et al. An oral SARS-CoV-2 M(pro) inhibitor clinical candidate for the treatment of COVID-19. *Science* **2021**, *374*, 1586–1593. [CrossRef]
52. Mok, N.Y.; Maxe, S.; Brenk, R. Locating sweet spots for screening hits and evaluating pan-assay interference filters from the performance analysis of two lead-like libraries. *J. Chem. Inf. Model.* **2013**, *53*, 534–544. [CrossRef]
53. Feng, B.Y.; Shoichet, B.K. A detergent-based assay for the detection of promiscuous inhibitors. *Nat. Protoc.* **2006**, *1*, 550–553. [CrossRef]
54. O'Donnell, H.R.; Tummino, T.A.; Bardine, C.; Craik, C.S.; Shoichet, B.K. Colloidal Aggregators in Biochemical SARS-CoV-2 Repurposing Screens. *J. Med. Chem.* **2021**, *64*, 17530–17539. [CrossRef] [PubMed]
55. Zhao, Y.; Cheng, W.; Hua, B.; Wang, S.; Yang, D. Effects of Gossypol Acetate on Proliferation and Apoptosis in Lymphoblastoid Cell Line and Primary ALL and CLL Cells. *Blood* **2005**, *106*, 4405. [CrossRef]
56. Ferenczyova, K.; Kalocayova, B.A.-O.; Bartekova, M.A.-O. Potential Implications of Quercetin and its Derivatives in Cardioprotection. *Int. J. Mol. Sci.* **2020**, *21*, 1585. [CrossRef] [PubMed]
57. Khaerunnisa, S.; Kurniawan, H.; Awaluddin, R.; Suhartati, S.; Soetjipto, S. Potential inhibitor of COVID-19 main protease (Mpro) from several medicinal plant compounds by molecular docking study. *Preprints* **2020**, 2020030226. [CrossRef]
58. Chaves, O.A.; Fintelman-Rodrigues, N.; Wang, X.; Sacramento, C.Q.; Temerozo, J.R.; Ferreira, A.C.; Mattos, M.; Pereira-Dutra, F.; Bozza, P.T.; Castro-Faria-Neto, H.C.; et al. Commercially Available Flavonols Are Better SARS-CoV-2 Inhibitors than Isoflavone and Flavones. *Viruses* **2022**, *14*, 1458. [CrossRef]
59. Chaves, O.A.; Lima, C.R.; Fintelman-Rodrigues, N.; Sacramento, C.Q.; de Freitas, C.S.; Vazquez, L.; Temerozo, J.R.; Rocha, M.E.N.; Dias, S.S.G.; Carels, N.; et al. Agathisflavone, a natural biflavonoid that inhibits SARS-CoV-2 replication by targeting its proteases. *Int. J. Biol. Macromol.* **2022**, *222*, 1015–1026. [CrossRef] [PubMed]

Article

In Silico and In Vitro Evaluation of Some Amidine Derivatives as Hit Compounds towards Development of Inhibitors against Coronavirus Diseases

Ahmed H. E. Hassan [1,*], Selwan M. El-Sayed [1], Mizuki Yamamoto [2], Jin Gohda [2], Takehisa Matsumoto [3], Mikako Shirouzu [3], Jun-ichiro Inoue [4], Yasushi Kawaguchi [2,5], Reem M. A. Mansour [1], Abtin Anvari [6] and Abdelbasset A. Farahat [6,7,*]

[1] Department of Medicinal Chemistry, Faculty of Pharmacy, Mansoura University, Mansoura 35516, Egypt; salwanmahmoud@mans.edu.eg (S.M.E.-S.); reemmansour@std.mans.edu.eg (R.M.A.M.)
[2] Research Center for Asian Infectious Diseases, Institute of Medical Science, The University of Tokyo, Tokyo 108-8639, Japan; mizuyama@g.ecc.u-tokyo.ac.jp (M.Y.); jgohda@g.ecc.u-tokyo.ac.jp (J.G.); ykawagu@g.ecc.u-tokyo.ac.jp (Y.K.)
[3] Drug Discovery Structural Biology Platform Unit, RIKEN Center for Biosystems Dynamics Research, Kanagawa 230-0045, Japan; takehisa.matsumoto@riken.jp (T.M.); mikako.shirouzu@riken.jp (M.S.)
[4] Infection and Advanced Research Center (UTOPIA), The University of Tokyo Pandemic Preparedness, Tokyo 108-8639, Japan; jun-i@g.ecc.u-tokyo.ac.jp
[5] Division of Molecular Virology, Department of Microbiology and Immunology, The Institute of Medical Science, The University of Tokyo, Tokyo 108-8639, Japan
[6] Master of Pharmaceutical Sciences Program, California Northstate University, 9700 W Taron Dr., Elk Grove, CA 95757, USA; abtin.anvari8815@cnsu.edu
[7] Department of Pharmaceutical Organic Chemistry, Faculty of Pharmacy, Mansoura University, Mansoura 35516, Egypt
* Correspondence: ahmed_hassan@mans.edu.eg (A.H.E.H.); abdelbasset.farahat@cnsu.edu (A.A.F.)

Citation: Hassan, A.H.E.; El-Sayed, S.M.; Yamamoto, M.; Gohda, J.; Matsumoto, T.; Shirouzu, M.; Inoue, J.-i.; Kawaguchi, Y.; Mansour, R.M.A.; Anvari, A.; et al. In Silico and In Vitro Evaluation of Some Amidine Derivatives as Hit Compounds towards Development of Inhibitors against Coronavirus Diseases. *Viruses* **2023**, *15*, 1171. https://doi.org/10.3390/v15051171

Academic Editor: Simone Brogi

Received: 9 March 2023
Revised: 4 May 2023
Accepted: 12 May 2023
Published: 15 May 2023

Abstract: Coronaviruses, including SARS-CoV-2, SARS-CoV, MERS-CoV and influenza A virus, require the host proteases to mediate viral entry into cells. Rather than targeting the continuously mutating viral proteins, targeting the conserved host-based entry mechanism could offer advantages. Nafamostat and camostat were discovered as covalent inhibitors of TMPRSS2 protease involved in viral entry. To circumvent their limitations, a reversible inhibitor might be required. Considering nafamostat structure and using pentamidine as a starting point, a small set of structurally diverse rigid analogues were designed and evaluated in silico to guide selection of compounds to be prepared for biological evaluation. Based on the results of in silico study, six compounds were prepared and evaluated in vitro. At the enzyme level, compounds **10–12** triggered potential TMPRSS2 inhibition with low micromolar IC_{50} concentrations, but they were less effective in cellular assays. Meanwhile, compound **14** did not trigger potential TMPRSS2 inhibition at the enzyme level, but it showed potential cellular activity regarding inhibition of membrane fusion with a low micromolar IC_{50} value of 10.87 μM, suggesting its action could be mediated by another molecular target. Furthermore, in vitro evaluation showed that compound **14** inhibited pseudovirus entry as well as thrombin and factor Xa. Together, this study presents compound **14** as a hit compound that might serve as a starting point for developing potential viral entry inhibitors with possible application against coronaviruses.

Keywords: antiviral agents; coronaviruses; viral entry

1. Introduction

Coronaviruses (CoVs) are a family of viruses known to cause respiratory, enteric, and neurogenic diseases. To date, seven human coronaviruses (hCoVs) are known. The outbreaks of severe acute respiratory syndrome (SARS-CoV) in 2002, Middle East respiratory syndrome (MERS-CoV) in 2012, and COVID-19 (SARS-CoV-2) by the end of 2019 highlight the serious threats of coronaviruses infections. Health emergencies were declared as a result of epidemics and/or pandemics caused by these viruses. SARS-CoV emerged in China and spread to 29 countries, while MERS-CoV was identified in Jordan and spread in 27 countries with 80% of the reported cases in Saudi Arabia [1,2]. In addition, the new SARS-CoV-2 that emerged in China and causes COVID-19 has spread globally. As of March 2020, the World Health Organization (WHO) declared COVID-19 as a global pandemic. It resulted in more than 660 million reported COVID-19 cases and more than six million reported deaths. The real number is estimated to be much higher as many cases were not documented. Despite devising vaccines against SARS-CoV-2, they do not provide enough protection but reduce the risk of developing severe symptoms. Furthermore, such vaccines become less effective against the emerging variants as SARS-CoV-2 continuously mutates. Therefore, even vaccinated people are still prone to catch infections especially when exposed to the emerging viral mutant variants such as omicron, delta and other variants harboring diverse mutations particularly within spike (s) proteins [3–5]. Unfortunately, at least 65 million people suffer long COVID which is a post-COVID syndrome characterized by the persistence of SARS-CoV-2 symptoms for weeks or even months after catching infection [6–8]. Despite some therapeutic agents targeting viral proteins having been developed, such as those targeting main protease, resistant strains of the continuously mutating virus might evolve [9,10]. To circumvent these insufficiencies of the currently available therapeutics targeting the viral proteins that are subjected to mutations, it might be beneficial to address novel agents relying instead on host-based processes required or exploited for viral infection [11–13].

In principle, infection by coronaviruses including SARS-CoV, MERS-CoV, SARS-CoV-2, and influenza A viruses involves a common initial process of viral entry into the cell that starts with the binding of the viral S protein to the angiotensin-converting enzyme-2 (ACE2) receptor on the host cells. Subsequently, proteolysis of the S protein by the action of host proteases leads to viral fusion with the cell and entry of the viral RNA-encoded genome. In this process, transmembrane protease serine-2 (TMPRSS2) plays a crucial role [14]. Only subsequent to entry, can the virus recruit normal cellular functions for production of its proteins and genome to replicate. Accordingly, developing inhibitors of the fusion and entry steps would protect against new cell infections and would stall the propagation and replication of the virus. Intercepting such a conserved host-dependent process could offer more advantages than targeting the continuously mutating viral proteins. In lieu, inhibitors of the proteolytic activity of TMPRSS2 could be promising inhibitors of viral fusion and entry, and could be promising host-based targeted therapy against coronaviruses including SARS-CoV, MERS-CoV, SARS-CoV-2, and influenza A viruses. Such agents would be expected to circumvent limitation of viral resistance because of viral mutations [15–18].

2. Materials and Methods

2.1. In Silico Docking Study

A molecular docking study was performed using the crystal structure of TMPRSS2 deposited in Protein Data Bank (PDB: 7MEQ) employing standard computational protocols as described in Supplementary Materials [19–27].

2.2. Chemistry

Compounds **5**, **11**, and **12** were reported previously [28,29]. Compounds **8–10** were prepared following well-established chemical reactions as detailed in the Supplementary Materials.

2.3. Cells and Materials

A pair of previously described 293FT-based reporter cell lines that constitutively express individual split reporters (DSP1-7 and DSP8-11 proteins, Supplementary Figure S1a) were maintained in Dulbecco's modified Eagle's medium (DMEM) containing 10% fetal bovine serum (FBS) and 1 µg/mL puromycin. TMPRSS2 expressing VeroE6 cells (VeroE6-TMPRSS2) were maintained in Eagle's minimum essential medium (EMEM) containing

15% fetal bovine serum (FBS) as described previously [30]. See Supplementary Figure S2 for recombinant enzymes and substrates used in enzyme assays.

2.4. Cell Fusion Assay Using Dual Split Proteins (DSP)

As shown in Supplementary Figure S1b, a DSP-assay was employed to perform a SARS-CoV-2 S protein/TMPRSS2/ACE2 assay for a quantitative investigation of ACE2/TMPRSS2-dependant SARS-CoV-2 S protein-mediated membrane fusion (S2TA assay). Co-transfection assay (CoTF assay) was also applied to investigate TMPRSS2-independent inhibition of membrane fusion (Supplementary Figure S1c). All tests were performed according to the reported protocol [31]. Briefly, for the S2TA assay, effector cells expressing S protein with DSP8-11 and target cells expressing ACE2 and TMPRSS2 with DSP1-7 were seeded in 10 cm plates and incubated overnight. Cells were treated with 6 µM EnduRen (Promega, WI, USA), a substrate for Renilla luciferase (RL), for 2 h. To test the effect of the inhibitor, 1 µL of the compound dissolved in DMSO was added to the 384-well plates (Greiner Bioscience, Frickenhausen, Germany). Next, 50 µL of each single cell suspension (effector and target cells) was added to the 384-well plates using a Multidrop dispenser (Thermo Fisher Scientific, MA, USA). After incubation at 37 °C in 5% CO_2 for 4 h, RL activity was measured using a Centro xS960 luminometer (Berthold, Bad Wildbad, Germany). For the CoTF assay, cells expressing DSP1-7 and DSP8-11 were used to evaluate the effect of compounds on TMPRSS2-independent inhibition of cell fusion or reduction in measured luminescence and fluorescence that might arise from effect on DSP reassociation.

2.5. Enzyme Assays

In these assays, TMPRSS2 and thrombin were used at final concentrations of 0.3 and 3 nM, respectively. Factor Xa was used at a final concentration of 3 or 12 nM. Peptides with a fluorescence quenching pair (Dabcyl and Edans) at both ends were used as substrates for each enzyme (see Supplemental Figure S2). To test the effect of compounds on enzyme activity, enzyme and compound were mixed in 90 µL of assay buffer (20 mM Tris-HCl (pH 8.0), 150 mM NaCl) at 1.25 times the final concentration in a 96-well plate. After incubation for 30 min at room temperature, 80 µL of the mixture was added to 20 µL of 50 µM substrate solution prepared from the assay buffer. Fluorescence was read every 5 min at room temperature using CLARIOstar Plus (BMG LABTECH, Ortenberg, Germany) at an excitation wavelength of 340 nm and an emission wavelength of 490 nm. The IC_{50} values of the compound were calculated using Prism version 9.0 (GraphPad Software, CA, USA).

2.6. Pseudovirus Assay

To produce replication-deficient vesicular stomatitis virus (VSV), BHK cells expressing T7 RNA polymerase were transfected with T7 promoter-driven expression plasmids for VSV proteins (pBS-N/pBS-P/pBS-L/pBS-G) and pΔG-Luci (a plasmid encoding VSV genomic RNA lacking the G gene and encoding firefly luciferase) as described previously [32,33]. At 48 h post-transfection, the supernatants were harvested. The 293T cells were then transfected with an expression plasmid for S protein or VSV G using calcium phosphate precipitation. At 16 h post-transfection, cells were infected with replication-deficient VSV at a multiplicity of infection (MOI) of 1. At 2 h post-infection, cells were washed and incubated for another 16 h before supernatants containing pseudovirus were harvested. For the infection assay, VeroE6-TMPRSS2 cells were seeded in 96-well plates (2×10^4 cells/well) and incubated overnight. Cells were pretreated with inhibitors for 1 h prior to pseudovirus infection. Luciferase activity was measured 16 h after infection using the Bright-Glo Luciferase Assay System or ONE-Glo Luciferase Assay System (Promega) and the Centro xS960 luminometer (Berthold, Bad Wildbad, Germany).

2.7. Cell Toxicity Assay

To test the toxicity of the compounds, VeroE6-TMPRSS2 cells were treated with the compounds and cell viability was analyzed using the Celltiter-Glo luminescent cell viability assay (G7570, Promega) 24 h after treatment according to the manufacturer's protocol.

3. Results

3.1. Design Rational

Hit discovery and confirmation is an indispensable first step in early drug discovery [34–38]. While a hit compound would show the desired type of activity, its activity might be of low potential and/or it can possess undesirable effects. Later on, in the drug discovery and development pipeline, hits might be developed into lead compounds that are subjected to optimization steps before achieving a preclinical agent. Unless a hit compound is discovered, it is hard to proceed in this drug discovery pipeline. Such a hit compound might be discovered through a variety of strategies that might utilize high-throughput screening, fragment-based methods, focused libraries, or repurposing molecules, as well as others. In lieu, it might be desirable to identify some hit TMPRSS2 inhibitor compounds to be advanced later in the drug discovery pipeline of new antiviral agents against coronaviruses.

Repurposing is one the drug discovery strategies that has been successfully applied [39–42]. Briefly, in such a strategy, compounds or drugs that have been studied for a certain therapeutic use would be re-investigated for a different therapeutic purpose. Adopting this method, repurposing of the anticoagulant drug nafamostat (**1**, Figure 1) and the anti-acute pancreatitis camostat (**2**, Figure 1) showed that they can potentially inhibit TMPRSS2-dependent viral entry, and they underwent clinical trials as possible COVID-19 treatments [15,17,43–45]. Nevertheless, they suffered drawbacks and limitations that stalled further advancement. Both nafamostat (**1**) and camostat (**2**) are irreversible covalent TMPRSS2 inhibitors where the ester moiety undergoes nucleophilic attack by Ser441 amino acid after the initial binding step of the drug within the binding site. While the ester moiety enables this second step of establishing a covalent bond with the TMPRSS2, it is also responsible for metabolic liability of nafamostat (**1**) and camostat (**2**) proved by their short plasma half-life of less than 1 and 23.1 min, respectively [15,46,47]. In addition, covalent inhibitors are notorious for high risk and toxicities because of off-target interactions [48]. Accordingly, rapid metabolic deactivation and off-target inhibition consequences could be major obstacles for development of covalent TMPRSS2 inhibitors despite their potent activity. Despite less potency, it might be desirable to identify reversible TMPRSS2 inhibitor hit compounds that would possibly open a gateway for development of a common treatment for SARS-CoV-2, SARS-CoV, MERS-CoV, and influenza A viruses.

Analysis of structural features of nafamostat (**1**) and camostat (**2**) shows that, in addition to the common central ester group, they contain a common left 4-guanidinophenyl moiety that might play a critical role in binding to the target protein [43,49,50]. Referring to X-ray crystal structure (PDB ID: 7MEQ), this 4-guanidinophenyl moiety is found covalently bound as a result of nucleophilic attack on ester moiety by the Ser441 amino acid of TMPRSS2. While nafamostat (**1**) possesses an amidinonaphthyl fragment on the right of the central ester moiety, camostat (**2**) has a phenyl moiety bearing a longer aliphatic substituent. Despite this right moiety being a leaving moiety that is absent in the X-ray crustal structure, it might affect the first step of ligand binding prior to the subsequent step of establishing the covalent binding. This might be supported by the found higher potency of nafamostat relative to camostat [51]. Guanidine and amidine are two closely related functional groups that differ by only one nitrogen atom. In literature, several amide-based peptidomimetics incorporating amidine or guanidine functionalities were reported as TMPRSS2 inhibitors, such as 3-amidinophenylalanyl-derived inhibitor **3** [52,53] and ketobenzothiazole-containing diguanidine inhibitor **4** [54,55]. However, it is known that amide-based peptidomimetics, despite being potent inhibitors, have disappointing activity because of their metabolic liabilities and poor pharmacokinetics [56,57]. In this regard, the small molecule non-peptidomimetic inhibitors (AKA class C and D mimetics) offer more advantages [58–62]. Interestingly, it was found that the small molecules pentamidine (**5**) and propamidine (**6**), in which the right amidinoaryl feature was conserved in the form of 4-amidinophenyl moiety and the left 4-guanidinophenyl moiety was converted also into 4-amidinophenyl moiety while the covalent binding-responsible and metabolically labile central ester moiety was replaced by flexible alkyl chains of variable lengths, were found to possess TMPRSS2 inhibitory activity [50,51].

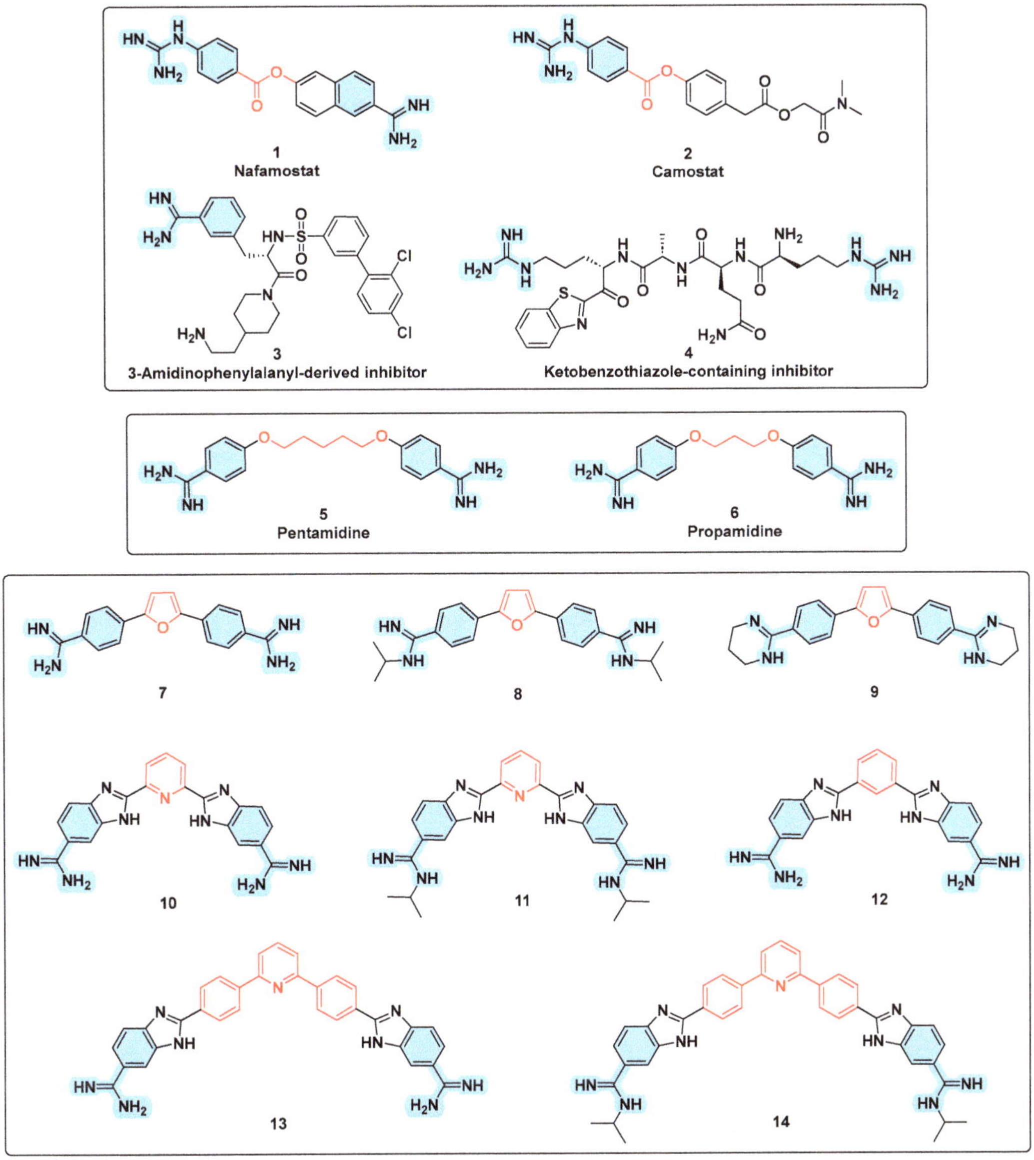

Figure 1. Chemical structure of nafamostat (**1**), camostat (**2**), 3-amidinophenylalanyl-derived inhibitor (**3**), ketobenzothiazole-containing diguanidine inhibitor (**4**), pentamidine (**5**), propamidine (**6**), and eight investigated compounds as TMPRSS2 inhibitors (**7–14**).

Deployment of in silico methods for drug discovery and design prior to synthesis and biological evaluation enables efficient utilization of limited resources as it helps to eliminate the likely-to-fail ligands and directs synthetic and biological evaluation efforts towards possibly promising compounds. In this regard, it might be helpful to establish a small focused library of compounds based on available data for in silico investigation prior to synthesis and in vitro testing [37]. Considering the structural features of nafamostat (**1**), camostat (**2**), amidinophenylalanyl-derived inhibitor **3**, and ketobenzothiazole-containing diguanidine inhibitor **4**, coupled with those of pentamidine (**5**) and propamidine (**6**), a small, focused library of eight compounds (**7–14**, Figure 1) was designed for in silico study prior to synthesis and evaluation. While the covalent binding-responsible and metabolically labile central ester moiety no longer exists in pentamidine (**5**) and propamidine (**6**), the flexibility of their central alkyl chain translates into probability of multiple conformers. In fact, the desired biological activity might be associated with some conformers while other conformers do not contribute to the desired bioactivity and, even worse, might trigger undesirable effects and/or toxicity. Therefore, rigidification of flexible moieties might be helpful if the new structure is conformationally locked in a configuration that mimics the conformer mediating the desired activity. Accordingly, the six-membered phenyl or pyridinyl as well as the five-membered furanyl aromatic moieties were introduced as rigid central moieties replacing the flexible alkyl chains of pentamidine (**5**) and propamidine (**6**). As biological evaluation showed that the size of the central moiety impacts the activity of pentamidine (**5**) and propamidine (**6**) where pentamidine turned out to be more active than propamidine [51], compounds incorporating the larger three rings-based rigid system 2,6-bisphenylpyridine, were considered as a central moiety to investigate the size impact. In addition to unsubstituted amidines, the effect of substituent introduction as well as incorporation of the amidine moiety into a cyclic moiety such as tetrahydropyrimidine were also considered. Furthermore, the aromatic ring bearing the amidine group in the terminal right and left moieties varied between the monocyclic phenyl moiety and the heterobicyclic benzimidazolyl moiety. A literature search showed that most of these postulated structurally diverse compounds were previously synthesized and explored as antiparasitic agents [28,29,63,64]. Accordingly, their re-investigation as TMPRSS2 inhibitors might be considered as a repurposing effort. The established small library was subjected to in silico study followed by in vitro evaluations to confirm their activity as possible hit compounds for development of therapeutics against coronaviruses diseases.

3.2. In Silico Evaluation

Prior to preparation and evaluation of members of the established small library, compounds were subjected to in silico evaluation to advance only the structures predicted to possess the desired bioactivity, and eliminate those likely not active. Accordingly, a molecular docking study was conducted using the reported crystal structure of TMPRSS2 (PDB ID: 7MEQ) and the docked poses and binding interactions of these compounds within the substrate binding site of TMPRSS2 were examined. The reported crystal structure of TMPRSS2 showed that the left 4-guanidinobenzoyl fragment of nafamostat was covalently bound through the hydrolyzed ester moiety to Ser441, which is one of the TMPRSS2 catalytic triad residues (Asp435, Ser441, and His296). This fragment of nafamostat establishes a network of favorable interactions that involve the hydrogen bond between its carbonyl and Gln438 and Gly439 residues in addition to hydrogen bonding interactions between the guanidine functionality with Asp435, Ser436, Gly464 and Gly472 residues. In fact, the presence of a guanidinium moiety in a structure of proposed inhibitor was found to be crucial to the binding with TMPRSS2 [17,65].

While the investigated eight compounds could dock into the substrate binding site (Figure 2), binding scores of compounds **7–9** possessing the five-membered furan ring as central moiety coupled with phenyl moiety as the aromatic ring bearing the amidine group in the terminal right and left moieties were, in general, relatively lower than compounds **10–14** possessing the six-membered phenyl or pyridine containing central moieties coupled with benzimidazole as the aromatic ring bearing the amidine group in the terminal right and left moieties (Table 1). Amongst furan-based compounds **7–9**, only compound **7** having unsubstituted amidines could establish an interaction with one of the catalytic triad amino acid residues (Asp435). Meanwhile, compounds **8** and **9**, in which the amidine is substituted or incorporated within a cycle, despite a higher binding score relative to compound **7**, could not establish any interaction with the catalytic triad. Interestingly, in silico results showed that compounds **10–14**, possessing the six-membered phenyl or pyridine containing central moieties and moieties coupled with benzimidazole as the aromatic ring bearing the amidine groups, could establish one or more favorable interactions with the catalytic triad amino acid residues (Table 1). Binding scores of compounds **11** and **14**, possessing isopropyl-substituted amidine moieties, were higher than corresponding compounds **10** and **13** possessing unsubstituted amidine moieties. Furthermore, compounds **13** and **14** with the larger three rings-based system 2,6-bisphenylpyridine central moiety showed higher binding scores relative to corresponding compounds **10** and **11** having the smaller pyridine ring as central moiety. As the catalytic triad residues (Asp435, Ser441, and His296) are crucial for activity of TMPRSS2, a predicted interaction with at least one of these residues would translate into a significant inhibitory activity, and the established interactions by compounds **7** and **10–14** with triad residues and other amino acids were further scrutinized. The catalytic Asp435 was the most frequent triad residue involved in favorable interactions with compounds **7** and **10–14**; specifically, with the amidine moieties. It is noteworthy to mention that such interaction with Asp435 was amongst the potential interactions detected in the X-ray crystal structure of TMPRSS2 with the nafamostat fragment (PDB ID: 7MEQ). However, interaction with Asp435 was not predicted for compound **14**, having the highest docking score, but instead interactions with the other two catalytic triad residues Ser441 and His296 were predicted. Favorable interactions with Ser441 were also predicted for compounds **12** and **13** while favorable interaction with His296 was the only triad interaction predicted for compound **11**. Meanwhile, Asp435 was the sole triad residue involved in two favorable interactions with compounds **10** and one favorable interaction with compound **7**. In addition to interactions with triad residues, favorable interactions of the amidine moiety with Ser436 (for compounds **7**, **12**, **13**, and **14**) and Gly464 (for compounds **7** and **10**) were predicted. Such interactions with Ser436 and Gly464 were amongst the potential detected interactions for the guanidine moiety of nafamostat fragment in the X-ray crystal structure (PDB ID: 7MEQ).

Based on these results, compounds **8** and **9**, which missed all interactions with the catalytic triad residue were excluded from further biological evaluation. As compounds **7** and **10–14** were predicted to establish a more or less diverse interaction pattern with one or more catalytic triad residues (one Asp435 interaction for compound **7**, two Asp435 interaction for compound **10**, Asp435 and Ser441 interactions for compounds **12** and **13**, Ser441 and His296 interactions for compound **14**, and one His296 interaction for compound **11**) coupled with calculated favorable binding scores, they were advanced to biological evaluation to assess their inhibitory activity on TMPRSS2.

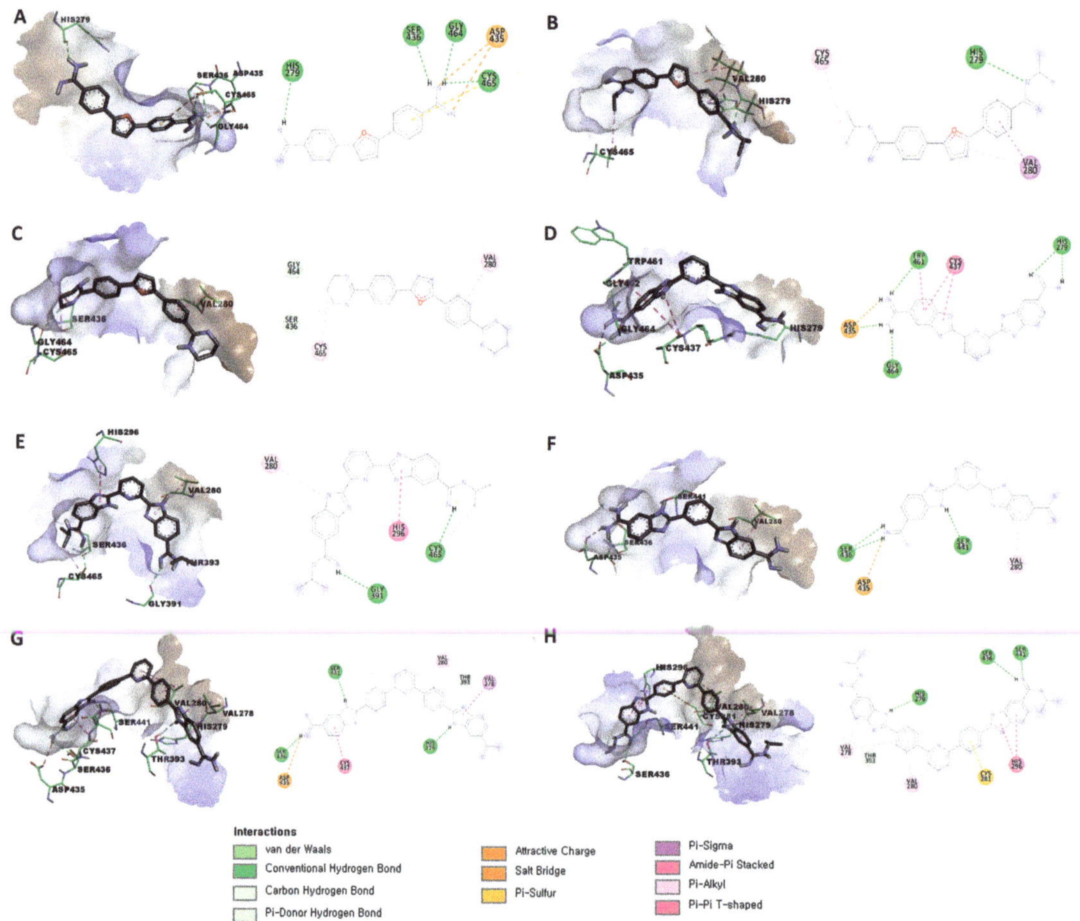

Figure 2. Predicted binding modes and interactions of compounds (**7–14**) within the binding site of TMPRSS2 (PDB ID: 7MEQ): (**A**) predicted pose of compound **7** (left) and schematic representation (right) of established interactions within the active site of TMPRSS2; (**B**) predicted pose of compound **8** (left) and schematic representation (right) of established interactions within the active site of TMPRSS2; (**C**) predicted pose of compound **9** (left) and schematic representation (right) of established interactions within the active site of TMPRSS2; (**D**) predicted pose of compound **10** (left) and schematic representation (right) of established interactions within the active site of TMPRSS2; (**E**) predicted pose of compound **11** (left) and schematic representation (right) of established interactions within the active site of TMPRSS2; (**F**) predicted pose of compound **12** (left) and schematic representation (right) of established interactions within the active site of TMPRSS2; (**G**) predicted pose of compound **13** (left) and schematic representation (right) of established interactions within the active site of TMPRSS2; (**H**) predicted pose of compound **14** (left) and schematic representation (right) of established interactions within the active site of TMPRSS2.

Table 1. The docking scores and type of binding interactions of the target compounds (**7–14**) with the crystal structure of TMPRSS2 (PDB ID: 7MEQ).

Compound	Binding Score [1]	Established Interactions
7	−5.7172	Three hydrogen bonds between (Gly464, Cys465, and Ser436) and the protons of one amidine group. Hydrogen bond between His279 with a proton of the other amidine group. Attractive charge between Asp435 and one amidine group. π-sulfur interaction between Cys465 and the phenyl (linker) group.
8	−6.5412	Hydrogen bond between His279 with protons of one amidine group. π-σ interaction between Val280 and the phenyl (linker) group. π-alkyl interaction between Val280 and the furan central ring. Alkyl interaction between Cys465 and isopropyl substitution on one amidine group.
9	−6.0349	C-H bond between Ser436 and protons of the tetrahydropyrimidine ring. C-H bond between Gly464 and protons of the tetrahydropyrimidine ring. π-alkyl interaction between Val280 and the phenyl (linker) group. Alkyl interaction between Cys465 and the tetrahydropyrimidine ring.
10	−6.2871	Salt bridge between Asp435 and one amidine group. Three hydrogen bonds between (Asp435, Gly464, and Trp461) and the protons of one amidine group. Two hydrogen bonds between His279 and the other amidine group. Two amide-π stacked interactions between Cys437 and the benzimidazole ring. Amide-π stacked interactions between Trp461 and the benzimidazole ring.
11	−7.4058	Hydrogen bond between Cys465 with protons of one amidine group. Hydrogen bond between Gly391 with protons of the other amidine group. Alkyl interaction between Cys465 and isopropyl substitution on one amidine group. π-π T-shaped interaction between His296 and the benzimidazole ring. π -alkyl interaction between Val280 and the benzimidazole ring.
12	−6.4987	Salt bridge between Asp435 and one amidine group. Two hydrogen bonds between Ser436 and (the same) amidine group. Hydrogen bond between Ser441 with a proton of the benzimidazole ring. π-alkyl interaction between Val280 and the benzimidazole ring.
13	−7.6965	Salt bridge between Asp435 and one amidine group. Hydrogen bond between Ser436 and (the same) amidine group. Hydrogen bond between Ser441 with a proton of the benzimidazole ring. Hydrogen bond between His279 with a proton of the other benzimidazole ring. π-donor hydrogen bond between Thr393 with a proton of the benzimidazole ring. π-alkyl interaction between Val278 with the (same) benzimidazole ring. π-σ interaction between Val278 with the (same) benzimidazole ring. π-alkyl interaction between Val280 and the phenyl (linker) group. π-alkyl interaction between Val280 and the pyridine central ring. Amide-π stacked interaction between Cys437 and the benzimidazole ring.
14	−8.4438	Two hydrogen bonds between Ser441 and Ser436 with a proton of one amidine group. Hydrogen bond between His279 with a proton of the benzimidazole ring. Two π-alkyl interaction between Val278 with the (same) benzimidazole ring. π-π interaction between His296 with the (other) benzimidazole ring. π-alkyl interaction between Val278 with the phenyl (linker) group. π-alkyl interaction between Val280 with the phenyl (linker) group. π-alkyl interaction between Val280 with the pyridine central ring. π-donor hydrogen bond between Thr393 with a proton of the benzimidazole ring. π-sulfur interaction between Cys281 with the phenyl (linker) group.

[1] Binding score was calculated in kcal/mol.

3.3. Chemistry

Synthesis of compounds **7**, **13**, and **14** was reported previously [28,29]. Synthesis of compounds **10–12** is outlined in Scheme 1. 4-Amidino-1,2-phenylenediamine derivatives **16** were prepared from commercially available 3,4-diaminobenzonitrile by applying Pinner reaction conditions [66]. Bis-Aldehydes **15** were reacted with 4-amidino-1,2-phenylenediamine derivatives **16** in ethanol in the presence of sodium metabisulphite to furnish the crude final compounds [67]. After sodium hydroxide treatment, crude compounds were converted to the hydrochloride salt by stirring with ethanolic HCl to furnish the desired compounds **10–12** as HCl salts.

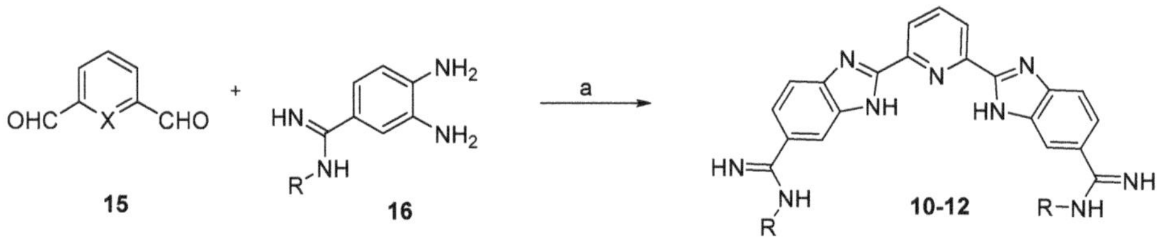

	10	**11**	**12**
X =	N	N	CH
R =	H	isopropyl	H

Scheme 1. Reagents and conditions: (a) (i) $Na_2S_2O_5$, EtOH-H_2O, reflux; (ii) NaOH/H_2O, EtOH/HCl.

3.4. In Vitro Evaluations

3.4.1. TMPRSS2 Inhibition Assay

Based on in silico calculations, compounds **7** and **10–14** that could establish interactions with one or more of the catalytic triad residues were advanced for in vitro evaluation to assess their inhibitory activity on TMPRSS2. For comparison, pentamidine and nafamostat were used as reference standard compounds. While nafamostat, the covalent irreversible inhibitor showed less than 1 nM IC_{50} value, pentamidine exhibited nearly 40 µM IC_{50}. Interestingly, the low micromolar IC_{50} values of compounds **12** and **10** were comparable to the reference pentamidine (Table 2), suggesting their potential inhibitory activity against TMPRSS2. Structurally, both compounds **12** and **10** have 6-amidinobenzimidazol-2-yl moieties bearing no substituents on the amidine fragment, but compound **12** has the six-membered phenyl moiety as the molecule's central moiety instead of the six-membered pyridine moiety of compound **10**. In comparison, compound **11** having the same structure as compound **10** but having N-isopropyl substituents at the amidine moieties, possessed a considerably lower potency relative to compound **10**. The results showed that increasing the size of the central moiety of compounds **10** and **11** from the one ring-based pyridine ring to the three rings-based 2,6-bisphenylpyridine system in compounds **13** and **14** resulted in a high decrease in TMPRSS2 inhibitory activity. Thus, compounds **13** and **14** showed high micromolar IC_{50} values (Table 2). Comparing results of compounds **13** and **14** considering their structures, it can be inferred that introduction of N-isopropyl substituents at the amidine moieties (compound **14**) resulted in lowering the potency relative to the N-unsubstituted structure (compound **13**). Similarly, compound **7** having the smaller 4-amidinophenyl moiety as left and right moieties coupled with the five-membered furan ting as a central moiety possessed a low potency reflected by the found high micromolar IC_{50} value (Table 2).

Table 2. Results of in vitro TMPRSS2 enzymatic assay, the two DSP-assays (S2TA and CoTF), and cellular viability for the selected set of compounds **7**, **10–14** and the reference nafamostat and pentamidine.

Compound	TMPRSS2 IC$_{50}$ (µM)	S2TA Assay		CoTF Assay IC$_{50}$ (µM)
		% Inhibition [1]	IC$_{50}$ (µM)	
7	82.7	39.91	>100	>100
10	5.57	89.53	16.93	37.84
11	>10 [2]	44.17	>100	>100
12	4.89	55.42	>100	>100
13	72.4	93.98	13.70	37.00
14	>100	97.33	10.87	30.97
Pentamidine	3.98	71.30	32.27	>100
Nafamostat	<0.001	89.32	0.007115	>100

[1] % Inhibition at 100 µM concentration. [2] Exact IC$_{50}$ value could not be determined because of the compound's fluorescence interference at higher concentrations.

3.4.2. Cell Fusion Assays

To assess the inhibitory effect of compounds **7** and **10–14**, the previously developed SARS-CoV-2 S protein/TMPRSS2/ACE2 assay (S2TA assay) for quantitation of the impact on ACE2/TMPRSS2-dependant SARS-CoV-2 S protein-mediated membrane fusion utilizing "Dual Split Protein (DSP)" reporter complex [31] was used (Supplementary Figure S1). As used target cells encompass DSP1-7 while effector cells encompass DSP8-11, fusion of effector cells with target cells will bring DSP together, resulting in reassociation to form functional Renilla luciferase (RL) and green fluorescent protein (GFP), which can be detected by luminescence or fluorescence. As tested compounds might elicit TMPRSS2-independent effects on membrane fusion, or affects reassociation of DSP proteins and, thus, the measured luminescence or fluorescence signals, an assay employing cells co-transfected with DSP1-7 and DSP8-11 was employed (CoTF assay) to account for the inhibition of membrane fusion mediated by targets other than TMPRSS2. In addition to compounds **7** and **10–14**, pentamidine and nafamostat were used as reference standard compounds. As compounds **7**, **10–14** and pentamidine exhibited micromolar IC$_{50}$ values, their percent inhibition values were first assessed at 100 µM concentration. The results are summarized in Table 2.

The reference compounds pentamidine and nafamostat triggered an inhibition percent of 71.30 and 89.32%, respectively, at 100 µM concentration, and an IC$_{50}$ value of 32.27 µM and 7.1 nM, respectively, in S2TA assay. It was interesting to find that compounds **10**, **13** and **14** triggered high inhibition percentages of 89.53, 93.98, and 97.33%, respectively, which were relatively higher than pentamidine, Table 2. However, the TMPRSS2 enzymatic assay showed that compound **10** exhibited a comparable IC$_{50}$ value to that of pentamidine, while compounds **13** and **14** were impotent TMPRSS2 inhibitors. Since the employed DSP assay evaluated the inhibitory effect not only on TMPRSS2 but also on various mechanisms required for membrane fusion of SARS-CoV-2 in cells, it might be possible that compounds **13** and **14** inhibited membrane fusion through other mechanisms. Noteworthy, it was reported that multiple membrane serine proteases can replace TMPRSS2 for membrane fusion [68]. Nevertheless, DSP assay can be used to evaluate the effects of compounds in cellular assay with high throughput and without the use of infectious viral particles, and various SARS-CoV-2 inhibitors have been found using this assay [30,31,69]. Although compound **10** has the one ring-based central moiety while compounds **13** and **14** have the larger three rings-based central moiety, all of them share the presence of a pyridine ring in the central moiety. Despite predicted close in silico binding scores and the measured close IC$_{50}$ value for TMPRSS2 inhibition, compound **12** possessing phenyl-based central moiety showed modest inhibition percent in DSP assay compared with the corresponding compound **10** possessing pyridine-based central moiety that showed potential activity (Table 2). This result reinforces the inferred contribution of mechanisms other than TMPRSS2 to the outcome of the DSP assay. In the case of compound **7** possessing the five-membered furan ring as the central moiety which was an impotent TMPRSS2 inhibitor (Table 2), it showed a low inhibition percent of 39.91%. Meanwhile, compound **11** which differed from compound **10** only by having isopropyl-substituted amidine moieties and possessed lower TMPRSS2 potency relative to compound **10**, was revealed to elicit a low inhibitory activity of 44.17% (Table 2). While IC_{50} values of low ineffective compounds **7**, **11** and **12** were above 100 μM, compounds **10**, **13** and **14** which triggered high inhibition percentages elicited better IC_{50} values than the 32.27 μM IC_{50} value of the reference reversible TMPRSS2 inhibitor pentamidine. Thus, compounds **10**, **13** and **14** possessed the more potent IC_{50} values of 10.87, 13.70, and 16.93, respectively. Next, evaluation of compounds **7** and **10–14**, pentamidine and nafamostat in CoTF assay, revealed that all of them had more than 100 μM IC_{50} values except for compounds **14**, **13** and **10** that showed IC_{50} values of 30.97, 37.00 and 37.84 μM, respectively.

3.4.3. Pseudovirus Entry and Cell Viability Assays

To evaluate the capability of the tested compounds to inhibit viral entry into cells, SARS-CoV-2 pseudovirus infection was assessed in the presence of compounds **7** and **10–14**, pentamidine and nafamostat. Initially, their percent inhibition values were first assessed at 100 μM concentration before assessing their IC_{50} values. Cellular viability was assessed to confirm that inhibition results were not because of cell death. The results are summarized in Table 3.

Table 3. Results of SARS-CoV-2 pseudovirus entry assays and cell viability of compounds **7**, **10–14** and the reference nafamostat and pentamidine.

Compound	SARS-CoV-2 Pseudovirus		Cell Viability Inhibition IC_{50} (μM)
	% Inhibition [1]	IC_{50} (μM)	
7	2.66	>100	>100
10	19.57	>100	>100
11	27.02	>100	>100
12	17.07	>100	>100
13	20.53	>100	>100
14	77.44	83.66	>100
Pentamidine	69.48	45	>100
Nafamostat	87.34	0.06	>100

[1] % Inhibition at 100 μM concentration.

As shown in Table 3, neither the tested compounds, pentamidine or nafamostat showed potential cytotoxic activity, as IC_{50} values for inhibition of cellular viabilities were more than 100 μM. While the standard pentamidine showed 69.48% inhibition of pseudovirus infection at 100 μM concentration and an IC_{50} value of 45 μM, only compound **14** showed a considerable 77.44% inhibition at 100 μM concentration and IC_{50} value of 83.66 μM. All other tested compounds possessed low percentage inhibition values of less than 21% at the 100 μM concentration. As compound **14** had a disappointing >100 IC_{50} value for TMPRSS2 in enzymatic assay but considerable activities in cellular DSP and pseudovirus entry assays, it might be inferred that its inhibitory activity could be mediated through molecular targets other than TMPRSS2. Meanwhile, the low micromolar TMPRSS2 inhibitory activity found for compounds **10–12** at the enzymatic level is not sufficient to trigger considerable activity at the cellular level. However, these compounds might serve as hit starting point compounds for development of more potential inhibitors.

3.4.4. Thrombin and Factor Xa Enzyme Inhibition Assays

Coagulopathy is a serious complication of coronavirus infections including SARS-CoV-2, SARS-CoV-1, MERS-CoV and others [70,71]. Formation of blood clots might result in thromboembolism that could be life-threatening. In this regard, a compound that inhibits thrombin or factor Xa might be double-edged. It might minimize the risk of thromboembolism, yet it might also be associated with a bleeding risk. As for the overlap of TMPRSS2, thrombin and factor Xa substrates [72], compounds **7** and **10–14**, pentamidine and nafamostat were subjected to enzymatic assays for the inhibition of thrombin and factor Xa to assess their possible inhibitory effects. The outcome is summarized in Table 4.

Table 4. Results of thrombin and factor Xa enzyme inhibition assays by compounds **7, 10–14** and the reference nafamostat and pentamidine.

Compound	IC_{50} (μM)	
	Thrombin	**Factor Xa**
7	79.1	30.9
10	0.921	5.93
11	>1 [1]	>1 [1]
12	0.862	34.1
13	0.929	4.89
14	2.25	>10 [1]
Pentamidine	1.51	6.22
Nafamostat	0.0341	1.74

[1] Exact IC_{50} value could not be determined because of the compound's fluorescence interference at higher concentrations.

While nafamostat inhibited thrombin with a nanomolar IC_{50} value of 34.1 nM, it also inhibited factor Xa but with a low micromolar IC_{50} value of 1.74 μM. Meanwhile, pentamidine might be relatively less capable of triggering bleeding while still inhibiting thrombin and factor Xa with a low micromolar IC_{50} value of 1.51 and 6.22 μM, respectively. Amongst evaluated compounds, compound **14**, which was the most effective inhibitor of viral infection despite it not being a potential TMPRSS2 inhibitor, possessed comparable IC_{50} values to those of pentamidine for the inhibition of thrombin and factor Xa (Table 4). Compound **13** had greater potential to inhibit thrombin and factor Xa relative to compound **14** but was still comparable to pentamidine. Structurally, compound **13** was relevant to compound **14** but had unsubstituted amidine moieties. Meanwhile, compounds **10, 11** and **12**, which had the most potential amongst tested compounds to inhibit TMPRSS2, showed submicromolar to low micromolar IC_{50} values for thrombin inhibition which were still comparable to pentamidine. However, compounds **10** and **11**, having pyridine ring as a central moiety, were greater potential inhibitors of factor Xa relative to compound **13** having the central phenyl ring instead of the pyridine. In comparison with compounds **10–14**, compound **7**, having the furan ring as a central moiety with the smaller 4-amidinophenyl moiety as left and right moieties, lacked potential thrombin inhibitory activity and also had 5-fold less potential factor Xa inhibitor relative to pentamidine (Table 4).

4. Discussion

Coronaviruses are notorious for triggering outbreaks of epidemic viral infections which are characterized by being contagious and morbid such as MERS-CoV, SARS-CoV, and SARS-CoV-2. Despite the attempted development of vaccines and therapeutics targeting viral proteins, continuous viral mutation raises major issues that challenge such efforts. As coronaviruses require a conserved host-based mechanism for viral entry, targeting this mechanism rather than other viral-dependent mechanisms might offer common and effective tools against coronaviruses. TMPRSS2 is involved in this process of host-dependent mechanism of viral entry into the lung epithelial cell in vitro [73] and in vivo [74]. While nafamostat and camostat were discovered as potential TMPRSS2 inhibitors, they are metabolically labile, covalent inhibitors and, furthermore, failed clinical studies [75–77]. To circumvent limitations of covalent inhibitors and metabolic liability of nafamostat and camostat, development of a reversible TMPRSS2 inhibitor lacking the central ester group responsible for metabolic instability and irreversible covalent inhibition might be a promising strategy despite the anticipated lower potency. Considering similarity of the structural features of pentamidine with nafamostat, camostat, ketobenzothiazole-containing and amidinophenylalanine-derived peptidomimetic inhibitors, pentamidine might offer a suitable starting point to develop potential TMPRSS2 inhibitors. However, the central alkyl chain of pentamidine renders it a flexible molecule with multiple conformers. Structure rigidification via replacement of the flexible moieties by cyclic fragments would result in a beneficial conformational lock. To identify new hit molecules, a small set of eight compounds (**7–14**) were designed with diverse structural features considering size of the introduced cyclic central moiety and the terminal aryl moiety bearing the amidine moiety in addition to the absence/presence of substituents at the amidine moiety or incorporation into a ring. Considering that the utility of conducting in silico studies in initial steps to guide the selection of structures more likely to possess desired activity and to reduce the workload and discovery costs, in silico study was conducted for compounds (**7–14**) using the reported crystal structure of TMPRSS2 (PDB ID: 7MEQ). In silico results predicted that compounds **7–9** incorporating the five-membered furan ring as the central moiety were less able to establish interactions with residues of the catalytic triad than compounds **10–14** possessing central moieties incorporating a six-membered ring. Meanwhile, compounds **14** and **13** were also able to establish interactions with catalytic triad residues with calculated good binding scores. Based on interactions and binding scores, in silico study enabled the filtering-out of compounds **8** and **9** from further consideration. Consequently, compounds **7** and **10–14** were advanced for preparation and in vitro evaluation. The found in vitro low TMPRSS2 inhibitory activity for TMPRSS2 inhibition by compound **7** possessing the five-membered furan ring-based central moiety was in agreement with its predicted in silico low binding score. In vitro assay for TMPRSS2 inhibition confirmed the influential role for the size of the central moiety as compounds **10–12**, possessing one six-membered-based central moieties, had potential low micromolar activity comparable to pentamidine, while compounds **14** and **13**, possessing the larger three rings-based rigid system 2,6-bisphenylpyridine as central moiety, showed low TMPRSS2 inhibitory activity. Meanwhile, the presence of *N*-isopropyl substituents at the amidine moieties had a negative influence on TMPRSS2 inhibitory activity. As cellular activity is very important, a known experimental model from the literature for quantification of the inhibitory activity of a compound on membrane fusion and cell entry employed using DSP was addressed [31]. Surprisingly, compounds **14**, **13**, despite not being potential TMPRSS2 inhibitors, were the most active compounds triggering high inhibition percentages and low micromolar IC_{50} values in the conducted cell-based assay. Meanwhile, compound **10**, possessing potentially low micromolar IC_{50} for TMPRSS2 inhibition in enzymatic assay was less active, and compounds **11** and **12**, possessing also potential TMPRSS2 inhibitory activity were of low cellular activity. Considering that this assay evaluated the inhibitory effect not only on TMPRSS2 but also on various mechanisms involved for membrane fusion of SARS-CoV-2 in cells, coupled with the reported finding that multiple membrane serine proteases can replace TMPRSS2 for membrane fusion [68], it might be inferred that compounds **14** and **13** might act by mechanisms other than TMPRSS2 inhibition. These findings emphasize the importance of evaluating the effects of compounds in cellular assay as it can help to identify potential compounds with different molecular targets. Structure analysis considering these results revealed that compounds **14**, **13**, and **11**, sharing the presence of the six-membered pyridine ring in their central moiety, and **14** and **13**, having the larger 2,6-bis(phenyl)pyridine moiety, were the most active. Meanwhile, phenyl-based central moiety afforded the less potent compound **12**. To check for activity that might arise from inhibition of molecular targets other than TMPRSS2, as well as checking for reduction in measured luminescence and fluorescence that might arise from the effect on DSP reassociation, a CoTF assay was performed. Compounds **14**, **13** and **10** showed IC_{50} values of 30.97, 37.00 and 37.84 µM in the CoTF assay, respectively. IC_{50} values for the CoTF assay were higher than the IC_{50} values for the TMPRSS2-dependent membrane fusion assay (S2TA assay), indicating that these compounds have inhibitory effects on membrane fusion, but their inhibitory activity is possibly accompanied by TMPRSS2-independent inhibition of membrane fusion or effects on reassociation of reporter DSP proteins. Therefore, the compounds' effects on pseudovirus entry and cell viability assays were assessed. All tested compounds showed no potential cytotoxic activity. The results showed again that compound **14**, lacking potential TMPRSS2 inhibition, was the most active inhibitor for pseudovirus entry amongst the tested compounds, while other compounds including the most effective TMPRSS2 inhibitors **10–12** exhibited low inhibitory activity for pseudovirus entry. This reinforces the conclusion that compound **14** has an inhibitory activity on viral entry not associated with TMPRSS2 inhibition. As coagulopathy and formation of blood clots is a serious complication of coronaviruses infections, inhibition of thrombin and factor Xa might have some benefits. However, inhibition of thrombin and factor Xa also bears bleeding risks. Consequently, thrombin and factor Xa inhibition by tested compounds was checked. Except for compound **7** that did not show potential inhibition of thrombin and factor Xa, all other tested compounds triggered thrombin and factor Xa inhibition comparable to the drug pentamidine. Similar to the systemically-used drug pentamidine, monitoring bleeding risks should be considered upon the use of this class of compounds. In conclusion, compound **14** might serve as a hit compound that might require further development into lead compounds against cellular entry of coronaviruses.

Supplementary Materials: The following supporting information can be downloaded at: https://www. mdpi.com/article/10.3390/v15051171/s1, new compounds synthesis procedures. Figure S1: Cell-based membrane fusion assay for SARS-Cov-2 S protein using the DSP reporter; Figure S2: Enzymes and substrates for enzyme assays.

Author Contributions: Conceptualization, A.H.E.H. and A.A.F.; methodology, A.H.E.H. and A.A.F.; validation, S.M.E.-S., R.M.A.M. and A.A.; formal analysis, M.Y., J.G., T.M., M.S., J.-i.I. and Y.K.; investigation, A.H.E.H., S.M.E.-S., R.M.A.M. and A.A.; resources, A.H.E.H. and A.A.F.; data curation, S.M.E.-S., R.M.A.M. and A.A.; writing—original draft preparation, A.H.E.H. and S.M.E.-S.; writing —review and editing, M.Y., J.G., J.-i.I. and Y.K.; visualization, A.H.E.H. and S.M.E.-S.; supervision, A.A.F.; project administration, A.H.E.H. and A.A.F. All authors have read and agreed to the published version of the manuscript.

Funding: This work was supported, in part, by the Japanese Society for the Promotion of Science (18K15235, 20K07610 and 23K06571 to M.Y.) and from the Japan Agency for Medical Research and Development (AMED) [Program of Japan Initiative for Global Research Network on Infectious Diseases (JGRID) JP22wm0125002 to Y.K., Japan Program for Infectious Diseases Research and Infrastructure JP22wm0325052 to M.Y., Research Program on Emerging and Re-emerging Infectious Diseases JP21fk0108562 to M.Y. and Research on Development of New Drugs JP22ak0101165 to M.Y.].

Data Availability Statement: No data were used in this study.

Conflicts of Interest: The authors declare no conflict of interest.

References

1. Chang, L.; Yan, Y.; Wang, L. Coronavirus Disease 2019: Coronaviruses and Blood Safety. *Transfus. Med. Rev.* **2020**, *34*, 75–80. [CrossRef] [PubMed]
2. Zhu, Z.; Lian, X.; Su, X.; Wu, W.; Marraro, G.A.; Zeng, Y. From SARS and MERS to COVID-19: A brief summary and comparison of severe acute respiratory infections caused by three highly pathogenic human coronaviruses. *Respir. Res.* **2020**, *21*, 224. [CrossRef] [PubMed]
3. Notarte, K.I.; Catahay, J.A.; Velasco, J.V.; Pastrana, A.; Ver, A.T.; Pangilinan, F.C.; Peligro, P.J.; Casimiro, M.; Guerrero, J.J.; Gellaco, M.M.L.; et al. Impact of COVID-19 vaccination on the risk of developing long-COVID and on existing long-COVID symptoms: A systematic review. *EClinicalMedicine* **2022**, *53*, 101624. [CrossRef] [PubMed]
4. Zheng, C.; Shao, W.; Chen, X.; Zhang, B.; Wang, G.; Zhang, W. Real-world effectiveness of COVID-19 vaccines: A literature review and meta-analysis. *Int. J. Infect. Dis.* **2022**, *114*, 252–260. [CrossRef]
5. Iacopetta, D.; Ceramella, J.; Catalano, A.; Saturnino, C.; Pellegrino, M.; Mariconda, A.; Longo, P.; Sinicropi, M.S.; Aquaro, S. COVID-19 at a Glance: An Up-to-Date Overview on Variants, Drug Design and Therapies. *Viruses* **2022**, *14*, 573. [CrossRef]
6. Davis, H.E.; McCorkell, L.; Vogel, J.M.; Topol, E.J. Long COVID: Major findings, mechanisms and recommendations. *Nat. Rev. Microbiol.* **2023**, *21*, 133–146. [CrossRef]
7. Fernández-de-las-Peñas, C. Long COVID: Current definition. *Infection* **2022**, *50*, 285–286. [CrossRef]
8. Raveendran, A.V.; Jayadevan, R.; Sashidharan, S. Long COVID: An overview. *Diabetes Metab. Syndr. Clin. Res. Rev.* **2021**, *15*, 869–875. [CrossRef]

9. Parigger, L.; Krassnigg, A.; Schopper, T.; Singh, A.; Tappler, K.; Köchl, K.; Hetmann, M.; Gruber, K.; Steinkellner, G.; Gruber, C.C. Recent changes in the mutational dynamics of the SARS-CoV-2 main protease substantiate the danger of emerging resistance to antiviral drugs. *Front. Med.* **2022**, *9*, 1061142. [CrossRef]
10. Jiao, Z.; Yan, Y.; Chen, Y.; Wang, G.; Wang, X.; Li, L.; Yang, M.; Hu, X.; Guo, Y.; Shi, Y.; et al. Adaptive Mutation in the Main Protease Cleavage Site of Feline Coronavirus Renders the Virus More Resistant to Main Protease Inhibitors. *J. Virol.* **2022**, *96*, e00907–e00922. [CrossRef]
11. Kumari, M.; Lu, R.-M.; Li, M.-C.; Huang, J.-L.; Hsu, F.-F.; Ko, S.-H.; Ke, F.-Y.; Su, S.-C.; Liang, K.-H.; Yuan, J.P.-Y.; et al. A critical overview of current progress for COVID-19: Development of vaccines, antiviral drugs, and therapeutic antibodies. *J. Biomed. Sci.* **2022**, *29*, 68. [CrossRef] [PubMed]
12. Lei, S.; Chen, X.; Wu, J.; Duan, X.; Men, K. Small molecules in the treatment of COVID-19. *Signal Transduct. Target. Ther.* **2022**, *7*, 387. [CrossRef] [PubMed]
13. Mei, M.; Tan, X. Current Strategies of Antiviral Drug Discovery for COVID-19. *Front. Mol. Biosci.* **2021**, *8*, 671263. [CrossRef]
14. Mahoney, M.; Damalanka, V.C.; Tartell, M.A.; Chung, D.h.; Lourenço, A.L.; Pwee, D.; Mayer Bridwell, A.E.; Hoffmann, M.; Voss, J.; Karmakar, P.; et al. A novel class of TMPRSS2 inhibitors potently block SARS-CoV-2 and MERS-CoV viral entry and protect human epithelial lung cells. *Proc. Natl. Acad. Sci. USA* **2021**, *118*, e2108728118. [CrossRef] [PubMed]
15. Zhu, H.; Du, W.; Song, M.; Liu, Q.; Herrmann, A.; Huang, Q. Spontaneous binding of potential COVID-19 drugs (Camostat and Nafamostat) to human serine protease TMPRSS2. *Comput. Struct. Biotechnol. J.* **2021**, *19*, 467–476. [CrossRef] [PubMed]
16. Gil, C.; Ginex, T.; Maestro, I.; Nozal, V.; Barrado-Gil, L.; Cuesta-Geijo, M.Á.; Urquiza, J.; Ramírez, D.; Alonso, C.; Campillo, N.E.; et al. COVID-19: Drug Targets and Potential Treatments. *J. Med. Chem.* **2020**, *63*, 12359–12386. [CrossRef]
17. Fraser, B.J.; Beldar, S.; Seitova, A.; Hutchinson, A.; Mannar, D.; Li, Y.; Kwon, D.; Tan, R.; Wilson, R.P.; Leopold, K.; et al. Structure and activity of human TMPRSS2 protease implicated in SARS-CoV-2 activation. *Nat. Chem. Biol.* **2022**, *18*, 963–971. [CrossRef]
18. Wettstein, L.; Kirchhoff, F.; Münch, J. The Transmembrane Protease TMPRSS2 as a Therapeutic Target for COVID-19 Treatment. *Int. J. Mol. Sci.* **2022**, *23*, 1351. [CrossRef]
19. Hassan, A.H.E.; Kim, H.J.; Park, K.; Choi, Y.; Moon, S.; Lee, C.H.; Kim, Y.J.; Cho, S.B.; Gee, M.S.; Lee, D.; et al. Synthesis and Biological Evaluation of O6-Aminoalkyl-Hispidol Analogs as Multifunctional Monoamine Oxidase-B Inhibitors towards Management of Neurodegenerative Diseases. *Antioxidants* **2023**, *12*, 1033. [CrossRef]
20. Hassan, A.H.E.; Wang, C.Y.; Lee, H.J.; Jung, S.J.; Kim, Y.J.; Cho, S.B.; Lee, C.H.; Ham, G.; Oh, T.; Lee, S.K.; et al. Scaffold hopping of N-benzyl-3,4,5-trimethoxyaniline: 5,6,7-Trimethoxyflavan derivatives as novel potential anticancer agents modulating hippo signaling pathway. *Eur. J. Med. Chem.* **2023**, *256*, 115421. [CrossRef]
21. Gulia, K.; Hassan, A.H.E.; Lenhard, J.R.; Farahat, A.A. Escaping ESKAPE resistance: In vitro and in silico studies of multifunctional carbamimidoyl-tethered indoles against antibiotic-resistant bacteria. *R. Soc. Open Sci.* **2023**, *10*, 230020. [CrossRef] [PubMed]
22. Elkamhawy, A.; Paik, S.; Ali, E.M.H.; Hassan, A.H.E.; Kang, S.J.; Lee, K.; Roh, E.J. Identification of Novel Aryl Carboxamide Derivatives as Death-Associated Protein Kinase 1 (DAPK1) Inhibitors with Anti-Proliferative Activities: Design, Synthesis, In Vitro, and In Silico Biological Studies. *Pharmaceuticals* **2022**, *15*, 1050. [CrossRef] [PubMed]
23. Hassan, A.H.E.; Kim, H.J.; Gee, M.S.; Park, J.-H.; Jeon, H.R.; Lee, C.J.; Choi, Y.; Moon, S.; Lee, D.; Lee, J.K.; et al. Positional scanning of natural product hispidol's ring-B: Discovery of highly selective human monoamine oxidase-B inhibitor analogues downregulating neuroinflammation for management of neurodegenerative diseases. *J. Enzyme Inhib. Med. Chem.* **2022**, *37*, 768–780. [CrossRef] [PubMed]
24. Lee, H.-H.; Shin, J.-S.; Chung, K.-S.; Kim, J.-M.; Jung, S.-H.; Yoo, H.-S.; Hassan, A.H.E.; Lee, J.K.; Inn, K.-S.; Lee, S.; et al. 3′,4′-Dihydroxyflavone mitigates inflammatory responses by inhibiting LPS and TLR4/MD2 interaction. *Phytomedicine* **2023**, *109*, 154553. [CrossRef] [PubMed]
25. Hassan, A.H.E.; Park, K.T.; Kim, H.J.; Lee, H.J.; Kwon, Y.H.; Hwang, J.Y.; Jang, C.-G.; Chung, J.H.; Park, K.D.; Lee, S.J.; et al. Fluorinated CRA13 analogues: Synthesis, in vitro evaluation, radiosynthesis, in silico and in vivo PET study. *Bioorg. Chem.* **2020**, *99*, 103834. [CrossRef]
26. El-Demerdash, A.; Al-Karmalawy, A.A.; Abdel-Aziz, T.M.; Elhady, S.S.; Darwish, K.M.; Hassan, A.H.E. Investigating the structure–activity relationship of marine natural polyketides as promising SARS-CoV-2 main protease inhibitors. *RSC Adv.* **2021**, *11*, 31339–31363. [CrossRef]
27. Hassan, A.H.E.; Lee, K.-T.; Lee, Y.S. Flavone-based arylamides as potential anticancers: Design, synthesis and in vitro cell-based/cell-free evaluations. *Eur. J. Med. Chem.* **2020**, *187*, 111965. [CrossRef]
28. Paul, A.; Nanjunda, R.; Kumar, A.; Laughlin, S.; Nhili, R.; Depauw, S.; Deuser, S.S.; Chai, Y.; Chaudhary, A.S.; David-Cordonnier, M.-H.; et al. Mixed up minor groove binders: Convincing A·T specific compounds to recognize a G·C base pair. *Bioorg. Med. Chem. Lett.* **2015**, *25*, 4927–4932. [CrossRef]
29. Anbazhagan, M.; Boykin, D.W.; Stephens, C.E. Direct Conversion of Amidoximes to Amidines via Transfer Hydrogenation. *Synthesis* **2003**, *2003*, 2467–2469. [CrossRef]
30. Yamamoto, M.; Gohda, J.; Kobayashi, A.; Tomita, K.; Hirayama, Y.; Koshikawa, N.; Seiki, M.; Semba, K.; Akiyama, T.; Kawaguchi, Y.; et al. Metalloproteinase-Dependent and TMPRSS2-Independent Cell Surface Entry Pathway of SARS-CoV-2 Requires the Furin Cleavage Site and the S2 Domain of Spike Protein. *mBio* **2022**, *13*, e00519–e00522. [CrossRef]

31. Yamamoto, M.; Kiso, M.; Sakai-Tagawa, Y.; Iwatsuki-Horimoto, K.; Imai, M.; Takeda, M.; Kinoshita, N.; Ohmagari, N.; Gohda, J.; Semba, K.; et al. The Anticoagulant Nafamostat Potently Inhibits SARS-CoV-2 S Protein-Mediated Fusion in a Cell Fusion Assay System and Viral Infection In Vitro in a Cell-Type-Dependent Manner. *Viruses* **2020**, *12*, 629. [CrossRef] [PubMed]

32. Tani, H.; Komoda, Y.; Matsuo, E.; Suzuki, K.; Hamamoto, I.; Yamashita, T.; Moriishi, K.; Fujiyama, K.; Kanto, T.; Hayashi, N.; et al. Replication-Competent Recombinant Vesicular Stomatitis Virus Encoding Hepatitis C Virus Envelope Proteins. *J. Virol.* **2007**, *81*, 8601–8612. [CrossRef] [PubMed]

33. Tani, H.; Shiokawa, M.; Kaname, Y.; Kambara, H.; Mori, Y.; Abe, T.; Moriishi, K.; Matsuura, Y. Involvement of Ceramide in the Propagation of Japanese Encephalitis Virus. *J. Virol.* **2010**, *84*, 2798–2807. [CrossRef] [PubMed]

34. Hassan, A.H.E.; Mahmoud, K.; Phan, T.-N.; Shaldam, M.A.; Lee, C.H.; Kim, Y.J.; Cho, S.B.; Bayoumi, W.A.; El-Sayed, S.M.; Choi, Y.; et al. Bestatin analogs-4-quinolinone hybrids as antileishmanial hits: Design, repurposing rational, synthesis, in vitro and in silico studies. *Eur. J. Med. Chem.* **2023**, *250*, 115211. [CrossRef]

35. Hassan, A.H.E.; Phan, T.-N.; Choi, Y.; Moon, S.; No, J.H.; Lee, Y.S. Design, Rational Repurposing, Synthesis, In Vitro Evaluation, Homology Modeling and In Silico Study of Sulfuretin Analogs as Potential Antileishmanial Hit Compounds. *Pharmaceuticals* **2022**, *15*, 1058. [CrossRef]

36. Hassan, A.H.E.; Phan, T.-N.; Yoon, S.; Lee, C.J.; Jeon, H.R.; Kim, S.-H.; No, J.H.; Lee, Y.S. Pyrrolidine-based 3-deoxysphingosylphosphorylcholine analogs as possible candidates against neglected tropical diseases (NTDs): Identification of hit compounds towards development of potential treatment of *Leishmania donovani*. *J. Enzyme Inhib. Med. Chem.* **2021**, *36*, 1922–1930. [CrossRef]

37. Elkamhawy, A.; Paik, S.; Hassan, A.H.E.; Lee, Y.S.; Roh, E.J. Hit discovery of 4-amino-N-(4-(3-(trifluoromethyl)phenoxy)pyrimidin-5-yl)benzamide: A novel EGFR inhibitor from a designed small library. *Bioorg. Chem.* **2017**, *75*, 393–405. [CrossRef]

38. Hassan, A.H.E.; Phan, T.-N.; Moon, S.; Lee, C.H.; Kim, Y.J.; Cho, S.B.; El-Sayed, S.M.; Choi, Y.; No, J.H.; Lee, Y.S. Design, synthesis, and repurposing of O6-aminoalkyl-sulfuretin analogs towards discovery of potential lead compounds as antileishmanial agents. *Eur. J. Med. Chem.* **2023**, *251*, 115256. [CrossRef]

39. Farag, A.K.; Hassan, A.H.E.; Ahn, B.S.; Park, K.D.; Roh, E.J. Reprofiling of pyrimidine-based DAPK1/CSF1R dual inhibitors: Identification of 2,5-diamino-4-pyrimidinol derivatives as novel potential anticancer lead compounds. *J. Enzyme Inhib. Med. Chem.* **2020**, *35*, 311–324. [CrossRef]

40. Elkamhawy, A.; Hassan, A.H.E.; Paik, S.; Sup Lee, Y.; Lee, H.-H.; Shin, J.-S.; Lee, K.-T.; Roh, E.J. EGFR inhibitors from cancer to inflammation: Discovery of 4-fluoro-N-(4-(3-(trifluoromethyl)phenoxy)pyrimidin-5-yl)benzamide as a novel anti-inflammatory EGFR inhibitor. *Bioorg. Chem.* **2019**, *86*, 112–118. [CrossRef]

41. Kang, S.; Lee, J.M.; Jeon, B.; Elkamhawy, A.; Paik, S.; Hong, J.; Oh, S.-J.; Paek, S.H.; Lee, C.J.; Hassan, A.H.E.; et al. Repositioning of the antipsychotic trifluoperazine: Synthesis, biological evaluation and in silico study of trifluoperazine analogs as anti-glioblastoma agents. *Eur. J. Med. Chem.* **2018**, *151*, 186–198. [CrossRef] [PubMed]

42. Hassan, A.H.E.; Yoo, S.Y.; Lee, K.W.; Yoon, Y.M.; Ryu, H.W.; Jeong, Y.; Shin, J.-S.; Kang, S.-Y.; Kim, S.-Y.; Lee, H.-H.; et al. Repurposing mosloflavone/5,6,7-trimethoxyflavone-resveratrol hybrids: Discovery of novel p38-α MAPK inhibitors as potent interceptors of macrophage-dependent production of proinflammatory mediators. *Eur. J. Med. Chem.* **2019**, *180*, 253–267. [CrossRef] [PubMed]

43. Fujimoto, K.J.; Hobbs, D.C.F.; Umeda, M.; Nagata, A.; Yamaguchi, R.; Sato, Y.; Sato, A.; Ohmatsu, K.; Ooi, T.; Yanai, T.; et al. In Silico Analysis and Synthesis of Nafamostat Derivatives and Evaluation of Their Anti-SARS-CoV-2 Activity. *Viruses* **2022**, *14*, 389. [CrossRef] [PubMed]

44. Vardhan, S.; Sahoo, S.K. Virtual screening by targeting proteolytic sites of furin and TMPRSS2 to propose potential compounds obstructing the entry of SARS-CoV-2 virus into human host cells. *J. Tradit. Complement. Med.* **2022**, *12*, 6–15. [CrossRef] [PubMed]

45. Yan, Y.; Yang, J.; Xiao, D.; Yin, J.; Song, M.; Xu, Y.; Zhao, L.; Dai, Q.; Li, Y.; Wang, C.; et al. Nafamostat mesylate as a broad-spectrum candidate for the treatment of flavivirus infections by targeting envelope proteins. *Antivir. Res.* **2022**, *202*, 105325. [CrossRef]

46. Midgley, I.; Hood, A.J.; Proctor, P.; Chasseaud, L.F.; Irons, S.R.; Cheng, K.N.; Brindley, C.J.; Bonn, R. Metabolic fate of 14C-camostat mesylate in man, rat and dog after intravenous administration. *Xenobiotica* **1994**, *24*, 79–92. [CrossRef]

47. Tsukagoshi, S. Pharmacokinetics studies of nafamostat mesilate (FUT), a synthetic protease inhibitor, which has been used for the treatments of DIC and acute pancreatitis, and as an anticoagulant in extracorporeal circulation. *Gan Kagaku Ryoho* **2000**, *27*, 767–774.

48. González-Bello, C. Designing Irreversible Inhibitors—Worth the Effort? *ChemMedChem* **2016**, *11*, 22–30. [CrossRef]

49. Hempel, T.; Raich, L.; Olsson, S.; Azouz, N.P.; Klingler, A.M.; Hoffmann, M.; Pöhlmann, S.; Rothenberg, M.E.; Noé, F. Molecular mechanism of inhibiting the SARS-CoV-2 cell entry facilitator TMPRSS2 with camostat and nafamostat. *Chem. Sci.* **2021**, *12*, 983–992. [CrossRef]

50. Huang, X.; Pearce, R.; Omenn, G.S.; Zhang, Y. Identification of 13 Guanidinobenzoyl- or Aminidinobenzoyl-Containing Drugs to Potentially Inhibit TMPRSS2 for COVID-19 Treatment. *Int. J. Mol. Sci.* **2021**, *22*, 7060. [CrossRef]

51. Peiffer, A.L.; Garlick, J.M.; Wu, Y.; Soellner, M.B.; Brooks, C.L.; Mapp, A.K. TMPRSS2 inhibitor discovery facilitated through an in silico and biochemical screening platform. *bioRxiv* **2021**. [CrossRef]

52. Meyer, D.; Sielaff, F.; Hammami, M.; Böttcher-Friebertshäuser, E.; Garten, W.; Steinmetzer, T. Identification of the first synthetic inhibitors of the type II transmembrane serine protease TMPRSS2 suitable for inhibition of influenza virus activation. *Biochem. J.* **2013**, *452*, 331–343. [CrossRef] [PubMed]

53. Pilgram, O.; Keils, A.; Benary, G.E.; Müller, J.; Merkl, S.; Ngaha, S.; Huber, S.; Chevillard, F.; Harbig, A.; Magdolen, V.; et al. Improving the selectivity of 3-amidinophenylalanine-derived matriptase inhibitors. *Eur. J. Med. Chem.* **2022**, *238*, 114437. [CrossRef] [PubMed]

54. Colombo, É.; Désilets, A.; Duchêne, D.; Chagnon, F.; Najmanovich, R.; Leduc, R.; Marsault, E. Design and Synthesis of Potent, Selective Inhibitors of Matriptase. *ACS Med. Chem. Lett.* **2012**, *3*, 530–534. [CrossRef]

55. Beaulieu, A.; Gravel, É.; Cloutier, A.; Marois, I.; Colombo, É.; Désilets, A.; Verreault, C.; Leduc, R.; Marsault, É.; Richter, M.V. Matriptase Proteolytically Activates Influenza Virus and Promotes Multicycle Replication in the Human Airway Epithelium. *J. Virol.* **2013**, *87*, 4237–4251. [CrossRef]

56. Patrick, G.L. Plasmepsins as targets for antimalarial agents. In *Antimalarial Agents*; Patrick, G.L., Ed.; Elsevier: Amsterdam, The Netherlands, 2020; pp. 217–270.

57. Nanjappan, S.K.; Surendran, S.; Paul, D. Pharmacokinetics and pharmacodynamics of peptidomimetics. In *Peptide and Peptidomimetic Therapeutics*; Qvit, N., Rubin, S.J.S., Eds.; Academic Press: Cambridge, MA, USA, 2022; pp. 195–211.

58. Trabocchi, A. Principles and applications of small molecule peptidomimetics. In *Small Molecule Drug Discovery*; Trabocchi, A., Lenci, E., Eds.; Elsevier: Amsterdam, The Netherlands, 2020; pp. 163–195.

59. Pelay-Gimeno, M.; Glas, A.; Koch, O.; Grossmann, T.N. Structure-Based Design of Inhibitors of Protein–Protein Interactions: Mimicking Peptide Binding Epitopes. *Angew. Chem. Int. Ed.* **2015**, *54*, 8896–8927. [CrossRef]

60. Li Petri, G.; Di Martino, S.; De Rosa, M. Peptidomimetics: An Overview of Recent Medicinal Chemistry Efforts toward the Discovery of Novel Small Molecule Inhibitors. *J. Med. Chem.* **2022**, *65*, 7438–7475. [CrossRef]

61. Tamanini, E.; Buck, I.M.; Chessari, G.; Chiarparin, E.; Day, J.E.H.; Frederickson, M.; Griffiths-Jones, C.M.; Hearn, K.; Heightman, T.D.; Iqbal, A.; et al. Discovery of a Potent Nonpeptidomimetic, Small-Molecule Antagonist of Cellular Inhibitor of Apoptosis Protein 1 (cIAP1) and X-Linked Inhibitor of Apoptosis Protein (XIAP). *J. Med. Chem.* **2017**, *60*, 4611–4625. [CrossRef]

62. Kumar, R.; Bavi, R.; Jo, M.G.; Arulalapperumal, V.; Baek, A.; Rampogu, S.; Kim, M.O.; Lee, K.W. New compounds identified through in silico approaches reduce the α-synuclein expression by inhibiting prolyl oligopeptidase in vitro. *Sci. Rep.* **2017**, *7*, 10827. [CrossRef]

63. Trent, J.O.; Clark, G.R.; Kumar, A.; Wilson, W.D.; Boykin, D.W.; Hall, J.E.; Tidwell, R.R.; Blagburn, B.L.; Neidle, S. Targeting the Minor Groove of DNA: Crystal Structures of Two Complexes between Furan Derivatives of Berenil and the DNA Dodecamer d(CGCGAATTCGCG)2. *J. Med. Chem.* **1996**, *39*, 4554–4562. [CrossRef]

64. Das, B.P.; Boykin, D.W. Synthesis and antiprotozoal activity of 2,5-bis(4-guanylphenyl)furans. *J. Med. Chem.* **1977**, *20*, 531–536. [CrossRef] [PubMed]

65. Moumbock, A.F.; Tran, H.T.; Lamy, E.; Günther, S. BC-11 is a covalent TMPRSS2 fragment inhibitor that impedes SARS-CoV-2 host cell entry. *Arch. Pharm.* **2022**, *356*, e2200371. [CrossRef] [PubMed]

66. Paul, A.; Kumar, A.; Nanjunda, R.; Farahat, A.A.; Boykin, D.W.; Wilson, W.D. Systematic synthetic and biophysical development of mixed sequence DNA binding agents. *Org. Biomol. Chem.* **2017**, *15*, 827–835. [CrossRef] [PubMed]

67. Depauw, S.; Lambert, M.; Jambon, S.; Paul, A.; Peixoto, P.; Nhili, R.; Marongiu, L.; Figeac, M.; Dassi, C.; Paul-Constant, C.; et al. Heterocyclic Diamidine DNA Ligands as HOXA9 Transcription Factor Inhibitors: Design, Molecular Evaluation, and Cellular Consequences in a HOXA9-Dependant Leukemia Cell Model. *J. Med. Chem.* **2019**, *62*, 1306–1329. [CrossRef]

68. Fuentes-Prior, P. Priming of SARS-CoV-2 S protein by several membrane-bound serine proteinases could explain enhanced viral infectivity and systemic COVID-19 infection. *J. Biol. Chem.* **2021**, *296*, 100135. [CrossRef]

69. Kandeel, M.; Yamamoto, M.; Tani, H.; Kobayashi, A.; Gohda, J.; Kawaguchi, Y.; Park, B.K.; Kwon, H.-J.; Inoue, J.-i.; Alkattan, A. Discovery of New Fusion Inhibitor Peptides against SARS-CoV-2 by Targeting the Spike S2 Subunit. *Biomol. Ther.* **2021**, *29*, 282–289. [CrossRef]

70. Giannis, D.; Ziogas, I.A.; Gianni, P. Coagulation disorders in coronavirus infected patients: COVID-19, SARS-CoV-1, MERS-CoV and lessons from the past. *J. Clin. Virol.* **2020**, *127*, 104362. [CrossRef]

71. Mackman, N.; Antoniak, S.; Wolberg, A.S.; Kasthuri, R.; Key, N.S. Coagulation Abnormalities and Thrombosis in Patients Infected with SARS-CoV-2 and Other Pandemic Viruses. *Arterioscler. Thromb. Vasc. Biol.* **2020**, *40*, 2033–2044. [CrossRef]

72. Kastenhuber, E.R.; Mercadante, M.; Nilsson-Payant, B.; Johnson, J.L.; Jaimes, J.A.; Muecksch, F.; Weisblum, Y.; Bram, Y.; Chandar, V.; Whittaker, G.R.; et al. Coagulation factors directly cleave SARS-CoV-2 spike and enhance viral entry. *eLife* **2022**, *11*, e77444. [CrossRef]

73. Hoffmann, M.; Kleine-Weber, H.; Schroeder, S.; Krüger, N.; Herrler, T.; Erichsen, S.; Schiergens, T.S.; Herrler, G.; Wu, N.-H.; Nitsche, A.; et al. SARS-CoV-2 Cell Entry Depends on ACE2 and TMPRSS2 and Is Blocked by a Clinically Proven Protease Inhibitor. *Cell* **2020**, *181*, 271–280.e278. [CrossRef]

74. Iwata-Yoshikawa, N.; Kakizaki, M.; Shiwa-Sudo, N.; Okura, T.; Tahara, M.; Fukushi, S.; Maeda, K.; Kawase, M.; Asanuma, H.; Tomita, Y.; et al. Essential role of TMPRSS2 in SARS-CoV-2 infection in murine airways. *Nat. Commun.* **2022**, *13*, 6100. [CrossRef] [PubMed]

75. Quinn, T.M.; Gaughan, E.E.; Bruce, A.; Antonelli, J.; O'Connor, R.; Li, F.; McNamara, S.; Koch, O.; MacKintosh, C.; Dockrell, D.; et al. Randomised controlled trial of intravenous nafamostat mesylate in COVID pneumonitis: Phase 1b/2a experimental study to investigate safety, Pharmacokinetics and Pharmacodynamics. *eBioMedicine* **2022**, *76*, 103856. [CrossRef]

76. Kinoshita, T.; Shinoda, M.; Nishizaki, Y.; Shiraki, K.; Hirai, Y.; Kichikawa, Y.; Tsushima, K.; Shinkai, M.; Komura, N.; Yoshida, K.; et al. A multicenter, double-blind, randomized, parallel-group, placebo-controlled study to evaluate the efficacy and safety of camostat mesilate in patients with COVID-19 (CANDLE study). *BMC Med.* **2022**, *20*, 342. [CrossRef]
77. Gunst, J.D.; Staerke, N.B.; Pahus, M.H.; Kristensen, L.H.; Bodilsen, J.; Lohse, N.; Dalgaard, L.S.; Brønnum, D.; Fröbert, O.; Hønge, B.; et al. Efficacy of the TMPRSS2 inhibitor camostat mesilate in patients hospitalized with COVID-19-a double-blind randomized controlled trial. *eClinicalMedicine* **2021**, *35*, 100849. [CrossRef] [PubMed]

Article

Targeting SARS-CoV-2 Macrodomain-1 to Restore the Innate Immune Response Using In Silico Screening of Medicinal Compounds and Free Energy Calculation Approaches

Anwar Mohammad [1,*,†], Eman Alshawaf [1,†], Hossein Arefanian [2], Sulaiman K. Marafie [1], Abbas Khan [3], Dong-Qing Wei [3], Fahd Al-Mulla [4,5] and Jehad Abubaker [1]

[1] Department of Biochemistry and Molecular Biology, Dasman Diabetes Institute, Dasman 15462, Kuwait; eman.alshawaf@dasmaninstitute.org (E.A.); sulaiman.marafie@dasmaninstitute.org (S.K.M.); jehad.abubakr@dasmaninstitute.org (J.A.)

[2] Department of Immunology and Microbiology, Dasman Diabetes Institute, Dasman 15462, Kuwait; hossein.arefanian@dasmaninstitute.org

[3] Department of Bioinformatics and Biological Statistics, School of Life Sciences and Biotechnology, Shanghai Jiao Tong University, Shanghai 200240, China; abbaskhan@sjtu.edu.cn (A.K.); dqwei@sjtu.edu.cn (D.-Q.W.)

[4] Department of Genetics and Bioinformatics, Dasman Diabetes Institute, Dasman 15462, Kuwait; fahd.almulla@dasmaninstitute.org

[5] Translational Research Department, Dasman Diabetes Institute, Dasman 15462, Kuwait

* Correspondence: anwar.mohammad@dasmaninstitute.org; Tel.: +965-22242999 (ext. 2450)

† These authors contributed equally to this work.

Citation: Mohammad, A.; Alshawaf, E.; Arefanian, H.; Marafie, S.K.; Khan, A.; Wei, D.-Q.; Al-Mulla, F.; Abubaker, J. Targeting SARS-CoV-2 Macrodomain-1 to Restore the Innate Immune Response Using In Silico Screening of Medicinal Compounds and Free Energy Calculation Approaches. *Viruses* **2023**, *15*, 1907. https://doi.org/10.3390/v15091907

Academic Editor: Simone Brogi

Received: 17 July 2023
Revised: 31 August 2023
Accepted: 5 September 2023
Published: 12 September 2023

Abstract: Among the different drug targets of SARS-CoV-2, a multi-domain protein known as NSP3 is a critical element of the translational and replication machinery. The macrodomain-I, in particular, has been reported to have an essential role in the viral attack on the innate immune response. In this study, we explore natural medicinal compounds and identify potential inhibitors to target the SARS-CoV-2–NSP3 macrodomain-I. Computational modeling and simulation tools were utilized to investigate the structural-dynamic properties using triplicates of 100 ns MD simulations. In addition, the MM/GBSA method was used to calculate the total binding free energy of each inhibitor bound to macrodomain-I. Two significant hits were identified: 3,5,7,4′-tetrahydroxyflavanone 3′-(4-hydroxybenzoic acid) and 2-hydroxy-3-O-beta-glucopyranosyl-benzoic acid. The structural-dynamic investigation of both compounds with macrodomain-I revealed stable dynamics and compact behavior. In addition, the total binding free energy for each complex demonstrated a robust binding affinity, of ΔG -61.98 ± 0.9 kcal/mol for Compound A, while for Compound B, the ΔG was -45.125 ± 2.8 kcal/mol, indicating the inhibitory potential of these compounds. In silico bioactivity and dissociation constant (K_D) determination for both complexes further validated the inhibitory potency of each compound. In conclusion, the aforementioned natural products have the potential to inhibit NSP3, to directly rescue the host immune response. The current study provides the basis for novel drug development against SARS-CoV-2 and its variants.

Keywords: SARS-CoV-2; NSP3; macrodomain-I; medicinal compounds; computational biology

1. Introduction

The severe acute respiratory syndrome coronavirus-2 (SARS-CoV-2) is the causative agent of COVID-19, which the World Health Organization (WHO) declared a global pandemic in March 2020. The SARS-CoV-2 virus has a 29.8 kb positive-sense single-stranded RNA genome with 14 open reading frames (ORFs) encoding 29 proteins that include four structural proteins (Envelope (E), Membrane (M), Nucleocapsid (N), and Spike (S) protein), 16 nonstructural proteins (NSPs), and nine accessory proteins [1,2]. Despite the proofreading capacity during replication, the rapid spread of the SARS-CoV-2 virus has led to a high rate of mutations in the viral proteins. This is evident from the emergence of new variants since the start of the pandemic, such as D614G [3–5], Alpha (B.1.1.7) [6–8], Beta (B.1.351) [9], Gamma (P.1) [10], Delta (B.1.617.2) [11], Epsilon (B.1.427), Eta (B.1.525), Iota (B.1.526) [12], Kappa (B.1.617.1) [13], Mu (B.1.621), Zeta (P.2), and Omicron (B.1.1.529, BA.1–BA.5 lineages) [14–17], that have been reported around the world [18]. Furthermore, these SARS-CoV-2 variants and strains have evolved and are more efficient in host cell entry and evasion of the immune system.

Advances in clinical research have led to a better understanding of SARS-CoV-2, facilitating the rapid development of a range of vaccines that present considerable efficacy against severe COVID-19 infection. Nonetheless, the emergence of SARS-CoV-2 variants has compromised the efficacy of the developed vaccines [19]. One effective method to address this issue would involve identifying novel small-molecule drug-like ligands against potent viral protein sites to serve as therapeutic agents for treating and managing SARS-CoV-2 infection. Therefore, drug discovery target studies have investigated several potential drug targets, such as papain-like protease (PL^{Pro}), S-protein, RNA-dependent RNA polymerase (RdRp), and the nsp10–nsp16 complex [20–22], in addition to the main protease (Mpro), which was the principal explored target of SARS-CoV-2 [23,24]. Macrodomain I (Mac-I) of the NSP3 is an attractive therapeutic target for treating SARS-CoV-2 [25,26]. This conserved macrodomain lies within NSP3, the largest membrane-associated cysteine protease encoded by the SARS-CoV-2 genome. Mac-I disrupts the innate immune response to increase viral pathogenesis and virulence, making Mac-I an appealing drug target [25–28].

NSP3 is a multi-domain protein with three Mac-I domains and two SUD-M-like domains critical in the SARS-CoV-2 translational and replication machinery. Mac-I, also known as X-domain, is highly conserved, with an ADP-ribose (ADPr)-binding site reported in many viruses [29,30], and plays an essential role in the viral attack on the innate immune response. Mac-I functions by hydrolyzing mono-ADP-ribose from target proteins by reversing the activity of ADP-ribosyltransferases [28,31–33] to counteract the host anti-viral response of ADP-ribosylation [28]. As a consequence, the role of Mac-I in ADP-ribosylation is associated with the degree of viral pathogenicity [28,34]. Therefore, blocking the ADPr binding to Mac-I would decrease viral pathogenicity, as reported for infectious bronchitis virus (IBV) [29].

Mac-I subverts the host immune response by interfering with the IFN pathway and dysregulating the signal transducer and activator of transcription 1 (STAT1) [35,36], suggesting a contribution to the cytokine storm phenomenon [37–39]. Consequently, STAT1 is an in vivo target of SARS-CoV-2 Mac-I, which precisely counteracts its mono-ADP-ribosylation with human PARP14 [39]. Additionally, previous studies have reported that targeting Mac-I attenuates NSP3 activity, leading to a significant reduction in viral pathogenesis and an increase in interferon response and viral neutralization [40].

Considering the importance of Mac-I in SARS-CoV-2 pathogenesis, it is a potential druggable target for new anti-viral compounds. In a previous report, Brosey et al. screened PARGi drugs and reported them as effective candidates against Mac-I in SARS-CoV-2 [40]. Further efforts are needed to discover novel broad-spectrum drugs that could efficiently interact with Mac-I to reduce or block ADPr binding. Drugs with such characteristics would debilitate SARS-CoV-2 virulence and help increase the sensitivity of the innate immune response. Molecular screening and fragment-based drug design studies would help discover fragment binders that function as effective Mac-I inhibitors. In this regard, there is a growing interest in employing computational studies to identify novel small-molecule inhibitors. Studies of the chemical composition of natural products are focused on secondary metabolites, mainly polyphenols that include flavonoids, quercetin, and bibenzyl [41]. Some secondary metabolites exhibit diverse biological activities with potential medicinal use [42]. These metabolites are flavonoids, which are reported as effective antioxidants and potent enzyme inhibitors [43].

This report targets the NSP3–Mac-I domain by screening the MPD3 database with 2295 phytochemicals and the East African Natural Compounds Database (EANCDB), which includes 1875 compounds, to inhibit the Mac-I–ADPr binding site. Furthermore, the conformational stability and dynamic features of Mac-I bound to the selected compounds were tested by subjecting each complex to 100 ns molecular dynamic (MD) simulations and the molecular mechanics–generalized Born surface area (MM/GBSA) to extract free binding energies. This study may provide a basis for in vitro and in vivo experiments for novel drug development, which could inhibit the binding of ADPr to Mac-I and NSP3 activity in SARS-CoV-2 research.

2. Materials and Methods

2.1. Structures, Sequence Retrieval, and Modeling

The Protein Databank (http://www.rcsb.org/ accessed on 21 December 2022) was used to retrieve the X-ray crystal structure of Mac-I using PDB ID 6W02 [39,44]. The Mac-I structure was prepared and minimized using Chimera and AMBER simulation packages using the FF14SB force field. To screen targets against ligand-specific Mac-I, we utilized compounds from the MPD3 database and the East African Natural Compounds Database (EANCDB) [45–47].

2.2. Molecular Screening of Medicinal Compound Databases

The MPD3 database with 2295 phytochemicals and 1875 EANCDB compounds were retrieved, prepared, and filtered to meet Lipinski rules before the screening [48,49]. The active site information was based on the available X-ray crystal structure of Mac-I–ADPr (PDB ID: 6W02) for screening. The "structure-based screening module" available online on Mcule (https://mcule.com/dashboard/ accessed on 21 December 2022) was used to screen the MPD3 and EANCDB databases. In addition, Lipinski filtration was employed for the top-scoring compounds, after which the AutoDock Vina algorithm was used for screening purposes, and the top-selected compounds were chosen for the induced-fit docking (IFD) approach to remove false positive results [50].

2.3. Molecular Dynamic Simulation (MD)

The top complexes were subjected to molecular simulation using AMBER20 by adding water around each complex (OPC water model). The OPC water model has been developed to accurately reproduce various properties of water, including structure, dynamics, and thermodynamic properties. It has been extensively validated against experimental data, providing reliable results in many applications. Moreover, the OPC model is transferable across various conditions, including different temperatures and pressures. The drugs were extracted from the proteins and parameterized using GAFF. For the whole protein simulation, FF19SB was employed and then subjected to minimization [51–53]. For minimization, we applied weak harmonic restraints to the protein backbone atoms (Cα, C, N, and O) while keeping the solvent and ions unrestrained, helping maintain the protein's secondary structure during minimization. The initial energy minimization used algorithms such as steepest descent or conjugate gradient, which help to relax the system and eliminate close contacts. The following energy minimization step was conducted without restraints on the entire protein and solvent, allowing for further relaxation and optimization of the structure. Each minimization was run for 6000 and 4000 steps, respectively. The heating and equilibration of each complex were performed, followed by the production of 100 ns. A linear heating method was employed, gradually increasing the system temperature over a specified time period to ensure a smooth transition to the desired simulation temperature. Heating was applied to efficiently raise the system temperature, allowing for rapid equilibration while avoiding abrupt changes that could lead to structural distortions. The equilibration process started from 0 K and was gradually raised to the desired simulation temperature of 300 K, ensuring an appropriate starting point and enabling the system to reach the target temperature for subsequent simulations. Positional restraints were applied to specific atoms or groups during the equilibration phase to keep them fixed, ensuring stability while allowing other parts of the system to adapt to the changing temperature and relax into a suitable configuration. The equilibration phase was carried out for 50 nanoseconds, providing sufficient time for the system to relax, reach equilibrium, and establish stable interactions at the desired temperature. The constant pH method was used with a solvent pH set to 7 since we wanted to emulate a simulation at physiological pH. Constant pH MD simulation allows the protonation state of ionizable groups in a protein to change during the simulation according to the local electrostatic environment and the actual pH of the solution [54,55]. The protonation state of amino acid continuously changes, where the pKa values of the ionizable groups can then be obtained from the distributions of the protonation states across the time of the MD simulation. The amino acids with two atoms carrying a proton are aspartate (Asp) and glutamate (Glu). The protonation/deprotonation percentages of Asp and Glu are presented in the Supplementary Table S1.

For each complex, the simulation was run three times to achieve accuracy and confirm the reproducibility of the results. The long-range electrostatic interactions were treated with the particle mesh Ewald algorithm with a 10.0 Å cutoff distance, while the covalent bonds were treated with the SHAKE algorithm [52,56]. Finally, the CPPTRAJ package was used to analyze the trajectories, and PMEMD.cuda was used for running the simulations [57].

2.4. Binding Free Energy Calculations

Estimating free energy for the interacting small molecules and the target receptor is the most widely used practice to determine accurate binding strengths. It is employed for diverse macromolecule sets such as protein–ligand, protein–protein, or protein–RNA/DNA to precisely estimate the interacting energy [53–56]. Thus, to determine the binding free energy, the top two hits were subjected to a molecular simulation MM/GBSA approach employed using the simulation trajectory [58]. Along with the total binding energy (G), van der Waal (vdW), electrostatic energy, generalized Born (GB), and ESURF were estimated.

2.5. Determination of Dissociation Constant and Bioactivity for the Top Hits

Quantifying the binding strength by estimating the dissociation constant (KD) using PRODIGY-LIG (PRODIGY for LIGands) and in silico bioactivity prediction against various classes of druggable proteins informs the selection of the final small molecule with Molinspiration cheminformatics [59].

3. Results

3.1. Macrodomain I Structural Modeling

The SARS-CoV-2 Nsp3 consists of 1945 amino acids with ten functional domains (Figure 1A), with the Nsp3–Macro domains contributing significantly to inhibit the innate immune response. The Mac-I domain is 169 residues in length, highly conserved [34], and plays an essential role in counteracting host-mediated anti-viral ADPr signaling. The hydrolase activity enables it to remove ADPr from target proteins, and this biochemical feature is directly associated with the SARS-CoV-2 pathogenicity level [34]. The Mac-I domain is an attractive drug target, identifying specific small-molecule inhibitors that would rescue and support the host immune innate response. In our study, we utilized the X-ray crystal structure of the Mac-I domain (PDB ID: 6W02) (Figure 1B) to identify natural compounds from the MPD3 and EANCDB natural product databases that can disrupt the ADPr interactions with Mac-I.

3.2. Discovery of Small-Molecule Inhibitors by Screening Large Libraries

The Mac-I-ADPr binding site was targeted with a multi-step computational screening approach using the Mcule [60] and AutoDock Vina docking tools. Initially, for MPD3 and EANCDB, 4170 compounds were retrieved and subjected to ADMET analysis, whereby 2153 compounds obeyed Lipinski's rule of five. By setting the docking score threshold to ≥ -5 kcal/mol, the MPD3 database presented 30 compounds with docking scores ranging between -8.2 and -10.6 kcal/mol, while the EANCDB showed 112 compounds' docking scores between -6.6 and -10.0 kcal/mol [27]. To further narrow down the selection, the docking threshold was increased to -9.46 kcal/mol, corresponding to the reported score for

ADPr docking to Mac-I [27]. The increased threshold resulted in three compounds from MPD3 and eight from the EANCDB database with docking scores higher than −9.46 kcal/mol. The eleven compounds were re-docked with AutoDock Vina with four compounds, resulting in docking scores to Mac-I higher than −9.46 kcal/mol (Table 1). Furthermore, for the top two compounds—3,5,7,4′-tetrahydroxyflavanone 3′-(4-hydroxybenzoic acid) (Compound A) and 2-hydroxy-3-O-beta-glucopyranosyl-benzoic acid (Compound B)—complexes with Mac-I underwent MD simulations to measure their conformational dynamics and stability.

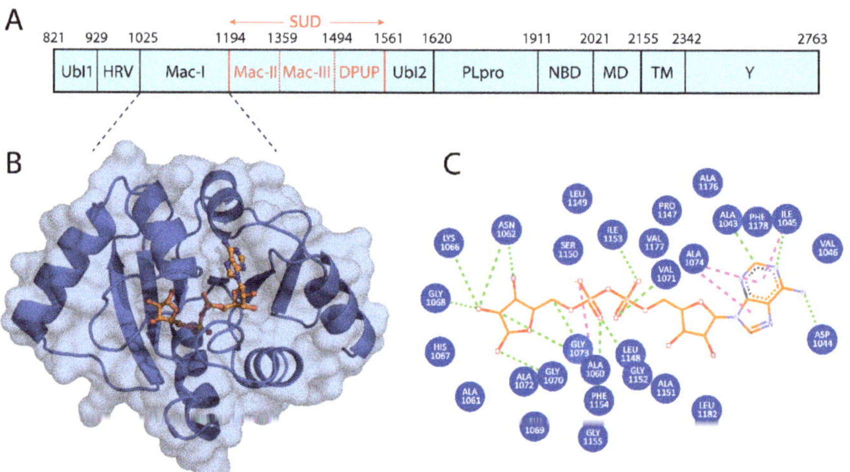

Figure 1. (**A**) Domain structure of NSP3. (**B**) Cartoon representation of Mac-I (blue) with the bound ligand shown in stick form with orange carbons. (**C**) Two-dimensional structure of the drug showing its interaction with Mac-I. Hydrogen bonds are represented by green dashed lines, and pink dashed lines indicate hydrophobic interactions.

3.3. Binding Modes of the Selected Compounds

3,5,7,4′-Tetrahydroxyflavanone 3′-(4-hydroxybenzoic acid) (Compound A) is an extract from the moss species Hypnum cupressiforme, commonly known as cypress-leaved plait-moss. In general, Bryophyta (mosses) are reported to be rich in active metabolites exhibiting antioxidant, antimicrobial, as well as anti-viral properties (Table 2) [42,61]. Since Compound A is a polyphenolic compound extracted from Hypnum cupressiforme, it is expected to have anti-viral and antimicrobial activities. Compound A, with a docking score of −11.54 kcal/mol, formed three hydrophobic interactions, which included bonds with Ile1153 and two interactions with Phe1154. In addition, 11 hydrogen bonds were formed with residues Gly1068, Gly1070, Val1071, Ala1072, Ser1150, Ala1151, Gly1152, Ile1153, Phe1154, Phe1178, and Asp1179 (Figure 2). 2-hydroxy-3-O-beta-glucopyranosyl-benzoic acid (Compound B) is extracted from the dried stem and roots of the Strychnos cocculoides plant (Table 2). Strychnos cocculoides is widely distributed in tropical regions, and in Tanzanian folk medicine, the root and stem barks are used to treat fevers, stomach pain, and snake bites [62,63]. The roots are also widely used to alleviate eczema and treat infections [64]. The interaction pattern of Compound B with Mac-I resulted in two hydrophobic interactions, including bonds with Ile1153 and Phe1154, while nine hydrogen bonds involving residues Gly1068, Gly1070, Val1071, Ala1072, Leu1148, Ser1150, Ala1151, and Gly1152 (Figure 3) were formed.

Table 1. Top hits identified through muti-step screening and rescoring via the IFD method. The table presents the 2D structures, compound names, and docking scores of the top four.

2D Structure	Compound Name	IFD Score	Identifier
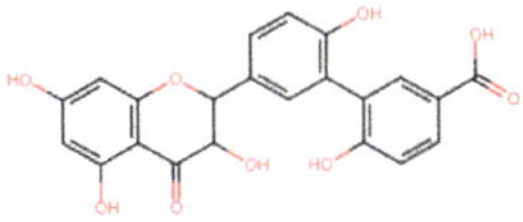	3,5,7,4′-Tetrahydroxy flavanone 3′-(4-hydroxybenzoic acid)	−11.54	A
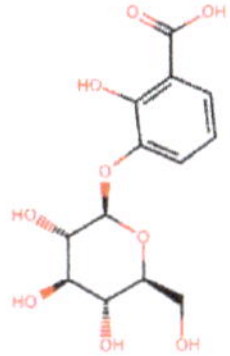	2-Hydroxy-3-O-beta-glucopyranosyl-benzoic acid	−10.0	B
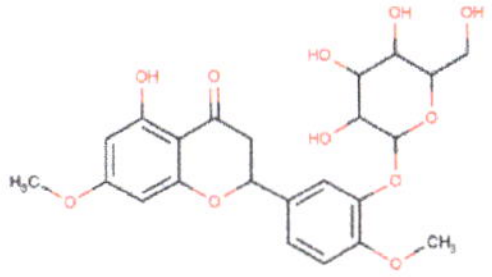	5,3′-Dihydroxy-7,4′-dimethoxyflavanone 3′-glucoside	−9.76	C
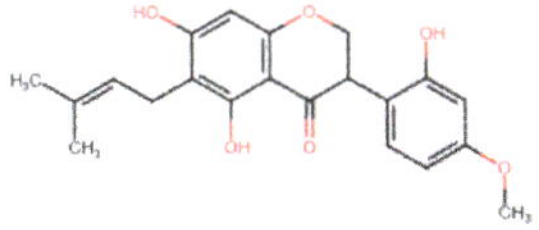	5,7,3′-Trihydroxy-2′,4′-dimethoxy-6-prenylisoflavanone	−9.66	D

Table 2. ADMET analysis results and biological and toxicological properties.

Compound	Identifier	MW. (g/mol)	Source	Molecule Class	Biological Activity	Lipinski Violation	AMES * Toxicity	Rat Oral LD50 (mol/kg)	Max. Tolerated Dose (Human) (log Mg/kg/day)	T. pyriformis Toxicity ** (log µg/L)	HBD	HBA	Rotatable Bonds No.	TPSA	Bioactivity
3,5,7,4′-Tetrahydroxy flavanone 3′-(4-hydroxybenzoic acid)	A	424.36	*Hypnum cupressiforme*	Flavonoid	Anti-viral	1	-	-	-	-	6	9	3	164.74 Å^2	0.24
2-hydroxy-3-O-beta-glucopyranosyl-benzoic acid	B	316.26	*Strychnos cocculoides*	Phenolic	Anti-inflammatory	1	No	2.305	1.294	0.285	6	9	4	156.91 Å^2	0.43

ADMET (Absorption, Distribution, Metabolism, Excretion, and Toxicity). * AMES toxicity test, in-vitro testing to assess the potential carcinogenic effect of chemicals. ** Tetrahymena pyriformis, the most commonly ciliated model, used for toxicological studies.

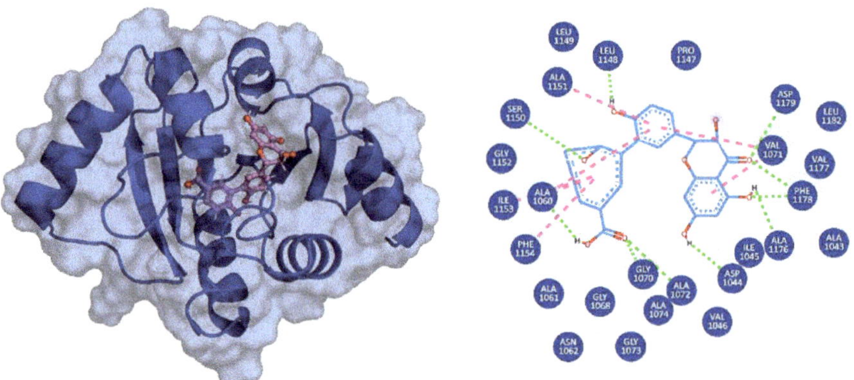

Figure 2. Interaction pattern of 2-hydroxy-3-O-beta-glucopyranosyl-benzoic acid (Compound B). (**A**) Cartoon representation of Mac-I (blue) and the bound ligand shown in stick form with pink carbon atoms. (**B**) Two-dimensional representation of the drug–Mac-I interactions. Hydrogen bonds are represented by green dashed lines; pink dashed lines indicate hydrophobic interactions.

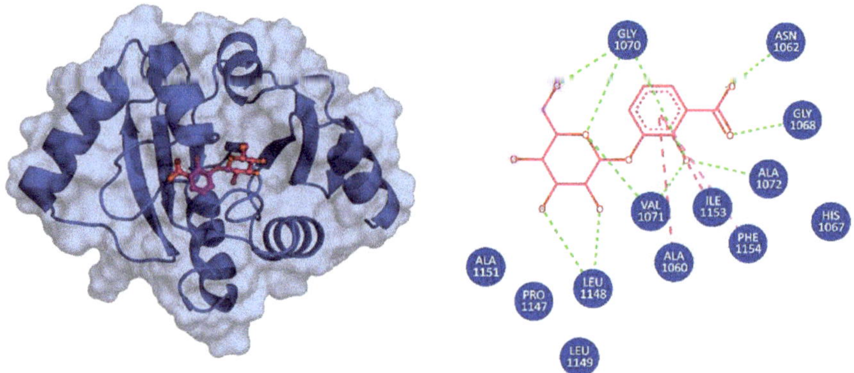

Figure 3. Interaction pattern of 3,5,7,4′-Tetrahydroxyflavanone 3′-(4-hydroxybenzoic acid) (Compound A). (**A**) Cartoon representation of Mac-I (blue) and the bound ligand shown in stick form with purple carbon atoms. (**B**) Two-dimensional representation of the drug–Mac-I interactions. Hydrogen bonds are represented by green dashed lines; pink dashed lines indicate hydrophobic interactions.

3.4. Dynamic Stability and Compactness Assessment

The conformational stability and dynamic environment of Mac-I bound to Compounds A and B were elucidated by running 100 ns MD simulations of the complexes, with the simulations being run in triplicate to ensure the accuracy of the reproducibility of the results. The root-mean-square deviation (RMSD) trajectories of the Cα-atoms demonstrated each system's dynamic stability and convergence (Figure 4A). The radius of gyration (Rg) indicates the structural compactness of the Mac-I–ligand complexes as a function of time (Figure 4B). The structural compactness of the interacting partners reveals essential information regarding the binding and unbinding events during the MD simulation.

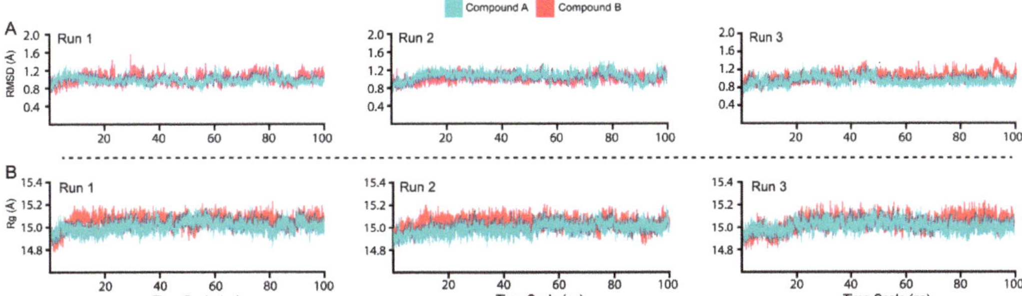

Figure 4. Dynamic stability and compactness assessment of ligands bound to Mac-I. (**A**) RMSD of both the ligands in complex with Mac-I. (**B**) Structural compactness in a dynamic environment.

The complexes of Mac-I with Compounds A and B demonstrated overall stable structures (Figure 4A). In the initial stages of Mac-I in complex with Compound A, the structure converged to 1.2 Å in the first 10 ns. From 10 ns onwards, the complex remained at equilibrium for the duration of the 100 ns simulation. Furthermore, Runs 1, 2, and 3 of Mac-I (Figure 4A turquoise) in complex with Compound A demonstrated similar RMSD atomic configurations. For Mac-I in complex with Compound B (Figure 4A magenta), the RMSD converged to 1.2 Å in the first 15 ns, after which the complex equilibrated and averaged 1.0 Å during the 100 ns simulation. The convergence of the second and third Mac-I–Compound B complex runs showed a similar atomic configuration to the first run, demonstrating the reliability of the MD simulation. For both Mac-I complexes with natural products, the RMSD was maintained with no structural perturbation, revealing a stable binding to the active site residues. Moreover, the average RMSD for both complexes was 1.0 Å, indicating that these ligands form a very stable complex with Mac-I and may inhibit the Mac-I interaction with ADPr, consequently reducing SARS-CoV-2 pathogenesis.

The Rg values of Mac-I binding to Compounds A and B averaged 15.0 Å in both duplicate runs (Figure 4B). This resulted from the stable binding of ligands with minimal unbinding events during the simulation, further corroborated by the RMSD results. The RG data indicate that Compounds A and B may bind Mac-I more favorably than ADPr. The structural compactness strongly aligns with the RMSD results, with no significant variations in the size of the MAC-I complex with Compounds A and B. Consequently, such robust binding indicates the favorable pharmacological properties of both molecules.

3.5. Estimation of Hydrogen Bonding and Residual Flexibility

Hydrogen bonds (H-bonds) contribute to protein–protein and protein–ligand binding. The determination of H-bonds is vital to ascertaining the intermolecular interactions between proteins and the selected ligands. The number of H-bonds formed between Mac-I and Compounds A and B can estimate the strength of the protein–ligand complex (Figure 5). The average number of H-bonds in the Compound A–Mac-I complex was 79 bonds, while the Compound B–Mac-I complex demonstrated an average of 77 hydrogen bonds. The higher number of hydrogen bonds in the Compound A–Mac-I complex than in the Compound B–Mac-I complex indicated a higher-affinity interaction and potentially more significant inhibitory effect. The H-bond estimation for Runs 2 and 3 revealed a similar bonding network with an average of 75 to 90 bonds for each complex. As such, demonstrating a stable conformation for each system could predict the native binding conformation, too. Overall, the results from each replica validate and reproduce the findings, thus showing the reliability of the results and the anti-viral potential of both compounds.

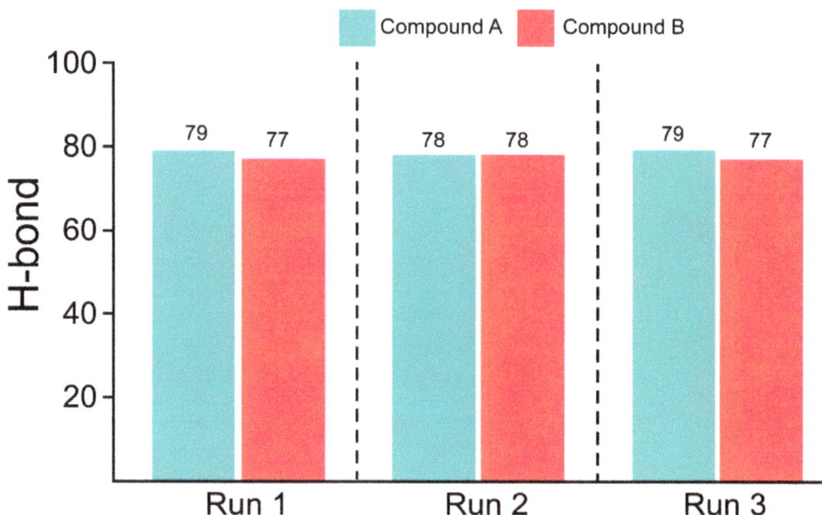

Figure 5. H-bonds of Mac-I bound to Compounds A (magenta) and B (fuchsia). The H-bonds were calculated for each run separately.

The root-mean square fluctuations (RMSFs) of the $C\alpha$ atoms demonstrate the flexibility and average position in a given conformation when the protein is in complex with a protein or ligand (Figure 6). The complexes demonstrated similar residual flexibility, except in regions 41–50, 95–110, and 125–135 of Mac-I, which showed a slight fluctuation. The higher fluctuations may have resulted from the higher conformational sampling in the Mac-I binding pocket. The Mac-I–Compound B complex demonstrated higher fluctuations between residues 40 and 45, with the Compound A–Mac-I complex showing higher fluctuations between residues 95–105 and 125–135. The results for Run 2 and 3 aligned with the results for Run 1, thus showing similar dynamic behavior during simulation. Our findings showed a very low mean RMSF, demonstrating that the residues of Mac-I in complex with Compounds A and B conformed to favorable energy minima [65]. These results are consistent with earlier findings describing low RMSF for the best compounds interacting with SARS-CoV-2 proteins [50].

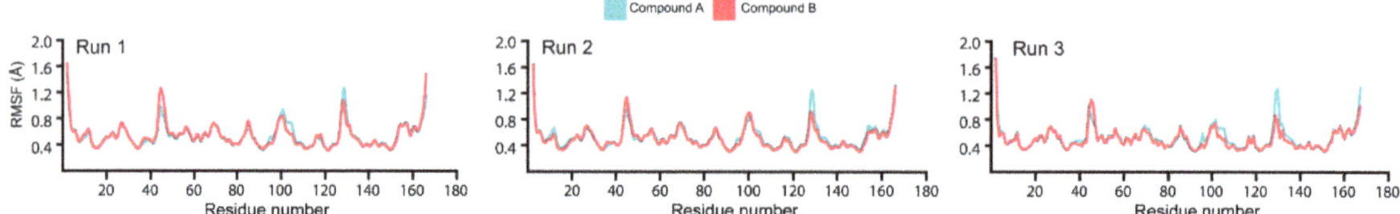

Figure 6. Residue flexibility assessment of Mac-I interacting with Compounds A and B to elucidate the dynamic environment.

3.6. Binding Free Energy Estimation

The MM/GBSA method is more robust in calculating binding energies than the classical docking scores. In addition, the MM/GBSA approach is computationally affordable compared with the costly alchemical-free energy methods. Previous studies have widely applied this approach to discover potential drug candidates for treating SARS-CoV-2. Henceforth, we employed the MM/GBSA method in the current study to estimate the binding free energy of Compound A and B complexes with Mac-I (Table 3). The MM/GBSA values were calculated for both MD simulation runs, and the average values were compared between Compounds A and B in complex with Mac-I.

Table 3. Binding free energy calculated as MM/GBSA of Compounds A and B with SD. All the values are presented in kcal/mol.

MM/GBSA	Compound A-Run 1	Compound A-Run 2	Compound A-Run 3	Average	SD	Compound B-Run 1	Compound B-Run 2	Compound B-Run 3	Average	SD
vdW	−71.16	−68.44	−64.37	−68.0	3.4	−50.96	−50.96	−51.67	−51.2	0.41
Electrostatic	−22.77	−21.72	−21.89	−22.1	0.6	−34.13	−28.76	−21.22	−28.0	6.49
ESURF	23.29	18.78	17.64	19.9	3.0	28.10	24.52	23.78	25.5	2.31
EGB	8.05	10.01	8.85	9.0	1.0	13.83	9.29	4.78	9.3	4.53
ΔG Bind	−62.59	−61.37	−59.77	−61.2	1.4	−43.16	−47.09	−44.33	−44.9	2.02

Our findings revealed average values of vdW (−68 ± 3.4), electrostatic (−21.1 ± 0.6), EGB (9.0 ± 1.0), ESURF (19.9 ± 3.0), and the total binding energy ΔG (−61.2 ± 1.4) kcal/mol for the Compound A–Mac-I complex. For the Compound B–Mac-I complex, the average values were vdW (−51.2 ± 0.41), electrostatic (−28 ± 6.4), EGB (9.3 ± 4.53), ESUF (25.5 ± 2.31), and total binding energy ΔG (−44.9 ± 2.02) kcal/mol. These findings demonstrate that Compound A, with an average ΔG of −61.98 ± 0.9 kcal/mol, is a more potent natural compound in blocking ADPr binding than Compound B, with an average ΔG of −45.125 ± 2.8 kcal/mol. To further corroborate the above findings, we calculated Mac-I's entropic values (ΔTS) [65] in complex with Compounds A and B, resulting in a ΔTS of −16.26 and −11.74, respectively. This indicates that Compound A demonstrates a tighter binding with the Mac-I domain, presenting it as a potential drug to bind Mac-I and inhibit NSP3 activity.

3.7. In Silico Bioactivity and K_D Estimation

The dissociation constant (KD) is the fundamental criterion to elucidate the binding properties of ligands to proteins. In silico PRODIGY-LIG (PRODIGY for LIGands), a post-MD simulation, and an MM/GBSA analysis were used to calculate the KD of Compounds A and B bound to Mac-I (Figure 7A). The Compound A complex with Mac-I showed a KD of −6.9 kcal/mol, whereas Mac-I–Compound B demonstrated a binding affinity of −5.8 kcal/mol. The stronger binding affinity of the latter correlates with the MM/GBSA data calculated from the MD simulations. In addition, in silico bioactivity levels for the complexes were found to be 0.39 and 0.44, respectively (Figure 7A), showing the strong potency of both compounds, with Compound A being the most potent.

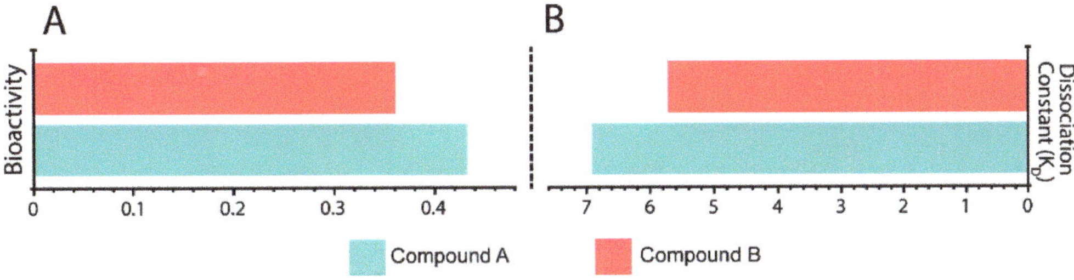

Figure 7. (**A**) In silico bioactivity results for 3,5,7,4′-tetrahydroxyflavanone 3′-(4-hydroxybenzoic acid)–Mac-I and 2-hydroxy-3-O-beta-glucopyranosyl-benzoic acid–Mac-I complexes. (**B**) K_D results for 3,5,7,4′ tetrahydroxyflavanone 3′-(4-hydroxybenzoic acid)–Mac-I and 2-hydroxy-3-O-beta-glucopyranosyl-benzoic acid–Mac-I complexes.

4. Conclusions

The current study used computational modeling and simulation tools to target the Mac-I domain of SARS-CoV-2. Screening of large libraries such as MPD3 and EANCDB identified two hits: 3,5,7,4′-tetrahydroxyflavanone 3′-(4-hydroxybenzoic acid) (Compound A) and 2-hydroxy-3-O-beta-glucopyranosyl-benzoic acid (Compound B). These drugs can potentially bind to Mac-I and inhibit NSP3 activity, thereby directly rescuing the host immune response. The current study provides a basis for novel drug development against SARS-CoV-2 and its variants.

Supplementary Materials: The following supporting information can be downloaded at: https://www.mdpi.com/article/10.3390/v15091907/s1, Table S1: Time-dependent distribution of protonation and deprotonation states of Aspartate and Glutamate residues.

Author Contributions: A.M. conceptualized the concept of this article. A.M. conducted the analysis, E.A. and H.A. were involved in the formal analysis, and A.M., S.K.M. and A.K. contributed to formulating the methodology. A.M., E.A. and H.A. participated in writing—original draft, and D.-Q.W., J.A. and F.A.-M. reviewed and edited the original draft writing. All authors have read and agreed to the published version of the manuscript.

Funding: This research received no external funding.

Institutional Review Board Statement: Not applicable.

Informed Consent Statement: Not applicable.

Data Availability Statement: Data availability upon request.

Acknowledgments: The authors of this study acknowledge the support provided by KFAS and DDI to facilitate this research.

Conflicts of Interest: The authors declare no conflict of interest.

References

1. Masters, P.S. Coronavirus genomic RNA packaging. *Virology* **2019**, *537*, 198–207. [CrossRef]
2. Masters, P.S. The molecular biology of coronaviruses. *Adv. Virus Res.* **2006**, *66*, 193–292. [CrossRef]
3. Mohammad, A.; Alshawaf, E.; Marafie, S.K.; Abu-Farha, M.; Abubaker, J.; Al-Mulla, F. Higher binding affinity of Furin to SARS-CoV-2 spike (S) protein D614G could be associated with higher SARS-CoV-2 infectivity. *Int. J. Infect. Dis.* **2020**, *103*, 611–616. [CrossRef]
4. Haddad, D.; John, S.E.; Mohammad, A.; Hammad, M.M.; Hebbar, P.; Channanath, A.; Nizam, R.; Al-Qabandi, S.; Al Madhoun, A.; Alshukry, A.; et al. SARS-CoV-2: Possible recombination and emergence of potentially more virulent strains. *PLoS ONE* **2021**, *16*, e0251368. [CrossRef]
5. Eaaswarkhanth, M.; Madhoun, A.A.; Al-Mulla, F. Could the D614 G substitution in the SARS-CoV-2 spike (S) protein be associated with higher COVID-19 mortality? *Int. J. Infect. Dis.* **2020**, *96*, 459–460. [CrossRef]
6. Yadav, P.D.; Nyayanit, D.A.; Sahay, R.R.; Sarkale, P.; Pethani, J.; Patil, S.; Baradkar, S.; Potdar, V.; Patil, D.Y. Isolation and characterization of the new SARS-CoV-2 variant in travellers from the United Kingdom to India: VUI-202012/01 of the B.1.1.7 lineage. *J. Travel Med.* **2021**, *28*, taab009. [CrossRef]
7. Rambaut, A.; Loman, N.; Pybus, O.; Barclay, W.; Barrett, J.; Carabelli, A.; Connor, T.; Peacock, T.; Robertson, D.L.; Volz, E. Preliminary genomic characterisation of an emergent SARS-CoV-2 lineage in the UK defined by a novel set of spike mutations. *Genom. Epidemiol.* **2020**, 1–5.
8. Mohammad, A.; Abubaker, J.; Al-Mulla, F. Structural modelling of SARS-CoV-2 alpha variant (B.1.1.7) suggests enhanced furin binding and infectivity. *Virus Res.* **2021**, *303*, 198522. [CrossRef]
9. Faria, N.R.; Mellan, T.A.; Whittaker, C.; Claro, I.M.; Candido, D.d.S.; Mishra, S.; Crispim, M.A.E.; Sales, F.C.S.; Hawryluk, I.; McCrone, J.T.; et al. Genomics and epidemiology of the P.1 SARS-CoV-2 lineage in Manaus, Brazil. *Science* **2021**, *372*, 815–821. [CrossRef]
10. Funk, T.; Pharris, A.; Spiteri, G.; Bundle, N.; Melidou, A.; Carr, M.; Gonzalez, G.; Garcia-Leon, A.; Crispie, F.; O'Connor, L.; et al. Characteristics of SARS-CoV-2 variants of concern B.1.1.7, B.1.351 or P.1: Data from seven EU/EEA countries, weeks 38/2020 to 10/2021. *Euro Surveill. Bull. Eur. Mal. Transm./Eur. Commun. Dis. Bull.* **2021**, *26*, 2100348. [CrossRef]
11. Cherian, S.; Potdar, V.; Jadhav, S.; Yadav, P.; Gupta, N.; Das, M.; Rakshit, P.; Singh, S.; Abraham, P.; Panda, S.; et al. SARS-CoV-2 Spike Mutations, L452R, T478K, E484Q and P681R, in the Second Wave of COVID-19 in Maharashtra, India. *Microorganisms* **2021**, *9*, 1542. [CrossRef]

12. Zhang, L.; Cui, Z.; Li, Q.; Wang, B.; Yu, Y.; Wu, J.; Nie, J.; Ding, R.; Wang, H.; Zhang, Y.; et al. Ten emerging SARS-CoV-2 spike variants exhibit variable infectivity, animal tropism, and antibody neutralization. *Commun. Biol.* **2021**, *4*, 1196. [CrossRef]
13. Yadav, P.D.; Sapkal, G.N.; Abraham, P.; Ella, R.; Deshpande, G.; Patil, D.Y.; Nyayanit, D.A.; Gupta, N.; Sahay, R.R.; Shete, A.M.; et al. Neutralization of Variant under Investigation B.1.617.1 With Sera of BBV152 Vaccinees. *Clin. Infect. Dis.* **2021**, *74*, 366–368. [CrossRef] [PubMed]
14. Hoffmann, M.; Krüger, N.; Schulz, S.; Cossmann, A.; Rocha, C.; Kempf, A.; Nehlmeier, I.; Graichen, L.; Moldenhauer, A.S.; Winkler, M.S.; et al. The Omicron variant is highly resistant against antibody-mediated neutralization: Implications for control of the COVID-19 pandemic. *Cell* **2022**, *185*, 447–456.e411. [CrossRef]
15. Karim, S.S.A.; Karim, Q.A. Omicron SARS-CoV-2 variant: A new chapter in the COVID-19 pandemic. *Lancet* **2021**, *398*, 2126–2128. [CrossRef] [PubMed]
16. Tegally, H.; Moir, M.; Everatt, J.; Giovanetti, M.; Scheepers, C.; Wilkinson, E.; Subramoney, K.; Makatini, Z.; Moyo, S.; Amoako, D.G.; et al. Emergence of SARS-CoV-2 Omicron lineages BA.4 and BA.5 in South Africa. *Nat. Med.* **2022**, *28*, 1785–1790. [CrossRef]
17. Khan, A.; Waris, H.; Rafique, M.; Suleman, M.; Mohammad, A.; Ali, S.S.; Khan, T.; Waheed, Y.; Liao, C.; Wei, D.-Q. The Omicron (B.1.1.529) variant of SARS-CoV-2 binds to the hACE2 receptor more strongly and escapes the antibody response: Insights from structural and simulation data. *Int. J. Biol. Macromol.* **2022**, *200*, 438–448. [CrossRef] [PubMed]
18. CDC. SARS-CoV-2 Variant Classifications and Definitions. Available online: https://www.cdc.gov/coronavirus/2019-ncov/variants/variant-classifications.html (accessed on 21 December 2022).
19. Ghazy, R.M.; Ashmawy, R.; Hamdy, N.A.; Elhadi, Y.A.M.; Reyad, O.A.; Elmalawany, D.; Almaghraby, A.; Shaaban, R.; Taha, S.H.N. Efficacy and Effectiveness of SARS-CoV-2 Vaccines: A Systematic Review and Meta-Analysis. *Vaccines* **2022**, *10*, 350. [CrossRef]
20. Wu, C.-R.; Yin, W.-C.; Jiang, Y.; Xu, H.E. Structure genomics of SARS-CoV-2 and its Omicron variant: Drug design templates for COVID-19. *Acta Pharmacol. Sin.* **2022**, *43*, 3021–3033. [CrossRef]
21. Mohammad, A.; Alshawaf, E.; Marafie, S.K.; Abu-Farha, M.; Al-Mulla, F.; Abubaker, J. Molecular Simulation-Based Investigation of Highly Potent Natural Products to Abrogate Formation of the nsp10–nsp16 Complex of SARS-CoV-2. *Biomolecules* **2021**, *11*, 573. [CrossRef] [PubMed]
22. Khan, A.; Ali, S.S.; Khan, M.T.; Saleem, S.; Ali, A.; Suleman, M.; Babar, Z.; Shafiq, A.; Khan, M.; Wei, D.-Q. Combined drug repurposing and virtual screening strategies with molecular dynamics simulation identified potent inhibitors for SARS-CoV-2 main protease (3CLpro). *J. Biomol. Struct. Dyn.* **2021**, *39*, 4659–4670. [CrossRef]
23. Chaves, O.A.; Fintelman-Rodrigues, N.; Wang, X.; Sacramento, C.Q.; Temerozo, J.R.; Ferreira, A.C.; Mattos, M.; Pereira-Dutra, F.; Bozza, P.T.; Castro-Faria-Neto, H.C.; et al. Commercially Available Flavonols Are Better SARS-CoV-2 Inhibitors than Isoflavone and Flavones. *Viruses* **2022**, *14*, 1458. [CrossRef]
24. Chaves, O.A.; Lima, C.R.; Fintelman-Rodrigues, N.; Sacramento, C.Q.; de Freitas, C.S.; Vazquez, L.; Temerozo, J.R.; Rocha, M.E.N.; Dias, S.S.G.; Carels, N.; et al. Agathisflavone, a natural biflavonoid that inhibits SARS-CoV-2 replication by targeting its proteases. *Int. J. Biol. Macromol.* **2022**, *222*, 1015–1026. [CrossRef] [PubMed]
25. Schuller, M.; Correy, G.J.; Gahbauer, S.; Fearon, D.; Wu, T.; Díaz, R.E.; Young, I.D.; Carvalho Martins, L.; Smith, D.H.; Schulze-Gahmen, U.; et al. Fragment binding to the Nsp3 macrodomain of SARS-CoV-2 identified through crystallographic screening and computational docking. *Sci. Adv.* **2021**, *7*, eabf8711. [CrossRef]
26. Yan, W.; Zheng, Y.; Zeng, X.; He, B.; Cheng, W. Structural biology of SARS-CoV-2: Open the door for novel therapies. *Signal Transduct. Target. Ther.* **2022**, *7*, 26. [CrossRef] [PubMed]
27. Hussain, I.; Pervaiz, N.; Khan, A.; Saleem, S.; Shireen, H.; Wei, D.-Q.; Labrie, V.; Bao, Y.; Abbasi, A.A. Evolutionary and structural analysis of SARS-CoV-2 specific evasion of host immunity. *Genes Immun.* **2020**, *21*, 409–419. [CrossRef]
28. Fehr, A.R.; Jankevicius, G.; Ahel, I.; Perlman, S. Viral Macrodomains: Unique Mediators of Viral Replication and Pathogenesis. *Trends Microbiol.* **2018**, *26*, 598–610. [CrossRef] [PubMed]
29. Leung, A.K.; McPherson, R.L.; Griffin, D.E. Macrodomain ADP-ribosylhydrolase and the pathogenesis of infectious diseases. *PLoS Pathog.* **2018**, *14*, e1006864. [CrossRef]
30. Frick, D.N.; Virdi, R.S.; Vuksanovic, N.; Dahal, N.; Silvaggi, N.R. Molecular Basis for ADP-Ribose Binding to the Mac1 Domain of SARS-CoV-2 nsp3. *Biochemistry* **2020**, *59*, 2608–2615. [CrossRef]
31. Alhammad, Y.M.O.; Fehr, A.R. The Viral Macrodomain Counters Host Antiviral ADP-Ribosylation. *Viruses* **2020**, *12*, 384. [CrossRef]
32. Fehr, A.R.; Channappanavar, R.; Jankevicius, G.; Fett, C.; Zhao, J.; Athmer, J.; Meyerholz, D.K.; Ahel, I.; Perlman, S. The Conserved Coronavirus Macrodomain Promotes Virulence and Suppresses the Innate Immune Response during Severe Acute Respiratory Syndrome Coronavirus Infection. *mBio* **2016**, *7*, 01721-16. [CrossRef]
33. Li, C.; Debing, Y.; Jankevicius, G.; Neyts, J.; Ahel, I.; Coutard, B.; Canard, B. Viral Macro Domains Reverse Protein ADP-Ribosylation. *J. Virol.* **2016**, *90*, 8478–8486. [CrossRef] [PubMed]
34. Han, W.; Li, X.; Fu, X. The macro domain protein family: Structure, functions, and their potential therapeutic implications. *Mutat. Res./Rev. Mutat. Res.* **2011**, *727*, 86–103. [CrossRef] [PubMed]
35. Lin, M.-H.; Chang, S.-C.; Chiu, Y.-C.; Jiang, B.-C.; Wu, T.-H.; Hsu, C.-H. Structural, biophysical, and biochemical elucidation of the SARS-CoV-2 nonstructural protein 3 macro domain. *ACS Infect. Dis.* **2020**, *6*, 2970–2978. [CrossRef]

36. Srinivasan, S.; Cui, H.; Gao, Z.; Liu, M.; Lu, S.; Mkandawire, W.; Narykov, O.; Sun, M.; Korkin, D. Structural genomics of SARS-CoV-2 indicates evolutionary conserved functional regions of viral proteins. *Viruses* **2020**, *12*, 360. [CrossRef]
37. Hoch, N.C. Host ADP-ribosylation and the SARS-CoV-2 macrodomain. *Biochem. Soc. Trans.* **2021**, *49*, 1711–1721. [CrossRef]
38. Molaei, S.; Dadkhah, M.; Asghariazar, V.; Karami, C.; Safarzadeh, E. The immune response and immune evasion characteristics in SARS-CoV, MERS-CoV, and SARS-CoV-2: Vaccine design strategies. *Int. Immunopharmacol.* **2021**, *92*, 107051. [CrossRef]
39. Claverie, J.M. A Putative Role of de-Mono-ADP-Ribosylation of STAT1 by the SARS-CoV-2 Nsp3 Protein in the Cytokine Storm Syndrome of COVID-19. *Viruses* **2020**, *12*, 646. [CrossRef]
40. Brosey, C.A.; Houl, J.H.; Katsonis, P.; Balapiti-Modarage, L.P.F.; Bommagani, S.; Arvai, A.; Moiani, D.; Bacolla, A.; Link, T.; Warden, L.S.; et al. Targeting SARS-CoV-2 Nsp3 macrodomain structure with insights from human poly(ADP-ribose) glycohydrolase (PARG) structures with inhibitors. *Prog. Biophys. Mol. Biol.* **2021**, *163*, 171–186. [CrossRef] [PubMed]
41. Pandey, K.B.; Rizvi, S.I. Plant polyphenols as dietary antioxidants in human health and disease. *Oxid. Med. Cell. Longev.* **2009**, *2*, 270–278. [CrossRef]
42. Lunić, T.M.; Oalđe, M.M.; Mandić, M.R.; Sabovljević, A.D.; Sabovljević, M.S.; Gašić, U.M.; Duletić-Laušević, S.N.; Božić, B.D.; Božić Nedeljković, B.D. Extracts Characterization and In Vitro Evaluation of Potential Immunomodulatory Activities of the Moss *Hypnum cupressiforme* Hedw. *Molecules* **2020**, *25*, 3343. [CrossRef]
43. Lin, D.; Xiao, M.; Zhao, J.; Li, Z.; Xing, B.; Li, X.; Kong, M.; Li, L.; Zhang, Q.; Liu, Y.; et al. An Overview of Plant Phenolic Compounds and Their Importance in Human Nutrition and Management of Type 2 Diabetes. *Molecules* **2016**, *21*, 1374. [CrossRef]
44. Rose, P.W.; Prlić, A.; Altunkaya, A.; Bi, C.; Bradley, A.R.; Christie, C.H.; Costanzo, L.D.; Duarte, J.M.; Dutta, S.; Feng, Z. The RCSB protein data bank: Integrative view of protein, gene and 3D structural information. *Nucleic Acids Res.* **2016**, gkw1000.
45. Webb, B.; Sali, A. Protein structure modeling with MODELLER. In *Structural Genomics*; Springer: New York, NY, USA, 2021; pp. 239–255.
46. Webb, B.; Sali, A. Comparative protein structure modeling using MODELLER. *Curr. Protoc. Bioinforma.* **2016**, *54*, 5.6.1–5.6.37. [CrossRef]
47. Pettersen, E.F.; Goddard, T.D.; Huang, C.C.; Meng, E.C.; Couch, G.S.; Croll, T.I.; Morris, J.H.; Ferrin, T.E. UCSF ChimeraX: Structure visualization for researchers, educators, and developers. *Protein Sci.* **2021**, *30*, 70–82. [CrossRef] [PubMed]
48. Simoben, C.V.; Qaseem, A.; Moumbock, A.F.; Telukunta, K.K.; Günther, S.; Sippl, W.; Ntie-Kang, F. Pharmacoinformatic investigation of medicinal plants from East Africa. *Mol. Inf.* **2020**, *39*, 2000163. [CrossRef]
49. Mumtaz, A.; Ashfaq, U.A.; ul Qamar, M.T.; Anwar, F.; Gulzar, F.; Ali, M.A.; Saari, N.; Pervez, M.T. MPD3: A useful medicinal plants database for drug designing. *Nat. Prod. Res.* **2017**, *31*, 1228–1236. [CrossRef] [PubMed]
50. Trott, O.; Olson, A.J. AutoDock Vina: Improving the speed and accuracy of docking with a new scoring function, efficient optimization, and multithreading. *J. Comput. Chem.* **2010**, *31*, 455–461. [CrossRef]
51. Salomon-Ferrer, R.; Case, D.A.; Walker, R.C. An overview of the Amber biomolecular simulation package. *Wiley Interdiscip. Rev. Comput. Mol. Sci.* **2013**, *3*, 198–210. [CrossRef]
52. Pearlman, D.A.; Case, D.A.; Caldwell, J.W.; Ross, W.S.; Cheatham, T.E., III; DeBolt, S.; Ferguson, D.; Seibel, G.; Kollman, P. AMBER, a package of computer programs for applying molecular mechanics, normal mode analysis, molecular dynamics and free energy calculations to simulate the structural and energetic properties of molecules. *Comput. Phys. Commun.* **1995**, *91*, 1–41. [CrossRef]
53. Case, D.A.; Cheatham, T.E., III; Darden, T.; Gohlke, H.; Luo, R.; Merz, K.M., Jr.; Onufriev, A.; Simmerling, C.; Wang, B.; Woods, R.J. The Amber biomolecular simulation programs. *J. Comput. Chem.* **2005**, *26*, 1668–1688. [CrossRef] [PubMed]
54. Dobrev, P.; Vemulapalli, S.P.B.; Nath, N.; Griesinger, C.; Grubmüller, H. Probing the Accuracy of Explicit Solvent Constant pH Molecular Dynamics Simulations for Peptides. *J. Chem. Theory Comput.* **2020**, *16*, 2561–2569. [CrossRef] [PubMed]
55. Swails, J.; McGee, D., Jr. The Amber Project. Available online: https://ambermd.org/tutorials/advanced/tutorial18/section2.php (accessed on 21 December 2022).
56. Salomon-Ferrer, R.; Götz, A.W.; Poole, D.; Le Grand, S.; Walker, R.C. Routine microsecond molecular dynamics simulations with AMBER on GPUs. 2. Explicit solvent particle mesh Ewald. *J. Chem. Theory Comput.* **2013**, *9*, 3878–3888. [CrossRef] [PubMed]
57. Roe, D.R.; Cheatham, T.E., III. PTRAJ and CPPTRAJ: Software for processing and analysis of molecular dynamics trajectory data. *J. Chem. Theory Comput.* **2013**, *9*, 3084–3095. [CrossRef]
58. Mishra, S.K.; Koča, J. Assessing the Performance of MM/PBSA, MM/GBSA, and QM–MM/GBSA Approaches on Protein/Carbohydrate Complexes: Effect of Implicit Solvent Models, QM Methods, and Entropic Contributions. *J. Phys. Chem. B* **2018**, *122*, 8113–8121. [CrossRef]
59. Vangone, A.; Schaarschmidt, J.; Koukos, P.; Geng, C.; Citro, N.; Trellet, M.E.; Xue, L.C.; Bonvin, A.M. Large-scale prediction of binding affinity in protein–small ligand complexes: The PRODIGY-LIG web server. *Bioinformatics* **2019**, *35*, 1585–1587. [CrossRef]
60. Kiss, R.; Sandor, M.; Szalai, F.A. http://Mcule.com: A public web service for drug discovery. *J Cheminform.* **2012**, *4*, P17. [CrossRef]
61. Yayıntaş, O.T.; Yılmaz, S.; Sökmen, M. Determination of antioxidant, antimicrobial and antitumor activity of bryophytes from Mount Ida (Canakkale, Turkey). *Indian J. Tradit. Knowl.* **2019**, *18*, 395–401.
62. Sunghwa, F.; Koketsu, M. Phenolic and bis-iridoid glycosides from *Strychnos cocculoides*. *Nat. Prod. Res.* **2009**, *23*, 1408–1415. [CrossRef]
63. Sitrit, Y.; Loison, S.; Ninio, R.; Dishon, E.; Bar, E.; Lewinsohn, E.; Mizrahi, Y. Characterization of monkey orange (*Strychnos spinosa* Lam.), a potential new crop for arid regions. *J. Agric. Food Chem.* **2003**, *51*, 6256–6260. [CrossRef]

64. Mwamba, C.K. *Monkey Orange: Strychnos cocculoides*, 1st ed.; International Centre for Underutilised Crops, Southampton University: Southampton, UK, 2006.
65. Genheden, S.; Ryde, U. How to obtain statistically converged MM/GBSA results. *J. Comput. Chem.* **2010**, *31*, 837–846. [CrossRef] [PubMed]

Communication

Applications of the Microscale Thermophoresis Binding Assay in COVID-19 Research

Damian T. Nydegger [1,2], Jonai Pujol-Giménez [1,2,*], Palanivel Kandasamy [1,2], Bruno Vogt [1,2] and Matthias A. Hediger [1,2,*]

1 Department of Nephrology and Hypertension, Inselspital, University of Bern, Kinderklinik, Freiburgstrasse 15, 3010 Bern, Switzerland; damian.nydegger@unibe.ch (D.T.N.)
2 Department of Biomedical Research, Inselspital, University of Bern, Kinderklinik, Freiburgstrasse 15, 3010 Bern, Switzerland
* Correspondence: jonai.pujol@unibe.ch (J.P.-G.); matthias.hediger@unibe.ch (M.A.H.)

Abstract: As the COVID-19 pandemic progresses, new variants of SARS-CoV-2 continue to emerge. This underscores the need to develop optimized tools to study such variants, along with new coronaviruses that may arise in the future. Such tools will also be instrumental in the development of new antiviral drugs. Here, we introduce microscale thermophoresis (MST) as a reliable and versatile tool for coronavirus research, which we demonstrate through three different applications described in this report: (1) binding of the SARS-CoV-2 spike receptor binding domain (RBD) to peptides as a strategy to prevent virus entry, (2) binding of the RBD to the viral receptor ACE2, and (3) binding of the RBD to ACE2 in complex with the amino acid transporter SLC6A20/SIT1 or its allelic variant rs61731475 (p.Ile529Val). Our results demonstrate that MST is a highly precise approach to studying protein–protein and/or protein–ligand interactions in coronavirus research, making it an ideal tool for studying viral variants and developing antiviral agents. Moreover, as shown in our results, a unique advantage of the MST assay over other available binding assays is the ability to measure interactions with membrane proteins in their near-native plasma membrane environment.

Keywords: COVID-19; microscale thermophoresis; SLC6A20 amino acid transporter; antiviral agents

Citation: Nydegger, D.T.; Pujol-Giménez, J.; Kandasamy, P.; Vogt, B.; Hediger, M.A. Applications of the Microscale Thermophoresis Binding Assay in COVID-19 Research. *Viruses* **2023**, *15*, 1432. https://doi.org/10.3390/v15071432

Academic Editor: Simone Brogi

Received: 23 May 2023
Revised: 16 June 2023
Accepted: 21 June 2023
Published: 25 June 2023

1. Introduction

Coronavirus disease 2019 (COVID-19) is caused by the severe acute respiratory syndrome coronavirus 2 (SARS-CoV-2). This enveloped positive-strand RNA virus, belonging to the group of β-coronaviruses that infect mammals, is characterized by club-like spike glycoproteins projecting from its surface [1,2]. The disease emerged in December 2019, with the first cases observed in Wuhan, China, followed by global spread, resulting in the largest pandemic so far this century [3–5]. Upon entry into the respiratory tract, one of the major sites of infection, SARS-CoV2 uses its spike protein receptor binding domain (RBD) to bind to the angiotensin-converting enzyme-2 (ACE2) receptor, which is located on the surface of lung epithelial cells. Binding is followed by priming of the spike protein through the peptidase TMPRSS2, which in turn triggers endocytosis and stimulates virus production and viral spread. The most common symptoms of COVID-19 are fever, dry cough, dyspnea, myalgia, and fatigue. Although SARS-CoV-2 infection can be asymptomatic, the disease is fatal in others, particularly the elderly or people who are immunocompromised [4,6]. As of April 2023, there have been more than 760 million confirmed cases and nearly 7 million deaths, according to the World Health Organization [7].

SARS-CoV-2 was the third outbreak of coronaviruses this century. Previously, severe acute respiratory syndrome coronavirus (SARS-CoV) emerged in 2002 and Middle East respiratory syndrome (MERS) in 2012. These repeated outbreaks highlight the importance of having reliable tools at hand to quickly investigate emerging viruses.

Here, we present and discuss the various applications of a binding assay known as microscale thermophoresis (MST) and review its suitability for coronavirus research. Briefly, MST depends on the movement of molecules in a temperature gradient (Figure 1). The rate of this movement is sensitive to changes in size, charge, and hydration shell. As a result of the binding of two molecules, the movement rate changes, which can be detected by MST. To measure changes in movement, one of the binders must be fluorescently labeled. To determine the binding affinity, a serial dilution of the ligand is generated and assessed in the capillaries where the binding experiments take place. For each capillary, the initial fluorescence is measured, then the infrared laser is activated to generate a local temperature gradient within the capillary (Figure 1a). This causes the labeled binder (e.g., labeled RBD) to move out of the focal area of the fluorescence detector, resulting in a decrease in fluorescence (Figure 1b). After switching off the IR laser, the labeled binder moves back into the focal area and the fluorescence returns to its initial state. A protein (e.g., RBD) that is bound to a ligand (e.g., a peptide) will exhibit a different rate of motion in response to laser exposure than an unbound protein because its size, charge, and/or hydration shell will change after binding. For each dilution, fluorescence is measured at the same time point, resulting in an affinity curve which is then used to determine the dissociation constant (K_D) (Figure 1c) [8–13].

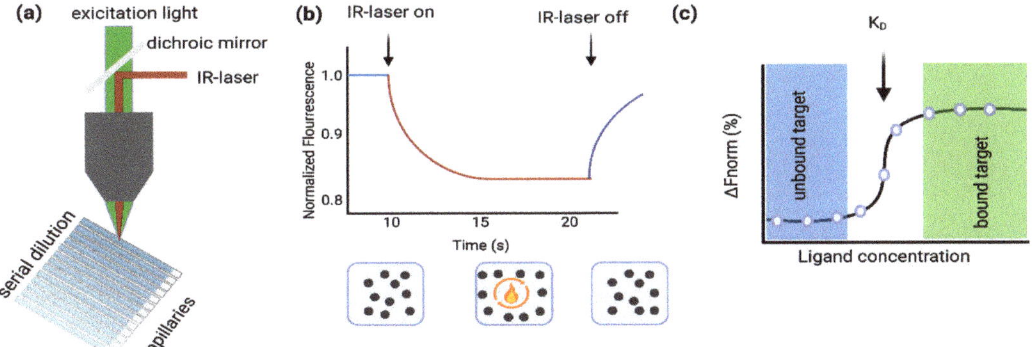

Figure 1. (**a**) Schematic representation of the MST assay. (**b**) (Upper panel.) Representative MST trace showing fluorescence measurement upon IR stimulation (On/Off) over the indicated time period. (Lower panel.) Schematic representation of the response of the proteins in solution upon IR stimulation. (**c**) Representative fitting of the fluorescent values obtained for the different ligand concentrations at a given time point of the MST traces. The figure was created using the BioRender software (https://www.biorender.com).

There are a variety of alternative biochemical assays that can be used to determine the dissociation constant (K_D) of protein–ligand interactions such as thermal shift assay [14], surface plasmon resonance (SPR) [15], circular dichroism [14], isothermal titration calorimetry [16], small-angle X-ray scattering [17], and nuclear magnetic resonance spectrometry [18], among others. Given the wide variety of approaches that exist for affinity measurements, our goal was to select the method that offered the greatest advantages for our applications.

Importantly, the techniques mentioned above usually require purified proteins in relatively large quantities. But it is often costly to purchase these purified proteins, and the alternative of performing protein purification in-house can be challenging. In addition, membrane proteins often behave differently when extracted from their natural lipid environment. In this regard, a major advantage of the MST assay is that it allows measurements with crude cell lysates [19,20], thus avoiding harsh protein purification procedures and also greatly simplifying sample preparation. In addition, MST allows the retention of the natural lipid environment and potential protein binding partners. Thus, MST offers the possibility of performing affinity measurements in a native lipid bilayer environment, which distinguishes this approach from the other binding assays mentioned above.

In our study, three different MST approaches were evaluated as resources for coronavirus disease investigation. One approach was to use purified RBD protein and to determine the affinity between RBD and potential peptide ligands as a strategy to prevent viral RBD binding to ACE2. Another approach was to use cell lysates overexpressing the membrane protein receptor ACE2, which allows physiologically relevant measurements of the RBD-ACE2 interaction. In the third approach, we examined the effect of ACE2 interacting partners on RBD-ACE2 binding. For this purpose, we investigated the co-expression of ACE2 with amino acid transporter SLC6A20 or its allelic variant rs61731475 (p.Ile529Val).

For further information on this versatile MST binding assay, please refer to previous publications showing a variety of additional applications of this cutting-edge technique [9–12,21].

2. Materials and Methods

2.1. Labeling of Purified Proteins

RBD (recombinant SARS-CoV-2, S1 subunit protein) was obtained from RayBiotech, Inc., Peachtree Corners, GA (Lucerna-Chem #230-30162, Luzern, Switzerland) and labeled according to the protein labeling protocol of NanoTemper Technologies GmbH (Munich, Germany). The labeling protocol features the N-hydroxy succinimide (NHS) coupling of the fluorescent dye NT647 (RED-NHS 2nd Generation, NanoTemper # MO-L011). The fluorescence dye of the kit carries a reactive NHS-ester group that reacts with the primary amines (i.e., lysine residues) to form a covalent bond. The labeling and the subsequent MST experiments were performed in PBS-TR (PBS + 0.05% Triton X100).

2.2. Cell Culture and Lysate Preparation

A HEK293 cell line stably expressing pcDNA3-ACE2 (WT)-8his (Addgene # 149268; Addgene Europe, Teddington, UK) was generated using geneticin (g418) selection. Prior to MST measurements, cells were washed with PBS, removed with a cell scraper and collected by centrifugation. The resulting pellets were suspended in PBS-T (PBS + 0.05% Tween-20) and snap-frozen in liquid nitrogen for 1 min. This was followed by homogenization in a Teflon-glass homogenizer. Subsequently, the samples were centrifuged at $1500 \times g$ for 5 min and the upper milky fractions, which contained membranes and vesicles enriched in the overexpressed protein, were collected and used for His-tag labeling.

2.3. His-Tag Labeling of Cell Lysate

Labeling of cell lysates was performed according to the protocol of the His-tag labeling kit RED-tris-NTA 2nd Generation (NanoTemper # MO-L018). The optimal protein concentration for labeling the cell lysate was determined by titrating the cell lysate against a constant dye concentration (25 nM) on MST. The optimal protein concentration used corresponded to the concentration at which the curve began to saturate and was 0.56 mg/mL. As recommended in the protocol of NanoTemper, PBST-buffer (PBS + 0.05% Tween-20) was used for the labeling and measurements.

2.4. MST Measurements

The measurements were performed using the RED channel Monolith NT.115 MST device (NanoTemper Technologies GmbH) and the corresponding Monolith capillaries (NanoTemper; MO-K022). The data were collected with MO. Control v2 and evaluated using the MO. Affinity v2.3 Software (NanoTemper).

2.5. Generation of the SLC6A20 Variant I529V (Ile529Val; rs61731475)

For the co-expression experiments with the SLC6A20 amino acid transporter, we purchased the NM_020208 Myc-DDK-tagged-human ORF clone #RC215764 (OriGene; OriGene

EU, Herford, Germany). The I529V variant was generated by site-directed mutagenesis using the standard polymerase chain reaction (PCR)-based approach, as previously described [22]. The primer used to generate the SLC6A20_I529V variant had the sequence 5$'$ CCTGAGCGACTACGTCCTCACGGGGACCC 3$'$.

2.6. Amino Acid Sequences of the RBD-Binding Peptides That Were Tested with MST

NB001R: RRRRRRFFERHHMVGSCMRAFHQL (24 Residues)
NB001: FFERHHMVGSCMRAFHQL (18 Residues)
NB002: FAHMNWKMQWLQKWQQGK (18 Residues)

3. Results

SARS-CoV-2 relies on the receptor ACE2 to enter cells of the human body such as the epithelial cells of the lung. The first contact between the virus and the target cell is the binding of the RBD of the viral spike protein to the extracellular catalytic domain of ACE2. This entry pathway is an attractive target for the development of antiviral agents that could bind either the RBD of the viral spike protein or its receptor ACE2 to prevent viral infection. Given that cell membranes are complex structures, the binding between RBD and ACE2 is expected to be influenced by a number of factors. These include interacting proteins, including the amino acid transporter SLC6A20, which is being investigated in our laboratory.

In the present study, we present three examples to illustrate how MST can be employed to measure protein–ligand interactions that are relevant in COVID-19 research (Figure 2).

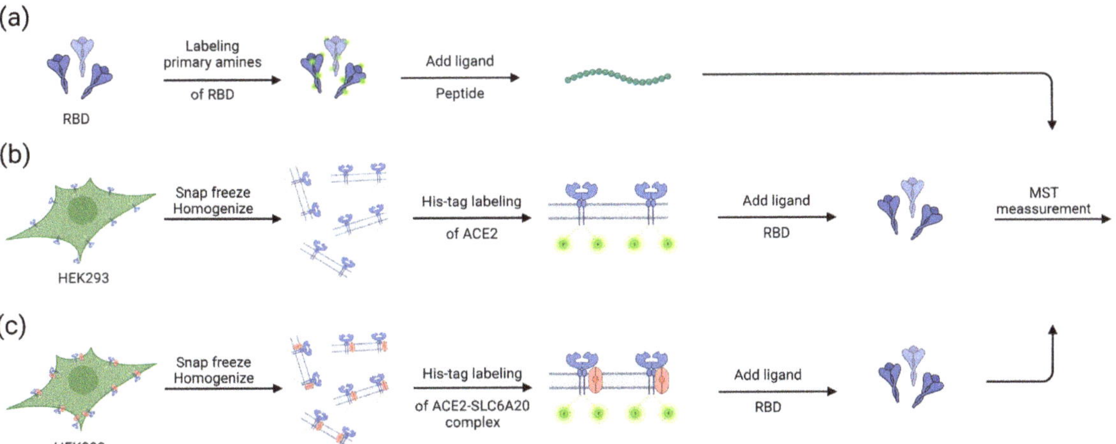

Figure 2. Applications of MST in coronavirus research. (**a**) The primary amines of purified RBD were labeled and a serial dilution of peptides that were predicted to bind RBD were added to perform MST. (**b**) C-terminal His-tag ACE2 is overexpressed in HEK293 cells, the ACE2 in the cell lysate was labeled with His-tag dye, and a serial dilution of RBD was added to perform MST. (**c**) His-tagged ACE2 and SLC6A20 were overexpressed in HEK293 cells, the ACE2 in the cell lysate was tagged with His-tag dye, and a serial dilution of RBD was added to perform MST. The figure was created using the BioRender software (https://www.biorender.com).

3.1. Determination of the Binding of SARS-CoV-2-S1-RBD to Antiviral Agents

Since the crystal structure of SARS-CoV-2 RBD bound to ACE2 is known [23,24], it is now possible to design peptides that could prevent the binding of the virus to ACE2. One strategy is to engineer peptides that can bind to the receptor binding domain of the viral spike protein in such a way that the ACE2 binding site is occupied and the virus cannot enter the cell. In silico peptide engineering can provide putative RBD-binding peptides, but binding needs to be validated experimentally, for which we used MST as a suitable assay for peptide screening. To confirm the applicability of MST for this purpose, a series of putative RBD-binding peptides were tested with MST. The peptides were kindly provided by Reymond and Gunasekera (see Acknowledgments), who had developed them using in silico methods. With these experiments, we aimed to validate the binding efficacy of the designed peptides to RBD, which is a first step towards assessing whether these peptides could prevent SARS-CoV-2 infection.

When working with nonfluorescent purified proteins (e.g., the spike protein), there are the following two labeling choices: (1) the unspecific labeling procedure, whereby the RED-NHS 2nd generation dye binds randomly to primary amides (i.e., lysine residues), and (2) the specific His-tag labeling procedure which involves labeling to the specific His-tag site of the expressed protein of interest. Both labeling methods are well established.

Nonspecific labeling of RBD was used in order to test the binding of peptides to RBD (Figure 2a). As expected from the in silico predictions, our MST results (Figure 3) confirm the binding of peptide NB001R to the spike protein, with a K_D of 2.08 µM ($\pm$0.43 µM) (Figure 3a). The measurements for the different dilutions tested are shown in Figure 3b. They are based on the MST traces (i.e., the IR-triggered fluorescence changes) shown in Figure 3c. Similarly, the binding of two other peptides, NB001 and NB002, predicted to bind RBD, were tested using MST and binding was confirmed (NB001, K_D = 1.08 µM ($\pm$0.51 µM) and NB002, K_D = 0.94 µM ($\pm$0.54 µM) (Supplementary Figure S1)). This demonstrates that MST is a suitable tool for screening the binding ability of peptides to RBD.

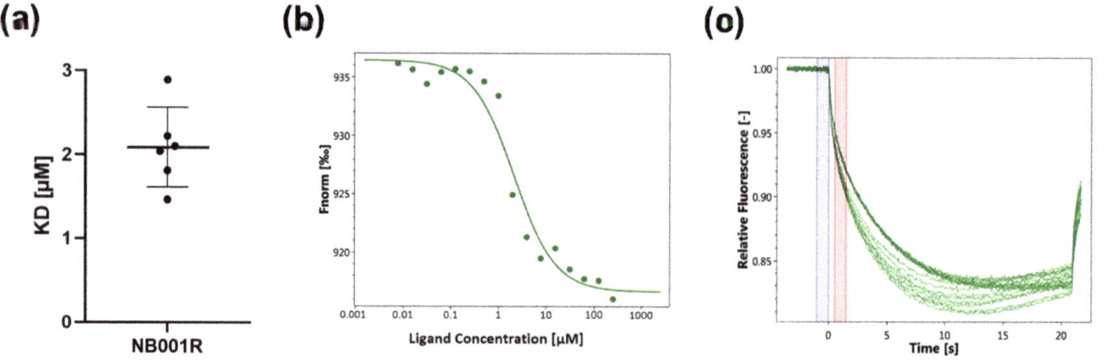

Figure 3. Binding of SARS-CoV-2-S1-RBD to peptide NB001R. (**a**) The K_D of NB001R and RBD is 2.08 µM. A serial dilution of the peptide from 1 mM–30 nM was used. (**b**) Representative dose–response curve. The red bar indicates the selected time point for the dose response used to determine the K_D (**c**) MST traces.

3.2. Determination of the Binding of SARS-CoV-2-S1-RBD to His-Tag-Labeled ACE2

While binding studies of purified proteins are an appropriate strategy to assess the interaction of viral spike proteins with potential ligands, a more sophisticated strategy is required when the ligand of interest is a transmembrane protein such as ACE2. Remarkably, in our studies, MST allowed us to assess the binding of transmembrane proteins to ligands using crude cell lysates. This success is likely due to the fact that the natural lipid environment of the proteins under study is largely preserved in our crude cell lysate preparations. For this type of experiment, nonspecific labeling of the protein of interest would not be a reliable option because it would label any lysine-containing proteins present in the cell lysate. An alternative approach would be to generate a cell line that overexpresses a recombinant GFP or RFP version of the protein of interest. However, the insertion of a bulky fluorescent tag could interfere with the expression or function of the target protein. For that reason, we preferred the His-tag labeling approach. For this purpose, a stable HEK293 cell line overexpressing ACE2 containing a His-tag attached to the Cterminus was generated, and the His-tag dye was mixed directly with the cell lysates for labeling. A serial dilution of RBD was added to the different capillaries prior to the MST measurements. As shown in Figure 4a, the K_D of membrane-embedded ACE2 to RBD was 37.2 nM ($\pm$10.7 nM). A representative dose–response curve is shown in Figure 4b and MST traces are shown in Figure 4c. Our results demonstrate that this approach allows for accurate and reliable binding measurements in the natural lipid environment without the need to prepare purified protein.

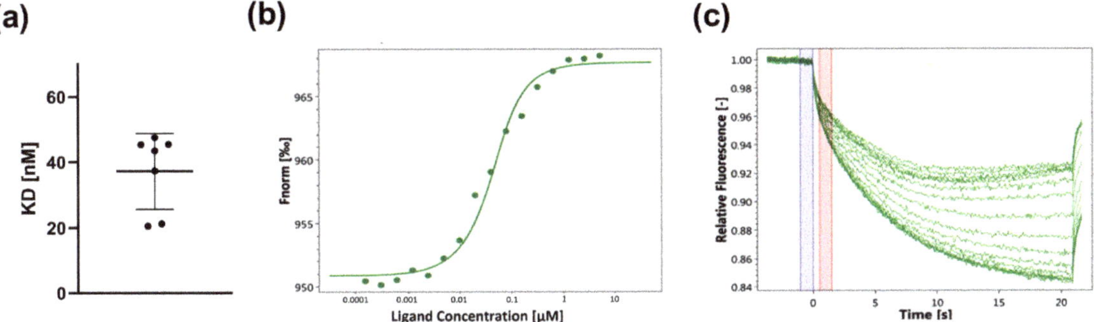

Figure 4. Binding of SARS-CoV-2-S1-RBD to His-tag-labeled ACE2. (**a**) The K_D of RBD and ACE2 in its natural lipid environment is 37 nM. A serial dilution of RBD from 5 µM–0.15 nM was used. Each K_D value was calculated from dose–response curves obtained in independent experiments using the stably transfected HEK293 ACE2 His-tag cell line. (**b**) Representative dose–response curve. (**c**) MST traces.

3.3. Determination of the Binding of SARS-CoV-2-S1-RBD to the ACE2-SLC6A20 Heterodimeric Complex in the Native Lipid Bilayer Environment

Often, proteins in the cell membrane are not expressed as monomers but form homo-multimers or heteromultimers with other proteins. It is well-known that complexation of proteins can alter the ability of protein monomers to interact with their environment [25–27]. Since proteins are usually purified as monomers, this aspect cannot be investigated using a conventional binding procedure, as ACE2 forms a complex with SLC6A20 (Figure 5c). However, MST allows working with crude membrane extracts in which a native lipid environment is present and protein complexes are still intact, making it an ideal tool to overcome this limitation. This protein complex is of particular interest because SLC6A20 has been shown to have an impact on the outcome of SARS-CoV2 infection [28,29]. In addition, variants of SLC6A20 are associated with diabetes mellitus [30], which we believe may also impact the infection efficiency [31]. And recently, the SLC6A20 variant I529V (Ile529Val; rs61731475) was indeed reported to affect the SARS-CoV-2 clinical outcome in Italian families [32].

To assess whether the co-expression of SLC6A20 and/or variant I529V together with ACE2 affects RBD binding, we wanted to investigate whether our MST method could answer this question. For this experiment, the stable cell line expressing ACE2 was transfected with either an empty vector (control), SLC6A20_WT, or SLC6A20_I529V. As described previously, measurements were performed with cell lysates after His-tag labeling. The results were normalized to the control condition (empty vector) to determine the fold change of K_D for complexes of ACE2 with SLC6A20_WT and SLC6A20_I529V. Interestingly, as shown in Figure 5a, the ACE2-SLC6A20_WT complex exhibited slightly stronger binding to RBD compared to the control, with a fold change of 0.67 ($\pm$0.12), whereas for the ACE2-SLC6A20_I529V complex there was somewhat weaker binding, with a fold change of 2.03 ($\pm$0.33). The representative dose–response curves and MST traces are shown in Figure 5b.

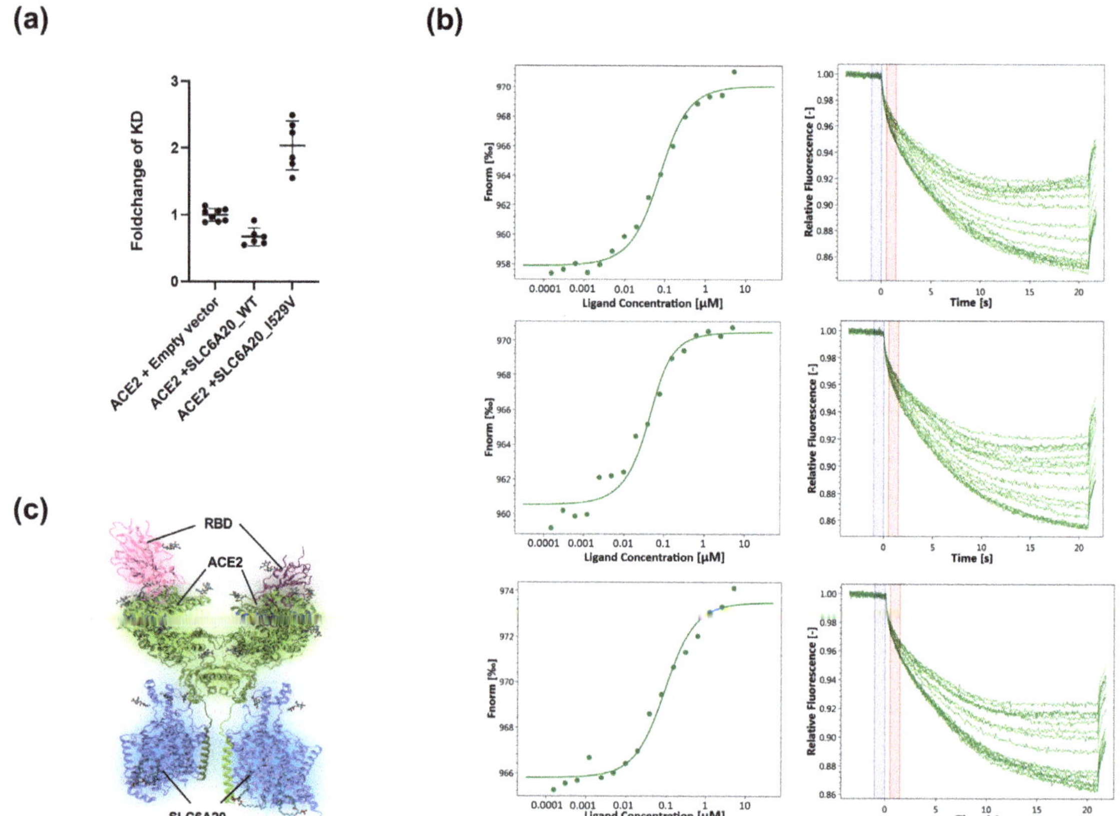

Figure 5. Binding of SARS-CoV-2-S1-RBD to the ACE2-SLC6A20 heterodimeric complex. (**a**) Impact of SLC6A20_WT and SLC6A20_I529V on the binding of ACE2 to RBD. The data shows a stronger binding of RBD to ACE2-SLC6A20 WT. The SLC20_I529V variant showed a reduction in binding affinity. A serial dilution of RBD from 5 µM to 0.15 nM was used. Three biological replicates with at least two technical replicates each were performed. (**b**) Dose response and MST traces. Top: ACE2 + empty vector; middle: ACE2 + SLC6A20_WT; bottom: ACE2 + SLC6A20_I529V. (**c**) Structure of ACE2-SLC6A20 complex binding to RBD based on PDB ID: 7Y75 (https://www.rcsb.org/structure/7Y75) (Shen, Y.P., Li, Y.N., Zhang, Y.Y., Yan, R.H., [33]). The figure was created using the BioRender software (https://www.biorender.com).

These experiments demonstrate that MST enables the study of protein–ligand interactions in the native lipid bilayer environment, including intact protein complexes, and thus provides in-depth insight into the binding behavior. Of course, the binding of ACE2 to RBD, as shown in Figure 5c, is only the first step of a complex infection process. Therefore, further studies are still required using different approaches, such as experiments with pseudoviruses or the original Wuhan SARS-CoV-2 virus and its variants, to assess whether SLC6A20 or its genetic variants affect the outcome of SARS-CoV-2 infection.

4. Discussion

In our study, we present several applications of MST that illustrate the versatility of this approach in coronavirus research and subsequent drug discovery: Binding of peptides to RBD that may ultimately prevent virus entry, and binding of RBD to ACE2, either alone or in complex with the amino acid transporter SLC6A20 or its allelic variant SLC6A20_I529V.

The K_D of membrane-embedded ACE2 for RBD was measured to be 37.2 nM ($\pm$10.7 nM). This value falls within a comparable range to the reported value (44.2 nM) based on measurements using the SPR assay [24]. Although similar results were obtained with both methods, we believe that MST has distinct advantages: While in SPR, purified RBD is immobilized on a sensor chip and the ligand, which is purified ACE2, is added at various concentrations to perform the measurements [24]; in MST, purified RBD is used for the dilutions and there is no need for purified ACE2, nor is the step of immobilizing RBD required. Thus, crude membrane extracts of a cell line transiently or stably expressing ACE2 can be used for the MST experiments without the need for further purification. His-tag labeling is performed within 30 min, and if ACE2-GFP was used as an alternative, no labeling was needed at all. Furthermore, working with crude cell lysates is more cost and time effective. In addition, as mentioned earlier, MST measurements were performed in the native environment of crude membrane extracts, where the associated lipids and proteins are likely still embedded in the membrane after cell rupture, allowing binding affinity to be measured at conditions closer to in vivo, which is in sharp contrast to the SPR method. In this context, our co-expression studies have shown that the presence of protein partners such as SLC6A20 can affect the binding affinity of ACE2 for RBD. Therefore, the absence of the additional partners present in the natural environment of ACE2 could distort the interpretation of the physiological significance of measurements using only purified proteins, which further highlights the advantage of the MST approach over other binding assays.

Because MST allows for the reliable examination of ligand interactions with membrane proteins in complex with other proteins, we were able to test whether the ACE2-SLC6A20-WT complex alters the binding affinity of RBD to ACE2 compared with ACE2 alone. Indeed, our experiments demonstrate that MST is capable of distinguishing subtle changes in binding affinity, which could be key to interpreting changes in the infectivity rate due to alterations in the ACE2-SLC6A20 complex. Since our experiments in fact show different K_D values for WT and I529V, this supports our hypothesis of a possible effect of this mutation on the infectivity of the SARS-CoV2 capacity, as previously hypothesized [28,29].

Interestingly, in contrast to our results, a recent study showed lower RBD binding for the ACE2-SLC6A20_WT complex with a higher K_D value of 63.23 nM [33], whereas the K_D value for ACE2 alone (43.64 nM) was fairly consistent with the K_D value obtained in our study by MST under the same conditions (37 nM, see Figure 4). The difference in ACE2-SLC6A20_WT binding may be attributed to the different binding method used in the latter study, namely flow cytometry. During measurement and incubation with RBD, the cells are still intact in flow cytometry, which means that only ACE2 expressed at the membrane can bind RBD. Interestingly, the co-expression of ACE2 and SLC6A20_WT was shown to result in a significant decrease in surface expression of the ACE2-SLC6A20-WT complex compared with ACE2 alone in the latter study. Consequently, a dramatic 2.1-fold decrease in maximal binding strength was observed for the ACE2-SLC6A20-WT complex. While K_D values should be independent of maximal binding levels, a lower signal-to-noise ratio could hinder proper curve fitting and K_D determination, which in this case could potentially explain the differences in K_D between MST and flow cytometry methods. In contrast to flow cytometry, our MST measurements showed similar maximal binding levels for all overexpressed constructs (Figure 5). This is likely a consequence of using cell lysates for MST that contain a mixture of surface membrane and cytosolic components, whereas flow cytometry relies only on proteins expressed on the surface membrane. Finally, it is worth mentioning that, in the same study, the K_D for the binding of the ACE2 N-terminal peptidase domain to RBD was also measured via biolayer interferometry, yielding a K_D value of 18.4 $\pm$ 0.03 nM, which is lower than that determined by MST or flow cytometry. This result once again highlights the differences that might arise as a consequence of using purified proteins instead of proteins in their native cellular environment in the binding study.

It is also worth noting that another amino acid transporter member of the SLC6 family, SLC6A19, forms a complex with ACE2 in enterocytes of the small intestine. There, SLC6A19 and its genetic variants could influence SARS-CoV-2 infectivity across the intestinal barrier [23,34,35]. Further experiments are needed to clarify whether and how SLC6 amino acid transporters and their allelic variants affect COVID-19 infectivity in different epithelial tissues.

The strength of MST in virus research is further evidenced by several recent studies. In one of them, the nonstructural protein 9, an RNA-binding protein essential for viral replication of SARS-CoV, was the subject [36]. While we measured protein–protein and peptide–protein interactions, the affinity of a protein–ssDNA interaction was determined in this study. This further highlights the broad applicability of MST. Another report shows the protein–protein interaction of nonstructural protein 15, expressed in MERS, with other nonstructural proteins [37]. Recently, a paper was published in which MST was used to screen pan-coronaviral major protease inhibitors. Similar to our initial experiments, but on a much larger scale, in silico experiments revealed promising inhibitors and confirmed binding affinity with MST [38].

In summary, it can be concluded that, based on our findings on membrane proteins, MST is a straightforward method to detect protein–protein and protein–peptide interactions based on changes in molecular weight and hydration shell. While the MST method is known to be well-suited for measuring the interactions of purified soluble proteins, for membrane proteins with alternative hydrophobic and hydrophilic domains, purification of the required amounts of high-quality concentrated material is challenging, and the resulting protein–detergent complexes compromise the binding events. Therefore, the strength of our MST approach lies in its ability to perform measurements in crude cell lysates where membrane proteins can retain their near-natural environment. Moreover, the influence of co-expressed membrane proteins that form complexes with ACE2 such as SLC6A20 can be readily studied using our MST approach. The successful use of MST with crude membrane extracts has also been demonstrated for other SLC solute carriers such as the H^+-coupled oligopeptide transporter PepT1/SLC15A1 [19] and the lysosomal SLC15A4 peptide/histidine transporter (Hediger et al., unpublished data). This highlights the versatility of the MST approach for studying membrane proteins, even when expressed in intracellular membranes. Moreover, the herein-presented approach is likely applicable to other coronaviruses, including emerging virus variants.

Supplementary Materials: The following supporting information can be downloaded at: https://www.mdpi.com/article/10.3390/v15071432/s1, Supplementary Figure S1: Binding of SARS-CoV-2-S1-RBD to the peptides NB001 and NB002. (a) The KD for the binding of NB001 to RBD is 1.08 µM, and that for NB002 is 0.94 µM. The following serial dilutions were used for the peptides: NB001R, 1 mM–30 nM; NB001 and NB002, 250 µM–7.6 nM. (b) Representative dose response curves (c) MST traces. The red bar indicates the selected time point for the dose response used to determine the KD.

Author Contributions: D.T.N.: performance of experiments, experimental design, paper writing; J.P.-G.: project strategy, supervision, experimental design, interpretation of the data, paper writing; P.K.: project strategy, supervision, experimental design; B.V.: project strategy, supervision; M.A.H.: project strategy, supervision, experimental design, interpretation of the data, paper writing. All authors have read and agreed to the published version of the manuscript.

Funding: This work was supported by the National Research Program NRP 78 "COVID-19" of the Swiss National Science Foundation (grant number NRP78 4078P0_198281; PIs Matthias Hediger and Bruno Vogt).

Institutional Review Board Statement: Not applicable.

Informed Consent Statement: Not applicable.

Data Availability Statement: All available data are included in the Section 3 and Supplementary Data.

Acknowledgments: The authors thank Jean-Louis Reymond and Kapila Gunasekera, Department of Chemistry, Biochemistry, and Pharmaceutical Sciences, University of Bern, for providing the peptides NB001, NB001R, and NB002.

Conflicts of Interest: The authors declare no conflict of interest.

References

1. Wang, Y.; Grunewald, M.; Perlman, S. Coronaviruses: An Updated Overview of Their Replication and Pathogenesis. In *Methods in Molecular Biology*; Humana: New York, NY, USA, 2020; Volume 2203, pp. 1–29. [CrossRef]
2. Maier, H.J.; Bickerton, E. Coronaviruses Methods and Protocols. In *Methods in Molecular Biology*; Humana: New York, NY, USA, 2020; Volume 2203, ISBN 9781071608999.
3. Chilamakuri, R.; Agarwal, S. COVID-19: Characteristics and therapeutics. *Cells* **2021**, *10*, 206. [CrossRef] [PubMed]
4. Sharma, A.; Farouk, I.A.; Lal, S.K. COVID-19: A Review on the Novel Coronavirus Disease. *Viruses* **2021**, *13*, 202. [CrossRef]
5. Khan, M.; Adil, S.F.; Alkhathlan, H.Z.; Tahir, M.N.; Saif, S.; Khan, M.; Khan, S.T. COVID-19: A Global Challenge with Old History, Epidemiology and Progress So Far. *Molecules* **2021**, *26*, 39. [CrossRef]
6. Chen, Y.; Klein, S.L.; Garibaldi, B.T.; Li, H.; Wu, C.; Osevala, N.M.; Li, T.; Margolick, J.B.; Pawelec, G.; Leng, S.X. Aging in COVID-19: Vulnerability, immunity and intervention. *Ageing Res. Rev.* **2021**, *65*, 101205. [CrossRef] [PubMed]
7. WHO Coronavirus (COVID-19) Dashboard. WHO Coronavirus (COVID-19) Dashboard With Vaccination Data. Available online: https://covid19.who.int/ (accessed on 7 April 2023).
8. Seidel, S.A.I.; Dijkman, P.M.; Lea, W.A.; Van Den Bogaart, G.; Jerabek-willemsen, M.; Lazic, A.; Joseph, J.S.; Srinivasan, P.; Baaske, P.; Simeonov, A.; et al. Microscale thermophoresis quantifies biomolecular interactions under previously challenging conditions. *Methods* **2023**, *59*, 301–315. [CrossRef] [PubMed]
9. El Deeb, S.; Al-Harrasi, A.; Khan, A.; Al-Broumi, M.; Al-Thani, G.; Alomairi, M.; Elumalai, P.; Sayed, R.A.; Ibrahim, A.E. Microscale thermophoresis as a powerful growing analytical technique for the investigation of biomolecular interaction and the determination of binding parameters. *Methods Appl. Fluoresc.* **2022**, *10*, 042001. [CrossRef]
10. Magnez, R.; Bailly, C.; Thuru, X. Microscale Thermophoresis as a Tool to Study Protein Interactions and Their Implication in Human Diseases. *Int. J. Mol. Sci.* **2022**, *23*, 7672. [CrossRef]
11. Jerabek-Willemsen, M.; Wienken, C.J.; Braun, D.; Baaske, P.; Duhr, S. Molecular interaction studies using microscale thermophoresis. *Assay Drug Dev. Technol.* **2011**, *9*, 342–353. [CrossRef]
12. Asmari, M.; Ratih, R.; Alhazmi, H.A.; El Deeb, S. Thermophoresis for characterizing biomolecular interaction. *Methods* **2018**, *146*, 107–119. [CrossRef]
13. Jerabek-Willemsen, M.; André, T.; Wanner, R.; Roth, H.M.; Duhr, S.; Baaske, P.; Breitsprecher, D. MicroScale Thermophoresis: Interaction analysis and beyond. *J. Mol. Struct.* **2014**, *1077*, 101–113. [CrossRef]
14. Chatzikyriakidou, Y.; Ahn, D.H.; Nji, E.; Drew, D. The GFP thermal shift assay for screening ligand and lipid interactions to solute carrier transporters. *Nat. Protoc.* **2021**, *16*, 5357–5376. [CrossRef] [PubMed]
15. Homola, J. Present and future of surface plasmon resonance biosensors. *Anal. Bioanal. Chem.* **2003**, *377*, 528–539. [CrossRef] [PubMed]
16. Ghai, R.; Falconer, R.J.; Collins, B.M. Applications of isothermal titration calorimetry in pure and applied research-survey of the literature from 2010. *J. Mol. Recognit.* **2012**, *25*, 32–52. [CrossRef]
17. Zhang, J.; Jones, C.P.; Amaré, A.R.F. Biochimica et Biophysica Acta Global analysis of riboswitches by small-angle X-ray scattering and calorimetry. *BBA-Gene Regul. Mech.* **2014**, *1839*, 1020–1029. [CrossRef]
18. Becker, W.; Bhattiprolu, K.C.; Zangger, K. Investigating Protein—Ligand Interactions by Solution Nuclear Magnetic Resonance Spectroscopy. *ChemPhysChem* **2018**, *19*, 895–906. [CrossRef] [PubMed]
19. Clémençon, B.; Lüscher, B.P.; Hediger, M.A. Establishment of a novel microscale thermophoresis ligand-binding assay for characterization of SLC solute carriers using oligopeptide transporter PepT1 (SLC15 family) as a model system. *J. Pharmacol. Toxicol. Methods* **2018**, *92*, 67–76. [CrossRef]
20. Romain, M.; Thiroux, B.; Tardy, M.; Quesnel, B.; Thuru, X. Measurement of Protein-Protein Interactions through Microscale Thermophoresis (MST). *Bio-Protocol* **2020**, *10*, e3574. [CrossRef]
21. Dijkman, P.M.; Lea, W.A.; Gmbh, N.T.; Lazic, A.; Gmbh, N.T. Microscale Thermophoresis Quantifies Biomolecular Interactions under Previously Challenging Conditions. *Methods* **2014**, *59*, 301–315.
22. Li, J.; Li, C.; Xiao, W.; Yuan, D.; Wan, G.; Ma, L. Site-directed mutagenesis by combination of homologous recombination and DpnI digestion of the plasmid template in *Escherichia coli*. *Anal. Biochem.* **2008**, *373*, 389–391. [CrossRef]
23. Yan, R.; Zhang, Y.; Li, Y.; Xia, L.; Guo, Y.; Zhou, Q. Structural basis for the recognition of SARS-CoV-2 by full-length human ACE2. *Science* **2020**, *367*, 1444–1448. [CrossRef]
24. Shang, J.; Ye, G.; Shi, K.; Wan, Y.; Luo, C.; Aihara, H.; Geng, Q.; Auerbach, A.; Li, F. Structural basis of receptor recognition by SARS-CoV-2. *Nature* **2020**, *581*, 221–224. [CrossRef]
25. Thompson, J.L.; Mignen, O.; Shuttleworth, T.J. The ARC Channel—An Endogenous Store-Independent Orai Channel. *Curr. Top. Membr.* **2013**, *71*, 125–148. [CrossRef] [PubMed]

26. Rutz, S.; Deneka, D.; Dittmann, A.; Sawicka, M.; Dutzler, R. Structure of a volume-regulated heteromeric LRRC8A/C channel. *Nat. Struct. Mol. Biol.* **2022**, *30*, 52–61. [CrossRef] [PubMed]

27. Borroto-Escuela, D.O.; Fuxe, K. Oligomeric Receptor Complexes and Their Allosteric Receptor-Receptor Interactions in the Plasma Membrane Represent a New Biological Principle for Integration of Signals in the CNS. *Front. Mol. Neurosci.* **2019**, *12*, 230. [CrossRef] [PubMed]

28. Semiz, S. SIT1 transporter as a potential novel target in treatment of COVID-19. *Biomol. Concepts* **2021**, *12*, 156–163. [CrossRef]

29. Kasela, S.; Daniloski, Z.; Bollepalli, S.; Jordan, T.X.; tenOever, B.R.; Sanjana, N.E.; Lappalainen, T. Integrative approach identifies SLC6A20 and CXCR6 as putative causal genes for the COVID-19 GWAS signal in the 3p21.31 locus. *Genome Biol.* **2021**, *22*, 1–10. [CrossRef]

30. Ling, Y.; Van Herpt, T.T.W.; van Hoek, M.; Dehghan, A.; Hofman, A.; Uitterlinden, A.G.; Jiang, S.; Lieverse, A.G.; Bravenboer, B.; Lu, D.; et al. A genetic variant in SLC6A20 is associated with Type 2 diabetes in white-European and Chinese populations. *Diabet. Med.* **2014**, *31*, 1350–1356. [CrossRef]

31. Severe Covid-19 GWAS Group. Genomewide Association Study of Severe COVID-19 with Respiratory Failure. *N. Engl. J. Med.* **2020**, *383*, 1522–1534. [CrossRef] [PubMed]

32. Azzarà, A.; Cassano, I.; Paccagnella, E.; Tirindelli, M.C.; Nobile, C.; Schittone, V.; Lintas, C.; Sacco, R.; Gurrieri, F. Genetic variants determine intrafamilial variability of SARS-CoV-2 clinical outcomes in 19 Italian families. *PLoS ONE* **2022**, *17*, e0275988. [CrossRef] [PubMed]

33. Shen, Y.; Wang, J.; Li, Y.; Zhang, Y.; Tian, R.; Yan, R. Structures of ACE2–SIT1 recognized by Omicron variants of SARS-CoV-2. *Cell Discov.* **2022**, *8*, 123. [CrossRef]

34. Stevens, B.R.; Ellory, J.C.; Preston, R.L. B0AT1 Amino Acid Transporter Complexed With SARS-CoV-2 Receptor ACE2 Forms a Heterodimer Functional Unit: In Situ Conformation Using Radiation Inactivation Analysis. *Function* **2021**, *2*, zqab027. [CrossRef] [PubMed]

35. Bröer, S. The role of the neutral amino acid transporter B0AT1 (SLC6A19) in Hartnup disorder and protein nutrition. *IUBMB Life* **2009**, *61*, 591–599. [CrossRef] [PubMed]

36. Zeng, Z.; Deng, F.; Shi, K.; Ye, G.; Wang, G.; Fang, L.; Xiao, S.; Fu, Z.; Peng, G. Dimerization of Coronavirus nsp9 with Diverse Modes Enhances Its Nucleic Acid Binding Affinity. *J. Virol.* **2018**, *92*, e00692-18. [CrossRef] [PubMed]

37. Zhang, L.; Li, L.; Yan, L.; Ming, Z.; Jia, Z.; Lou, Z.; Rao, Z. Structural and Biochemical Characterization of Endoribonuclease Nsp15 Encoded by Middle East Respiratory Syndrome Coronavirus. *J. Virol.* **2018**, *92*, e00893-18. [CrossRef]

38. Shahhamzehei, N.; Abdelfatah, S.; Efferth, T. In Silico and In Vitro Identification of Pan-Coronaviral Main Protease Inhibitors from a Large Natural Product Library. *Pharmaceuticals* **2022**, *15*, 308. [CrossRef]

*viruses*

MDPI

Article

Antiviral Activity of an Indole-Type Compound Derived from Natural Products, Identified by Virtual Screening by Interaction on Dengue Virus NS5 Protein

Leidy Lorena García-Ariza [1,*], Natalia González-Rivillas [1], Cindy Johanna Díaz-Aguirre [1], Cristian Rocha-Roa [2], Leonardo Padilla-Sanabria [1,*] and Jhon Carlos Castaño-Osorio [1]

[1] Grupo de Inmunología Molecular GYMOL, Universidad del Quindío, Armenia 630001, Quindío, Colombia; ngonzalezr@uqvirtual.edu.co (N.G.-R.); cjdiaza@uqvirtual.edu.co (C.J.D.-A.); jhoncarlos@uniquindio.edu.co (J.C.C.-O.)
[2] Grupo de Parasitología Molecular GEPAMOL, Universidad del Quindío, Armenia 630001, Quindío, Colombia
* Correspondence: llgarcia@uniquindio.edu.co (L.L.G.-A.); lpadilla@uniquindio.edu.co (L.P.-S.)

Abstract: Dengue is an acute febrile illness caused by the Dengue virus (DENV), with a high number of cases worldwide. There is no available treatment that directly affects the virus or the viral cycle. The objective of this study was to identify a compound derived from natural products that interacts with the NS5 protein of the dengue virus through virtual screening and evaluate its *in vitro* antiviral effect on DENV-2. Molecular docking was performed on NS5 using AutoDock Vina software, and compounds with physicochemical and pharmacological properties of interest were selected. The preliminary antiviral effect was evaluated by the expression of the NS1 protein. The effect on viral genome replication and/or translation was determined by NS5 production using DENV-2 Huh-7 replicon through ELISA and viral RNA quantification using RT-qPCR. The *in silico* strategy proved effective in finding a compound (M78) with an indole-like structure and with an effect on the replication cycle of DENV-2. Treatment at 50 µM reduced the expression of the NS5 protein by 70% and decreased viral RNA by 1.7 times. M78 is involved in the replication and/or translation of the viral genome.

Keywords: Dengue virus; natural compounds; NS5 protein; antiviral activity

Citation: García-Ariza, L.L.; González-Rivillas, N.; Díaz-Aguirre, C.J.; Rocha-Roa, C.; Padilla-Sanabria, L.; Castaño-Osorio, J.C. Antiviral Activity of an Indole-Type Compound Derived from Natural Products, Identified by Virtual Screening by Interaction on Dengue Virus NS5 Protein. *Viruses* **2023**, *15*, 1563. https://doi.org/10.3390/v15071563

Academic Editor: Juan De la Torre

Received: 13 May 2023
Revised: 7 July 2023
Accepted: 8 July 2023
Published: 17 July 2023

1. Introduction

Dengue virus (DENV) is a flavivirus transmitted by the bite of female mosquitoes of the genus Aedes spp., endemic in tropical and subtropical countries worldwide [1]. It is the causative agent of the infection known as Dengue or break bone fever [2]. Approximately 400 million cases [3] and 22,000 deaths occur worldwide each year due to Dengue [4]. According to the World Health Organization (WHO), the global incidence of Dengue has dramatically increased in the last decade, and approximately half of the world's population is at risk [5].

DENV has four genetically distinct serotypes (DENV 1–4). It is an enveloped virus with a single-stranded positive-sense RNA genome that encodes three structural proteins (capsid [C], pre-membrane [prM], and envelope [E]) and seven non-structural proteins (NS1, NS2A, NS2B, NS3, NS4A, NS4B, and NS5) [4,6], each of which perform different functions during the virus infectious cycle. The non-structural proteins are responsible for viral replication and host immune evasion. The NS5 protein plays an essential role in viral RNA replication, as the deletion of this protein from the viral genome inhibits replication [7], making it a promising pharmacological target [8,9]. This protein has two domains, the RNA-dependent RNA polymerase (RdRp) domain at the C-terminal end and the methyltransferase (Mtase) domain at the N-terminal end. The latter is responsible for protecting the RNA at the 5′ end of new viral genomes [1].

Despite the significant economic and social impact of this disease and the important advances made against Dengue, there is currently no effective antiviral therapy available [10–12]. Considering these limitations, it has become increasingly important to continue the search for molecules, compounds, or drugs that can inhibit enzymatic targets or essential processes for the replication cycle of the virus. The development and search for therapeutic molecules, such as direct-acting antivirals (DAA), has been shown to be a truly effective approach [2]. As a result, the use of computational techniques that have been employed in other research is considered a strategy of interest. Although DAA have not been approved for use in treatment of DENV [13], these have shown great promise in *in vitro* assays. Furthermore, bioactive agents from natural resources have laid a great foundation for the design of new therapeutic drugs [14], allowing a return to the use of traditional medicine to search for treatments for emerging diseases. Similarly, the innovation in the X-ray structures of several DENV proteins has allowed the development of *in silico* computational screening strategies [11]. Therefore, the execution of screenings from databases and docking analysis is promising when selecting an action target, such as important proteins in the virus's infectious cycle [15–19], with the viral polymerase NS5 protein standing out among these [7].

In this research, compounds derived from natural products were identified through virtual screening with an interaction on the NS5 protein. The *in vitro* antiviral effect of an indole-type compound, identified here as M78, was evaluated in a DENV-2 infected cellular model. This evaluation showed that the action is related to intervention during stages of replication and/or translation of the genome.

2. Materials and Methods

2.1. In Silico Assays

2.1.1. Virtual Screening of Natural Compound Derivatives on DENV NS5 Protein

The structures of NS5 proteins from the four DENV serotypes were obtained as described by García et al. [20]. The selected cavities for interaction corresponded to the substrate binding site of its natural substrate, S-adenosyl homocysteine (SAH), located in the methyltransferase (MTase) domain, and the entrance to the RNA tunnel, present in the RNA-dependent RNA polymerase (RdRp) domain. The ligands SAH and 68E, crystalized in the selected cavities, respectively [21], were subjected to re-docking onto their binding sites to find the coordinates and dimensions of the interaction boxes and obtain a theoretical value of the binding energy as a starting reference point for the selection of the best natural compounds with stronger binding on each region. The Root Mean Square Deviation (RMSD) was calculated as the most commonly used quantitative measure of similarity between two superimposed atomic coordinates [22]. In total, eight virtual screenings were performed (two per serotype). To perform this, the library of 190,090 natural compound derivatives available on the DrugDiscovery@TACC web portal (https://drugdiscovery.tacc.utexas.edu/#/) (accessed on 20 March 2019) [23] from the Texas Advanced Computing Center (TACC) was used. It is worth mentioning that all virtual screening calculations were executed using the AutoDock Vina 1.1 software [24], which is implemented in the DrugDiscovery@TACC web portal.

2.1.2. Selection of Compounds by Interaction on NS5 of DENV

The compounds were selected based on their ability to bind to NS5 in the four serotypes of DENV. Predictions of aqueous solubility were performed using the SwissADME web server (http://www.swissadme.ch/) [25]. This server delivers three predictions for this physicochemical descriptor with six possible outcomes: insoluble, poorly soluble, moderately soluble, soluble, highly soluble, and very highly soluble. We used a score of 0 for the descriptors of insoluble and poorly soluble, a score of 1 for the descriptors of moderately soluble and soluble, and a score of 2 for the descriptors of highly soluble and very highly soluble. Thus, only compounds that obtained a value of two or higher (by summing the scores of the three predictions provided by the SwissADME server) were selected for the next filter. A prediction was made of compliance or violation of the four Lipinski rules (Molecular weight $\leq$ 500, LogP $\leq$ 5, hydrogen bond acceptors $\leq$ 10, hydrogen bond donors $\leq$ 5) [26]; thus, only compounds that had a maximum of one violation were accepted for the next filter. This prediction was made using the SwissADME server. Prediction of possible toxicological risks using the ProTox-II web server (http://tox.charite.de/protox_II/) [27], such as hepatotoxicity, carcinogenicity, immunotoxicity, mutagenicity, and cytotoxicity, were carried out, and only those compounds that did not present toxicological risks after the prediction were selected. 3D visualizations of protein-ligand complexes were performed with the Chimera v1.13.1 program [28].

2.2. In Vitro Assays

2.2.1. Determination of the Cytotoxic Effect of Compounds on Huh-7 Cells

The compounds identified and selected through *in silico* assays were acquired through a synthesis service at MolPort (https://www.molport.com/shop/index). Subsequently, the cytotoxic effect on Huh-7 cells (ATCC HB 8065) was evaluated using the cell viability assay, using resazurin as a metabolic indicator. For this purpose, 15,000 cells were cultured per well in DMEM medium (Dulbecco's Modified Eagle Medium, Life Technologies 12100-046, New York, NY, USA), supplemented with 10,000 units/mL of penicillin/streptomycin, 20 mM L-glutamine, and 2% (*v/v*) heat-inactivated fetal bovine serum (Eurobio, CVFSVF00-01, Les Ulis, France) in 96-well Multiwell plates (Costar 3590, New York, NY, USA). The cell monolayer was allowed to stabilize for approximately 24 h at 37 °C and 5% CO_2. Subsequently, the cells were treated with the compounds at concentrations of 12.5, 25, 50, 100, 125, 250, and 500 μM. Cell viability was determined after 24 and 48 h of compound exposure, using resazurin at a final concentration of 44 μM, incubating for 2 h under the previously described conditions. Finally, absorbance was measured at 603 and 570 nm. The percentage of cell viability was calculated considering the difference between the absorbances for each treatment and the untreated cell control (CC), using the following formula:

$$\text{Cell viability (\%)} = [\text{Sample absorbance}/\text{Control absorbance}] \times 100 \qquad (1)$$

The mean cytotoxic concentration (CC_{50}) was also established, defined as the concentration at which cell viability decreases by 50%.

2.2.2. Antiviral Screening of Compounds on NS1 Protein Production in DENV-2 Infected Cells

For antiviral screening, 15,000 Huh-7 cells (ATCC HB 8065) were cultured per well in 96-well Multiwell plates (Costar 3590, New York, NY, USA) under the same conditions as described in item 2.2.1, and then these were infected with DENV-2 New Guinea for 2 h at a multiplicity of infection (MOI) of 1. The DENV-2 strain used was isolated and cultivated in C6/36 mosquito cells (ATCC® CRL-1660) and maintained in L-15 medium supplemented with 10% tryptose and 2% fetal bovine serum, incubated for seven days at 28 °C and stored at −80 °C. This strain was provided by the Biomedical Research Center of the University of Quindío, Colombia. After infection, the supernatant was removed, and the compounds were added at non-cytotoxic concentrations (between 40 and 100 μM, previously determined) and incubated for 24 h. Mycophenolic acid 20 μM [29] was used as an inhibition control, and 0.45% DMSO (vehicle) was used as a negative control. Subsequently, the cells were treated with 4% paraformaldehyde for 30 min and permeabilized for 5 min with 1X PBS, 0.5% Triton. The cells were then blocked with 5% fetal bovine serum in 0.05% PBS-Tween for 24 h at 4 °C. The primary Monoclonal anti-Dengue Virus NS1 antibody (SAB2702307 Sigma Aldrich, Saint Louis, MO, USA) (1:1000) was added and incubated for 1 h and 30 min at 37 °C. The secondary antibody, Anti-Mouse IgG (whole molecule) −Alkaline Phosphatase antibody produced in goats (A3562-Sigma Aldrich), was then added and incubated for 1 h at 37 °C. Finally, the alkaline phosphatase substrate pNPP (S0942-Sigma Aldrich) was added and incubated for 30 min. Then, NaOH 0.1 M solution was added, and absorbance was measured at 405 nm on an Epoch spectrophotometer. Absorbance values were transformed into percentages of NS1 production and compared to the viral control (VC), using the following formula:

$$\text{NS1 production (\%)} = [\text{Sample absorbance}/\text{Control absorbance}] \times 100 \qquad (2)$$

2.2.3. Determination of the Inhibitory Effect on NS5 Protein on the Expression Using ELISA Technique

To determine the effect of the compound on protein expression, the production of NS5 was evaluated using the Huh-7 cell line, which carries a DENV-2 subgenomic replicon. The replicon includes a luciferase reporter gene, a geneticin resistance gene, and the coding region of NS proteins (NS1 to NS5), allowing for stable expression of these proteins. These systems contain genetic elements necessary for autonomous genome replication in cells and have been useful for expressing viral genes in several flaviviruses, including DENV, WNV, YFV, and TBEV [30]. The cells were cultured in DMEM (Dulbecco's Modified Eagle Medium, Life Technologies 12100-046, New York, NY, USA), supplemented with 10,000 units/mL of penicillin/streptomycin, 20 mM L-glutamine, and 10% (*v/v*) heat-inactivated fetal bovine serum. Geneticin G418 (10131-035, Gibco, Grand Island, New York, NY, USA) was added at a final concentration of 0.2 mg/mL as a selection antibiotic for the transfected cells with the replicon. This cell line was provided by the Biomedical Research Center of the University of Quindío, Colombia. To begin, 15,000 Huh-7 DENV-2 replicon cells were cultured per well in 96-well Multiwell plates (Costar 3590, New York, NY, USA). After reaching 70–80% confluency, they were treated with compounds that had previously shown an effect on the viral cycle and were incubated for 24 h at 37 °C and 5% CO_2. NITD008 compound, an NS5 protein inhibitor [31], was used as a positive control, along with other respective controls. The cells were treated with 4% paraformaldehyde for 30 min, and permeabilized for 5 min with a 0.5% PBS 1X Triton X100 solution. Then, the cells were blocked with 5% fetal bovine serum in 0.05% PBS-Tween for 24 h at 4 °C. Then, the cells were treated with primary Anti-NS5 antibody produced in rabbits (SAB2700025 Sigma Aldrich) (1:10,000) and incubated for 1 h and 30 min at 37 °C. Then, goat anti-rabbit (whole molecule) alkaline phosphatase-conjugated antibody (A3687-Sigma Aldrich) diluted 1:30,000 was added and incubated for 1 h at 37 °C. Subsequently, alkaline phosphatase substrate (S0942-Sigma Aldrich®) was added for 1 h at 37 °C, followed by the addition of 0.1 M NaOH solution, and the absorbance was measured at 405 nm using an Epoch spectrophotometer. The absorbance values were transformed into NS5 production percentage and compared to the viral control, using the (2) formula. The IC_{50} was estimated through the dose-response curve, using the GraphPad Prism 6 software, and the selectivity index (SI) was also predicted by calculating the ratio between CC_{50}/IC_{50}.

2.2.4. Determination of the Inhibitory Effect on Viral RNA Production of DENV-2

After treatments with the selected compounds on the previously infected Huh-7 cells with DENV-2, as described above, total RNA extraction was performed using the TRIzol LS Reagent® (Lot.50867000) following the manufacturer's recommendations. The concentration and purity of the RNA were determined by the absorbance ratio at 260 nm and 280 nm, read on an Epoch spectrophotometer. Subsequently, the amplification of the NS5 protein gene was performed using the primers DENV 7764 Fwd 5′-CGTCGAGAGAAATATGGTCA CACC-3′ and DENV 7844 Rev 5′-CCACAATAGTATGACCAGCCT-3′. The endogenous GAPDH gene was amplified using the primers hGAPDH Fwd 5′-TGTTGCCATCAATGA CCCCTT-3′ and hGAPDH Rev 5′-CTCCACGACGTACTCAGCG-3′. RT-qPCR was performed using the Power SYBR® Green RNA-to-CTTM 1-Step kit (Ref 4389986, Applied Biosystems™), following the manufacturer's instructions, for a total reaction volume of 20 µL. As a negative amplification control, a reaction mixture without genetic material was included. The RT was performed at 48 °C for 30 min, enzyme activation at 95 °C for 10 min, denaturation for 40 cycles of 95 °C for 15 s, and annealing and extension at 60 °C for 1 min. The relative expression of this gene was calculated using the comparative CT method ($2^{-\Delta\Delta CT}$) [32], which makes several assumptions, including that the PCR efficiency is close to 1 and the PCR efficiency of the target gene is similar to that of the internal control gene, using the following equation:

$$2^{-\Delta\Delta CT} = [(\text{CT gene of interest} - \text{CT internal control}) \text{ Sample A} - (\text{CT gene of interest} - \text{CT internal control}) \text{ Sample B}]. \quad (3)$$

2.3. Statistical Analysis

In all cases, treatments were compared with their respective controls. A Shapiro–Wilk test was performed to evaluate data normality. Parametric data were evaluated using one-way ANOVA with multiple comparisons test through Dunne's t method. Non-parametric data were evaluated using Kruskal–Wallis test, with comparison test through Dunn's test. *t*-test was performed to compare two groups of parametric data, Tukey test to compare means, and Mann–Whitney U test for two groups of non-parametric data. A p-value < 0.05 was considered statistically significant. The analyses were performed using GraphPad Prism 6 software.

3. Results

3.1. The MTase and RdRp Domains Were Validated As Binding Sites

The crystallized structure of NS5 protein from serotype 3, PDB 5JJR, containing two ligands on the regions of interest of the MTase and RdRp domains (SAH and 68E, respectively), was used to re-dock these two compounds onto the NS5 protein of all DENV serotypes. The defined dimensions for all boxes were 24 Å in all axes (x, y, and z).

In Figure 1, the re-docking of the SAH and 68E ligands onto the NS5 protein of serotype 3 is presented. As shown, molecular docking was able to approximately reproduce the crystallographic pose of the control ligands. In the case of PDB 5JJR, a value of 1.2 Å was obtained for the RMSD between the crystallized SAH ligand and the predicted pose, and for the case of the 68E ligand present in the RdRp domain, a value of 1.7 Å was obtained for the RMSD. The calculated interaction energies for SAH was −7.6 kcal/mol, and for 68E was −8.6 kcal/mol.

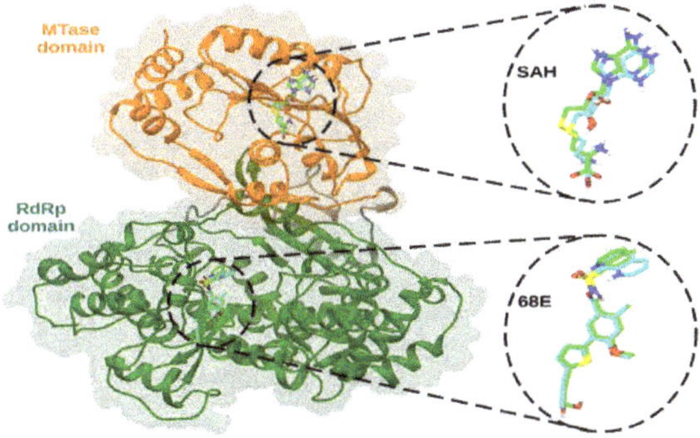

Figure 1. Molecular re-docking of co-crystallized ligands onto NS5 protein of serotype 3 of DENV (PDB 5JJR). SAH ligand in MTase domain, with an RMSD of 1.2 Å, and 68E ligand in RdRp domain, with an RMSD of 1.7 Å. The MTase and RdRp domains are represented in orange and dark green ribbons, respectively. The crystallized and predicted poses are shown in green and cyan sticks, respectively.

3.2. Selected Compounds by Interaction on MTase and RdRp of DENV NS5 Protein

Figure 2 shows the steps taken to select compounds from the Zinc Natural Compounds database that have interaction with the DENV NS5 protein.

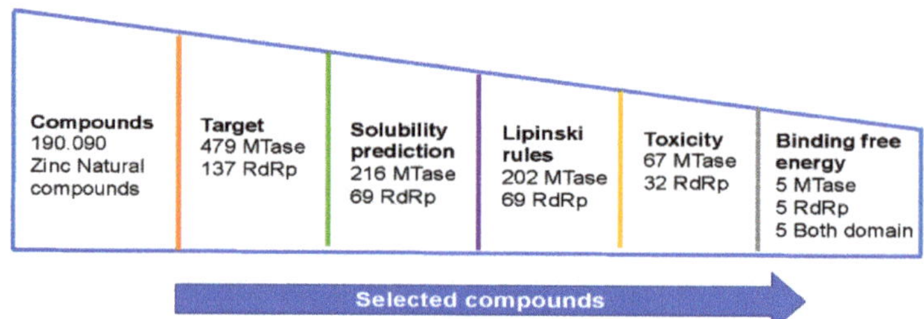

Figure 2. Stages and selection criteria for the selection of natural compounds with potential anti-DENV activity.

After selecting the binding sites and setting up the interaction boxes used in the compound search, virtual screening was performed. Initially, 190,090 natural compounds were docked to the two binding sites of interest (MTase and RdRp) in the four serotypes. This means that a total of eight virtual screenings and approximately 1,520,720 molecular dockings were performed on the NS5 protein models of DENV. The DrugDiscovery@TACC web portal [23] provided the top 1000 compounds for each screening, and the list was reduced to 8000 natural compounds (4000 for each domain). In order to identify compounds that may exhibit anti-Dengue effects, only compounds that interacted with the NS5 protein domains in all Dengue serotypes were selected. After verifying the compounds at each binding site, a total of 479 compounds were obtained in the MTase domain and 127 compounds were obtained in the RdRp domain in all four serotypes. Finally, physicochemical and toxicological property screening was performed. The solubility evaluation in aqueous systems allowed for the selection of 216 compounds for the MTase domain and 69 compounds for the RdRp domain. Then, the Lipinski rule compliance was checked, resulting in 202 compounds selected for the MTase domain and all previously identified compounds for the RdRp domain. The next analysis was based on the predictions of possible toxicological risks, which resulted in 67 compounds for the MTase domain and 32 compounds for the RdRp domain. The final selection of compounds was based on the binding energies. Five compounds were chosen for the MTase domain, five for the RdRp domain, and five with interaction in both domains (Table 1). The compounds were named according to the interaction site and the last two digits of the Zinc code. Based on the results, the compounds were acquired from Molport (https://www.molport.com/shop/index) (accessed on 30 August 2019) for synthesis and to begin the *in vitro* evaluation of the molecules' activity against DENV-2.

Table 1. Natural compounds identified through virtual screening with interaction in the NS5 protein of DENV.

Binding Site	Zinc Code	Assigned Name	Binding Energy (kcal/mol)			
			DENV-1	DENV-2	DENV-2	DENV-4
Mtase Domain	ZINC08790808	M08	−11.4	−11.4	−11.5	−11.5
	ZINC03839432	M32	−11.1	−11.1	−11.1	−11.2
	ZINC08791166	M66	−11.4	−11.5	−11.5	−11.4
	ZINC35485176	M76	−12.0	−12.1	−12.2	−12.0
	ZINC08790178	M78	−12.0	−12.1	−12.2	−12.0
RdRp Domain	ZINC02094107	R07	−9.8	−9.7	−9.8	−9.6
	ZINC04085432	R32	−10.1	−10	−10.5	−9.9
	ZINC04085246	R46	−9.8	−9.7	−10.1	−9.2
	ZINC12884853	R53	−9.8	−9.8	−10.1	−9.3
	ZINC20611155	R55	−9.8	−9.6	−9.8	−10.0

Table 1. *Cont.*

Binding Site	Zinc Code	Assigned Name	Binding Energy (kcal/mol)			
			DENV-1	DENV-2	DENV-2	DENV-4
Both domain	ZINC08790125	MR25	MTase −11.8	−11.9	−12.1	−11.9
			RdRp −10.1	−9.8	−10.0	−9.7
	ZINC08791241	MR41	MTase −11.2	−11.3	−11.4	−11.3
			RdRp −10.0	−9.9	−10.4	−9.4
	ZINC12885588	MR88	MTase −11.2	−11.1	−11.3	−11.2
			RdRp −10.4	−10.3	−10.8	−10.2
	ZINC04086794	MR94	MTase −11.1	−11.2	−11.4	−11.1
			RdRp −10.2	−10.4	−10.4	−10.2
	ZINC08791299	MR99	MTase −11.3	−11.3	−11.5	−11.3
			RdRp −10.2	−10.1	−10.4	−9.9

3.3. Cytotoxic Effect Evaluation of Compounds in Huh-7 Cells

The cytotoxic effect was evaluated for 10 of the 15 identified compounds (M66, M76, M78, R07, R32, R53, R55, MR25, MR41, and MR94). According to the results, among the compounds with interaction on the MTase domain, M66 caused a decrease in cell viability ($\geq 50\%$) at concentrations of 50 and 100 μM, with statistically significant differences when compared to the control cells (**** $p < 0.0001$). The estimated CC_{50} was 44.08 μM. For compound M76, no effect was observed on cell morphology or metabolism at the highest concentration evaluated (100 μM) after 24 h of treatment exposure; therefore, CC_{50} was not determined. On the other hand, after treatment with compound M78, cellular viability close to 100% was evidenced at concentrations of 12.5, 25, and 50 μM, while at 100 μM, damage to the monolayer and therefore loss of viability was observed, with statistically significant difference when compared to the control cells (**** $p < 0.0001$), finding a CC_{50} of 60.77 μM. Regarding the compounds identified with interaction on the RdRp domain, it was found that compounds R07, R32, and R55 did not reduce cell viability at any of the evaluated concentrations. On the other hand, treatment with compound R53 reduced cell viability at all evaluated concentrations, with evident cell death. The determined CC_{50} for the compound was 30 μM. On the other hand, for compounds with binding in both domains, CC_{50} of 70.57 μM was found for MR25 and CC_{50} of 46.22 μM for MR94, and no effect was evidenced for MR41 at any of the evaluated concentrations, so CC_{50} was considered >100 μM (Table 2).

Table 2. Mean cytotoxic concentration (CC_{50}) of evaluated compounds on Huh-7 cell line.

Compound	Structure	CC_{50} (μM)
M66		44.08
M76		>100

Table 2. *Cont.*

Compound	Structure	CC$_{50}$ (μM)
M78	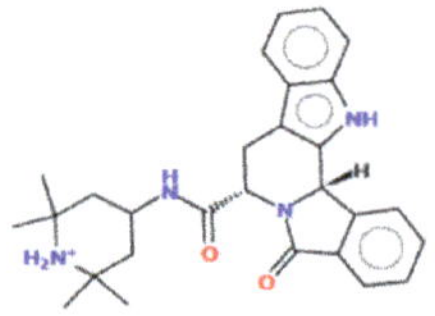	60.77
R07	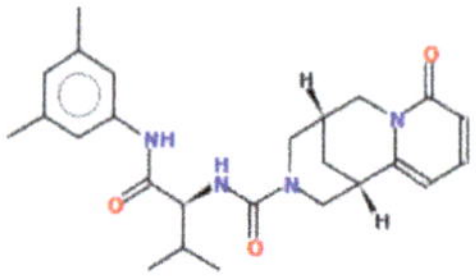	>100
R32	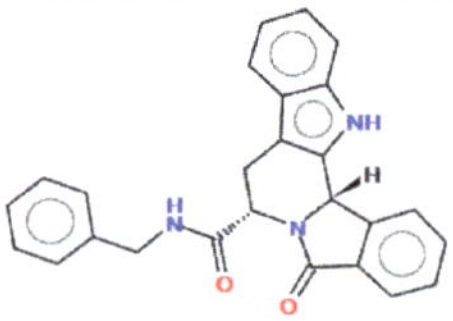	>100
R53	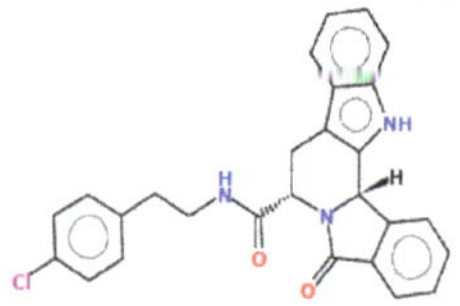	30
R55	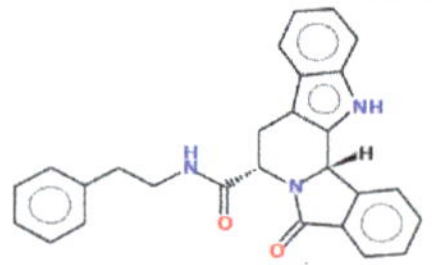	>100
MR25	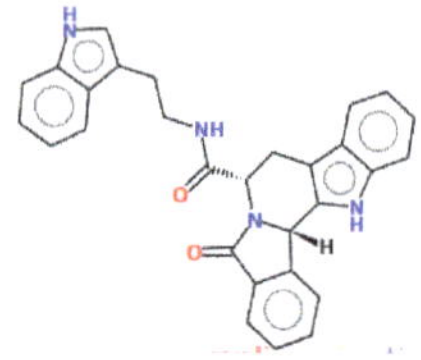	>100
MR41	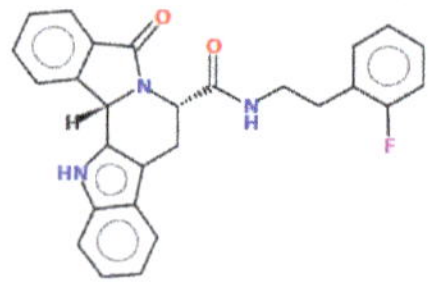	>100
MR94	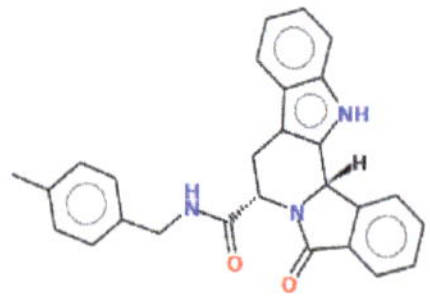	46.22

3.4. The Compounds Reduce the Production of NS1 Protein in Cells Infected with DENV-2

The antiviral activity of the compounds was determined for nine (9) out of the ten (10) compounds previously evaluated, excluding compound R53, which showed higher toxic effects with a CC_{50} of 30 μM. This first selection analysis was performed through an ELISA assay with detection of the DENV NS1 protein. According to these results, there was a reduction in NS1 production after treatment with eight out of nine compounds (M66, M76, M78, R07, R32, R55, MR25, MR41), with statistically significant differences when compared to the viral control (**** p <0.0001), except for treatment with MR94. It is worth noting that the addition of M78 and MR25 reduced the expression of this protein, with production rates close to 38% and 45%, respectively. These rates were lower than that presented by the inhibition control (mycophenolic acid), which was around 63% (Figure 3).

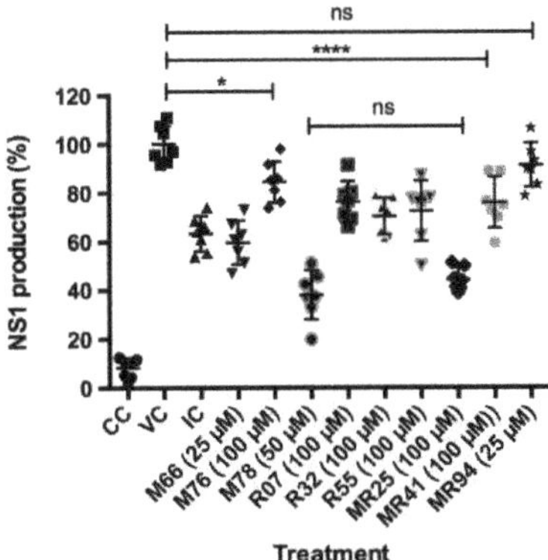

Figure 3. Percentage of NS1 protein production after treatment with compounds M66, M76, M78, R07, R32, R55, MR25, MR41, and MR94 in DENV-2 infected cells, as measured by ELISA. VC: Viral control, CC: Cell control, CI: Inhibition control (20 μM mycophenolic acid). Data are represented as mean and standard deviation (n = 7). One-way ANOVA analysis indicated statistically significant differences between the compounds and viral control (VC) (*** p < 0.001), (* p < 0.1; **** p < 0.0001), except for compound MR94. *t*-test indicated no difference between M78 and MR25 (ns: non-significant).

3.5. Compound M78 Affects NS5 Protein Production by Interfering with Genome Replication and/or Translation

The compound M78 was selected because it showed the greatest reduction in NS1 protein production when compared to the other compounds, decreasing expression by approximately 60%, according to the previously described results (Figure 3). M78 was evaluated at lower concentrations to assess its effect on NS5 expression. The ELISA results are presented in Figure 4, indicating the percentage of protein production and showing dose-dependent reduction with statistically significant differences when compared to viral control (VC) (*** p < 0.001). However, treatment with the compound at three different concentrations reduced protein production to values close to those observed in the positive inhibition control. Nevertheless, there were statistically significant differences between the three treatments, indicating a greater effect of M78 at 50 μM, where NS5 production was close to 30% compared to VC (100%). The IC_{50} was estimated, with a value of 24.61 μM, and according the determined CC_{50}, an SI of 2.5 was found.

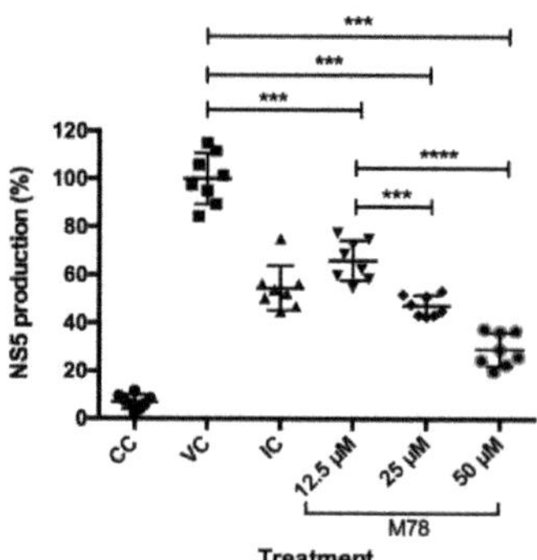

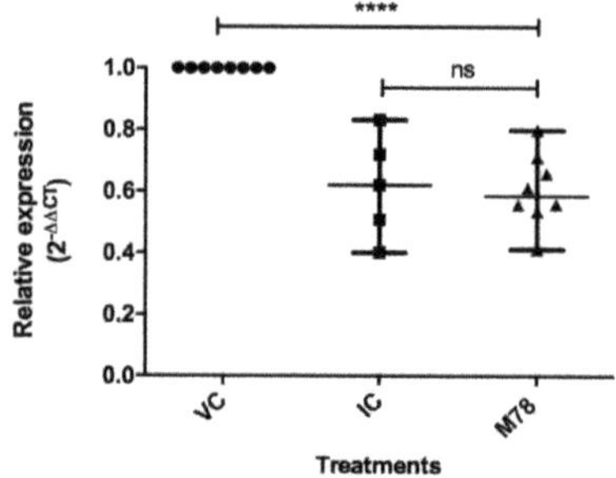

Figure 4. Effect of M78 compound on the production of DENV-2 viral protein NS5 in Huh-7 replicon cells using ELISA assay. CC: Cell control, VC: Viral control, IC: Inhibition control (20 µM Mycophenolic Acid). Data represent mean and standard deviation (n = 8). One-way ANOVA analysis indicates statistically significant differences between all concentrations of M78 evaluated in relation to viral control (*** $p < 0.001$). Tukey test indicates difference between the three M78 treatments, between 12.5 µM and 25 µM (*** $p < 0.001$), and between 12.5 µM and 50 µM (**** $p < 0.0001$).

3.6. The M78 Compound Affects The Production of DENV-2 Viral RNA

The results of the antiviral effect and protein expression after treatment with 50 µM of M78 led to the evaluation of the compound's action on viral RNA synthesis. The evaluation showed a decrease in the relative expression of the DENV-2 gene when compared to the viral control (Figure 5).

Figure 5. Effect of compound M78 (50 µM) on the production of DENV-2 viral RNA determined through relative expression levels, by RT-qPCR, normalized using GAPDH. VC: Viral control, IC: Inhibition control (20 µM Mycophenolic acid). Data represents the median and interquartile range (n = 8). Kruskal–Wallis test analysis shows statistically significant differences between treatment with compound M78 and inhibition control (IC) compared to viral control (VC) (*** $p < 0.001$). Mann–Whitney test indicates no difference between inhibition control (IC) and M78 (ns: non-significant).

Figure 5 shows the relative expression levels for VC, IC, and M78 treatment. The results indicate statistically significant difference between treatment with M78 and viral control (**** $p < 0.0001$). Based on this, it is possible to state that the expression of the evaluated gene was reduced 1.7 times due to the compound treatment, with similar behavior to that of the inhibition control.

4. Discussion

The use of bioinformatics tools for the study of molecular docking has become an important component in the drug discovery process [33]. Over the past two decades, computational technologies have played a crucial role in the development of antiviral drugs [34]. For DENV, the NS5 protein has been reported as an important target for the search for new molecules with inhibitory capacity of its function, knowing that it does not have homologues in the eukaryotic cell, which decreases the probability of toxic effects. It is highly conserved in flaviviruses, and within this protein, the thumb subdomain in RdRp plays a crucial role in assisting in the synthesis of viral RNA [2]. Likewise, the MTase domain has essential enzymatic activity in replication and positively influences polymerase activity [20,35]. Considering this, the molecules that bind to these sites may hinder conformational changes for RdRp activity [21]. Seeking compounds that bind in these regions remains an objective, since it has been demonstrated that the development of therapeutic molecules, such as direct-acting antivirals (DAA), is a truly effective approach [2].

The NS5 structures of the four serotypes were obtained from García et al. These models include most of the NS5 amino acids from all serotypes, and they are reliable and comparable based on validation of Z-score values against structures resolved by X-ray and NMR and torsion angles [20]. Molecular docking with natural compounds was performed on the SAM and RNA tunnel regions of NS5, after the re-docking of the SAH and 68E ligands on the NS5 protein of serotype 3. The RMSD values obtained between the computationally calculated and experimental poses for the control ligands SAH and 68E were 1.0 Å and 0.8 Å, respectively (Figure 1), presenting small values on an atomic scale. A value equal to 0 indicates identical structures, and as the two structures become different, its value increases [36], suggesting that the two compared structures are very close to having the same formation. In other words, this result supports the interaction coordinates used in the re-docking, indicating that they are suitable in the search for natural compounds with interaction in DENV NS5, known as the most conserved protein, and considered a promising pharmacological target due to its fundamental role in replication, viral RNA methylation, RNA polymerization, and evasion of the host immune system [8].

Molecular docking has become an essential component for drug development and represents an approach that can aid in therapeutic and pharmaceutical development. The purpose of evaluating the antiviral potential of compounds derived from natural products is aimed at their proven success in many pharmacological therapies. As such, and because they present antiviral properties, they may be an alternative target for drug development in order to combat the Dengue virus [37]. The binding energies of compounds on NS5 protein were found to be between -9.2 and -12 kcal/mol (Table 1), indicating the global minimum energy of the complex formed between ligand and receptor [38]. The molecular docking results suggest that all our compounds have a higher binding site affinity than their respective controls, based on re-docking (for SAH, this was -7.6 kcal/mol; for 68E this was -8.6 kcal/mol). This condition was considered as an initial criterion when choosing the compounds. The binding energy is often used to determine the affinity of biomolecular interactions and drug efficacy [39]; the more negative the binding affinity, the stronger the ligand-receptor interaction and the better molecular docking prediction [40]. Further, predictions of physicochemical and toxicological properties favored the selection of these compounds. Compliance with Lipinski's rules was established as an important criterion for this classification, as higher molecular weight is associated with a lower rate of permeability in lipid bilayer membranes, and a LogP less than five has approximately a 90% probability of being orally soluble. When a drug-like molecule satisfies the five fundamental principles, it exhibits greater pharmacokinetic properties and bioavailability [37], making it feasible to consider whether the compounds possess possible drug properties and can be used in the future as candidates [20,41]. Likewise, predicting possible risks favored ordering the compounds in order to evaluate the most promising ones *in vitro*. This allowed for the postulation of 15 candidate compounds for *in vitro* evaluation in DENV.

The experimental phase began with the determination of the cytotoxicity of 10 of the 15 identified compounds. The results showed that six of them had CC_{50} in Huh-7 cells above 100 μM (M76, R07, R32, R55, MR25 and MR41), while the CC_{50} of the others were less than or equal to that found for M78 (60.77 μM). No apparent cellular damage was found up to 50 μM (Supplementary Figure S1) so it was decided to evaluate these compounds at lower concentrations relative to the CC_{50} determined for each one (Table 2). According to the chemical structure of the compounds, eight out of ten evaluated *in vitro* contain a scaffold of five-ring named 7,8,13,13b-tetrahydro-5H-benz [1,2] indolizino [8,7-b] indole. Indole compounds possess pharmacological potential that has been used as an excellent scaffold in the discovery of antimicrobial drugs, anticancer agents, antihypertensive, antiproliferative, and anti-inflammatory agents [42]. The activity of these compounds is associated with the molecular interactions generated between the indole compound and the therapeutic target [43], with high affinity, which aids in the development of new biologically active compounds [44,45]. Medically useful or promising indole compounds span the entire structural spectrum, from simple indoles to highly complex indole alkaloids [43]. Recent studies have pointed out that these types of structures exhibit an effect against flaviviruses [46] such as DENV, and other viruses such as HIV and influenza virus [47], and that fused tricyclic derivatives of indoline and imidazolidinone have action on ZIKV and DENV infection [48].

The effect of the compounds against DENV was initially evaluated for nine out of the ten selected compounds (excluding R53) by measuring the expression of DENV-2 NS1 protein, which was used as a model because it is one of the most prevalent serotypes [49]. The results indicated that all compounds, except MR94, reduced NS1 expression when compared to the viral control (Figure 3). The presence of NS1 confirms Dengue infection and serves as evidence of successful viral replication [9,50]. Among the compounds, M78 showed better effects by decreasing protein production by over 60% (production close to 38%). In other studies, the detection of DENV-2 NS1 through ELISA also identified that hydroalcoholic extracts of leaves (UGL) and bark (UGB) from the medicinal species *Uncaria guinanensis* reduced the levels of this protein at concentrations of 0.5, 5, and 10 µg/mL [51]. Considering that NS1 is a multifunctional protein essential for virus production, and that, in infected cells, it is necessary for the formation of virus-induced membrane structures that serve as replication sites for DENV [52], the findings presented here support the concept of an antiviral action by the evaluated compounds.

On the other hand, it was found that compound M78 induces a reduction in NS5 expression, with a dose-dependent effect (Figure 4), and a predicted selectivity index (SI) of 2.5 µM, indicating that the compound is effective and selective at this concentration. This parameter is accepted to express the efficacy of a compound in inhibiting viral replication, although studies have shown that SI values < 10 have limited antiviral activity, as observed in the case of OA, a methylated flavone from *Oroxylum indicum*, with an SI of 2.66 against DENV-2 [53]. Furthermore, by using the replicon system, which allows studying aspects of viral replication due to the lack of structural genes [30], it is possible to consider that the intervention of M78 is directed towards viral replication and/or the expression of proteins associated with this process [54], which is also evidenced by the reduction in viral RNA copies (Figure 5), where a $2^{-\Delta\Delta CT}$ value less than 1 indicates a reduction in gene expression due to the treatment [32], estimating that such reduction was 1.7 times greater compared to the viral control. Based on the above, we can say that M78 intervenes in the viral cycle of DENV-2, although the mechanism through which this effect occurs requires further validation, using complementary techniques that allow for a deeper understanding of the role of this compound as an antiviral.

Other studies have reported the evaluation of compounds and plant extracts against DENV-2; among them, it has been indicated that the ethanolic extract of *A. calamus* root (Tatanan A) presented a similar effect to that found for M78, related to the intervention in the initial stage of viral replication, while inhibiting DENV-2 mRNA and protein levels [55]. On the other hand, it has also been published that hirsutine, an alkaloid from *Uncaria rhynchophylla* that shares structural similarity with M78 in relation to the indole nucleus, was identified as a potent anti-DENV compound in the four serotypes, inhibiting viral particle assembly, budding, or release step, but not translation and viral replication in the DENV lifecycle [56]. However, we have found a different dynamic for the compound M78, demonstrating that the compound intervenes in DENV-2 viral replication by acting on RNA synthesis and/or translation of the viral genome, causing a decrease in the production of viral proteins, as we have observed (Figures 4 and 5). Furthermore, considering that indole is a potent basic pharmacophore present in a wide variety of antiviral agents [45], it is also known that some indole derivatives have been effective and selective inhibitors of this virus replication [57]. In this sense, the design of antiviral drugs containing indole is useful for combating viral infections [42], and furthermore, its application is interesting if this type of compound is found in natural products, as compounds from these sources have prevented DENV from infiltrating the genome or act by reducing structural and non-structural proteins that are produced [58]. This supports our findings. The identification of this naturally occurring compound is interesting, as similar bioactive compounds with antiviral properties could be combined with existing therapies along with different administration methods to enhance their efficacy [59].

Furthermore, our results indicate that the *in silico* strategy used in this research to search for compounds against Dengue proved to be effective, allowing the identification of 15 compounds with interaction in NS5 from DENV-1 to DENV-4 out of a total of 190,090 evaluated natural compounds. Among them, compound M78 showed *in vitro* antiviral activity, highlighting the utility of this methodology for future studies in the identification of compounds targeting viral targets. The findings suggest that the natural compound M78 could be considered a candidate for DENV-2. M78 is involved in viral replication, and it is recommended to study its role in NS5 in more detail, as well as its action in pre-treatment and its virucidal effect against other Dengue serotypes and flaviviruses. It is important to note that its chemical structure, for which no other biological activity assays are currently known, possesses an indole core that could be associated with its antiviral effect, which increases the interest in further investigating this compound identified here through virtual screening.

Supplementary Materials: The following supporting information can be downloaded at: https://www.mdpi.com/article/10.3390/v15071563/s1, Figure S1: Cell viability of Huh-7 for compound M78, as measured by resazurin assay.

Author Contributions: Conceptualization, L.L.G.-A., L.P.-S. and J.C.C.-O.; methodology, N.G.-R., C.J.D.-A., C.R.-R. and L.L.G.-A.; data curation, N.G.-R., C.J.D.-A., C.R.-R. and L.L.G.-A.; writing—original draft preparation, L.L.G.-A.; writing—review and editing, L.L.G.-A., L.P.-S. and J.C.C.-O.; supervision, L.L.G.-A., L.P.-S. and J.C.C.-O.; project administration, L.L.G.-A., L.P.-S. and J.C.C.-O.; funding acquisition, L.L.G.-A., L.P.-S. and J.C.C.-O. All authors have read and agreed to the published version of the manuscript.

Funding: This research was funded by Ministerio de Ciencia, Tecnología e Innovación -Minciencias, project number 111380863020, convocatory 808-2018 and the APC was funded by Minciencias and Universidad del Quindío.

Institutional Review Board Statement: This project was approved by the research bioethics committee of the University of Quindío, in the minutes No. 16 09-05-2018 (9 May 2018).

Informed Consent Statement: Not applicable.

Data Availability Statement: The study did not report any data.

Acknowledgments: We are grateful to Diego Alejandro Molina Lara for advice on *in silico* essays, Grupo Gepamol, Centro de Investigaciones Biomédicas—Universidad del Quindío.

Conflicts of Interest: The authors declare no conflict of interest.

References

1. Uno, N.; Ross, T.M. Dengue virus and the host innate immune response. *Emerg. Microbes Infect.* **2018**, *7*, 167. [CrossRef] [PubMed]
2. Kumar, S.; Bajrai, L.H.; Faizo, A.A.; Khateb, A.M.; Alkhaldy, A.A.; Rana, R.; Azhar, E.I.; Dwivedi, V.D. Pharmacophore-Model-Based Drug Repurposing for the Identification of the Potential Inhibitors Targeting the Allosteric Site in Dengue Virus NS5 RNA-Dependent RNA Polymerase. *Viruses.* **2022**, *14*, 1827. [CrossRef] [PubMed]
3. Nazmi, A.; Dutta, K.; Hazra, B.; Basu, A. Role of pattern recognition receptors in flavivirus infections. *Virus Res.* **2014**, *185*, 32–40. [CrossRef] [PubMed]

4. Roy, S.K.; Bhattacharjee, S. Dengue virus: Epidemiology, biology, and disease aetiology. *Can. J. Microbiol.* **2021**, *67*, 687–702. [CrossRef]

5. Li, Y.; Dou, Q.; Lu, Y.; Xiang, H.; Yu, X.; Liu, S. Effects of ambient temperature and precipitation on the risk of dengue fever: A systematic review and updated meta-analysis. *Environ. Res.* **2020**, *191*, 110043. [CrossRef]

6. Simmons, C.P.; Farrar, J.J.; Van Vinh Chau, N.; Wills, B. Dengue. *N. Engl. J. Med.* **2012**, *366*, 423–1432. [CrossRef]

7. Watterson, D.; Modhiran, N.; Young, P.R. The many faces of the flavivirus NS1 protein offer a multitude of options for inhibitor design. *Antivir. Res.* **2016**, *130*, 7–18. [CrossRef]

8. Fernandes, P.O.; Chagas, M.A.; Rocha, W.R.; Moraes, A.H. Non-structural protein 5 (NS5) as a target for antiviral development against established and emergent flaviviruses. *Curr. Opin. Virol.* **2021**, *50*, 30–39. [CrossRef]

9. Obi, J.O.; Gutiérrez-Barbosa, H.; Chua, J.V.; Deredge, D.J. Current Trends and Limitations in Dengue Antiviral Research. *Trop. Med. Infect. Dis.* **2021**, *6*, 180. [CrossRef]

10. Mushtaq, M.; Naz, S.; Parang, K.; Ul-Haq, Z. Exploiting Dengue Virus Protease as a Therapeutic Target: Current Status, Challenges and Future Avenues. *Curr. Med. Chem.* **2021**, *28*, 7767–7802. [CrossRef]

11. Troost, B.; Smit, J.M. Recent advances in antiviral drug development towards dengue virus. *Curr. Opin. Virol.* **2020**, *43*, 9–21. [CrossRef]

12. Low, J.G.; Gatsinga, R.; Vasudevan, S.G.; Sampath, A. Dengue Antiviral Development: A Continuing Journey. *Adv. Exp. Med. Biol.* **2018**, *1062*, 319–332. [CrossRef]

13. Park, S.J.; Kim, J.; Kang, S.; Cha, H.J.; Shin, H.; Park, J.; Jang, Y.S.; Woo, J.S.; Won, C.; Min, D.H. Discovery of direct-acting antiviral agents with a graphene-based fluorescent nanosensor. *Sci Adv.* **2020**, *6*, eaaz8201. [CrossRef]

14. Sagaya Jansi, R.; Khusro, A.; Agastian, P.; Alfarhan, A.; Al-Dhabi, N.A.; Arasu, M.V.; Rajagopal, R.; Barcelo, D.; Al-Tamimi, A. Emerging paradigms of viral diseases and paramount role of natural resources as antiviral agents. *Sci. Total Environ.* **2021**, *759*, 143539. [CrossRef]

15. Kausar, M.A.; Ali, A.; Qiblawi, S.; Shahid, S.; Asrar Izhari, M.; Saral, A. Molecular docking based design of Dengue NS5 methyltransferase inhibitors. *Bioinformation* **2019**, *15*, 394–401. [CrossRef]

16. Qamar, M.T.U.; Maryam, A.; Muneer, I.; Xing, F.; Ashfaq, U.A.; Khan, F.A.; Anwar, F.; Geesi, M.H.; Khalid, R.R.; Rauf, S.A.; et al. Computational screening of medicinal plant phytochemicals to discover potent pan-serotype inhibitors against dengue virus. *Sci. Rep.* **2019**, *9*, 1433. [CrossRef]

17. Hariono, M.; Choi, S.B.; Roslim, R.F.; Nawi, M.S.; Tan, M.L.; Kamarulzaman, E.E.; Mohamed, N.; Yusof, R.; Othman, S.; Rahman, N.A.; et al. Thioguanine-based DENV-2 NS2B/NS3 protease inhibitors: Virtual screening, synthesis, biological evaluation and molecular modelling. *PLoS ONE* **2019**, *14*, e0210869. [CrossRef]

18. Yang, C.C.; Hu, H.S.; Lin, H.M.; Wu, P.S.; Wu, R.H.; Tian, J.N.; Wu, S.H.; Tsou, L.K.; Song, J.S.; Chen, H.W.; et al. A novel flavivirus entry inhibitor, BP34610, discovered through high-throughput screening with dengue reporter viruses. *Antivir. Res.* **2019**, *172*, 104636. [CrossRef]

19. Idrus, S.; Tambunan, U.; Zubaidi, A.A. Designing cyclopentapeptide inhibitor as potential antiviral drug for dengue virus ns5 methyltransferase. *Bioinformation* **2012**, *8*, 348–352. [CrossRef]

20. García-Ariza, L.L.; Rocha-Roa, C.; Padilla-Sanabria, L.; Castaño-Osorio, J.C. Virtual Screening of Drug-Like Compounds as Potential Inhibitors of the Dengue Virus NS5 Protein. *Front. Chem.* **2022**, *10*, 637266. [CrossRef]

21. Lim, S.P.; Noble, C.G.; Seh, C.C.; Soh, T.S.; El Sahili, A.; Chan, G.K.; Lescar, J.; Arora, R.; Benson, T.; Nilar, S.; et al. Potent Allosteric Dengue Virus NS5 Polymerase Inhibitors: Mechanism of Action and Resistance Profiling. *PLoS Pathog.* **2016**, *12*, e1005737. [CrossRef] [PubMed]

22. Kufareva, I.; Abagyan, R. Methods of protein structure comparison. *Methods Mol. Biol.* **2012**, *857*, 231–257. [CrossRef] [PubMed]

23. Viswanathan, U.; Tomlinson, S.M.; Fonner, J.M.; Mock, S.A.; Watowich, S.J. Identification of a novel inhibitor of dengue virus protease through use of a virtual screening drug discovery Web portal. *J. Chem. Inf. Model.* **2014**, *54*, 2816–2825. [CrossRef] [PubMed]

24. Trott, O.; Olson, A.J. AutoDock Vina: Improving the speed and accuracy of docking with a new scoring function, efficient optimization, and multithreading. *J. Comput. Chem.* **2010**, *31*, 455–461. [CrossRef] [PubMed]

25. Daina, A.; Michielin, O.; Zoete, V. SwissADME: A free web tool to evaluate pharmacokinetics, drug-likeness and medicinal chemistry friendliness of small molecules. *Sci. Rep.* **2017**, *7*, 42717. [CrossRef]

26. Lipinski, C.A.; Lombardo, F.; Dominy, B.W.; Feeney, P.J. Experimental and computational approaches to estimate solubility and permeability in drug discovery and development settings. *Adv. Drug. Deliv. Rev.* **2001**, *46*, 3–26. [CrossRef]

27. Banerjee, P.; Eckert, A.O.; Schrey, A.K.; Preissner, R. ProTox-II: A webserver for the prediction of toxicity of chemicals. *Nucleic Acids Res.* **2018**, *46*, W257–W263. [CrossRef]

28. Pettersen, E.F.; Goddard, T.D.; Huang, C.C.; Couch, G.S.; Greenblatt, D.M.; Meng, E.C.; Ferrin, T.E. UCSF Chimera—A visualization system for exploratory research and analysis. *J. Comput. Chem.* **2004**, *25*, 1605–1612. [CrossRef]

29. Diamond, M.S.; Zachariah, M.; Harris, E. Mycophenolic acid inhibits dengue virus infection by preventing replication of viral RNA. *Virology* **2002**, *304*, 211–221. [CrossRef]

30. Kato, F.; Hishiki, T. Dengue Virus Reporter Replicon is a Valuable Tool for Antiviral Drug Discovery and Analysis of Virus Replication Mechanisms. *Viruses* **2016**, *8*, 122. [CrossRef]

31. Yin, Z.; Chen, Y.L.; Kondreddi, R.R.; Chan, W.L.; Wang, G.; Ng, R.H.; Lim, J.Y.; Lee, W.Y.; Jeyaraj, D.A.; Niyomrattanakit, P.; et al. N-sulfonylanthranilic acid derivatives as allosteric inhibitors of dengue viral RNA-dependent RNA polymerase. *J. Med. Chem.* **2009**, *52*, 7934–7937. [CrossRef]

32. Schmittgen, T.D.; Livak, K.J. Analyzing real-time PCR data by the comparative C(T) method. *Nat. Protoc.* **2008**, *3*, 1101–1108. [CrossRef]

33. Stanzione, F.; Giangreco, I.; Cole, J.C. Use of molecular docking computational tools in drug discovery. *Prog. Med. Chem.* **2021**, *60*, 273–343. [CrossRef]

34. Gaurav, A.; Agrawal, N.; Al-Nema, M.; Gautam, V. Computational Approaches in the Discovery and Development of Therapeutic and Prophylactic Agents for Viral Diseases. *Curr. Top. Med. Chem.* **2022**, *22*, 2190–2206. [CrossRef]

35. Potisopon, S.; Ferron, F.; Fattorini, V.; Selisko, B.; Canard, B. Substrate selectivity of Dengue and Zika virus NS5 polymerase towards 2′-modified nucleotide analogues. *Antivir. Res.* **2017**, *140*, 25–36. [CrossRef]

36. Carugo, O.; Pongor, S. A normalized root-mean-square distance for comparing protein three-dimensional structures. *Protein Sci.* **2001**, *10*, 1470–1473. [CrossRef]

37. Halder, S.K.; Ahmad, I.; Shathi, J.F.; Mim, M.M.; Hassan, M.R.; Jewel, M.J.I.; Dey, P.; Islam, M.S.; Patel, H.; Morshed, M.R.; et al. A Comprehensive Study to Unleash the Putative Inhibitors of Serotype2 of Dengue Virus: Insights from an In Silico Structure-Based Drug Discovery. *Mol. Biotechnol.* **2022**, 1–14. [CrossRef]

38. Manish, C.; Ghosh, B.P. Chapter 8—Anti-Tubercular Drug Designing Using Structural Descriptors. In *Advances in Mathematical Chemistry and Applications*; Subhash, C.B., Restrepo, G., José, L.V., Eds.; Bentham Science Publishers: Sharjah, United Arab Emirates, 2015; pp. 179–190. ISBN 9781681080536.

39. Hata, H.; Phuoc Tran, D.; Marzouk Sobeh, M.; Kitao, A. Binding free energy of protein/ligand complexes calculated using dissociation Parallel Cascade Selection Molecular Dynamics and Markov state model. *Biophys. Physicobiol.* **2021**, *18*, 305–316. [CrossRef]

40. Fadlan, A.; Nusantoro, Y. The Effect of Energy Minimization on The Molecular Docking of Acetone-Based Oxindole Derivatives. *JKPK* **2021**, *6*, 69–77. [CrossRef]

41. Ahmad, N.; Rehman, A.U.; Badshah, S.L.; Ullah, A.; Mohammad, A.; Khan, K. Molecular dynamics simulation of zika virus NS5 RNA dependent RNA polymerase with selected novel non-nucleoside inhibitors. *J. Mol. Struct.* **2020**, *1203*, 127428. [CrossRef]

42. Dorababu, A. Indole—A promising pharmacophore in recent antiviral drug discovery. *RSC Med. Chem.* **2020**, *11*, 1335–1353. [CrossRef] [PubMed]

43. Norwood, V.M., 4th; Huigens, R.W., 3rd. Harnessing the Chemistry of the Indole Heterocycle to Drive Discoveries in Biology and Medicine. *Chembiochem* **2019**, *20*, 2273–2297. [CrossRef] [PubMed]

44. Sravanthi, T.V.; Manju, S.L. Indoles—A promising scaffold for drug development. *Eur. J. Pharm. Sci.* **2016**, *91*, 1–10. [CrossRef] [PubMed]

45. De, A.; Sarkar, S.; Majee, A. Recent advances on heterocyclic compounds with antiviral properties. *Chem. Heterocycl. Compd.* **2021**, *57*, 410–416. [CrossRef] [PubMed]

46. Giannakopoulou, E.; Pardali, V.; Frakolaki, E.; Siozos, V.; Myrianthopoulos, V.; Mikros, E.; Taylor, M.C.; Kelly, J.M.; Vassilaki, N.; Zoidis, G. Scaffold hybridization strategy towards potent hydroxamate-based inhibitors of *Flaviviridae* viruses and *Trypanosoma* species. *Medchemcomm* **2019**, *10*, 991–1006. [CrossRef]

47. Fikatas, A.; Vervaeke, P.; Meyen, E.; Llor, N.; Ordeix, S.; Boonen, I.; Bletsa, M.; Kafetzopoulou, L.E.; Lemey, P.; Amat, M.; et al. A Novel Series of Indole Alkaloid Derivatives Inhibit Dengue and Zika Virus Infection by Interference with the Viral Replication Complex. *Antimicrob. Agents Chemother.* **2021**, *65*, e0234920. [CrossRef]

48. Zhou, G.F.; Li, F.; Xue, J.X.; Qian, W.; Gu, X.R.; Zheng, C.B.; Li, C.; Yang, L.M.; Xiong, S.D.; Zhou, G.C.; et al. Antiviral effects of the fused tricyclic derivatives of indolassaine and imidazolidinone on ZIKV infection and RdRp activities of ZIKV and DENV. *Virus Res.* **2023**, *326*, 199062. [CrossRef]

49. Fang, Y.; Tambo, E.; Xue, J.B.; Zhang, Y.; Zhou, X.N.; Khater, E.I.M. Detection of DENV-2 and Insect-Specific Flaviviruses in Mosquitoes Collected from Jeddah, Saudi Arabia. *Front. Cell Infect. Microbiol.* **2021**, *11*, 626368. [CrossRef]

50. Nath, H.; Basu, K.; De, A.; Biswas, S. Dengue virus sustains viability of infected cells by counteracting apoptosis-mediated DNA breakage. *BioRxiv* **2020**. [CrossRef]

51. Mello, C.D.S.; Valente, L.M.M.; Wolff, T.; Lima-Junior, R.S.; Fialho, L.G.; Marinho, C.F.; Azeredo, E.L.; Oliveira-Pinto, L.M.; Pereira, R.C.A.; Siani, A.C.; et al. Decrease in Dengue virus-2 infection and reduction of cytokine/chemokine production by Uncaria guianensis in human hepatocyte cell line Huh-7. *Mem. Inst. Oswaldo Cruz.* **2017**, *112*, 458–468. [CrossRef]

52. Songprakhon, P.; Thaingtamtanha, T.; Limjindaporn, T.; Puttikhunt, C.; Srisawat, C.; Luangaram, P.; Dechtawewat, T.; Uthaipibull, C.; Thongsima, S.; Yenchitsomanus, P.T.; et al. Peptides targeting dengue viral nonstructural protein 1 inhibit dengue virus production. *Sci. Rep.* **2020**, *10*, 12933. [CrossRef]

53. Ratanakomol, T.; Roytrakul, S.; Wikan, N.; Smith, D.R. Oroxylin A shows limited antiviral activity towards dengue virus. *BMC Res. Notes* **2022**, *15*, 154. [CrossRef]

54. Kato, F.; Nio, Y.; Yagasaki, K.; Suzuki, R.; Hijikata, M.; Miura, T.; Miyazaki, I.; Tajima, S.; Lim, C.K.; Saijo, M.; et al. Identification of inhibitors of dengue viral replication using replicon cells expressing secretory luciferase. *Antivir. Res.* **2019**, *172*, 104643. [CrossRef]

55. Yao, X.; Ling, Y.; Guo, S.; Wu, W.; He, S.; Zhang, Q.; Zou, M.; Nandakumar, K.S.; Chen, X.; Liu, S. Tatanan A from the *Acorus calamus* L. root inhibited dengue virus proliferation and infections. *Phytomedicine* **2018**, *42*, 258–267. [CrossRef]
56. Hishiki, T.; Kato, F.; Tajima, S.; Toume, K.; Umezaki, M.; Takasaki, T.; Miura, T.H. An Indole Alkaloid of *Uncaria rhynchophylla*, Inhibits Late Step in Dengue Virus Lifecycle. *Front. Microbiol.* **2017**, *8*, 1674. [CrossRef]
57. Bardiot, D.; Koukni, M.; Smets, W.; Carlens, G.; McNaughton, M.; Kaptein, S.; Dallmeier, K.; Chaltin, P.; Neyts, J.; Marchand, A. Discovery of Indole Derivatives as Novel and Potent Dengue Virus Inhibitors. *J. Med. Chem.* **2018**, *61*, 8390–8401. [CrossRef]
58. Babbar, R.; Kaur, R.; Rana, P.; Arora, S.; Behl, T.; Albratty, M.; Najmi, A.; Meraya, A.M.; Alhazmi, H.A.; Singla, R.K. The Current Landscape of Bioactive Molecules against DENV: A Systematic Review. *Evid. Based Complement. Alternat. Med.* **2023**, *2023*, 2236210. [CrossRef]
59. Ghasemnezhad, A.; Ghorbanzadeh, A.; Sarmast, M.K.; Ghorbanpour, M. A review on botanical, phytochemical, and pharmacological characteristics of iranian junipers (*Juniperus* spp.). In *Plant-Derived Bioactives*; Springer: Singapore, 2020; pp. 493–508. [CrossRef]

viruses

MDPI

Article

The Anti-Dengue Virus Peptide DV2 Inhibits Zika Virus Both In Vitro and In Vivo

Maria Fernanda de Castro-Amarante [1,2], Samuel Santos Pereira [1], Lennon Ramos Pereira [1], Lucas Souza Santos [3], Alexia Adrianne Venceslau-Carvalho [1,2], Eduardo Gimenes Martins [1], Andrea Balan [3] and Luís Carlos de Souza Ferreira [1,2,*]

1 Laboratory of Vaccine Development, Institute of Biomedical Sciences, University of São Paulo, São Paulo 05508-000, Brazil
2 Scientific Platform Pasteur USP, São Paulo 05508-020, Brazil
3 Applied Structural Biology Laboratory, Institute of Biomedical Sciences, University of São Paulo, São Paulo 05508-000, Brazil
* Correspondence: lcsf@usp.br

Abstract: The C-terminal portion of the E protein, known as stem, is conserved among flaviviruses and is an important target to peptide-based antiviral strategies. Since the dengue (DENV) and Zika (ZIKV) viruses share sequences in the stem region, in this study we evaluated the cross-inhibition of ZIKV by the stem-based DV2 peptide (419–447), which was previously described to inhibit all DENV serotypes. Thus, the anti-ZIKV effects induced by treatments with the DV2 peptide were tested in both in vitro and in vivo conditions. Molecular modeling approaches have demonstrated that the DV2 peptide interacts with amino acid residues exposed on the surface of pre- and postfusion forms of the ZIKA envelope (E) protein. The peptide did not have any significant cytotoxic effects on eukaryotic cells but efficiently inhibited ZIKV infectivity in cultivated Vero cells. In addition, the DV2 peptide reduced morbidity and mortality in mice subjected to lethal challenges with a ZIKV strain isolated in Brazil. Taken together, the present results support the therapeutic potential of the DV2 peptide against ZIKV infections and open perspectives for the development and clinical testing of anti-flavivirus treatments based on synthetic stem-based peptides.

Keywords: Zika virus; peptide; antiviral; DV2 peptide; dengue virus; flavivirus

Citation: Castro-Amarante, M.F.d.; Pereira, S.S.; Pereira, L.R.; Santos, L.S.; Venceslau-Carvalho, A.A.; Martins, E.G.; Balan, A.; Souza Ferreira, L.C.d. The Anti-Dengue Virus Peptide DV2 Inhibits Zika Virus Both In Vitro and In Vivo. *Viruses* **2023**, *15*, 839. https://doi.org/10.3390/v15040839

Academic Editor: Simone Brogi

Received: 3 March 2023
Revised: 23 March 2023
Accepted: 23 March 2023
Published: 25 March 2023

1. Introduction

The Zika (ZIKV) and dengue (DENV) viruses belong to the Flaviviridae family, which also includes other clinically important viruses, such as yellow fever virus (YFV), West Nile virus (WNV), and Japanese encephalitis virus (JEV) [1,2]. DENV infection in humans has been detected in more than 100 countries, for which observers estimate about 390 million cases and 22,000 deaths by dengue disease worldwide each year [3,4]. On the other hand, around 1.5 million of ZIKV infections have been reported in more than 70 countries [5]. The Zika virus was probably introduced into Brazil in 2013 [6], where two years later it was related to the occurrence of neurologic damage in newborn babies [7]; it was and declared a global health emergency by the World Health Organization (WHO) in 2016 [8]. Even though ZIKV infections are usually associated with self-limiting symptoms, severe clinical outcomes may be observed, such as Congenital Zika Syndrome (CZS) and Guillain-Barré syndrome [9]. In Brazil, between 2015 and 2022, approximately 3707 cases of congenital ZIKV infection were confirmed, of which 1852 were classified as CZS. Additionally, 31,194 cases of severe dengue were reported in the period from 2020 to 2022 [10]. Thus, the high incidence of these infections over the years and the related clinical complications illustrate the importance and impact of these diseases on public health.

Several DENV and ZIKV vaccine candidates have been developed and tested under preclinical and clinical conditions [11,12]. Two tetravalent dengue vaccines have already been licensed, Dengvaxia® (Sanofi Pasteur) and QDENGA® (Takeda), which are based on chimaeric live-attenuated viruses [13]. Moreover, part of the anti-ZIKV vaccine strategies have evolved into phase I and II clinical trials (Clinical trials: NCT03008122, NCT03014089, NCT03110770, NCT02996461). The reported results demonstrated that formulations based on inactivated virus, DNA vaccine, and more recently, lipid nanoparticle-encapsulated mRNA-based vaccine encoding the prM-E antigens were well-tolerated and immunogenic, with induction of neutralizing antibodies in immunized individuals [14–19]. However, to date, none of the anti-ZIKV vaccines have been approved for use in humans. On the other hand, effective therapeutic strategies may reduce the lethality and morbidity related to severe forms of DENV and ZIKV infections. In this sense, different natural or synthetically produced antiviral candidates have also been tested against those arboviruses, including bioactive compounds from plant extracts, small molecules, or peptides [20–26]. Besides the high number of studies regarding the discovery of new compounds that have antiviral activity against those viruses, only a few have been tested in humans. Thus, so far, there are no specific drugs with proven efficacy against DENV, ZIKV, or other flaviviruses, and currently available therapies are only palliative [27].

Previous studies have demonstrated that synthetic peptides (<100 amino acid residues) have antiviral effects, and some of them show features compatible with pharmacological uses [28]. Furthermore, studies have demonstrated the inhibitory effects of synthetic peptides against different flaviviruses, such as DENV [29–33], JEV [34], WNV [35], and more recently, ZIKV [36]. Peptides derived from the proximal stem region of the DENV2 E protein inhibited viral infectivity under in vitro conditions. In addition, detailed analyses demonstrated that DENV2-inhibitory peptides act via a two-step mechanism. First, the peptides bind nonspecifically to the virion membrane at neutral pH using C-terminal hydrophobic residues; subsequently, at low pH into the endosome, the inhibitory peptides interact with the postfusion form of the E protein, blocking the membrane fusion step and, thus, the virus' infectivity. Specifically, a synthetic peptide named DV2, which was designed according to the domain III (EDIII) of the stem region of the E protein amino acid sequence, targets a late step of the viral fusion process and has been reported to inhibit the infectivity of all DENV serotypes [32,33]. Nonetheless, there is no additional data regarding the activity of the DV2 peptide on other flaviviruses, as well as on its protective effects in vivo under experimental conditions.

In the present study, we evaluated the interactions of the DV2 peptide with the ZIKV E protein using in silico analyses and biological effects data. Experimental data demonstrated that the DV2 peptide blocks ZIKV infectivity both in cultivated cells and under in vivo conditions, as demonstrated in immunodeficient mice subjected to lethal challenges with the virus, without any detected cytotoxic side effects.

2. Materials and Methods

2.1. Ethics Statement

The protocols were approved on 28 March 2016 by the Institutional Animal Care and Use Committee (CEUA) of the University of São Paulo (protocol number 22/2016) and were conducted according to the Ethical Principles of Animal Experimentation established by the Brazilian College of Animal Experimentation.

2.2. Cell Lines and Virus

The Vero CCL-81 cell line was cultured in DMEM supplemented with 10% fetal bovine serum (FBS) (Life Technologies). Aedes albopictus C6/36 cells were cultured in Leibovitz L-15 medium (Vitrocell, Campinas, SP, Brazil) supplemented with 2% FBS. ZIKV^{BR} strain (isolated from a clinical case during the 2015 Brazilian outbreak (isolate = "BeH823339", GenBank: KU729217) [37] was propagated in C6/36 cells, concentrated, and titrated as previously described [38,39].

2.3. Peptide Synthesis

DV2 (AWDFGSLGGVFTSIGKALHQVFGAIYGAA) and Ph-22 (ACNTIDPRHCGG GSAETVES)—a synthetic peptide previously generated in the laboratory with no reported activity against ZIKV—were purchased from Biomatik (Piscataway, NJ, USA), solubilized in dimethylsulfoxide (DMSO) to obtain stock solutions with end concentrations of 2.5 mM and 5 mM, respectively, and stored at $-20\ ^\circ$C until use. The purity of the peptides was higher than 95%.

2.4. Cytotoxicity Tests

The cytotoxic effects of the DV2 peptide were evaluated by measuring cell metabolic activity as well as cell viability. Cell proliferation was assessed using the reagent WST-1 (Roche) according to the manufacturer's instructions, and the number of live cells was determined by flow cytometry. Briefly, Vero CCL-81 cell monolayers established in 96-well plates were incubated at 37 $^\circ$C for 24 h with increasing concentrations of the peptide ranging from 0.75 to 24 µM. Equivalent volumes of DMSO were used as the control. The WST-1 reagent was added to the wells, which were then incubated for 1 h at 37 $^\circ$C. The absorbance was measured using a microplate reader at 440 nm. For flow cytometry analysis, live/dead aqua fluorescent reactive dye (Invitrogen) was added to the wells after a washing step for incubation for 30 min. at room temperature. The cells were then acquired on a LSR FortessaTM cytometer (BD, Franklin Lakes, NJ, USA). The data were analyzed using FlowJo software (version 10, Tree Star, San Carlo, CA, USA).

2.5. In Silico Docking of the DV2 Peptide

Prediction of the DV2 peptide-binding site on the ZIKV E protein in the postfusion conformation was performed using a three-dimensional model built with Modeller v10.1 [40] based on the structural coordinates from homologous DENV1 (PDB:3G7T) and DENV2 (PDB:1OK8). A total of 100 models were generated, and the selected structure was chosen on the basis of the DOPE score. Peptide docking was performed using CABS-dock v0.9.18 [41] with both conformations of the ZIKV E protein as docking pairings. Each simulation generated 10,000 models (from 10 independent trajectories), which were further reduced to 1000 models on the basis of low-energy selection. The 1000 top-scored models were then classified into 10 clusters, and the all-atom structures were derived by PD2 [42] from each cluster's medoid. The Zika E protein residues involved in the interaction with the DV2 peptide were identified using the Protein-Ligand Interaction Profiler (PLIP) server [43]. The PyMOL molecular graphics system (DeLano Scientific, San Carlos, CA, USA; DeLano 2002) was used for analyzing the structures and interactions of DV2 and the ZIKV E protein and for preparing the figures.

2.6. Virus Inhibition Assays

Different concentrations (0.75, 1.5, 3.0, 6.0, and 12 µM) of the DV2 or Ph22 peptides were incubated with ZIKV^{BR} strain (MOI of 1) for 30 min at 37 $^\circ$C and 5% CO2. The virus-peptide mixture was transferred to Vero CCL-81 cell monolayers (previously established in 96-flat well plates) and incubated for 1 h at 37 $^\circ$C and 5% CO2. After washing (three times) with PBS, fresh DMEM supplemented with 2% FBS was added to the cells and incubated for up to 24 h. The cell culture supernatants were harvested and stored at $-80\ ^\circ$C. The virus titers of the cell-free supernatants were determined by plaque assays (PFU/mL), as previously described [38].

2.7. Flow Cytometry Analyses of ZIKV-Infected Cells

The cell monolayers were washed twice with PBS and trypsinized using 1× Trypsin/EDTA (Gibco). Cells were fixed/permeabilized using a Cytofix/Cytoperm kit (BD Bioscience), according to the manufacturer's instructions, labeled with the 4G2 monoclonal antibody (Millipore) (10 µg/mL), and incubated with a rabbit anti-mouse IgG antibody coupled to AF488 (Thermo Fisher Scientific) (final dilution of 1:1000). Flow cytometry analyses were performed using an LSR FortessaTM cytometer (BD, Franklin Lakes, NJ, USA). The data were analyzed using FlowJo software (version 10, Tree Star, San Carlo, CA, USA) to determine the percentage of infected cells.

2.8. Mice Infection

The AG129 (IFNα/βR^{--}) (7–8 weeks old) mice used in this study were bred under specific pathogen-free conditions at the Isogenic Mouse Facility of the Microbiology Department, University of São Paulo, Brazil. ZIKV^{BR} preparations (100 PFU/animal) were incubated at 37 °C for 30 min. with 12 μM DV2 peptide (n = 5) or equivalent volumes of DMSO (n = 5), maintaining a final peptide-virus mixture volume of 50 μL injected per animal. Mice inoculated with ZIKV only (n = 4) and naive animals (n = 3) were used as the control groups. Naïve AG129 mice were s.c. inoculated with the peptide/DMSO-virus mixtures and monitored for up to 17 days. Morbidity signs were quantified on the basis of an arbitrary score scale (healthy, score 0; ruffled fur, score 1; paralysis, score 2; deformed spinal column, score 3; moribund, score 4) together with body weight measurements. Serum samples were collected on days 3, 5, 7, and 14 after infection and stored at −80 °C for analysis of viremia. A body weight reduction of 20% was used as a parameter for sacrificing animals in combination with morbidity evaluation (clinical score of 4).

2.9. Statistical Analysis

Statistical analyses were performed using Prism 6 (GraphPad Software Inc., LA Jolla, CA, USA). A t-test was used for two-group comparisons, whereas two-way ANOVA followed by Bonferroni correction was used when the data involved several groups and more than one variable (concentrations). The log-rank test (Mantel-Cox) was used to analyze the survival and morbidity data. Differences were considered significant when the p-value (p) was $\leq$0.05.

3. Results

3.1. In-Silico Docking Analyses of DV2 with the ZIKV E Protein

In silico docking analyses of the DV2 peptide were carried out with the postfusion conformation of the ZIKV E protein-binding site using the CABS-dock program. We used a structural model based on the coordinates of DENV1 and DENV2 proteins as the input for the program. The best docking complex was selected among those with the lowest energy parameters; it corresponded to the positioning of the DV2 peptide on a protein surface rich in hydrophobic and negatively charged amino acids, interacting with domains I and II (Figure 1). According to the docking simulation, the interaction of the peptide in the postfusion conformation mainly relies on residues from the conserved region (419–441) and involves hydrogen bonds, hydrophobic interactions, and salt bridges. These data are in accordance with previous results and models that describe the interaction of peptides derived from the membrane-proximal region of the dengue virus E protein in the postfusion conformation [32]. According to the PLIP program, 11 residues of the DV2 peptide participate in the interactions with ZIKV E protein, 8 in the conserved region, and only 3 among residues 441–447, reinforcing the notion that the conserved region is important for binding in the trimeric form. A list of the predicted interactions for all the residues is shown in Table S1 (Supplementary Materials).

3.2. In Vitro Cytotoxicity and Anti-ZIKV Effects of the DV2 Peptide

Based on the expected interactions among the amino acid residues of DV2 and ZIKV E proteins, we set up experiments to evaluate the in vitro effects of the peptide on ZIKV-infected cells. Initially, we determined the cytotoxic effects of DV2 on Vero CCL-81 cells incubated with the peptide for up to 24 h. The DV2 citotoxicity was evaluated measuring the number of live cells by staining them with a live/dead fluorescent and determining the cell metabolic activity using the WST-1 reagent. As shown in Figure 2, the DV2 peptide did not show any significant cytotoxicity at concentrations ranging from 0.7 to 24 µM. Similarly, the control (DMSO) tested at equivalent dilutions did not induce any significant cell death.

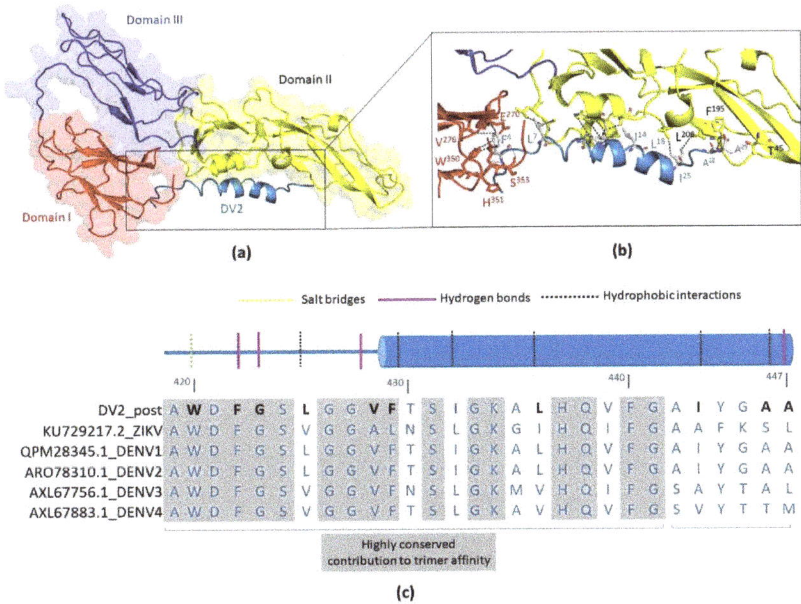

Figure 1. Prediction of DV2 peptide and ZIKV E protein interactions based on in silico docking analyses. The structural model of ZIKV E protein in postfusion conformation (based on the structural coordinates from PDBs 3G7T and 1OK8) was used for the analysis. The protein is shown in surface mode highlighting domain I, II, and III, displayed as red, yellow, and blue colored cartoons, respectively. The DV2 peptide is represented using a sky-blue-colored cartoon. (**a**) According to docking results, the peptide interacts with EDI and EDII, in the postfusion conformation. (**b**) Detailed view of the interacting residues between the ZIKV E protein and the DV2 peptide, predicted by the PLIP program. Residues are shown as colored sticks; sky blue has been used for the DV2 peptide, salmon for DI, and yellow for DII residues. (**c**) Representation of residues 420–447 of the DV2 peptide and their interactions with the ZIKV E protein. The amino acid sequence alignment of the DV2 peptide with the respective sequences of the envelope protein of ZIKV and DENV (types 1 to 4) is shown below. The residues of the peptide that interact with the postfusion conformation are highlighted in bold black.

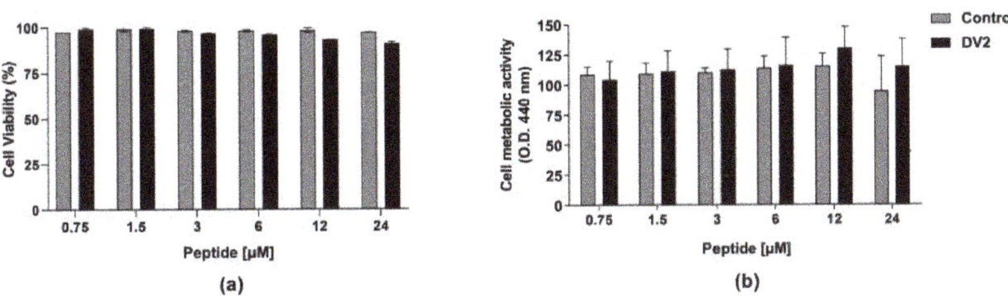

Figure 2. Cytotoxic effects of the DV2 peptide on eukaryotic cells. Vero CCL-81 cell monolayers were treated with increasing concentrations (0.75–24 µM) of the DV2 peptide for up to 24 h. Cell samples were exposed to the peptide solvent (DMSO) at the same volume/dilution (control). (**a**) Percentage of live cells obtained by flow cytometry analyses. (**b**) Cell proliferation measured by metabolic activity using the WST-1 reagent. Data are presented as mean ± SEM of live cells normalized to the respective control.

Next, we tested the in vitro antiviral effects of the DV2 peptide. Vero cells were infected with ZIKV (MOI of 1) in the presence of either the DV2 or the control Ph22 peptides, followed by detection of infected cells and infectious viral particles in the cell supernatants collected 24 h post infection (hpi). Notably, the DV2 peptide drastically reduced ZIKV infectivity at concentrations ranging from 3 to 12 µM, as measured by the number of infected cells (Figure 3a). The anti-ZIKV effects of DV2, compared to that of the control Ph22 peptide, reached 80% (3 µM) to 97.5% (12 µM) inhibition. Similarly, determination of the viable viral particles in the culture supernatants of the infected cells demonstrated that concentrations of 6 and 12 µM of DV2 reduced the number of PFU to values below the detection limit of the assay (Figure 3b). The inhibitory concentrations (IC50 and IC90) are listed in Table 1 and Figure S1 (Supplementary Materials).

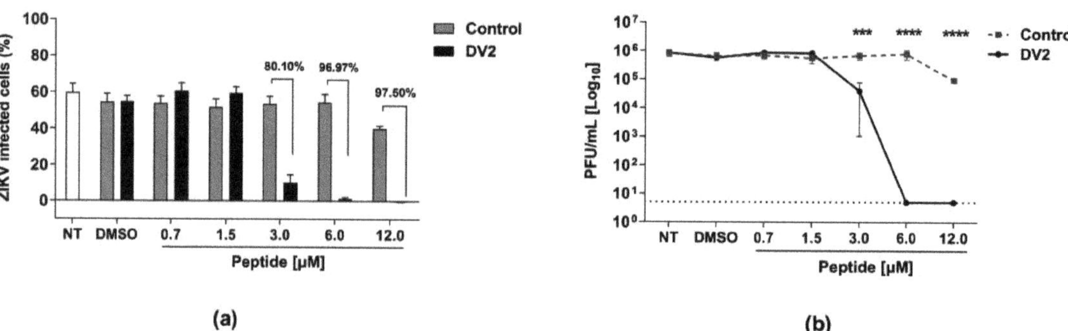

Figure 3. The DV2 peptide inhibits ZIKV infectivity in vitro cultured Vero cells. Vero cell monolayers were infected with ZIKV (MOI = 1.0) in the presence of the DV2 (DV2) or Ph22 (Control) peptides. The impact of the treatment with the DV2 peptide on virus infection was measured at 24 hpi. (**a**) Percentages of infected cells determined by flow cytometry. (**b**) Viable ZIKV titers (PFU/mL) determined by plaque assay using culture supernatants of infected cells. Cells incubated with viruses treated with Ph22, DMSO, or culture medium (NT) were used as negative controls. The data are presented as mean ± SEM (bars) from three independent experiments. Statistically significant differences were determined using two-way ANOVA with Bonferroni correction (*** $p < 0.001$ and **** $p < 0.0001$, compared to the DMSO control).

Table 1. DV2 peptide concentrations required to inhibit ZIKV infection by 50% (IC_{50}) or 90% (IC_{90}), determined by different methods.

	IC50 (µM)	IC90 (µM)
Flow Cytometry [1]	2.647	3.228
Plaque assay [2]	1.976	2.696

[1] ZIKV infected cells (%), [2] Viable ZIKV titers (PFU/mL).

3.3. Protective Effects of the DV2 Peptide in AG129 Mice Challenged with ZIKV

The in vivo anti-ZIKV effects of the DV2 peptide were determined in AG129 IFNα/βR^{-}/$^{-}$ mice following s.c. inoculation with a lethal dose of ZIKV^{BR}. As demonstrated in Figure 4, incubation of ZIKV with the DV2 peptide reduced morbidity signs in infected animals (Figure 4a–e). DV2 also prevented body weight loss and viremia (Figure 4f,g). Additionally, we measured the protection conferred by the DV2 peptide against mortality in ZIKV-infected mice. In contrast to mice inoculated with ZIKV particles in the presence of DMSO, 80% of mice infected with ZIKV exposed to 12 µM DV2 peptide survived the lethal challenge (Figure 4h). Taken together, these results support a strong anti-ZIKV effect of the DV2 peptide both in vitro and in vivo.

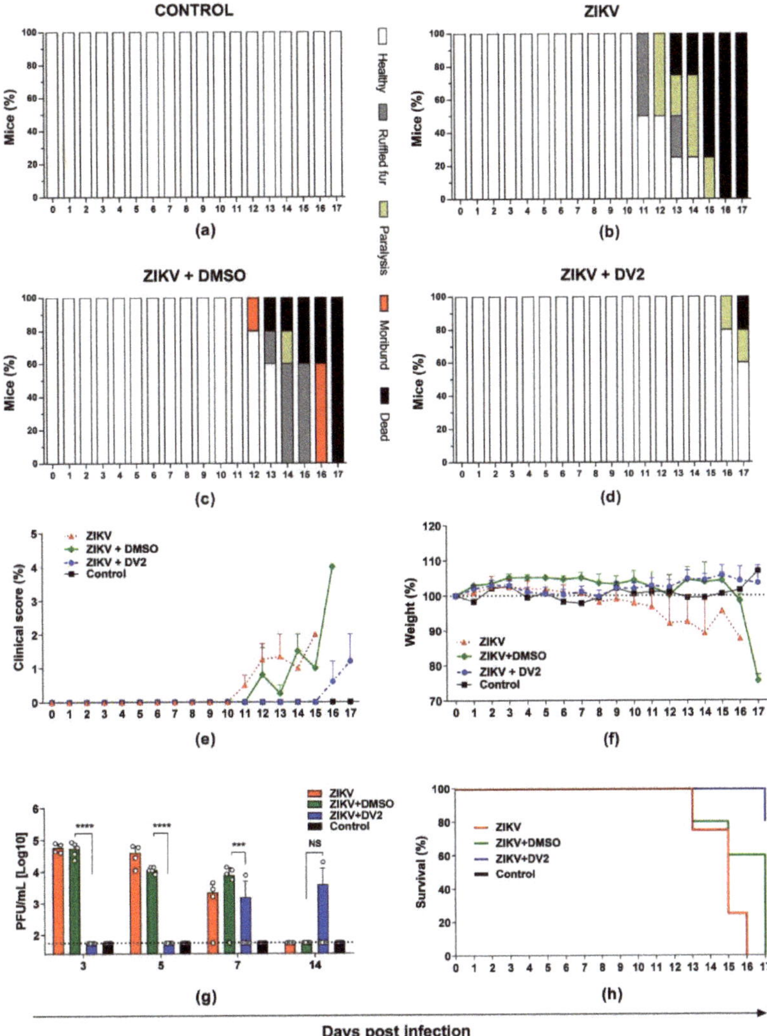

Figure 4. The DV2 peptide induces protection against ZIKV infection in AG129 mice. DV2 (12 μM) was incubated with 100 PFU ZIKV^{BR} for 30 min at 37 °C. The virus-peptide mixtures were s.c. inoculated in AG129 mice. The (**a–d**) morbidity, (**e**) clinical score, (**f**) weight changes, (**g**) viremia, and (**h**) survival were monitored in the infected animals for up to 17 days. The data are presented as means ±SD of the respective parameters in the different mice groups. Statistical significance was evaluated using two-way ANOVA followed by Bonferroni correction. Mantel-Cox test was used to analyze the survival results (*** $p < 0.001$; **** $p < 0.0001$; NS, not significant).

4. Discussion

The flavivirus E protein is a surface membrane protein involved in host cell receptor-binding functions and mediates the fusion of the viral envelope with endosomal membranes. The E protein is composed of three regions: domains I (EDI), II (EDII), and III (EDIII). EDII preserves the conserved hydrophobic fusion loop required for cell entry, whereas EDIII interacts with cellular receptors. The E protein is a key target for peptides capable of halting the infectivity of flaviviruses [44–46], particularly those derived from the C-terminal portion of the E protein [29,33,34,36]. This portion, known as stem, is a membrane-proximal region that plays important roles in the viral fusion process [47,48]. Since the stem region is conserved among flaviviruses [32], stem-based antiviral peptides are potential candidates for cross-inhibiting the infectivity of flaviviruses.

Here we showed that the DV2 stem-derived peptide (aa 419–447), previously reported to inhibit the four DENV serotypes [33], was capable of inhibiting ZIKV infection in cell culture and protected mice against an otherwise lethal challenge with the virus. Interestingly, despite the DV2 peptide being based on the DENV sequence, our in silico docking analyses showed that N- and C-terminal DV2 residues interact with ZIKV EDI and EDII in postfusion conformations. Moreover, this interaction is mainly based on residues from the conserved region (419–441) among the viruses. Accordingly, a two-step mechanism was initially proposed for the DV2-inhibitory activity against DENV infection. First, DV2 binds nonspecifically to the viral lipid layers using C-terminal hydrophobic residues, and, once in the endosome, it specifically interacts with the postfusion form of the E protein (N-terminal residues) and blocks viral endosome membrane fusion [33]. Thus, we assume that the ZIKV cross-inhibition effects observed in the presence of the DV2 peptide are similar to those reported for DENV.

The DV2 peptide was able to inhibit in vitro ZIKV infectivity by more than 95% at 6–12 μM. A previous study demonstrated that the DV2 peptide inhibited the four dengue types at different molar concentrations, DENV1, IC_{90} = 0.1 μM; DENV2, IC_{90} = 0.3 μM; DENV3, IC_{90} = 2 μM; and DENV4, IC_{90} = 0.7 μM. In contrast, peptides with sequences derived from other flaviviruses (YFV and WNV) did not inhibit DENV infection [33]. Here, the IC_{50} (1.98–2.65 μM) or IC_{90} (2.20–2.7 μM) values established for ZIKV inhibition were higher than those reported for the DENV types. However, IC parameters are difficult to compare as these have been shown to be dependent on the cell line and virus strain. Interestingly, despite the limitations mentioned above, the IC_{90} = 2.7 μM value shown here is close to the IC_{90} found by the authors for DENV3 inhibition, whose stem region has the highest similarity with that of ZIKV (76%) compared with the similarities observed with the stem regions of the other serotypes, DENV1 (52%), DENV2 (55%), and DENV4 (62%). Thus, our in vitro results demonstrate the cross-antiviral activity of a DENV2-derived peptide with ZIKV, highlighting the potential application of stem-derived peptides for the control of different flavivirus infections.

The cross-inhibitory activity of flavivirus stem-derived peptides has also been reported previously. In this context, a 33-residue peptide mimetic to the DENV2 sequence, named DN59 (412–444), was effective against the four DENV serotypes and WNV (IC_{50} = 2–5 μM). The authors suggested that DN59 interacts with the viral lipid membrane, disrupting the lipid bilayer and forming holes from where the viral RNA exits, irreversibly inactivating the virions [29]. Similarly, a ZIKV peptide, designated Z2 (aa 421–453), was shown to disrupt the ZIKV envelope membrane integrity, inhibiting different ZIKV strains, such as SZ01 (IC_{50} = 2.6 μM), FLR (IC_{50} = 4.0 μM), and MR766 (IC_{50} = 13.9 μM); Z2 also inhibited DENV2 (IC_{50} = 4.0 μM) and YFV 17D (IC_{50} = 5.0 μM) [36]. Another study described a stem-derived P5 peptide that inhibited JEV as well as ZIKV infections (IC_{50} = 3.27 μM). The authors also showed that the equivalent peptide ZP1 (424–445) derived from ZIKV inhibited ZIKV infection at lower concentrations (IC_{50} = 1.32 μM) than the JEV-derived peptide [34]. Thus, our data are in line with the literature and support the cross-inhibition effects of E protein stem-based peptides against different flaviviruses, raising perspectives for the development of anti-flavivirus therapies based on the use of synthetic peptides.

Some stem-based peptides have shown in vivo antiviral effects in animal models [34,36]. Here, we used a highly susceptible ZIKV infection animal model, IFNα/βR$^-$/$^-$AG129 mice, to evaluate in vivo DV2-induced protection against ZIKV [49]. Our results showed an important DV2-mediated protective effect against ZIKV infection based on different in vivo parameters, including viremia, morbidity, and mortality. Notably, 12 μM of the DV2 peptide protected most (80%) of the animals from death and prevented weight loss. To our knowledge, this is the first study to describe the in vivo antiviral effects of the DV2 peptide. In vivo anti-ZIKV protection has also been reported for the JEV E protein stem-derived peptide P5 [34]. The P5 peptide reduced viral load and histopathological damage in the brains of mice inoculated with ZIKV [34]. Alternatively, the Z2 peptide, derived from the ZIKV E protein stem region, conferred significant protection (75%) against death and neurologic symptoms among challenged mice [36]. Collectively, different synthetic peptides derived from the stem region of the E protein of different flaviviruses represent a potential therapeutic strategy against infection and, in the case of ZIKV, may avoid or reduce neurological damage in newborns associated with infection in pregnant women.

Supplementary Materials: The following supporting information can be downloaded at: https://www.mdpi.com/article/10.3390/v15040839/s1, Figure S1. Determination of DV2 peptide IC_{50} values against ZIKV through different techniques. Table S1. Prediction of non-covalent interactions between the ZIKV E protein and the synthetic DV2 peptide in the postfusion conformation according to the PLIP tool.

Author Contributions: Conceptualization, M.F.d.C.-A., S.S.P. and L.R.P.; methodology, M.F.d.C.-A., S.S.P., L.R.P., L.S.S., E.G.M. and A.B.; software, L.S.S., A.B.; validation, M.F.d.C.-A. and S.S.P.; formal analysis, M.F.d.C.-A., S.S.P., L.R.P. and L.S.S.; investigation, M.F.d.C.-A., S.S.P., A.A.V.-C. and L.R.P.; data curation, M.F.d.C.-A., L.R.P. and L.S.S.; writing—original draft preparation, M.F.d.C.-A., L.R.P. and A.B.; writing—review and editing, L.C.d.S.F. and A.B.; supervision, L.C.d.S.F.; project administration, E.G.M.; funding acquisition, L.C.d.S.F. and A.B. All authors have read and agreed to the published version of the manuscript.

Funding: This study was funded by Fundação de Amparo à Pesquisa do Estado de São Paulo (FAPESP), grant number 2016/20045-7; FAPESP, grant numbers 2018/14459-9 (M.F.C.A.), 2016/05570-8 (L.R.P.), 2022/01034-5 (A.A.V.B.C.), 2018/20162-9 (A.B.); Conselho Nacional de Desenvolvimento Científico e Tecnológico (CNPq), grant numbers 440409/2016-0 and 407592/2016-4. Coordenação de Aperfeiçoamento de Pessoal de Nível Superior (CAPES), grant numbers 116618/2016-01 and 130787/2016-01; CAPES, grant numbers 88887.600175/2021-00 (S.S.P.).

Institutional Review Board Statement: The study was conducted according to the Ethical Principles of Animal Experimentation established by the Brazilian College of Animal Experimentation, and approved by the Institutional Animal Care and Use Committee (CEUA) of the University of São Paulo (protocol number 22/2016) on 28 March 2016.

Informed Consent Statement: Not applicable.

Data Availability Statement: Not applicable.

Conflicts of Interest: The authors declare no conflict of interest.

References

1. Chong, H.Y.; Leow, C.Y.; Abdul Majeed, A.B.; Leow, C.H. Flavivirus Infection—A Review of Immunopathogenesis, Immunological Response, and Immunodiagnosis. *Virus Res.* **2019**, *274*, 197770. [CrossRef]
2. Daep, C.A.; Muñoz-Jordán, J.L.; Eugenin, E.A. Flaviviruses, an Expanding Threat in Public Health: Focus on Dengue, West Nile, and Japanese Encephalitis Virus. *J. Neurovirol.* **2014**, *20*, 539–560. [CrossRef]
3. Roy, S.K.; Bhattacharjee, S. Dengue Virus: Epidemiology, Biology, and Disease Aetiology. *Can. J. Microbiol.* **2021**, *67*, 687–702. [CrossRef] [PubMed]
4. Shepard, D.S.; Undurraga, E.A.; Halasa, Y.A.; Stanaway, J.D. The Global Economic Burden of Dengue: A Systematic Analysis. *Lancet Infect Dis.* **2016**, *16*, 935–941. [CrossRef] [PubMed]
5. Pielnaa, P.; Al-Saadawe, M.; Saro, A.; Dama, M.F.; Zhou, M.; Huang, Y.; Huang, J.; Xia, Z. Zika Virus-Spread, Epidemiology, Genome, Transmission Cycle, Clinical Manifestation, Associated Challenges, Vaccine and Antiviral Drug Development. *Virology* **2020**, *543*, 34–42. [CrossRef]
6. Faria, N.R.; Quick, J.; Claro, I.M.; Thézé, J.; de Jesus, J.G.; Giovanetti, M.; Kraemer, M.U.G.; Hill, S.C.; Black, A.; da Costa, A.C.; et al. Establishment and Cryptic Transmission of Zika Virus in Brazil and the Americas. *Nature* **2017**, *546*, 406–410. [CrossRef]
7. de Oliveira, W.K.; de França, G.V.A.; Carmo, E.H.; Duncan, B.B.; de Souza Kuchenbecker, R.; Schmidt, M.I. Infection-Related Microcephaly after the 2015 and 2016 Zika Virus Outbreaks in Brazil: A Surveillance-Based Analysis. *Lancet* **2017**, *390*, 861–870. [CrossRef] [PubMed]
8. Baud, D.; Gubler, D.J.; Schaub, B.; Lanteri, M.C.; Musso, D. An Update on Zika Virus Infection. *Lancet* **2017**, *390*, 2099–2109. [CrossRef]
9. Miner, J.J.; Diamond, M.S. Zika Virus Pathogenesis and Tissue Tropism. *Cell Host Microbe* **2017**, *21*, 134–142. [CrossRef]

10. França, G.V.A.D.; Pedi, V.D.; Garcia, M.H.D.O.; Carmo, G.M.I.D.; Leal, M.B.; Garcia, L.P. Síndrome congênita associada à infecção pelo vírus Zika em nascidos vivos no Brasil: Descrição da distribuição dos casos notificados e confirmados em 2015-2016. *Epidemiol. Serv. Saúde* **2018**, *27*, e2017473. [CrossRef]

11. Richner, J.M.; Diamond, M.S. Zika Virus Vaccines: Immune Response, Current Status, and Future Challenges. *Curr. Opin. Immunol.* **2018**, *53*, 130–136. [CrossRef]

12. Pinheiro-Michelsen, J.R.; Souza, R. da S.O.; Santana, I.V.R.; da Silva, P. de S.; Mendez, E.C.; Luiz, W.B.; Amorim, J.H. Anti-Dengue Vaccines: From Development to Clinical Trials. *Front. Immunol.* **2020**, *11*, 1252. [CrossRef]

13. Torres-Flores, J.M.; Reyes-Sandoval, A.; Salazar, M.I. Dengue Vaccines: An Update. *BioDrugs* **2022**, *36*, 325–336. [CrossRef] [PubMed]

14. Modjarrad, K.; Lin, L.; George, S.L.; Stephenson, K.E.; Eckels, K.H.; De La Barrera, R.A.; Jarman, R.G.; Sondergaard, E.; Tennant, J.; Ansel, J.L.; et al. Preliminary Aggregate Safety and Immunogenicity Results from Three Trials of a Purified Inactivated Zika Virus Vaccine Candidate: Phase 1, Randomised, Double-Blind, Placebo-Controlled Clinical Trials. *Lancet* **2018**, *391*, 563–571. [CrossRef]

15. Tebas, P.; Roberts, C.C.; Muthumani, K.; Reuschel, E.L.; Kudchodkar, S.B.; Zaidi, F.I.; White, S.; Khan, A.S.; Racine, T.; Choi, H.; et al. Safety and Immunogenicity of an Anti–Zika Virus DNA Vaccine. *N. Engl. J. Med.* **2021**, *385*, e35. [CrossRef]

16. Gaudinski, M.R.; Houser, K.V.; Morabito, K.M.; Hu, Z.; Yamshchikov, G.; Rothwell, R.S.; Berkowitz, N.; Mendoza, F.; Saunders, J.G.; Novik, L.; et al. Safety, Tolerability, and Immunogenicity of Two Zika Virus DNA Vaccine Candidates in Healthy Adults: Randomised, Open-Label, Phase 1 Clinical Trials. *Lancet* **2018**, *391*, 552–562. [CrossRef] [PubMed]

17. Stephenson, K.E.; Tan, C.S.; Walsh, S.R.; Hale, A.; Ansel, J.L.; Kanjilal, D.G.; Jaegle, K.; Peter, L.; Borducchi, E.N.; Nkolola, J.P.; et al. Safety and Immunogenicity of a Zika Purified Inactivated Virus Vaccine given via Standard, Accelerated, or Shortened Schedules: A Single-Centre, Double-Blind, Sequential-Group, Randomised, Placebo-Controlled, Phase 1 Trial. *Lancet Infect. Dis.* **2020**, *20*, 1061–1070. [CrossRef]

18. Han, H.-H.; Diaz, C.; Acosta, C.J.; Liu, M.; Borkowski, A. Safety and Immunogenicity of a Purified Inactivated Zika Virus Vaccine Candidate in Healthy Adults: An Observer-Blind, Randomised, Phase 1 Trial. *Lancet Infect. Dis.* **2021**, *21*, 1282–1292. [CrossRef]

19. Essink, B.; Chu, L.; Seger, W.; Barranco, E.; Le Cam, N.; Bennett, H.; Faughnan, V.; Pajon, R.; Paila, Y.D.; Bollman, B.; et al. The Safety and Immunogenicity of Two Zika Virus MRNA Vaccine Candidates in Healthy Flavivirus Baseline Seropositive and Seronegative Adults: The Results of Two Randomised, Placebo Controlled, Dose Ranging, Phase 1 Clinical Trials. *Lancet Infect. Dis.* **2023**. [CrossRef] [PubMed]

20. Lim, W.Z.; Cheng, P.G.; Abdulrahman, A.Y.; Teoh, T.C. The Identification of Active Compounds in *Ganoderma Lucidum Var. Antler* Extract Inhibiting Dengue Virus Serine Protease and Its Computational Studies. *J. Biomol. Struct. Dyn.* **2020**, *38*, 4273–4288. [CrossRef] [PubMed]

21. Norshidah, H.; Vignesh, R.; Lai, N.S. Updates on Dengue Vaccine and Antiviral: Where Are We Heading? *Molecules* **2021**, *26*, 6768. [CrossRef] [PubMed]

22. Diosa-Toro, M.; Troost, B.; van de Pol, D.; Heberle, A.M.; Urcuqui-Inchima, S.; Thedieck, K.; Smit, J.M. Tomatidine, a Novel Antiviral Compound towards Dengue Virus. *Antiviral Res.* **2019**, *161*, 90–99. [CrossRef] [PubMed]

23. Quintana, V.M.; Selisko, B.; Brunetti, J.E.; Eydoux, C.; Guillemot, J.C.; Canard, B.; Damonte, E.B.; Julander, J.G.; Castilla, V. Antiviral Activity of the Natural Alkaloid Anisomycin against Dengue and Zika Viruses. *Antiviral Res.* **2020**, *176*, 104749. [CrossRef]

24. Ruan, J.; Rothan, H.A.; Zhong, Y.; Yan, W.; Henderson, M.J.; Chen, F.; Fang, S. A Small Molecule Inhibitor of ER-to-Cytosol Protein Dislocation Exhibits Anti-Dengue and Anti-Zika Virus Activity. *Sci. Rep.* **2019**, *9*, 10901. [CrossRef]

25. Lee, S.H.; Kim, E.H.; O'neal, J.T.; Dale, G.; Holthausen, D.J.; Bowen, J.R.; Quicke, K.M.; Skountzou, I.; Gopal, S.; George, S.; et al. The Amphibian Peptide Yodha Is Virucidal for Zika and Dengue Viruses. *Sci. Rep.* **2021**, *11*, 602. [CrossRef] [PubMed]

26. Martins, I.C.; Ricardo, R.C.; Santos, N.C. Dengue, West Nile, and Zika Viruses: Potential Novel Antiviral Biologics Drugs Currently at Discovery and Preclinical Development Stages. *Pharmaceutics* **2022**, *14*, 2535. [CrossRef] [PubMed]

27. Low, J.G.H.; Ooi, E.E.; Vasudevan, S.G. Current Status of Dengue Therapeutics Research and Development. *J. Infect. Dis.* **2017**, *215*, S96–S102. [CrossRef]

28. Agarwal, G.; Gabrani, R. Antiviral Peptides: Identification and Validation. *Int. J. Pept. Res. Ther.* **2021**, *27*, 149–168. [CrossRef]

29. Lok, S.-M.; Costin, J.M.; Hrobowski, Y.M.; Hoffmann, A.R.; Rowe, D.K.; Kukkaro, P.; Holdaway, H.; Chipman, P.; Fontaine, K.A.; Holbrook, M.R.; et al. Release of Dengue Virus Genome Induced by a Peptide Inhibitor. *PLoS ONE* **2012**, *7*, e50995. [CrossRef]

30. Chew, M.-F.; Poh, K.-S.; Poh, C.-L. Peptides as Therapeutic Agents for Dengue Virus. *Int. J. Med. Sci.* **2017**, *14*, 1342–1359. [CrossRef]

31. Panya, A.; Bangphoomi, K.; Choowongkomon, K.; Yenchitsomanus, P. Peptide Inhibitors Against Dengue Virus Infection. *Chem. Biol. Drug Des.* **2014**, *84*, 148–157. [CrossRef] [PubMed]

32. Schmidt, A.G.; Yang, P.L.; Harrison, S.C. Peptide Inhibitors of Flavivirus Entry Derived from the E Protein Stem. *J. Virol.* **2010**, *84*, 12549–12554. [CrossRef] [PubMed]

33. Schmidt, A.G.; Yang, P.L.; Harrison, S.C. Peptide Inhibitors of Dengue-Virus Entry Target a Late-Stage Fusion Intermediate. *PLoS Pathog.* **2010**, *6*, e1000851. [CrossRef]

34. Chen, L.; Liu, Y.; Wang, S.; Sun, J.; Wang, P.; Xin, Q.; Zhang, L.; Xiao, G.; Wang, W. Antiviral Activity of Peptide Inhibitors Derived from the Protein E Stem against Japanese Encephalitis and Zika Viruses. *Antiviral Res.* **2017**, *141*, 140–149. [CrossRef]

35. Bai, F.; Town, T.; Pradhan, D.; Cox, J.; Ashish; Ledizet, M.; Anderson, J.F.; Flavell, R.A.; Krueger, J.K.; Koski, R.A.; et al. Antiviral Peptides Targeting the West Nile Virus Envelope Protein. *J. Virol.* **2007**, *81*, 2047–2055. [CrossRef] [PubMed]

36. Yu, Y.; Deng, Y.-Q.; Zou, P.; Wang, Q.; Dai, Y.; Yu, F.; Du, L.; Zhang, N.-N.; Tian, M.; Hao, J.-N.; et al. A Peptide-Based Viral Inactivator Inhibits Zika Virus Infection in Pregnant Mice and Fetuses. *Nat. Commun.* **2017**, *8*, 15672. [CrossRef]

37. Faria, N.R.; Azevedo, R.D.S.D.S.; Kraemer, M.U.G.; Souza, R.; Cunha, M.S.; Hill, S.C.; Thézé, J.; Bonsall, M.B.; Bowden, T.A.; Rissanen, I.; et al. Zika Virus in the Americas: Early Epidemiological and Genetic Findings. *Science* **2016**, *352*, 345–349. [CrossRef] [PubMed]

38. Pereira, L.R.; Alves, R.P.D.S.; Sales, N.S.; Andreata-Santos, R.; Venceslau-Carvalho, A.A.; Pereira, S.S.; Castro-Amarante, M.F.; Rodrigues-Jesus, M.J.; Favaro, M.T.D.P.; Chura-Chambi, R.M.; et al. Enhanced Immune Responses and Protective Immunity to Zika Virus Induced by a DNA Vaccine Encoding a Chimeric NS1 Fused with Type 1 Herpes Virus GD Protein. *Front Med. Technol.* **2020**, *2*, 604160. [CrossRef]

39. Andreata-Santos, R.; Alves, R.P.D.S.; Pereira, S.A.; Pereira, L.R.; Freitas, C.L.D.; Pereira, S.S.; Venceslau-Carvalho, A.A.; Castro-Amarante, M.F.; Favaro, M.T.P.; Mathias-Santos, C.; et al. Transcutaneous Administration of Dengue Vaccines. *Viruses* **2020**, *12*, 514. [CrossRef]

40. Šali, A.; Blundell, T.L. Comparative Protein Modelling by Satisfaction of Spatial Restraints. *J. Mol. Biol.* **1993**, *234*, 779–815. [CrossRef]

41. Kurcinski, M.; Pawel Ciemny, M.; Oleniecki, T.; Kuriata, A.; Badaczewska-Dawid, A.E.; Kolinski, A.; Kmiecik, S. CABS-Dock Standalone: A Toolbox for Flexible Protein–Peptide Docking. *Bioinformatics* **2019**, *35*, 4170–4172. [CrossRef] [PubMed]

42. Moore, B.L.; Kelley, L.A.; Barber, J.; Murray, J.W.; MacDonald, J.T. High-Quality Protein Backbone Reconstruction from Alpha Carbons Using Gaussian Mixture Models. *J. Comput. Chem.* **2013**, *34*, 1881–1889. [CrossRef] [PubMed]

43. Salentin, S.; Schreiber, S.; Haupt, V.J.; Adasme, M.F.; Schroeder, M. PLIP: Fully Automated Protein–Ligand Interaction Profiler. *Nucleic Acids Res* **2015**, *43*, W443–W447. [CrossRef] [PubMed]

44. Wei, J.; Hameed, M.; Wang, X.; Zhang, J.; Guo, S.; Anwar, M.N.; Pang, L.; Liu, K.; Li, B.; Shao, D.; et al. Antiviral Activity of Phage Display-Selected Peptides against Japanese Encephalitis Virus Infection in Vitro and in Vivo. *Antiviral Res.* **2020**, *174*, 104673. [CrossRef]

45. Mertinková, P.; Mochnáčová, E.; Bhide, K.; Kulkarni, A.; Tkáčová, Z.; Hruškovicová, J.; Bhide, M. Development of Peptides Targeting Receptor Binding Site of the Envelope Glycoprotein to Contain the West Nile Virus Infection. *Sci. Rep.* **2021**, *11*, 20131. [CrossRef]

46. Zoladek, J.; Burlaud-Gaillard, J.; Chazal, M.; Desgraupes, S.; Jeannin, P.; Gessain, A.; Pardigon, N.; Hubert, M.; Roingeard, P.; Jouvenet, N.; et al. Human Claudin-Derived Peptides Block the Membrane Fusion Process of Zika Virus and Are Broad Flavivirus Inhibitors. *Microbiol. Spectr.* **2022**, *10*, e0298922. [CrossRef]

47. Stiasny, K.; Kiermayr, S.; Bernhart, A.; Heinz, F.X. The Membrane-Proximal "Stem" Region Increases the Stability of the Flavivirus E Protein Postfusion Trimer and Modulates Its Structure. *J. Virol.* **2013**, *87*, 9933–9938. [CrossRef]

48. Hu, T.; Wu, Z.; Wu, S.; Chen, S.; Cheng, A. The Key Amino Acids of E Protein Involved in Early Flavivirus Infection: Viral Entry. *Virol. J.* **2021**, *18*, 136. [CrossRef]

49. Aliota, M.T.; Caine, E.A.; Walker, E.C.; Larkin, K.E.; Camacho, E.; Osorio, J.E. Characterization of Lethal Zika Virus Infection in AG129 Mice. *PLoS Negl. Trop. Dis.* **2016**, *10*, e0004682. [CrossRef]

viruses

MDPI

Article

Targeting the Human Influenza a Virus: The Methods, Limitations, and Pitfalls of Virtual Screening for Drug-like Candidates Including Scaffold Hopping and Compound Profiling

Thomas Scior [1,*], Karina Cuanalo-Contreras [1], Angel A. Islas [1,2] and Ygnacio Martinez-Laguna [2]

[1] Faculty of Chemical Sciences, Benemérita Universidad Autónoma de Puebla, Ciudad Universitaria, Colonia San Manuel, Puebla 72570, Mexico
[2] Vicerrectoría de Investigación y Estudios de Posgrado, Benemérita Universidad Autónoma de Puebla, Puebla 72592, Mexico
* Correspondence: tscior@gmail.com or thomas.scior@correo.buap.mx

Abstract: In this study, we describe the input data and processing steps to find antiviral lead compounds by a virtual screen. Two-dimensional and three-dimensional filters were designed based on the X-ray crystallographic structures of viral neuraminidase co-crystallized with substrate sialic acid, substrate-like DANA, and four inhibitors (oseltamivir, zanamivir, laninamivir, and peramivir). As a result, ligand–receptor interactions were modeled, and those necessary for binding were utilized as screen filters. Prospective virtual screening (VS) was carried out in a virtual chemical library of over half a million small organic substances. Orderly filtered moieties were investigated based on 2D- and 3D-predicted binding fingerprints disregarding the "rule-of-five" for drug likeness, and followed by docking and ADMET profiling. Two-dimensional and three-dimensional screening were supervised after enriching the dataset with known reference drugs and decoys. All 2D, 3D, and 4D procedures were calibrated before execution, and were then validated. Presently, two top-ranked substances underwent successful patent filing. In addition, the study demonstrates how to work around reported VS pitfalls in detail.

Keywords: influenza; neuraminidase inhibitors; noncompetitive inhibition; virtual screening; ligand docking; screening pitfalls; screening problems

check for updates

Citation: Scior, T.; Cuanalo-Contreras, K.; Islas, A.A.; Martinez-Laguna, Y. Targeting the Human Influenza a Virus: The Methods, Limitations, and Pitfalls of Virtual Screening for Drug-like Candidates Including Scaffold Hopping and Compound Profiling. *Viruses* **2023**, *15*, 1056. https://doi.org/10.3390/v15051056

Academic Editor: Simone Brogi

Received: 8 February 2023
Revised: 6 April 2023
Accepted: 12 April 2023
Published: 26 April 2023

1. Introduction

Despite world-wide vaccination efforts and "anti-flu" public health prevention campaigns (general hygiene and patient confinement), the influenza disease has never been controlled due to antigenic drifts or occasional abrupt shifting by gene mutations in the viruses—in addition to viruses occasionally crossing host–range barriers, thereupon expanding their genetic pools. Hence, as a severe setback, vaccines must be developed in advance of a forthcoming flu season based on predictions on previously known strains. Moreover, vaccine production time is a bottleneck, so vaccines are not readily available during the initial spread of a pandemic. As such, alternative treatments such as novel mini-antibodies have been proposed [1]. Human influenza A vaccines constitute the cornerstone of flu prevention while drugs have been developed for monotherapeutic purposes in the early stages of infection or as a co-medication, such as orally inhaled zanamivir (Relenza™) or orally administered oseltamivir (Tamiflu™). The latter suffered from complicated multi-step and resourceful synthetic preparation challenging production and delivery on time. These setbacks beg for simpler chemical moieties with broad anti-flu activity (against H1N1 and H5N1), as well as the possibility of structural derivatization.

Traditional experimental drug development techniques are based on laborious optimization cycles until favorable results are met. In contrast, computational molecular simulations—namely virtual screening, scaffold hopping, and drug candidate profiling—can save time, materials, and human resources. In particular, compound collections allow the inclusion of huge amounts of chemical substances—either already extant "real world" substances or non-existing molecules from virtual libraries. In this context, novel scaffolds against new influenza virus targets have emerged [2].

Three types of viral proteins are located on the influenza virus' surface: M2, HA, and NA. Primary sequence variants of the latter two define the subtypes, including human H1N1 or avian H5N1, etc. The human influenza A neuraminidase (protein name NA, gene name: na) is similar to the other viral surface glycoprotein hemagglutinin (HA), with complementary functions during the infection process. NA and HA protrude spike-like above the viral outer envelope, where four NA glycoproteins usually form homotetramers [3]. Mechanistic insight concerning virus-binding specificities to sialyloligosaccharides on human cells from clinical isolates was published as early as 1989 [4]. Of all nine existing NA subtypes, only influenza virus types A, B, and C circulate in the human body, specifically N1 from influenza A virus [5,6]. Originally H5N1 was an avian influenza subtype, but, after the 2009/10 "bird flu", this virus has been suspected to be on the brink of zoonosis, risking a viral pandemic spread among humans [7]. Thus, the N1 protein is a prime and timely target for antiviral screening.

Human influenza viral neuraminidase extracts sialic acid (N-acetylneuraminic acid) from the sugar chains of glycoproteins on the human cell membrane surfaces. During initial infection and subsequent reproduction cycles in the patients' mucosa cells, this viral sialidase has two essential functions: (i) it breaks down the mucin in the mucus layer, thereby facilitating viral entry into upper respiratory epithelial cells; (ii) it enables the dissemination—i.e., the "budding"—of newly reproduced virus particles from the host cells. During the last phase of budding, when the new viruses are still attached to host cell glycoproteins, galactose and sialic acid constitute the last two sugar monomers in the sialoglycan receptor on the human cell membrane surface. Both are linked by a glycosidic bond that is enzymatically cleaved by viral NA during budding when detaching offspring viruses from the human host cell. In contrast, viral HA binds to the terminal sialic acid residue of human sialoglycan receptors during the initial stage of infection to attach the incoming virus particle to the host cell surface. However, this HA-mediated virus–host binding can interfere during the late infection cycle when the multiplicated new outgoing offsprings are prevented from escaping the cell surface. Thus, the virus requires the enzymatic intervention of its neuraminidase to enable its separation from the host cell. The viral neuraminidase belongs to the glucosidase family (EC: 3.2.1.18), and has an aspartate and glutamate catalytic dyad in the active site that is responsible for cleaving the glycosidic bond between the hosts' sialic acid and galactose. Precisely, Glu277 attacks the glycosidic bond, forming a cationic intermediate (a carbenium-oxonium moiety) in noncovalent concert with the adjacent Tyr406 [8,9].

The present study describes not only the findings in details, but also the methods that we used to find drug-like molecules combining computational screening, docking, and profiling. We also discuss some of the limitations of virtual screening and the pitfalls to be avoided during virtual screening. Our approach is exemplified by a description of our discovery of small Fmol and AAmol molecules that are predicted to inhibit the budding of influenza viral particles. The target biomolecule belongs to the human influenza A H1N1 virus from the 2009 pandemic outbreak, also known as "Mexican flu" (A/California/04/2009 (H1N1)), as well as the H5N1 subtype, also known as "bird flu" (A/Viet Nam/1203/2004 (H5N1). In the meantime, clinical observation has revealed that H5N1 infection is less common in humans. The motivation for searching for a combination of virus types is intended to identify conserved scaffolds common to all N1 proteins that are unlikely to possess intrinsic target preferences or exhibit mutational resistance, i.e., complications which lie beyond our scope and are addressed by others [10,11]. Our study is embedded in the extant literature on ongoing drug research to find new scaffolds for targeting neuraminidase N1 [11,12].

Prior to virtual screening (VS), a pharmacophore model was created upon assessment of the common binding pattern. To this end, we inspected the known binding modes of four commercial influenza virus N1 inhibitors (oseltamivir, zanamivir, peramivir, and laninamivir). Sialic acid was also included in the study as it constitutes the natural component of host cell membranes, which is recognized as substrate by the viral neuraminidase enzyme. Also under scrutiny was structurally related DANA (Figure S1). In addition, a complete ADMET profile was generated for the two proposed drug-like hits on theoretical ground.

2. Materials and Methods

2.1. Computer Programs

The following computer programs and web-based general tools were used in this study: Vega ZZ [13], Pubchem at https://pubchem.ncbi.nlm.nih.gov/ (first accessed on 24 April 2013) [14], Autodock v. 4.2 with AutoDock tools (ADTs) [15], Discovery Studio v. 4.0 [16], and OpenBabel [17]. In addition, two specialized tools were used for the VS and ADMET profiling of VS hits: Molecular Operating Environment [18] and ADMET predictor v. 7.1 [19] at https://www.simulations-plus.com (first accessed on 29 August 2014).

The Brookhaven Protein Data Bank at http://www.rcsb.org/pdb (first accessed on 5 September 2013) [20] was visited to download the target structure N1 [21]. Moreover, we retrieved crystallized structures of the liganded N2 complex with sialic acid (PDB entry: 2BAT) [22], oseltamivir (PDB entry: 3CL0) [23], zanamivir (PDB entry: 3TI5) [24], laninamivir (PDB entry: 3TI4) [24], peramivir (PDB entry: 4MWV) [25], and substrate-like DANA (PDB entry: 2HTR) [26] as reference ligands with known binding modes and activities.

The water-accessible surface of the N1 active site was calculated at 4.5 Å and potential binding patterns with characteristic chemical properties (features) within the site were identified, i.e., surface locations to form hydrogen bonds, salt bridges, polar regions, and hydrophobic pockets.

The study design embraced four main procedures: (1) pharmacophore modeling, (2) drug-like dataset screening, (3) affinity docking, and (4) pharmacokinetic profiling.

1. To determine the pharmacophore patterns, we studied binding mode specificities and structure–activity relationships (SAR) of hitherto known 3D structure complexes between influenza virus neuraminidase targets and sialic acid substrates or four reference antiviral drugs (oseltamivir, zanamivir, laninamivir, and peramivir) to generate 1D, 2D, and 3D fingerprints used as filters during virtual screening (VS).
2. We also aimed to carry out VS on drug-like compounds by 1D, 2D, and 3D fingerprints. The method facilitates a fully automated (unsupervised) selection of drug-like candidates. To this end, 1-, 2-, or 3D filters have to be predefined or are built-in search tools [27–30]. The input data collection comprises a total of 660,961 small organic molecules (SOMs) [18]. It is composed of basic commercial structures for next-step experimental lead optimization and scaffold diversification upon identifying selected VS hits.
3. The ligand affinities were then calculated to target for the selected VS hits by molecular docking. The self- or back-docking of reference inhibitors compares their predicted poses and affinities with observed (crystallographic) data and validates the computational study, besides docking new molecules to target. The molecular affinities were compared to the viral neuraminidase target by molecular docking each of the following ligands: natural substrates, sialic acid, reference drugs, and VS hits. Interaction energies and affinities were quantitated by means of the inhibition constant (K_i), and the results of the computed affinities were compared with experimental K_i values of reference drugs from the literature.
4. For ADMET profiling, smile codes were created for the VS hits and ADMET data were assessed using ADMET Predictor software.

2.2. Virtual Screening

The search for new candidate drugs through VS can be carried out with in-house molecular databases or public virtual chemical libraries, both of which may store large amounts of molecules and biological metainformation [31]. Alternatively, researchers took a closer look at natural sources for agents against the flu [32]. Here, a commercial substance library with a total of 660,961 chemicals was screened [18]. To cope with the sheer number of data entries, a stepwise approach was established with four search levels by applying different data-type complexity and filtering conditions (Figure 1).

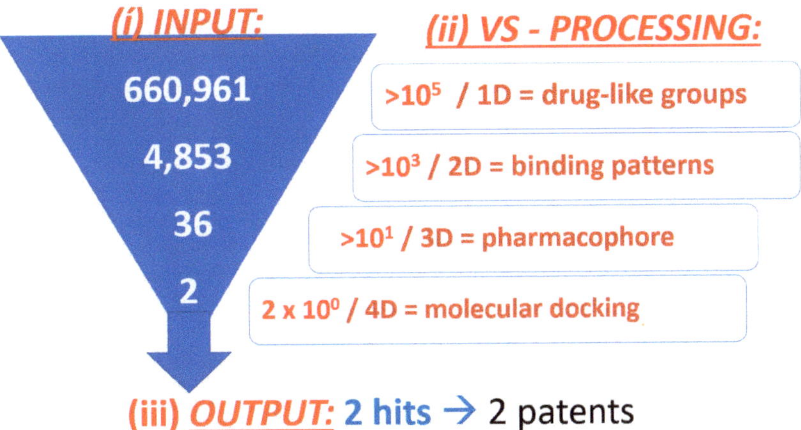

Figure 1. Schematic workflow of the virtual screen. This strategy filters over half a million substances of variable sizes and compositions. Positive and negative controls were manually added to the virtual library for screening. Due to funding limitations, only 2 out of 30 candidates could be experimentally examined for antiviral potency (0.005%, n = 660,961 initial entries). Both hits received patents as promising therapeutic leads. They constitute under-substituted scaffolds and are labeled AAmol and Fmol for short 10^3.

i.   Based on molecular overall features, thousands of chemical substances were eliminated by of their size (molecular weight), lipophilicity (log P), and toxicity (toxic groups, SMILES patterns). Such screening methods are termed one-dimensional (1D) filters.

ii.   All molecules which passed the 1D filter were filtered through topological searches for 2D binding patterns.

iii.   Utilizing active conformations of known ligands at the binding site, a pharmacophore 3D filter was designed and a conformational database of the remaining substances was searched for spatial matches (hits) of atoms, groups, or properties (acidic, basic, polar, nonpolar, ionic, H-bond etc.).

iv.   Finally, the few 3D filter hits were screened by docking simulations, also sometimes called 4D filtering.

Molecular fingerprints are 2D or 3D binding patterns with numeric thresholds to determine whether dataset molecules pass or fail these 2D or 3D filters. We used the chemometric feature known as the Tanimoto coefficient to group or cluster similar molecules in case they exceeded the established similarity threshold (cf. user guide [18]).

To verify if VS can successfully discriminate between target binders and non-binders, we applied retrospective virtual screening: we enriched our dataset with a small number of compounds with known antiviral activities (positive controls) or non-activities (negative controls, also known as decoys). A total of ten positive controls were included: 2-deoxy-2,3-didehydro-N-acetylneuraminic acid (DANA), oseltamivir, peramivir, zanamivir, laninamivir, sialic acid, BANA113, BANA106, BCX1898, and G20. The first six were extracted from the aforementioned PDB files whereas the remaining four were taken from a published source [33]. In addition, 23 negative controls were added from PubChem [14]. First, the following numerical descriptors were calculated for the positive controls: molecular weight; hydrogen bond acceptors (HBAs); hydrogen bond donors (HBDs); acidic, basic, and hydrophobic atoms; and rotatable bonds (RB). Next, we searched PubChem for compounds with descriptor values similar to the positive controls but without reported activities against the neuraminidases of the influenza A virus.

The preparation for retrospective and prospective 2D virtual screening required four steps: (i) the removal of binding-irrelevant molecular components; (ii) protonation/deprotonation at physiological pH; (iii) the calculation of Gasteiger partial charges; and (iv) the calculation of TGD-type fingerprints.

The preparation procedures used for the virtual compound library—the Molecular Operating Environment (MOE) database for retrospective and prospective 3D virtual screening—differed only in the last step to calculate ligand conformations, whereby a stochastic algorithm was used under the MOE's default force field MMFF94x [18]. Its applicability range not only covers drug-like molecules but also proteins. Three parameters were set as follows: (i) residual strain energy limits at the local minimum: 10 kcal/mol; (ii) the maximum number of conformations per molecule: 1000; and (iii) the root mean square distance (RMSD) threshold to remove duplicates: 0.15.

2.3. Molecular Docking

The target biomolecule was prepared for docking by removing undesired moieties from the crystal input structure using SPDBV. The protein's atom partial charges on all amino acid atoms were computed using the Gasteiger approach of ADT.

All ligand models for docking were preprocessed under Vega ZZ in order to assign correct bond orders, hybridizations, hydrogen atoms, atom types (Tripos), and partial charges (Gasteiger), and the protonation/deprotonation state at pH 7.

For flexible ligand docking, default program settings were used under the Lamarckian genetic algorithm, except for the number of runs (256), elitism (3), and the highest precision level (25,000,000). The numerous peptide bonds were held rigid in their natural trans configuration, i.e., they were not allowed to rotate.

2.4. ADMET Profiling

After screening and docking, pharmacokinetic characterization was carried out for the final candidates. Their experimentally confirmed antiviral activities are reported elsewhere [34].

Their 3D structures were converted into the SMILE code or mol file format under Vega ZZ, which were then used as input data for the ADMET predictor™ tool. The user manual describes the "expanded applicability domain" after including an in-house dataset from Bayer™. The improvements in the tool became evident "in prospective predictivity of S+ pKa certainly reflect [the tool´s] expanded applicability domain" [" … " cited from the ADMET predictor manual, 23 July 2014, version 7.1].

3. Results

3.1. Binding Pattern and Pharmacophore Modeling

The selected biomolecular target was the three-dimensional structures of the influenza A virus neuraminidase, corresponding to the viral strain A/Vietnam/1203/2004 (H5N1). As reference binders for target screening and docking, we retrieved the published crystal structures of four N1 inhibitor drugs in the complex with the target protein: zanamivir, oseltamivir, peramivir, and laninamivir (Table 1). In addition, the crystal complex of the N1 target with natural substrate sialic acid was also retrieved from the Protein Data Bank (PDB). Specifically, we inspected the following related PDB entries: 3CL2, 2HU0, 3B7E, 3CKZ, 2HTU, 2HTW, 2HTR, and 3NSS.

Table 1. Input data from the human influenza A virus for 3D model generation from experimentally observed crystal structures of the reference target protein and ligands. Asterisk (*): the target primary sequence corresponds to a single-point mutant Iso223Val (I223V) or (**) His274Tyr (H274Y). As an off-scope effect, these changing amino residues also give knowledge about published resistance mechanisms [21].

Extracted 3D-Structure	PDB Code	Types of Molecules and Viruses	Resolution (Å); Year	Ref.
Neuraminidase (target protein)	5NZ4	OS—liganded neuraminidase N1; unidentified strain; (*)	2.2; 2018	[21]
Sialic Acid (SA)	2BAT	SA—N2 complex; influenza A virus; A/Tokyo/3/1967 (H2N2)	2; 1992	[22]
Oseltamivir (OS)	3CL0	OS—N1 complex; influenza A virus; influenza A virus; A/Viet Nam/1203/2004 (H5N1) (**)	2.2; 2008	[23]
Zanamivir (ZA)	3TI5	ZA—N1 complex; influenza A virus; A/California/04/2009 (H1N1)	1.9; 2011	[24]
Peramivir (PE)	4MWV	PE—N9 complex; influenza H7N9 virus; human-infecting variant from avian origin	2.0; 2013	[25]
Laninamivir (LA)	3TI4	LA—N1 complex; influenza A virus (A/California/04/2009 (H1N1)	1.6; 2011	[24]
DANA	2HTR	DANA—N8 complex; influenza A virus (unspecified strain)	2.5; 2006	[26]

The fundamental tenet of (quantitative) structure–activity relationships declares that similar chemical structures reflect similar biological activities, although so-called activity cliffs create exceptions to the rule [35,36]. Therefore, we analyzed the binding modes at atomic scale of all five reference ligands from Table 1 (Figure 2).

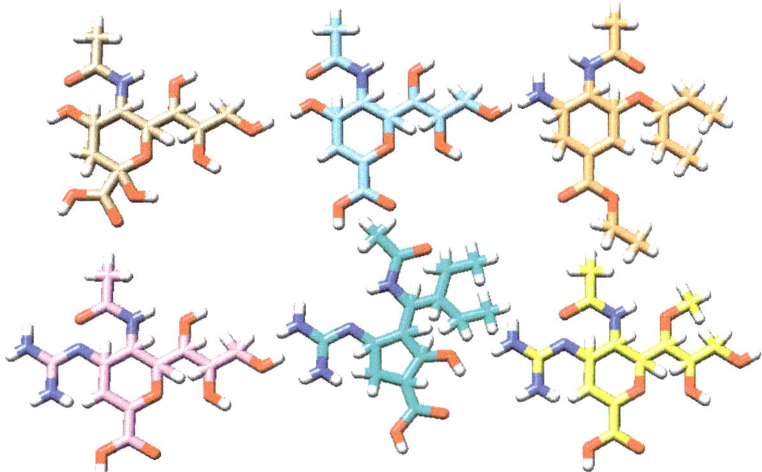

Figure 2. The endogenous substrate and five competitive reference ligands. The coloration and 2D orientation allows structural comparison by eye. Top row, left to right: sialic acid, DANA, and oseltamivir; bottom row, left to right: zanamivir, peramivir, and laninamivir. Color coding: white—H atoms, red—O atoms, blue—N atoms, and other colors—C atoms.

All reference inhibitor modes of binding are fairly similar to the natural substrate sialic acid (Table 2). In the Supplemental Material (SM), we document all details on ligand–receptor interaction patterns (Table S1), which are in line with literature describing essential residue active conformations for ligand recognition [37]. To be more precise, ionic (and polar) head group atoms of Glu276, Glu277, Arg292, Asn294, and Ser246 mark the location of a large cavity. A smaller hydrophobic pocket is surrounded by Arg224 and Ala248. A larger pocket encompasses hydrophilic and hydrophobic residues Glu119, Asp151, Arg152, and Glu227. A fourth, negatively charged pocket is composed of residues Arg118, Arg292, and Arg371 (Figure S1).

Table 2. Synopsis of neuraminidase amino acids which interact with the five reference binders and the two hits. Residue numbering scheme adopted from PDB entry 3CL0. Residues are grouped to reflect their spatial packing and ordered from ionic to nonpolar. Abbr.: 3OHprop = tri-hydroxy-propyl; BB = peptide backbone; IA = interaction.

Residue	Sial ac.	DANA	Osel	Zana	Pera	Lani	AAmol	Fmol
Arg118	-COO (-)	-COO (-)	-COO (-)	-COO (-)	-COO (-)	-COO (-)	BB amide	BB amide
Arg292	-COO (-)	-COO (-)	-COO (-)	-COO (-)	-COO (-)	-COO (-)	-COO (-)	-COO (-)
Arg371	-COO (-)	-COO (-)	-COO (-)	-COO (-)	-COO (-)	-COO (-)	acetamido	BB amide
Arg152	acetamido	acetamido	acetamido	acetamido	acetamido	acetamido	-COO (-)	-COO (-)
Arg224	-	tri-hydroxy-propyl	-	▪	▪		COO ( )	COO ( )
Glu276	tri-hydroxy-propyl	tri-hydroxy-propyl	-	tri-hydroxy-propyl	-	-	-	-
Glu277	-	-	-	guanidino	Guanidino	guanidino	-	-
Glu119	-	-	-NH3 (+)	guanidino	Guanidino	guanidino	-	-
Asp151; -CH- not OO	2-hydroxy on oxane ring;	-	-NH3 (+); none	guanidino; none	guanidino; 2-hydroxy on cyclopentane	guanidino; none	amido; -S-CH3	piperazinyl; none
Ser246	-	-	[no IA with alkyl]	tri-hydroxy-propyl	-	2,3-dihydroxy-1-methoxy propyl	-	-
Asn294	tri-hydroxy-propyl	same as left but Gly294 on N2 prot.	-	tri-hydroxy-propyl	-	2,3-dihydroxy—1-methoxy propyl	-	-
Tyr347			-COO (-)	-	-	-	acetamido	-
Tyr406	ether-O- in oxane	ether-O- in pyran	–	ether-O- in pyran	-	ether-O- in pyran	-	-
Val149 Ile 427 Pro431	-	-	-	-	-	-	phenyl	di-methyl-phenyl
Ala248	-	[no IA tri-hydroxy-propyl]	-alkyl	-	2-ethylbutyl	-	-	-

The X-ray structure of the neuraminidase subtype N2 with co-crystallized sialic acid indicates that the substrate binds the enzyme in a considerably deformed conformation due to strong ionic salt- and hydrogen-bonding energies exercised through the substrate's anionic carboxylate group. The latter is in contact with three cationic side chains of Arg118,

Arg282, and Arg371. The N-acetyl group, attached to ring atom C5, maintains polar and nonpolar contact with Arg152, Trp178, and Ile 222. These interactions help anchor the substrate to the active site. The C4-OH group of sialic acid hydrogen bonds with the negatively charged Glu119, Glu227, and Asp151. In docking simulations of the natural substrate sialic acid, we analyzed the pivotal interactions in the N1 target active site. Remarkably, there is a concert of strong salt bridges and elaborated networks of polar hydrogen bonds between the ligand's anionic carboxylate group and three cationic arginine residues (Arg118, Arg282, and Arg371). In addition, a fourth arginine Arg152 interacts with the substrate's acetamide group. The stabilizing noncovalent hydrogen bond is formed between anionic glutamate Glu276 and two hydroxyl groups on Carbon atomsC8 and C9 of the ligand´s triol side chain. Of note, Glu276 is conserved in N1 and N2.

It is evident that all five sialic acid analogues share an aliphatic ring, a carboxylic group, an acetamide, and hydroxyl groups or other oxygenated functions for hydrogen bond networking. These findings were merged with other findings into the 2D and 3D fingerprint (pharmacophore) models. Figure S1 exemplifies the graphical analysis for oseltamivir. Based on the calibration results, pharmacophore model 24 was chosen to perform the prospective 3D virtual screening. Stereochemical drawings of the reference ligands (Figure S2) were aligned (superposition) to generate 3D pharmacophore models (Figure 3; shown in detail in Figure S4). This allowed us to discriminate all 10 active molecules from the 23 inactive decoys. Moreover, model 24 obtained the best metrics score during evaluation.

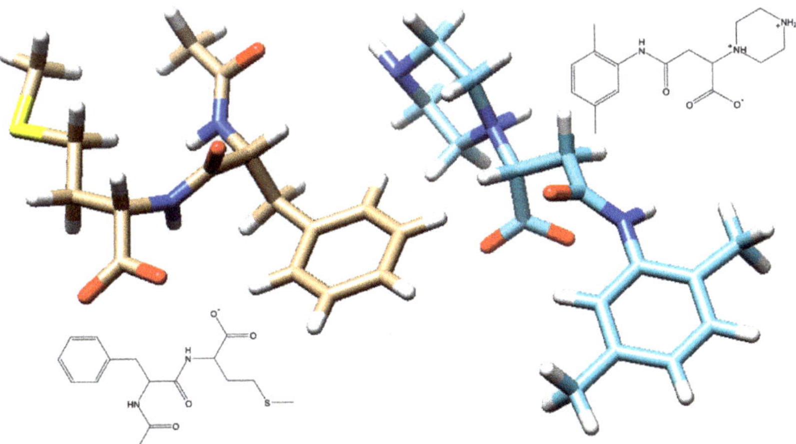

Figure 3. Three-dimensional stick models of the final hits with antiviral activity. AAmol (left) and Fmol (right) were selected among tens of final candidates from virtual screening. Fmol has two protonation siteson the tertiary and secondary amine groups on the piperazine ring. At physiological pH one is protonated while the carboxyl group is deprotonated. Two inlays show the structural 2D drawings of AAmol (below) and Fmol (above). The color coding for H, O, N, and S atoms: white, red, blue, or yellow, respectively. Carbon atoms are beige (AAmol) and light blue (Fmol).

3.2. D Filtering (2D Fingerprint Design)

Several fingerprint models were constructed from the two-dimensional features of the following ligands: sialic acid (2BAT), oseltamivir (3CL0), peramivir (2HTU), zanamivir (2B7E), and DANA (2HTR). The fingerprint filter is two-dimensional, so the conformations of the molecules and the RMSD between them were not included. The molecules were processed, as indicated in the Methods section. Subsequently, the 2D typed graph distance (TGD) fingerprints of the chosen molecules were calculated.

During retrospective virtual screening, all fingerprint models were constructed and underwent repeated rounds until an optimized outcome from the control test set was achieved. Finally, one best-scoring model was selected, i.e., that with the highest hit rate to discriminate between the active and inactive compounds.

3.3. D Filtering (3D Fingerprint Design)

To determine the essential set of 3D structural features to include in the pharmacophore models, the reference ligands underwent superposition operation, the RMSD values were calculated, and the overlapping features were deemed pivotal. These "frequently conserved" ligand substructures were merged to formulate the pharmacophoric models such that the several initial models were extracted from the co-crystallized ligands (with PDB codes): sialic acid (2BAT), oseltamivir (3CL0, 2HU0), peramivir (2HTU), zanamivir (2B7E), and sialic-acid-related DANA (2HTR). Since 3D substructures had been extracted from (3D) crystal data, all binding-relevant features could be located in 3D spaces (i.e., hydrophobic, anionic, and cationic chemical substructures) or hydrogen bond donor or acceptor (HBD or HBA, respectively) groups. Again, pharmacophore models were repeatedly evaluated against the enriched dataset by retrospective virtual screening. The model with the best activity vs. nonactivity discrimination capacity (model number 24) was finally selected.

3.4. Prospective 2D VS

The final 2D fingerprint model was applied to screen the 2D data-type version of the MOE database [18]. This approach reduced the original ~10^5 chemical entities of the MOE database to a smaller set by a 2D filter, which runs many thousand-fold faster than 3D filtering (Figure 1).

3.5. Prospective 3D VS

Pharmacophore model 24 was applied as a 3D filter to screen the 2D VS hits in order to obtain a final set of ~10^1 candidates after clustering. To this end, the conformation of the lowest RMSD value with respect to its underlying pharmacophore model was computed for each 3D hit. Specifically, the TAT-type fingerprint of each 3D hit was calculated. The hits were lumped together using the Tanimoto coefficient, and the molecules that exceeded an arbitrary 85% similarity threshold to other candidates (with an identity match set to 100%) were not included for further study.

Fully automated multi-step VS led to a considerable data size reduction (Figure 1). After screening, two molecules of the 30 remaining were of limited presence in academic literature, and thus were worthy of further scrutiny. Hence, the molecular docking of 30 candidates against the viral target led to two high-scoring candidates (Fmol and AAmol). We kept their nicknames: AA stands for two amino acids and F stands for the other (dimethyl-) phenyl group (written in Spanish as "Fenil") (Figure 3). In collaboration with a biomedical research center (CIBIOR, Metepec, PU, Mexico), they were experimentally validated as active inhibitors and published together with other inhibitors from studies on drug repurposing or structure–activity relationships [34].

AAmol, with the chemical name (N-acetyl-phenylalanyl)-methionine, possesses two amide bonds. It is a dipeptide derivative (Phe and Met), has a total charge of −1 as it takes a deprotonated carboxylate form in water at pH 7, and interacts with arginine residues at the binding site. Its molecular mass is 337 Daltons; with an estimated log P = 1.2; a polar surface area of 183 Å^2; a non-PSA of 441.9 Å^2; a HBA = 4; HBD = 2; highly flexible conformations (12 rotatable bonds); and a commercial vendor: Chembridge # 6429718. A recent review outlined the fundamental principles concerning peptide inhibitors against the flu [38]. One 24-residue-long oligopeptide binds NA with a micromolar inhibition constant (Ki = 0.29 mM), and a much smaller octapeptide was designed as a strong nanomolar binder at the active site where oseltamivir appears, showing one-digit micromolar inhibition activity in the cell tests. In this context, AAmol, with only two amide bonds, is an even smaller H1N1 NA inhibitor than all the other reported peptides (4 to 24 residues long) [38].

Fmol, or 3-[(2,5-dimethyl phenyl) carbamoyl]-2-(piperazin-1-yl) propanoic acid, has two rings and also contains a monoanionic carboxylate group to interact with the cationic active site arginine residues. The piperazine ring possesses two potential protonation sites and exists as a di-cationic species in strongly acidifying media, but the monocationic form dominates under cellular conditions. With a mono-anionic and cationic center, it is a sort of zwitterion at pH 7, with a molecular weight = 306 Daltons; a log P = -3.4; a polar surface area = 142 $Å^2$; a non-PSA of 427.9 $Å^2$; a PSA-to-non-PSA ratio = 1:3; a HBA = 4; a HBD = 3; flexible conformations (seven rotatable bonds); and a commercial vendor: Life Chemicals F1278-0516.

3.6. Virtual Library Performance under Fingerprint Model Number 24

Starting with 660,961 entries in the MOE database, the 2D screening yielded 4853 hits (0.7% of the drug library), whose subsequent conformational calculations were then further virtually screened through the 3D fingerprint filter (for details, cf. Tables S2–S6 concerning dataset test characteristics, control structures and decoys, dataset enrichment metrics, further metrics on positional fit by RMSD, as well as pharmacophore modeling and thus the associated performance).

In order to score and evaluate the post-3D-filter hits they were docked against the influenza A virus N1 neuraminidase receptor. For this post-screening ranking by docking scores only a numeric evaluation of the hits was requested. Specifically, an inspection of conformational space and steric requirements was not intended for that stage. The study did not focus on the search for favorable positions or conformations (since the 3D filter yields very specific steric, electronic, and spatial conformations), but rather on the ranking of selected hits.

Successful self- or back-docking tests gave reason to assume that the blind docking of unknown active conformations of the hit ligands will be faithful and biochemically meaningful (Figure S1 and Table S1). Validation took place with reference ligands extracted from liganded crystal structures (with the following PDB codes): sialic acid (2BAT), oseltamivir (3CL0), oseltamivir (2HUO), peramivir (2HTU), zanamivir (2B7E), and substrate analog DANA (2HTR). The hits were blind-docked and ranked. The computed inhibition constant of oseltamivir (K_i = 0.007 µM) lies within the experimental range reported with PDB entry 3CL0 (0.0001 to 0.008 µM). We add another proof of concept that uses K_i values for our docking approach (for details on docking precision, see the Discussion section below). The docked ligand successfully reproduces specific contacts from the experimental structure—between the acetamido function and conserved arginine Arg152—despite the changes in docked overall positions which could be expected from scaffold hopping in our analyses (Figure S1) [39].

3.7. Ligand–Target Docking

The 3D models were curated in the Vega ZZ program [13], and their ionization states were estimated. A total charge of -1 can be attributed to both AAmol and Fmol (each with one carboxylate group), while a zwitter form ($+1/-1$) was ascribed to Fmol, i.e., an additional tertiary ammonium next to the carboxylate function. At pH 7, the piperazinyl ring is a mixture of monoaninic > neutral >> dianionic species following the rule of thumb for the dissociation of (diluted) weak acids or bases: pKa-pH yields the % concentration for neutral bases (or deprotonated corresponding acidic forms, respectively). At that stage of work, partial charges were loaded by the Gasteiger approach under Vega ZZ and the Tripos force field atom types assigned prior to saving in the mol2 file format.

All five reference models could be successfully docked back into their crystal poses (Tables 1 and 2). For sialic acid (PDB entry: 2BAT), no experimental binding affinities have been published, but experimental affinity data of the four others have been included in their PDB entries. The experimental binding constants ranged from 0.1 to 817 nM for oseltamivir (PDB entry: 3CL0) and from 0.5 to 12 nM for zanamivir (PDB entry: 3TI5). Their computed best-scoring values from our docking studies lie in the same one-to-three-digit nanomolar range (7.5 nM to 363.7 nM). Reported experimental IC$_{50}$ values of peramivir (PDB entry: 4 MWV) and laninamivir (PDB entry: 3TI4) were 0.4 nM and 0.947 nM, respectively, which can be compared to their computed nanomolar inhibitory constants of 109.9 nM and 743.1 nM For the two hits under scrutiny, AD4 found the following affinities: (i) AAmol with a lower one-to-two-digit micromolar range; and (ii) Fmol with an upper nanomolar range, with the lowest K$_i$ of 0.1 mM and an average value of 0.8 mM for the most populated, best-scoring (first) cluster. For more details, see the Discussion section below. The final poses of the blind docking of AAmol and Fmol were compared to the reference complexes (Figures 4 and 5).

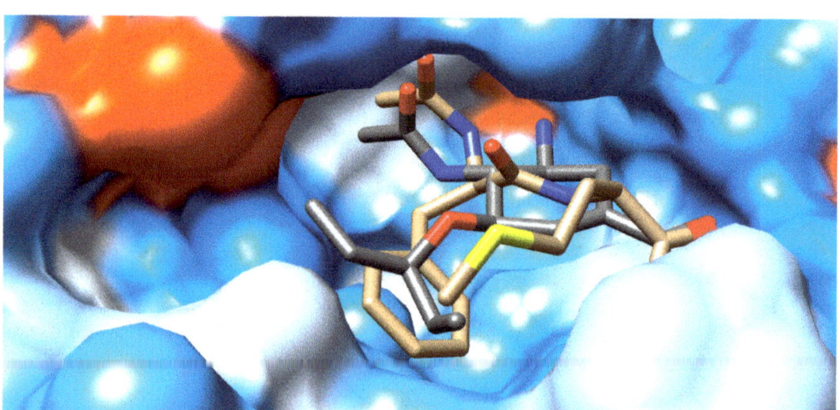

Figure 4. Structural binding model of AAmol (beige), with co-crystalized oseltamivir (grey) for reference. The cavity entry lies to the right hand coming from the foreground. Color coding for the stick models: red, blue, and yellow for O, N, and S atoms, respectively; beige or grey for carbon atoms of AAmol or oseltamivir, respectively; hydrogen atoms are omitted. The water-accessible protein surface is also colored: light blue for hydrophilic (polar) residues, orange for hydrophobic (nonpolar), and white for intermediate polarity.

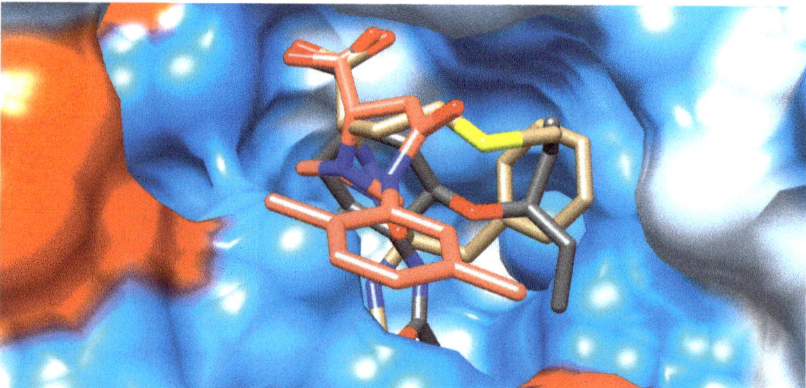

Figure 5. Structural models of binding modes for AAmol (beige), Fmol (pink), and co-crystalized oseltamivir (grey) for reference. Top-down view into the binding cavity. The viewing angle from Figure 4 is tilted 90° for orthogonal viewing. Coloration is shown in Figure 4.

Figure 4 illustrates the equivalent role of three structural features which are shared by AAmol and oseltamivir. They effectively occupy the same cavity locations: (i) acetamido; (ii) alkyl-ether (-CH$_2$-O-CH$_2$-) and alkyl thioeter (-CH$_2$-S-CH$_2$-); and (iii) anionic carboxylate groups (-COO$^{(-)}$). In the figure, both acetamide side chains (i) lie to the rear (topmost). The methyl group on AAmol (beige) is not visible, occluded by the acetamido group of oseltamivir which lies in the front. In the middle, both O/S-ether bridges (ii) are fully visible. The methyl-thioether group belongs to AAmol in the front (foremost, sulfur atom in yellow). Intriguingly, the three atoms of both carboxyl groups (iii) are in almost perfect superposition, presumably reflecting their role in ligand recognition as they help bind to the same cationic residues (Table 2), though only one oxygen atom from each of the COO$^{(-)}$ groups can be seen (two overlapping rightmost red O atoms). CH-rich (aliphatic) substructures on both ligands squeeze to the left into the hydrophobic area. The extended lipophilic pocket remains unoccupied by both binders (cf. Val149, Ile 427, and Pro431 in Table 2). Figure 5 complements Figure 4 in additionally presenting Fmol and a top-down view on the entire carboxyl group. Their three atoms (-COO) appear perfectly aligned (Table 2). In the mid-section, the alkyl ether (oseltamivir) and thioether (AAmol) are fully visible as well. To the cavity's side wall, N- and O-free alkyl parts on all ligands meet with more neutral residues (white surface), while the deeper, more hydrophilic rear (bluish surface) provides for more affinity for N- and O-rich substructures. The cavity entry coincides with the viewer's perspective. Fmol's dimethylphenyl ring extends between two hydrophobic areas in the forefront. In this way, it partly occupies the extended lipophilic pocket with Val149, Ile 427, and Pro431 (Table 2).

The interacting amino acids at the binding site of N1 were inspected in the 3D models with the docked poses and displayed in a 2D scheme (Figures 6 and 7). In particular, AAmol's docking shows that its acetamido group—which is part of the pharmacophore model—is correctly recognized by the Arg152 of the N1 protein target (Figure 6) [39]. Intriguingly, in Fmol, this amide bond is stabilized by a strong pi-electronic resonance effect since its nitrogen atom also belongs to the aromatic system of the dimethyl phenyl aniline substructure. Considering the latter as a sidechain and the remaining molecule part as the scaffold, the amide group orientation appears reversed to better fit into the cavity (Figures 5 and 7).

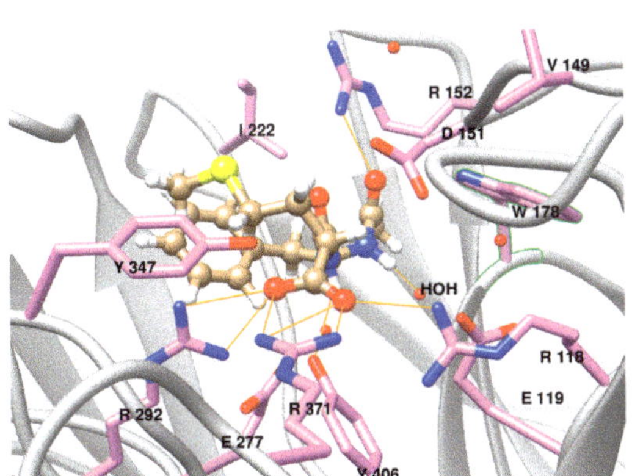
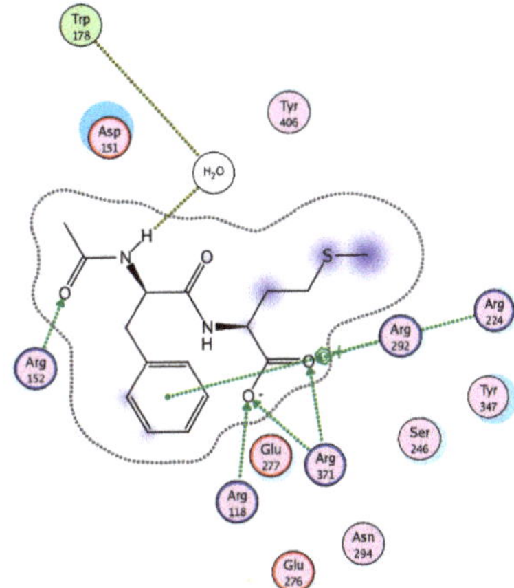

Figure 6. Schematic representation (**left**) and graphical display (**right**) of N1 binding site with docked AAmol (beige ball-and-stick model). The cavity entry lies to the foreground left. Color code for the stick models: pink, red, and blue indicates C, O, and N atoms, respectively; grey ribbons represent the protein backbone; residue hydrogen atoms are omitted; additional colors are shown in Figure 4.

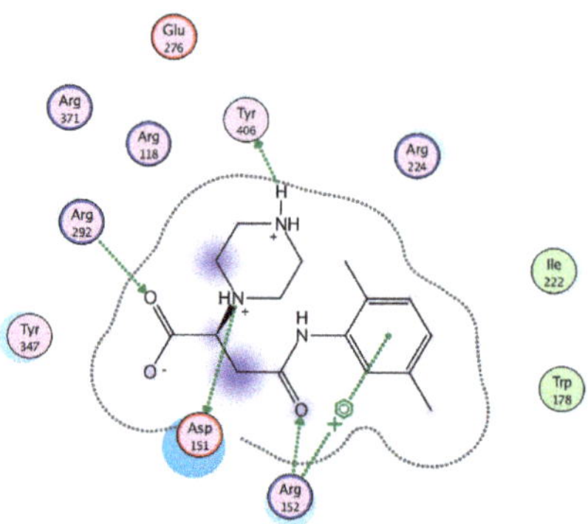

Figure 7. Graphical display of the N1 binding site with Fmol (pink). The cavity entry lies to the foreground left. Ile222 and Trp178 flank the hydrophobic dimethylphenyl side chain of the ligand to the right in the left-hand panel or to the left in the right-hand panel, respectively. Color coding is shown in Figures 4 and 6.

3.8. ADMET Profiling

This theoretical pharmacokinetic profiling step was carried out as a post-screening evaluation of hits for early candidate attrition. In the context of long and costly drug research and development projects (R&D), in silico profiling studies are routinely carried out (Table 3).

Table 3. Estimated pharmacokinetic data for AAmol and Fmol. Abbreviations: BBB, blood–brain barrier; CYP, cytochrome P450 enzymes; MRTD, id, atom number of substrate; lethal doses; perm, permeability; sol, solubility % (m/m); Vd, volume of distribution; w, water. Asterisk (*): even in its neutral form, Fmol remains hydrophilic, so its log P value becomes negative. Its intramolecular prototropy causes a zwitterionic form with total charges of +1 and −1 formally summating to zero. In contrast, AAmol can appear in a buffered solution as an undissociated (neutral) species with a positive log p value whichreflects its overall lipophilicity.

Name	Acidic pKa	MlogP (Neutral Form)	logD (ionized)	Perm Skin	Solu w
AAmol	3.8	1.0	−1.6	6.85	0.9
Fmol	11.3; 4.0	−1.6 (*)	−1.4 (*)	0.01	4.6
Name	**pH in w**	**BBB_Filter**	**Vd in L/Kg**	**RuleOf5**	**CYP_1A2**
AAmol	3.24	Low	0.22	0	No (96%)
Fmol	6.89	Low	0.54	0	No (96%)
Name	**CYP_2C8**	**CYP_2C8 (id)**	**CYP_2C9**	**CYP_risk**	**TOX_MRTD**
AAmol	Yes (73%)	S19(992); C20(869); C4(828)	No (56%)	0	Above_3.16
Fmol	No (92%)	NonSubstrate	No (98%)	0	Above_3.16
Name	**TOX_hERG**	**TOX_ER**	**TOX_rat**	**TOX_skin**	**TOX_biodeg**
AAmol	No (95%)	Nontoxic	2066.07	Nonsensit. (75%)	No (63%)
Fmol	No (95%)	Nontoxic	941.78	Nonsensit. (85%)	No (96%)

4. Discussion

4.1. Implications and Limitations for Drug Screening

Computational molecular simulation methods have ushered a new area of drug R&D in academia and industry. In silico approaches are popular, but currently generalized and in need of improvement. They are not intended to replace in vitro, ex vivo, or in vivo studies, but, rather, they should be understood as complementary tools within the drug discovery cycle. Pharmacophore modeling—specifically, molecular dataset screening for hits and blind docking of hitherto unknown target binders—is one such in silico approach which was validated herein by elaborated protocols, namely (i) analyzing binding patterns of published crystal complexes, (ii) fine-tuning of 2D and 3D filters and dataset enrichment with active controls and decoys, and (iii) back docking trials of crystalized ligand-target complexes. Bioassay results have been published to confirm the micromolar antiviral activities of our final candidates.

In general terms, VS saves time, material, human, and financial resources. However, this cutting-edge approach requires pre-existing information on drugs, compound libraries, and biomolecular receptors on an atomic scale. In short, as with any screen, the results are only as comprehensive as its inputs.

Although the filters used during VS had been selected after testing in the presence of known active and inactive control structures in order to perform a more selective screen, not finding hits with nanomolar target affinity does not necessarily jeopardize success, as high potency ought not to be expected in VS campaigns in general. Pragmatically, the ultimate goal of VS is to find hits or lead compounds for further research and validation by means of distinguishing between non-binders and binders—not necessarily between strong and weak binders (cf. pitfall (i) below).

To the best of our abilities, this VS study concept is designed to avoid known pitfalls [40]. In particular, we worked around the following setbacks:

Pitfall (i): the success criteria of VS are insufficiently defined; solution (i): focusing on new scaffolds rather than high target affinities, which would be improved in a subsequent step of scaffold derivatization (drug profiling with design, synthesis, and testing).

Pitfall (ii): variable water-mediated binding interactions; solution (ii): in analogy with the situation illustrated in Figure S1 Panel B with reference ligand zanamivir in interaction with a water moiety, we present the water-mediated AAmol interaction in Figure 6 with the graphical display of AAmol as a N1 binder.

Pitfall (iii): the rigorous and prospective validation of VS protocols; solution (iii): challenging our pharmacophore models with positive and negative control molecules in a virtual test library (Table S3).

Pitfall (iv): overcautious approaches to 'drug-likeness'; solution (iv): disregarding the "rule of five", which is a retro-perspective result of averaged values on a drug sample (statistics), though it is not a prospective rule which must be followed (i.e., selection bias). As a direct result, Fmol is an unprecedent case of an aromatic-ring-bearing compound that is not seen on hitherto known N1 inhibitors. Moreover, the acetamido side chain still conserved on AAmol was incorporated into a larger substructure and its orientation was inverted to enhance cavity fitting (Figure 3).

Pitfall (v): one-at-a-time approach; solution (v): limited binding pattern complexity (2D or 3D filter definition) upon the application of a single active compound for pharmacophore generation.

Pitfall (vi): meaningless binding pattern selection for pharmacophore design; solution (vi): testing the filter capacity to discriminate between active and inactive control molecules. The former should appear in the hit list VS, while the latter should not, as also shown in solution (iii).

Pitfall (vii): the data size and structural variety do not reflect an appropriate variety of chemical spaces; solution (vii): the VS aimed to find new under-substituted scaffolds. Thus, possible hits with structural variations were searched in a chemical landscape of over half a million simple small organic compounds, as also shown in solution (i).

4.2. Implications and Limitations for Ligand–Target Docking

Upon agonist or substrate binding, certain receptor types—such as nuclear steroid hormones, glucocorticoid receptors, or Cyp P450 enzymes—unfold side chains, backbones, or even domains in a phenomena called an induced fit [41]. In response to AD4's treatment of flexible ligands and rigid body receptors, authors have proposed to act of docking apo-forms and various liganded complexes to account for spatial differences [42–44]. Others have used molecular dynamics to adopt new conformations by educated guessing [45,46]. None of the cited solutions were followed here for a two-fold reasoning: (i) the hits are all smaller in size than the six reference ligands; and (ii) the binding patterns of hits are merely a subset of those observed in the corresponding crystal complexes (Table 1). As such, no conformational rearrangements were needed.

AD4 computes free binding energies (ΔG) as a crude approximation from linear scaling on rotatable bonds of the entropy term (ΔS). The tool converts these thermodynamic estimates into inhibition constants (K_i) by applying the thermodynamics formula $K_i = e\hat{}[\Delta G/(R^*T)\,]$. Moreover, K_i values are not always close to IC_{50} values, which depend on the substrate concentration in the assay. The estimated K_i value for the endogenous sialic acid substrate lies in a two-digit micromolar range. Referring to the reference compounds, their micromolar K_i values all lie in one-digit ranges of the best-scoring clusters.

For AAmol, the experimentally determined inhibitor constant was published with a K_i value of 15 mM [34]. The best value in our docking simulation for AAmol was $K_i = 0.001$ mM, with an average value for K_i of 0.12 mM, which is approximately a 100-fold overestimation of potency. With regard to calculation precision vs. experimental data, the docking program tutorial states that the numeric results are merely crude estimates in a wider 100-fold imprecision range (corroborated by private messages from Autodock scientist Prof. Dr. Stefano Forli, Dept. of Comp Biol, Scripps Research Institute, La Jolla, CA, USA)—a fact which is overlooked all too often by published AD4 studies which claim that their numeric result lies in excellent keeping with assay data as a misleading "proof of concept". Our proof-of-concept stems from the fact that the acetamido group on AAmol is recognized, as reported by the crystal complexes.

For Fmol, only a IC_{50} value was measured against N2, not N1. Given the experimental settings, the IC_{50} data of Fmol are not directly comparable to the K_i values. Of note, IC_{50} does not reflect directly binding affinities, but could be set on an equal footing by the Cheng–Prusoff equation, which is, unluckily, in need of additional experimentally determined parameters.

4.3. Implications and Limitations for ADMET Modeling

The predicted numerical results from ADMET profiling are crude estimates. The program's artificial neural network architecture has come under criticism due to variable degrees of reliability. The latter depends on the applicability range of each ADMET parameter, which, in turn, is given by the training set for calibration, i.e., adjusting the outcome to some experimental (measured) endpoint. To complicate reliability, even closely related structures—e.g., warfarin and phenprocoumon—fall short of expectations (an unpublished pitfall during our phenprocoumon research [47]). One explanation for ill-behaving cases (pitfalls) is that overfitting tends to increment the prediction power, and in the same time narrows the applicability range to only those entries of the underlying training set used during data fitting. As a direct consequence, a trade-off must be arranged for an "in-between solution. Stray outliers lie in the nature of this method. At the essence of the pros and cons discussion here, it is safe to presume that VS results do not foresee pharmacokinetic issues with absorption, distribution, metabolism, excretion, or toxicity.

5. Conclusions

More than half a million small organic compounds from a commercial dataset were virtually filtered in four stages. One-, two-, and three-dimensional pattern filters were utilized and verified by enriching the dataset with molecules of known activity or inactivity, until the screen could distinguish between the active and inactive controls. A final (4D) filter was imposed with a ligand–enzyme docking for the best-performing hits from the virtual screen. Two such top-ranked substances underwent successful patent filing (details on the experimental validation having been published elsewhere). In line with the present study design and choice of input data, novel lead compounds were used with simple under-substituted scaffolds of micromolar target affinities, representing a promising future of ongoing R&D cycles of synthetic derivation and biochemical testing to develop stronger and more specific target binders.

6. Patents

The two concluded national patents have the following identification numbers: (i) for AAmol: MX/E/2017/039727, IMPI no. 352708 (2018); (ii) for Fmol: MX/E/2017/034353, IMPI no. 352709 (2018).

Supplementary Materials: The following supporting information can be downloaded at: https://www.mdpi.com/article/10.3390/v15051056/s1. Figure S1: Reference ligand interaction. Figure S2: Stereochemical drawing of natural substrates with the five reference ligands. Figure S3: Ligand superposition for pharmacophore construction. Figure S4: Pharmacophore model. Table S1: Oseltamivir interaction. Table S2: Library test characteristics. Table S3: Control structures. Table S4: Enrichment factors. Table S5: Positional evaluation (RMSD). Table S6: Pharmacophore models. Table S7: Performance evaluation [48,49].

Author Contributions: Conceptualization, Y.M.-L. and T.S.; methodology, T.S.; software, T.S. and K.C.-C.; validation, K.C.-C. and T.S.; formal analysis, A.A.I.; investigation, T.S., K.C.-C. and A.A.I.; resources, A.A.I.; data curation, A.A.I.; writing—original draft preparation, T.S. and K.C.-C.; writing—review and editing, T.S.; visualization, T.S., K.C.-C. and A.A.I.; supervision, T.S.; project administration, Y.M.-L. and A.A.I.; funding acquisition, Y.M.-L. All authors have read and agreed to the published version of the manuscript.

Funding: This research received no external funding.

Institutional Review Board Statement: Not applicable.

Informed Consent Statement: Not applicable.

Data Availability Statement: Experimental results published at https://www.mdpi.com/1420-3049/25/18/4248 and at https://doi.org/10.3390/molecules25184248 (accessed on 17 April 2023). Patent document for AAmol at https://vidoc.impi.gob.mx/visor?usr=SIGA&texp=SI&tdoc=E&id=MX/a/2014/001768 (accessed on 20 April 2021). Patent document for Fmol at https://vidoc.impi.gob.mx/visor?usr=SIGA&texp=SI&tdoc=E&id=MX/a/2014/001759 (accessed on 20.04.2021).

Acknowledgments: We are very grateful for the personal support and scientific advice given to T.S. by Thierry Langer, Faculty of Life Sciences Department of Pharmaceutical Sciences, Univeristy Vienna, Austrich, as well as Prestwick Pharmaceuticals, Illkirch, F-67 Strasbourg, France, with assistant Fabrice Garrido. For patent filing management, we thank Martín Pérez Santos, Transferencia Tecnologica, DitCo, BUAP, Puebla, Mexico. We also wish to thank graduate student John H. Gardiner of the Timothy C. Mereidth Lab at Penn State University for English language revision thanks to his linguistic talent and proclivity to grammatical and lexical issues.

Conflicts of Interest: The authors declare no conflict of interest.

References

1. Lee, Y.; Hoang, P.; Kim, D.; Ayun, R.; Luong, Q.; Na, K.; Kim, T.; Oh, Y.; Kim, W.; Lee, S. A Therapeutically Active Minibody Exhibits an Antiviral Activity in Oseltamivir-Resistant Influenza-Infected Mice via Direct Hydrolysis of Viral RNAs. *Viruses* **2022**, *14*, 1105. [CrossRef]
2. Dufrasne, F. Baloxavir Marboxil: An Original New Drug against Influenza. *Pharmaceuticals* **2022**, *15*, 28. [CrossRef]
3. Su, B.; Wurtzer, S.; Rameix-Welti, M.; Dwyer, D.; van der Werf, S.; Naffakh, N.; Clavel, F.; Labrosse, B. Enhancement of the Influenza A Hemagglutinin (HA)-Mediated Cell-Cell Fusion and Virus Entry by the Viral Neuraminidase (NA). *PLoS ONE* **2009**, *4*, e8495. [CrossRef]

4. Rogers, G.N.; D'Souza, B.L. Receptor binding properties of human and animal H1 influenza virus isolates. *Virology* **1989**, *173*, 317–322. [CrossRef]
5. Rossignol, J.; La Frazia, S.; Chiappa, L.; Ciucci, A.; Santoro, M. Thiazolides, a new class of anti-influenza molecules targeting viral hemagglutinin at post-translational level. *J. Biol. Chem.* **2009**, *284*, 29798–29808. [CrossRef]
6. Monto, A. Epidemiology and Virology of Influenza Illness. *Am. J. Manag. Care* **2000**, *6* (Suppl. S5), S255–S264.
7. Yamada, S.; Suzuki, Y.; Suzuki, T.; Le, M.Q.; Nidom, C.A.; Sakai-Tagawa, Y.; Mu-ramoto, Y.; Ito, M.; Kiso, M.; Horimoto, T.; et al. Haemagglutinin mutations responsible for the binding of H5N1 influenza A viruses to human-type receptors. *Nature* **2006**, *444*, 378–382. [CrossRef] [PubMed]
8. Kosik, I.; Yewdell, J.W. Influenza Hemagglutinin and Neuraminidase: Yin-Yang Proteins Coevolving to Thwart Immunity. *Viruses* **2019**, *11*, 346. [CrossRef] [PubMed]
9. Du, W.; Guo, H.; Nijman, V.S.; Doedt, J.; van der Vries, E.; van der Lee, J.; Li, Z.; Boons, G.J.; van Kuppeveld, F.J.M.; de Vries, E.; et al. The 2nd sialic acid-binding site of influenza A virus neuraminidase is an important determinant of the hemagglutinin-neuraminidase-receptor balance. *PLoS Pathog.* **2019**, *15*, e1007860. [CrossRef]
10. Kurt, M.; Ercan, S.; Pirinccioglu, N. Designing new drug candidates as inhibitors against wild and mutant type neuraminidases: Molecular docking, molecular dynamics and binding free energy calculations. *J. Biomol. Struct. Dyn.* **2022**, *24*, 1–15. [CrossRef] [PubMed]
11. Jeyaram, R.A.; Anu Radha, C. N1 neuraminidase of H5N1 avian influenza a virus complexed with sialic acid and zanamivir—A study by molecular docking and molecular dynamics simulation. *J. Biomol. Struct. Dyn.* **2022**, *40*, 11434–11447. [CrossRef]
12. Colombo, C.; Podlipnik, C.; Lo Presti, L.; Niikura, M.; Bennet, A.J.; Bernardi, A. Design and synthesis of constrained bicyclic molecules as candidate inhibitors of influenza a neuraminidase. *PLoS ONE* **2018**, *28*, e0193623. [CrossRef] [PubMed]
13. Pedretti, A.; Mazzolari, A.; Gervasoni, S.; Fumagalli, L.; Vistoli, G. The VEGA suite of programs: An versatile platform for cheminformatics and drug design projects. *Bioinformatics* **2021**, *37*, 1174–1175. [CrossRef] [PubMed]
14. Bolton, E.E.; Wang, Y.; Thiessen, P.A.; Bryant, S.H. PubChem: Integrated platform of small molecules and biological activities. In *Annual Reports in Computational Chemistry*; Elsevier: Amsterdam, The Netherlands, 2008; Volume 4, pp. 217–241.
15. Morris, G.M.; Huey, R.; Lindstrom, W.; Sanner, M.F.; Belew, R.K.; Goodsell, D.S.; Olson, A.J. AutoDock4 and AutoDockTools4: Automated docking with selective receptor flexibility. *J. Comput. Chem.* **2009**, *30*, 2785–2791. [CrossRef]
16. Seiler, K.P.; George, G.A.; Happ, M.P.; Bodycombe, N.E.; Carrinski, H.A.; Norton, S.; Brudz, S.; Sullivan, J.P.; Muhlich, J.; Serrano, M.; et al. ChemBank: A small-molecule screening and cheminformatics resource database. *Nucleic Acids Res.* **2007**, *36*, D351–D359. [CrossRef]
17. Guha, R.; Howard, M.T.; Hutchison, G.R.; Murray-Rust, P.; Rzepa, H.; Steinbeck, C.; Wegner, J.; Willighagen, E.L. The Blue Obelisk-interoperability in chemical informatics. *J. Chem. Inf. Model.* **2006**, *46*, 991–998. [CrossRef]
18. Molecular Operating Environment Software (MOE). Available online: http://www.chemcomp.com (accessed on 26 July 2018).
19. ADMET Predictor™ Tool. Available online: https://www.simulations-plus.com (accessed on 29 August 2014).
20. Berman, H.M.; Westbrook, J.; Feng, Z.; Gilliland, G.; Bhat, T.N.; Weissig, H.; Shindyalov, I.N.; Bourne, P.E. The Protein Data Bank. *Nucleic Acids Res.* **2000**, *28*, 235–242. [CrossRef] [PubMed]
21. Pokorná, J.; Pachl, P.; Karlukova, E.; Hejdánek, J.; Rezáčová, P.; Machara, A.; Hudlický, J.; Konvalinka, J.; Kožíšek, M. Kinetic, Thermodynamic, and Structural Analysis of Drug Resistance Mutations in Neuraminidase from the 2009 Pandemic Influenza Virus. *Viruses* **2018**, *10*, 339. [CrossRef]
22. Varghese, J.N.; Laver, W.G.; Colman, P.M. Structure of the influenza virus glycoprotein antigen neuraminidase at 2.9 a resolution. *Nature* **1983**, *303*, 35–40. [CrossRef] [PubMed]
23. Collins, P.J.; Haire, L.F.; Lin, Y.P.; Liu, J.; Ruseell, R.J.; Walker, P.A.; Skehel, J.J.; Martin, S.R.; Hay, A.J.; Gamblin, S. Crystal structures of oseltamivir-resistant influenza virus neuraminidase mutants. *Nature* **2008**, *453*, 1258–1261. [CrossRef]
24. Vavricka, C.J.; Li, Q.; Wu, Y.; Qi, J.; Wang, M.; Liu, Y.; Gao, F.; Liu, J.; Feng, E.; He, J.; et al. Structural and functional analysis of laninamivir and itsoctanoate prodrug reveals group specificmechanisms for influenza NA inhibition. *PLoS Pathog.* **2011**, *7*, e1002249. [CrossRef] [PubMed]
25. Wu, Y.; Bi, Y.H.; Vavricka, C.J.; Sun, X.M.; Zhang, Y.F.; Gao, F.; Zhao, M.; Xiao, H.X.; Qin, C.F.; He, J.H.; et al. Characterization of two distinct neuraminidases from avian-origin hu-man-infecting H7N9 influenza viruses. *Cell Res.* **2013**, *23*, 1347–1355. [CrossRef]
26. Russell, R.; Haire, L.; Stevens, D.; Collins, P.J.; Lin, Y.P.; Blackburn, G.M.; Hay, A.J.; Gamblin, S.J.; Skehel, J.J. The structure of H5N1 avian influenza neuraminidase suggests new opportunities for drug design. *Nature* **2006**, *443*, 45–49. [CrossRef]
27. Shoichet, B.K. Virtual screening of chemical libraries. *Nature* **2004**, *432*, 862–865. [CrossRef] [PubMed]
28. Wright, D.L.; Anderson, A.C. The Design and Docking of Virtual Compound Libraries to Structures of Drug Targets. *Curr. Comput.-Aided Drug Des.* **2005**, *1*, 103–127.
29. Klebe, G.; Grädler, U.; Grüneberg, S.; Krämer, O.; Gohlke, H. *Methods and Principles in Medicinal Chemistry*; Wiley-VCH Verlag GmbH: Weinheim, Germany, 2008; pp. 207–227.
30. Kapetanovic, I. Computer-Aided Drug Discovery and Development (CADDD): In sili-co-chemico-biological approach. *Chem. Biol. Interact.* **2008**, *171*, 165–176. [CrossRef] [PubMed]
31. Scior, T.; Bernard, P.; Medina-Franco, J.L.; Maggiora, G.M. Large compound databases for structure-activity relationships studies in drug discovery. *Mini-Rev. Med. Chem.* **2007**, *7*, 851–860. [CrossRef]

32. Eichberg, J.; Maiworm, E.; Oberpaul, M.; Czudai-Matwich, V.; Lüddecke, T.; Vilcinskas, A.; Hardes, K. Antiviral Potential of Natural Resources against Influenza Virus Infections. *Viruses* **2022**, *14*, 2452. [CrossRef]
33. Steindl, T.; Langer, T. Influenza virus neuraminidase inhibitors: Generation and comparison of structure-based and common feature pharmacophore hypotheses and their application in virtual screening. *J. Chem. Inf. Comput. Sci.* **2004**, *44*, 1849–1856. [CrossRef]
34. Márquez-Domínguez, L.; Reyes-Leyva, J.; Herrera-Camacho, I.; Santos-López, G.; Scior, T. Five Novel Non-Sialic Acid-Like Scaffolds Inhibit In Vitro H1N1 and H5N2 Neuraminidase Activity of Influenza a Virus. *Molecules* **2020**, *25*, 4248. [CrossRef]
35. Scior, T.; Medina-Franco, J.L.; Do, Q.T.; Martínez-Mayorga, K.; Yunes Rojas, J.A.; Bernard, P. How to recognize and work-a-round pitfalls in QSAR studies: A critical review. *Curr. Med. Chem.* **2009**, *16*, 4297–4313. [CrossRef] [PubMed]
36. Medina-Franco, J.L.; Martínez-Mayorga, K.; Bender, A.; Marín, R.M.; Giulianotti, M.A.; Pinilla, C.; Houghten, R.A. Characterization of activity landscapes using 2D and 3D similarity methods: Consensus activity cliffs. *J. Chem. Inf. Model.* **2009**, *49*, 477–491. [CrossRef] [PubMed]
37. Bai, Y.; Jones, J.C.; Wong, S.-S.; Zanin, M. Antivirals Targeting the Surface Glycoproteins of Influenza Virus: Mechanisms of Action and Resistance. *Viruses* **2021**, *13*, 624. [CrossRef] [PubMed]
38. Agamennone, M.; Fantacuzzi, M.; Vivenzio, G.; Scala, M.C.; Campiglia, P.; Superti, F.; Sala, M. Antiviral Peptides as Anti-Influenza Agents. *Int. J. Mol. Sci.* **2022**, *23*, 11433. [CrossRef]
39. Sartori, A.; Dell'Amico, L.; Battistini, L.; Curti, C.; Rivara, S.; Pala, D.; Kerry, P.S.; Pelosi, G.; Casiraghi, G.; Rassu, G.; et al. Synthesis, structure and inhibitory activity of a stereoisomer of oseltamivir carboxylate. *Org. Biomol. Chem.* **2014**, *12*, 1561–1569. [CrossRef] [PubMed]
40. Scior, T.; Bender, A.; Tresadern, G.; Medina-Franco, J.L.; Martínez-Mayorga, K. Recognizing pitfalls in virtual screening: A critical review. *J. Chem. Inf. Model.* **2012**, *52*, 867–881. [CrossRef]
41. Quiroga, I.; Scior, T. Induced fit for cytochrome P450 3A4 based on molecular dynamics. *ADMET DMPK* **2019**, *11*, 252–266. [CrossRef]
42. Nabuurs, S.B.; Wagener, M.; De Vlieg, J. A flexible approach to induced fit docking. *J. Med. Chem.* **2007**, *50*, 6507–6518. [CrossRef]
43. Sano, E.; Li, W.; Yuki, H.; Liu, X.; Furihata, V.; Kobayashi, K.; Chiba, K.; Neya, S.; Hoshino, T. Mechanism of the decrease in catalytic activity of human cytochrome P450 2C9 polymorphic variants investigated by computational analysis. *J. Comput. Chem.* **2010**, *15*, 2746–2758. [CrossRef]
44. Sheng, Y.; Chen, Y.; Wang, L.; Liu, G.; Li, W.; Tang, Y. Effects of protein flexibility on the site of metabolism prediction for CYP2A6 substrates. *J. Mol. Graph. Model.* **2014**, *54*, 90–99. [CrossRef]
45. Naceri, S.; Marc, D.; Camproux, A.C.; Flatters, D. Influenza a Virus NS1 Protein Structural Flexibility Analysis According to Its Structural Polymorphism Using Computational Approaches. *Int. J. Mol. Sci.* **2022**, *4*, 1805. [CrossRef] [PubMed]
46. Xiao, W.; Wang, D.; Shen, Z.; Li, S.; Li, H. Multi-Body Interactions in Molecular Docking Program Devised with Key Water Molecules in Protein Binding Sites. *Molecules* **2018**, *23*, 2321. [CrossRef] [PubMed]
47. Quiroga, I.; Meléndez, F.; Atonal-Salvador, K.; Kammerer, B.; Scior, T. Identification a New Site of Metabolism for Phenprocoumon by Modeling it's CYP2C9 Hydroxylation Pattern. *SAJ Pharm. Pharmacol.* **2018**, *5*, 1–12.
48. Kirchmair, J.; Wolber, G.; Laggner, C.; Langer, T. Comparative performance assessment of the conformational model generators 397 omega and catalyst: A large-scale survey on the retrieval of protein-bound ligand conformations. *J. Chem. Inf. Model.* **2006**, *46*, 1848–1861.
49. Shaillay Kumar, D. Script for Computing Tanimoto Coefficient. QSARWorld—Free Online Resource for QSAR Modeling. Available online: http://www.qsarworld.com/virtual-workshop.php (accessed on 10 April 2023).

Article

The Antiviral Compound PSP Inhibits HIV-1 Entry via PKR-Dependent Activation in Monocytic Cells

Eduardo Alvarez-Rivera [1], Madeline Rodríguez-Valentín [2] and Nawal M. Boukli [1,*]

[1] Biomedical Proteomics Facility, Department of Microbiology and Immunology, Universidad Central del Caribe School of Medicine, Bayamón, PR 00960, USA
[2] Department of Microbiology, Houston Lee Moffitt Cancer Center and Research Institute, Tampa, FL 33612, USA
* Correspondence: nawal.boukli@uccaribe.edu; Tel.: +1-(787)-798-3001 (ext. 2080)

Abstract: Actin depolymerization factor (ADF) cofilin-1 is a key cytoskeleton component that serves to lessen cortical actin. HIV-1 manipulates cofilin-1 regulation as a pre- and post-entry requisite. Disruption of ADF signaling is associated with denial of entry. The unfolded protein response (UPR) marker Inositol-Requiring Enzyme-1α (IRE1α) and interferon-induced protein (IFN-IP) double-stranded RNA- activated protein kinase (PKR) are reported to overlap with actin components. In our published findings, *Coriolus versicolor* bioactive extract polysaccharide peptide (PSP) has demonstrated anti-HIV replicative properties in THP1 monocytic cells. However, its involvement towards viral infectivity has not been elucidated before. In the present study, we examined the roles of PKR and IRE1α in cofilin-1 phosphorylation and its HIV-1 restrictive roles in THP1. HIV-1 p24 antigen was measured through infected supernatant to determine PSP's restrictive potential. Quantitative proteomics was performed to analyze cytoskeletal and UPR regulators. PKR, IRE1α, and cofilin-1 biomarkers were measured through immunoblots. Validation of key proteome markers was done through RT-qPCR. PKR/IRE1α inhibitors were used to validate viral entry and cofilin-1 phosphorylation through Western blots. Our findings show that PSP treatment before infection leads to an overall lower infectivity. Additionally, PKR and IRE1α show to be key regulators in cofilin-1 phosphorylation and viral restriction.

Keywords: PSP; HIV; proteomics; cofilin-1; PKR; IRE1α; UPR

Citation: Alvarez-Rivera, E.; Rodríguez-Valentín, M.; Boukli, N.M. The Antiviral Compound PSP Inhibits HIV-1 Entry via PKR-Dependent Activation in Monocytic Cells. *Viruses* **2023**, *15*, 804. https://doi.org/10.3390/v15030804

Academic Editor: Simone Brogi

Received: 2 March 2023
Revised: 16 March 2023
Accepted: 17 March 2023
Published: 22 March 2023

1. Introduction

The amount of globally infected individuals (38.4 million as of 2021) has surpassed those who have access to antiretroviral therapy (28.7 million in 2021) according to the statistics of the World Health Organization (WHO) and the United Nations Programme on HIV/AIDS. The corresponding amount has been slowly increasing throughout the years [1,2]. This renders at least 25% of the HIV-1 infected worldwide population without access to highly active antiretroviral treatment (HAART). While HAART is used to counteract the devastating effects of HIV-1, it remains an expensive solution [3,4]. Due to these documented facts, new treatment therapies co-working with HAART are required to combat HIV-1, as well as to ensure new preventative methods.

The success of traditional therapies by the 2015 Nobel Prize for an antimalarial treatment from Chinese herbs has renewed the interest in developing new drugs with immunomodulatory properties [5]. To that effect, a body of research demonstrates examples of mushrooms that can be used as a great source of natural compounds [6]. It undertakes this task by increasing the immunomodulatory response described by ancient Chinese history [7]. It has been demonstrated by our research group that *Coriolus versicolor* polysaccharide peptide (PSP) possesses anti-HIV properties by lowering viral replication by an average of 61% in a toll-like receptor 4 (TLR4)-dependent manner [8].

HIV-1 targets CD4+ cells with CXCR4/CCR5 co-receptors such as monocytes/macrophages, dendritic, and T-helper cells [9–12]. HIV-1 interaction with CD4 remains an obstacle to viral entry. HIV-CD4/CXCR4/CCR5 activates Rho family GTPases to exchange GDP for GTP [13,14]. This leads to downstream phosphorylation of ROCK/LIMK-1 to inactivate actin-depolymerization factor (ADF) cofilin-1. This process is referred to as actin polymerization (Figure 1A). The strengthening/bundling of cortical actin filaments results in co-receptor clustering, increasing the probability of interaction towards the HIV-CD4 complex. At this stage, cytoskeleton remodeling effectively blocks entry by acting as a barrier [15] (Figure 1A). The HIV-CXCR4/CCR5 complex proceeds to recruit phosphatases to activate cofilin-1. This promotes the breakdown of actin filaments, effectively eliminating this barrier by actin depolymerization [16,17] (Figure 1B). Among the phosphatases regulating cofilin-1 activation are the slingshot homolog family: SSH1, SSH2, and SSH3 [18,19]. The exact signaling that HIV-1 uses to recruit these phosphatases remains elusive. HIV co-receptors are G-protein coupled receptors to which it holds influence towards the Gα subunit [20]. This affinity sets the stages of dephosphorylation and activation resulting in successful infection (Figure 1B). Interference with either cofilin-1 or its phosphorylated state from any of the pre-requisite steps leads to denial of entry [21,22].

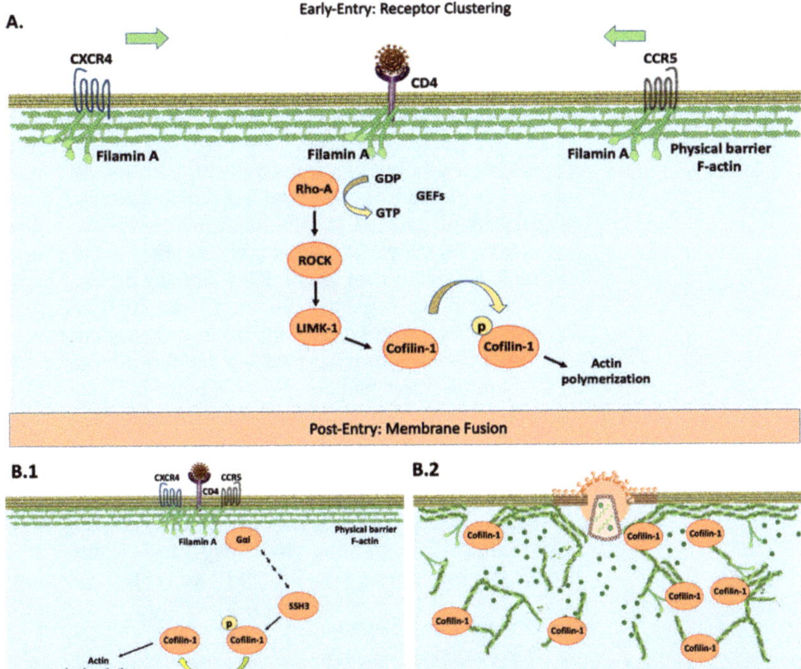

Figure 1. Diagram depicting HIV-1 entry during infection in CD4+ cells. (**A**) Early interaction between HIV-1 and CD4 receptor initiates activation of guanine nucleotide exchange factors (GEFs) and subsequently downstream signaling through RhoA/ROCK/LIMK-1. This results in phosphorylation and inactivation of cofilin-1. Actin dynamics will shift towards the polymerization state leading to CXCR4/CCR5 co-receptor clustering as the first requirement for viral entry. Filamin A serves as a linkage factor between CD4 and co-receptors. (**B.1**) HIV-1 associates with CXCR4/CCR5 in proximity and triggers downstream dephosphorylation and activation of cofilin-1. This process is carried out by G-protein coupled receptor signaling, specifically through Gα subunit-mediated phosphatases such as SSH3. (**B.2**) Cofilin-1 will begin the breakdown of actin filaments and subsequently lead to viral fusion as the final entry requirement.

The cytoskeleton contributes towards morphology, trafficking, migration, and adhesion [23–25], and ultimately acts as a deciding factor for HIV-1 entry [16,18,21,22,26–28]. Among its regulators, the actin binding protein (ABP) gelsolin is considered a strong influencer over actin dynamics. It holds the same function as cofilin-1, promoting the breakdown of actin filaments [29,30]. In contrast to its ADF counterpart, it is independent of HIV-1 signaling. The literature analysis shows that gelsolin overexpression interferes with the pre-fusion cascade, essentially blocking HIV-1 entry [31,32]. This unique property makes gelsolin a strong candidate as a restriction inducer. Although precise biological mechanisms remain elusive, evidence suggests that the endoplasmic reticulum (ER) plays a vital role in cytoskeleton regulation [33–36]. Aggregations of unfolded proteins due to cellular stress lead to an unfolded protein response (UPR). This characteristic is attributed to glucose-regulated protein 78 (GRP78) [37,38]. GRP78 up-regulation triggers ER stress by the Protein Kinase R-like ER kinase (PERK), activating transcription factor 6α (ATF6) and inositol requiring enzyme 1α (IRE1α) to decide cell fate. A short or prolonged UPR induces survival or apoptotic ER stress respectively [37–39], which overlaps with actin filaments and indirectly dictates the polymerization state [40–43]. Therefore, treatments that can interfere with HIV-1 depolymerization/polymerization status are pivotal for arresting entry at the pre- or post-fusion steps respectively.

The UPR aims at restoring cellular homeostasis following physiological stress exerted on the ER. This also invokes direct control of cytoskeleton re-arrangement in response to invading microorganisms [44,45]. In recent years, studies have surged to understand novel UPR overlapping signaling events. This survival mechanism has been shown to have an influence on the production of type I interferons (IFN) to ward off infections [46]. In a similar manner, it has been postulated that the interferon-induced protein (IFN-IP) double-stranded RNA-activated protein kinase (PKR) can interact directly with upstream regulators of cofilin-1 [47,48]. PKR has also been researched to be closely linked with the UPR [49]. Specifically, a study has shown the RNase activity of IRE1α can directly activate PKR downstream of TLR4 signaling [50]. The unique properties of PKR range from an immune response against viral infection to a survival regulator [51–53], but most importantly it has been linked to the phosphorylation of cofilin-1 (Ser3) [54].

The current study reveals, for the first time, the anti-HIV restrictive properties of PSP by influencing cofilin-dependent phosphorylation through PKR and IRE1α activation. Additionally, the identification of de-regulated cytoskeletal and UPR proteins were unveiled. Our findings establish a distinctive pattern of PSP before infection occurs as well as exerting control over cytoskeletal components. We show that the overexpression of PKR induced by PSP leads to a direct correlation to IRE1α and cofilin-1. These observations are based on our data that inhibition of either IRE1α or PKR rendered the cell vulnerable to infection by significantly increasing viral entry. These results provide the first insights into PSP's restrictive roles through UPR, IFN-IP, and cytoskeletal markers.

2. Materials and Methods

2.1. Cell Culture

THP1-Blue-CD14 derives from the human monocytic THP1 cell line (InvivoGen, San Diego, CA, USA). These cells carry a reporter plasmid expressing a secreted embryonic alkaline phosphatase (SEAP) and express all TLRs; however, they only respond to ligands that are specific for TLR2, TLR1/2, TLR2/6, TLR4, TLR5, and TLR8. Cells were cultured in Roswell Park Memorial Institute (RPMI) 1640 medium (ATCC, Manassas, VA, USA), supplemented with 100 U/mL penicillin and 100 µg/mL streptomycin, in addition to 10% heat-inactivated fetal bovine serum (ATCC, Manassas, VA, USA). Cells were maintained in T-75 cm^2 culturing flasks in a humidified incubator at 37 °C and 5% CO_2.

2.2. PSP Extraction and Treatment

PSP supplement tablets consist of the following: 28% polysaccharide-to-peptide ratio with 60.23 mg/g beta-1,3/1,6-glucan (Mushroom Science, Eugene, OR, USA). To achieve the extracted working treatment, each tablet is diluted in hot/boiling water (approximately 90–100 °C) and centrifuged at $2060 \times g$ for 5 min. This cycle is repeated until the supernatant is free of insoluble residues. Ethanol 80% was used to separate the solution into two phases and PSP was collected from the light-brown layer. This was followed by washing with absolute ethanol, centrifugation, and drying with a refrigerator vapor trap (Thermo Fisher Scientific, Waltham, MA, USA) at -105 °C. The extracted treatment can be used directly or stored at -20 °C. A concentration of 200 µg/mL was used and defined after the assessment of cell activation and cytotoxicity for a 6 day period as previously published [8]. PSP was added twice during the allotted time (at days 0 and 3) for a period of 72 h, with fresh RPMI 1640 culture media to achieve the desired exposure concentration. Cells were maintained in T-75 cm^2 cultured flasks, in a humidified incubator at 37 °C and 5% CO_2.

2.3. HIV Stocks and Propagation

HIV-1ME46 (NIH AIDS Reagent Program, Bethesda, MD, USA) virus was propagated with ex vitro cell culture using peripheral blood mononuclear cells (PBMC) as recommended by the NIH HIV Reagent program's protocol. Healthy PBMCs were extracted from blood donors using heparin tubes and subjected to centrifuging at $572 \times g$ for 12 min. Histopaque (Sigma-Aldrich, Saint Louis, MO, USA) was used to separate PBMCs from plasma and cell debris and subsequently washed with PBS 1X. Cells were cultured in RPMI 1640 (ATCC, Manassas, VA, USA) and incubated in T-75 cm^2 flasks in a humidified incubator at 37 °C, 5% CO_2. PBMCs were stimulated with 5 µg/mL of phytohemagglutinin and subjected to 200 µL of HIV-1ME46 after 3 days of stimulation. Infection lasted for a total period of 9 days with supplementation of fresh media every 3 days. PBMCs were centrifuged and HIV particles were measured by analyzing p24 antigens in cell culture supernatant through quantitative reverse transcription polymerase chain reaction (RT-qPCR).

2.4. HIV Infection

After reaching approximately 80% confluency, THP1-Blue-CD14 were centrifuged and re-suspended in RPMI 1640 culture medium. Cells were passaged into a T-75 cm^2 flask with fresh media and supplemented with 8 µg/mL hexadimethrine bromide (Sigma-Aldrich, Saint Louis, MO, USA) to enhance the infection process. Acute infection was achieved by using a multiplicity of infection (MOI) of 1.4 from dual-tropism HIV-1ME46 (NIH AIDS Reagent Program, Bethesda, MD, USA) per 1 million healthy cells. HIV-1-infected-THP1 were incubated in a humidified atmosphere at 37 °C and 5% CO_2 for a period of 24 h and subsequently washed with fresh culture media. To guarantee equal infection among cells with or without PSP treatment, infection was carried out in the same flask and then divided into their respective groups.

2.5. Viral Load Analysis

After 6 days of continuous exposure to PSP, cells were cultured in 12-well plates (Thermo Fisher Scientific, Waltham, MA, USA) using an optimal density of 3×10^6 cells. Subsequently, cells were acutely infected with HIV-1ME46 (NIH AIDS Reagents Program, Bethesda, MD, USA) with an MOI of 1.4 for a period of 24 h. Control group is composed of THP1 monocytic cells infected with HIV-1. Hexadimethrine bromide (Sigma-Aldrich, Saint Louis, MO, USA) at a concentration of 8 µg/mL was used for enhancement of infection. HIV particles were measured by analyzing HIV p24 antigens in cell culture supernatant after 24 h of infection through RT-qPCR using a COBAS 6800 system (Roche Diagnostics Corporation, Indianapolis, IN, USA). HIV-specific primers were supplemented in the HIV-1 Master Mix Reagent 2 kit (Roche Diagnostics Corporation, Indianapolis, IN, USA). Denaturalization, annealing, extension steps, and cycle numbers were carried out according to Roche Diagnostic's pre-determined protocols. Viral load calculations were determined as the percentage of total infection used relative to treated and infected cells.

2.6. IRE1α and PKR Inhibition Assays

During inhibition assays, THP1-Blue-CD14 cells were subjected twice (days 0 and 3) to 5-h treatment of either 56.09 nM Imidazolo-oxindole PKR inhibitor C16 (Sigma-Aldrich, Saint Louis, MO, USA) or 221.8 nM IRE1α Inhibitor III, 4μ8C (Sigma-Aldrich, Saint Louis, MO, USA) prior to PSP addition and infection. Cells were later exposed twice to PSP (200 μg/mL) during the same time frame for a total of six days. Dimethyl sulfoxide (DMSO) at 0.5% was used as a drug diluent during experiments and represents the vehicle group.

2.7. Quantitative Proteomics Experiments by Isobaric Labeling Described Below

2.7.1. Protein Extraction and Digestion

Cells were centrifuged at $321 \times g$ for 7 min, after 6 days of continuous PSP exposure (200 μg/mL). THP1-Blue-CD14 were lysed using a combination of ultrasonic and SDS lysis buffer composed of 2% SDS w/v and 250 mM NaCl. Additionally, PhosSTOP (Roche, Madison, WI, USA) phosphatase inhibitors, 2 mM sodium vanadate, an EDTA-free protease inhibitor cocktail (Promega, Madison, WI, USA), and 50 mM HEPES adjusted at pH 8.5 were added to the mixture for protein effective membrane lysis and protein preservation. The following were added for each lysate: 5 mM of dithiothreitol (DTT) and 14 mM of iodoacetamide in the dark for 30 min with the purpose to serve as a reducing and alkylation agent respectively. Proteins were extracted by methanol and chloroform precipitation. Subsequently, washes composed of ice-cold acetone were used. Each pellet was left to dry and afterwards resuspended in a combination of 8 M Urea and 50 mM HEPES (pH 8.5). Protein concentrations were measured using a bicinchoninic acid assay kit (Thermo Fisher Scientific, Waltham, MA, USA) and with a spectrophotometer using the Molecular Devices VersaMax Absorbance Microplate Reader (GMI Trusted Laboratory Solutions, Ramsey, MN, USA) prior to digestion through proteases. Samples were then diluted with 4 M Urea, and digested with LysC (Wako, Japan) in a 1:50 enzyme-to-protein ratio overnight. Further dilutions were carried out with 1.5 M Urea concentration the next day. To finalize the digestive process, Trypsin (Promega, Madison, WI, USA) was added to a ratio of 1:100 (enzyme to protein) for 6 h at 37 °C. Afterwards, 200 μL of 20% formic acid (FA) (adjusted to pH 2.0) was used to acidify the samples. Each was subjected to C18 solid-phase extraction (SPE) (Sep-Pak, Waters, Milford, MA, USA).

2.7.2. Tandem Mass Tagging (TMT) Labeling

TMT isobaric labeling was performed using the 6-plex TMT kit (Thermo Fisher Scientific, Waltham, MA, USA). TMT reagents at a concentration of 0.8 mg were dissolved in 40 μL of dry acetonitrile (Micro BCA, Thermo Fisher Scientific, Waltham, MA, USA), and 10 μL were added for each 0.2 μg of peptides. These were further dissolved in 100 μL of 200 mM HEPES, pH 8.5. After 1 h of incubation at room temperature, the reaction was quenched by adding 8 μL of 5% hydroxylamine. Lastly, each labeled peptide was combined, acidified with 20 μL of 20% FA (pH~2.0), and concentrated via C18 SPE on Sep-Pak cartridges (50 mg bed volume).

2.7.3. Reverse-Phase High-Performance Liquid Chromatography (HPLC)

TMT labeled HIV-infected-THP1 (controls and PSP treated) extracted peptides were subjected to basic-pH reverse phase fractionation. Peptides were solubilized in buffer composed of 5% ACN, 50 mM ammonium bicarbonate adjusted to pH 8.0. These samples were stably separated through HPLC using an Agilent 300Extend-C18 column (Specification: 5 μm particles, 4.6 mm i.d. and 220 mm in length). Additionally, the flow rate of the samples was measured using an Agilent 1100 binary pump for HPLC. This was equipped with a degasser and a photodiode array detector (Thermo Fisher Scientific, Waltham, MA, USA), with a 45 min linear gradient ranging from 8–35% acetonitrile in 10 mM ammonium bicarbonate pH 8.0 (flow rate parameters: 0.8 mL/min). The peptide mixtures were separated into a total of 96 fractions as a result of this. Subsequently, said fractions were consolidated into two sets of samples in a checkerboard manner, acidified with 10 μL of 20% FA, and vacuum dried. Each sample was re-dissolved in 5% FA, desalted via StageTip, dried via vacuum centrifugation, and reconstituted for LC–MS/MS analysis.

2.7.4. Triple Stage Mass Spectrometry (MS3)

During mass spectrometry (MS) experiments, all spectra were acquired on an Oribtrap Fusion (Thermo Fisher Scientific, Waltham, MA, USA) coupled to an Easy-nLC 1000 (Thermo Fisher Scientific, Waltham, MA, USA) ultrahigh pressure liquid chromatography (UHPLC) pump. Peptides were separated on a 100 μm inner diameter column composed of 0.5 cm of Magic C4 resin (5 μm, 100 Å, Michrom Bioresources, Auburn, CA, USA), followed by 25 cm of Sepax Technologies GP-C18 resin (1.8 μm, 120 Å, Newark, DE, USA) with a gradient ranging between 3–25% (ACN, 0.125% FA) in a time course of approximately 180 min.

For all experiments, the instrument was operated in the data-dependent mode. First-stage MS1 spectra were collected at a resolution of 120,000 using an automated gain control (AGC) target of 200,000 and a maximum injection time of 100 ms. A selection of the 10 most intense ions was used for the second stage MS2 approach. Precursors were filtered out according to relative charge state and monoisotopic peak assignments. These were later excluded using a dynamic window of 75 s $\pm$ 10 ppm and isolated with a quadrupole mass filter set to a width of 0.5 m/z.

During the second-stage MS2 spectra, the Orbitrap was operated at 60,000 resolutions, consisting of an AGC target of 50,000 and a maximum injection time of 250 ms. Precursors were fragmented by high-energy collision dissociation (HCD) at a normalized collision energy (NCE) of 37.5%.

Lastly, the MS3 approach was performed using the collected spectra during MS2 at an AGC of 4000, maximum injection time of 150 ms, and collision energy of 35%. The same Orbitrap parameters as for the MS2 method were used during MS3, with the exception that the HCD collision energy was increased to 55% to ensure maximal TMT reporter ion yield. The synchronous-precursor-selection (SPS) was enabled to include up to 3, 6, or 10 MS2 fragment ions in the MS3 scan.

2.7.5. MS3 Data Processing

All data relating to thermo."raw" files were converted into a readable file format (mz.XML) through a compilation of in-house software. This was also used to correct monoisotopic m/z measurements and incorrect peptide charge states that may have surged during the implementation of these methods. The distribution of MS2 spectra was performed using the SEQUEST algorithm. Each proteome experiment utilized the Human UniProt database to establish the accession number, gene name, and functions. For every experiment, reverse protein sequences were included for common contaminants such as human keratins. During SEQUEST analysis, a 50 ppm precursor ion tolerance was performed, while requiring each peptide's N/C terminus to have trypsin protease specificity and allowing up to two missed cleavages. MS2 spectra assignment false discovery rate (FDR) of less than 1% was achieved by applying the target-decoy database search strategy.

2.7.6. Determination of TMT Reporter Ion Intensities and Quantitative Data Analysis

For quantification, a 0.03 m/z (6-plex TMT) window centered on the theoretical m/z value of each reporter ion was queried for the nearest signal intensity. Reporter ions were adjusted to correct for the isotopic impurities of the different TMT reagents (manufacturer specifications). The signal-to-noise values for all peptides were summed within each TMT channel, where each was scaled according to the inter-channel differences to account for differences in sample handling. A total minimum sum of signal-to-noise values greater than 100 and isolation purity greater than 50% was required for each sample.

2.8. Western Blot

A concentration of 30 µg of protein samples extracted with RIPA buffer (Thermo Fisher Scientific, Waltham, MA, USA) was used for SDS-PAGE and transferred to a polyvinylidene difluoride membrane. A protease and phosphatase inhibitor cocktail (Thermo Fisher Scientific, Waltham, MA, USA) was used during the protein extraction. Protein concentrations were measured spectrophotometrically using the Molecular Devices VersaMax Absorbance Microplate Reader (GMI Trusted Laboratory Solutions, Ramsey, MN, USA). Bovine serum albumin (Fisher Scientific, Hampton, NH, USA) 5% in 1X TBST served as the blocking buffer for each membrane. Primary antibody incubation was done overnight at 4 °C in a shaker. The following primary antibodies were used at a 1:1000 dilution: cofilin-1, p-cofilin-1 (Ser 3), gelsolin, PKR, IRE1α, and GRP78, (Cell Signaling Technologies, Danvers, MA, USA), pPKR (Thr 446, Abcam, Cambridge, UK). Anti β-Actin (Cell Signaling Technologies, Danvers, MA, USA) was used as a loading control. Horseradish peroxidase (HRP)-conjugated secondary antibodies (Cell Signaling Technologies, Danvers, MA, USA) were used at a dilution of 1:10,000. Protein bands were visualized by chemiluminescence using the SuperSignal West Femto Maximum Sensitivity Substrate Kit (Thermo Fisher Scientific, Waltham, MA, USA). All images were analyzed with ImageJ image processing program version 1.52a (National Institute of Health, Bethesda, MD, USA).

2.9. Quantitative Reverse Transcription PCR

RNA was extracted from cell pellets using the RNeasy Plus Mini Kit (Qiagen, Hilden, Germany) following the manufacturer's protocols. RNA quality and concentration were quantified spectrophotometrically with a NanoDrop 1000 spectrophotometer (Thermo Fisher Scientific, Waltham, MA, USA). cDNA was reverse transcribed from 1 µg of total RNA using the iScript cDNA synthesis kit (Bio-Rad, Hercules, CA, USA). Samples were then processed through RT-qPCR for identification of cofilin-1, gelsolin, and PKR genes by means of SYBR green assays (Bio-Rad, Hercules, CA, USA) using 500 nM of primers (Sigma-Aldrich, Saint Louis, MO, USA). Amplification was carried out in a Bio-Rad CFX96 Touch real time PCR detection system (Bio-Rad, Hercules, CA, USA) using the following parameters: 15 s at 95 °C, 1 min at the gene-specific annealing temperature, and 1 min at 72 °C for a total of 40 cycles. The gene-specific primers were as follows: cofilin-1 forward, 5$'$-GCTTCTTCTTGATGGCGTCCTTG-3$'$ and cofilin-1 reverse 5$'$-GTCACCTGTGGCTTTGCTGT-3$'$; gelsolin forward, 5$'$-CTACCCAGGATGAGGTCGCT-3$'$ and gelsolin reverse, 5$'$-GTGCCGCCCTTGTAGATGA-3$'$; PKR forward, 5$'$-TCCAACACTG-GACTCCCTTC-3$'$; PKR reverse, 5$'$-TACTGGGGGCATATGGGTAA-3$'$. The gene expression level was defined as the threshold cycle number (CT). The mean fold changes in expression of the target genes were calculated using the comparative CT method (RU, 2$\Delta\Delta$CT). All data were normalized through the quantity of RNA input of 18S forward, 5$'$-GGCCCTGTAATTGGAATGAGTC-3$'$ and 18S reverse, 5$'$-CCAAGATCCAACTACGAGCTT-3$'$, serving as the endogenous control and for normalization.

2.10. MTT Cell Viability Assay for IC50 Determination

The half inhibitory concentration (IC50) for Imidazolo-oxindole PKR inhibitor C16 (Sigma-Aldrich, Saint Louis, MO, USA) and IRE1α Inhibitor III, 4µ8C (Sigma-Aldrich, Saint Louis, MO, USA) were determined using Thiazolyl Blue Tetrazolium Bromide (Sigma-Aldrich, Saint Louis, MO, USA) MTT cell cytotoxicity assays. THP1-Blue-CD14 cells were seeded at a density of 1×10^4 cells per well using a 96-well plate model in biological and technical triplicates for a period of 24 h. The next day, cells were exposed to either C16 or 4µ8C inhibitor at days 0 and 3 for a total of 6 days to simulate PSP treatments. The range of concentrations used for C16 and 4µ8C were 0–200 nM and 0–300 nM respectively. On the sixth day, 5 mg/mL of MTT was added to each sample using 1X PBS which served as the diluent as stated by Sigma-Aldrich protocols. After 4 h of incubation, the formazan crystals were dissolved using 100% DMSO. The absorption of the formazan solution was measured using the Molecular Devices VersaMax Absorbance Microplate Reader (GMI

Trusted Laboratory Solutions, Ramsey, MN, USA) at a wavelength of 570 nm. Cell viability was calculated as the percentage of THP1-BLUE-CD14 treated cells relative to positive and negative controls. MTT IC50 results can be found in Figure S1.

2.11. Statistical Analysis

All data sets are expressed as mean $\pm$ S.E.M. for biological triplicates. Statistical analysis for in vitro studies was performed by using one way analysis of variance (ANOVA) with post hoc Tukey for multiple comparison groups or unpaired t-test as appropriate for each experiment, using GraphPad PRISM v9.5.0 (GraphPad Software, San Diego, CA, USA) statistical power software. A p-value of $p \leq 0.05$ was considered as statistically significant.

3. Results

3.1. Isobaric TMT Labeling Quantitative Proteomics Profiling Revealed Differential Regulated Proteins Associated with Cytoskeletal and ER Stress Function Categories in THP1-Blue-CD14 Cells Treated with PSP

Untreated and PSP (200 µg/mL) treated (24 h) THP1 cells were used in the differential protein expression analysis. Relative high-throughput estimation of cellular protein abundances and quantitation have been achieved using stable isotope chemical labeling referred to as TMT labeling of proteins or peptides using MS. This quantitative proteomics approach has been widely used in our laboratory [55,56] for its high multiplexing capacity and deep proteome coverage. Proteins relating to cytoskeletal re-arrangement and ER stress UPR process with p-values ≤ 0.05 (p-value adjusted as false discovery rate) were deemed as statistically significant and selected for Supplementary Tables S1 and S2 respectively. The detailed information for accession number, gene symbol, number of spectra counts, protein name, average expression in control relating to treatment, fold change, and p-value are included in Tables S1 and S2. Due to the infeasibility to discuss all identified proteins in the MS data, the selection criteria are based on the significance of the fold change for each category of signaling markers.

A total of 111 proteins were associated with cytoskeletal functions and were identified as significantly expressed in PSP-treated cells. From these, 54 upregulated and 57 downregulated proteins make up the total. Among the differentially expressed proteins, the key factor of HIV entry involved in cofilin-1 activation, SSH3 was found to be downregulated (-1.46-fold). Interestingly, the interferon-induced, double-stranded RNA-activated protein kinase (EIF2AK2), which corresponds to PKR, is found significantly upregulated (2.44 fold). Actin-binding regulators tropomyosin and tropomodulin are identified as up-regulated (1.48 and 1.53 folds respectively). These two proteins need special mention since they are involved in sensing and modulating cofilin's depolymerization activity [57–59]. In addition, seven guanine nucleotide exchange factors (GEFs) pertaining to the Rho, Ras, and Rab families (Table S1) involved in regulating the polymerization state of actin filaments are found de-regulated. An adapter and regulator of actin cytoskeletal, switch associated protein 70 (SWAP70), had a significant decrease in expression (-1.26). This marker has been proposed as a diagnostic marker for HIV-1 infection due to its presence in the membrane surface of HIV-positive cells [60]. Additionally, this protein has close ties with the GEF family, modulating their activity [61]. This correlates to the decrease in fold value of the Ras-specific guanine nucleotide release factor RalGPS2 (-1.43). Cytoskeletal and transport motor proteins such as myosin phosphatase Rho-interactin (MPRIP), unconventional myosin-XVIIIa (MYO18A), and tubulin-specific chaperone C (TBCC) have been listed in downregulated states: -1.24-, -1.28- and -1.24-fold respectively. Meanwhile, unconventional myosin-Ig (MYO1G) is upregulated by 1.34-fold. Phosphatidylinositol 3,4,5-triphosphate dependent Rac exchanger-1 (PREX1) is associated with the Rho family GTPases of Rac, Ral, and Rho. Additionally, it is a downstream cytoskeletal modulator of CXCR4 co-receptors relating to viral membrane fusion [35]. This protein is identified as significantly downregulated (-1.38) in the presence of PSP. This regulation correlates with its association with the Rho GTPase-activating protein 15 and 17 (-1.29 and -1.37-fold respectively). Among other HIV-1 entry factors, α-actinin was found to be downregulated

(−1.20-fold). This protein has been known as a marker for virion fusion. [62]. Other facilitators of viral entry, such as microtubule-actin cross linking factor 1 (isoforms 1/2/3/5) are determined to be downregulated by −1.29-fold in treated cells. These cytoskeletal components are known for their internalization of HIV-1 capsid interactions during the early fusion steps [63].

Moreover, MS3 analysis has revealed 28 deregulated proteins (21 up- and seven downregulated) relating to UPR (Table S2). Among these, associated UPR and ER stress markers such as protein niban (FAM129A), ER degradation-enhancing alpha-mannosidase-like protein 1 (EDEM1), and heat shock protein 90-beta (HSP90AB1) were upregulated by 1.35-, 1.25- and 1.16-fold respectively. These markers are known for targeting translational regulation, misfolded proteins, and chaperone activity respectively [64–66]. In correlation with the UPR and EDEM1, F-box only protein 6 (FBXO6) is involved in the ER-associated degradation pathway for misfolded proteins. FBXO6 has also been associated with the inhibition of chronic ER stress, making it a suitable pro-survival marker [67]. FBOX06 has been significantly present in an upregulated state corresponding to 1.71-fold. PSP also extends its UPR signaling by positively modulating pro-survival chaperones. A clear indicator of this involves the upregulation (1.31-fold) of heat shock protein 105 kDa (HSPH1), which is involved in alleviating protein accumulation [68]. The effect of PSP also targets markers that possesses duality roles, such as heat-shock protein beta-1 (HSPB1). The overall levels of this marker were decreased by 1.14-fold. This protein has been associated as a cytoskeletal regulator and molecular chaperone [69]. Most notably, PSP treatment led to the significant upregulation of the ER survival chaperone protein disulfide isomerase (PDI) by a 1.17-fold change. Lastly, the UPR marker serine/threonine-protein kinase/endoribonuclease IRE1α (ERN1) is significantly overexpressed (1.41-fold).

3.2. PSP Treatment Leads to an Increase in Viral Restriction

Our research group has previously published a significant internal viral inhibition corresponding to 61% for PSP-treated HIV-infected THP1 and 35.99% PBMC [8]. To further widen the knowledge of these studies, the external antiviral effect of PSP was evaluated in THP1 cells. With the goal of determining PSP's efficacy as an anti-HIV-1 restrictive agent, THP1 cells were supplemented twice with 200 μg/mL of PSP over a six-day period before succumbing to HIV-1. Infection was carried out for a total time lapse of 24 h on the sixth day relative to treated cells. Hexadimethrine bromide at a concentration of 5 μg/mL was added as an enhancer of infection and a cationic polymer which carries the effect of neutralizing charge repulsions between viral and cell surface membranes. HIV-1 entry was measured by analyzing the concentration of HIV-1 core p24 antigen through RT-qPCR in the cell culture supernatant. Viral load analysis revealed a significant reduction in HIV-mediated entry by an average of 26% relative to 85.67% in the control group, with a mean difference of 59.67 ± 9.735 (Figure 2). This result correlates to a restriction average of 74% in contribution to the effects of PSP.

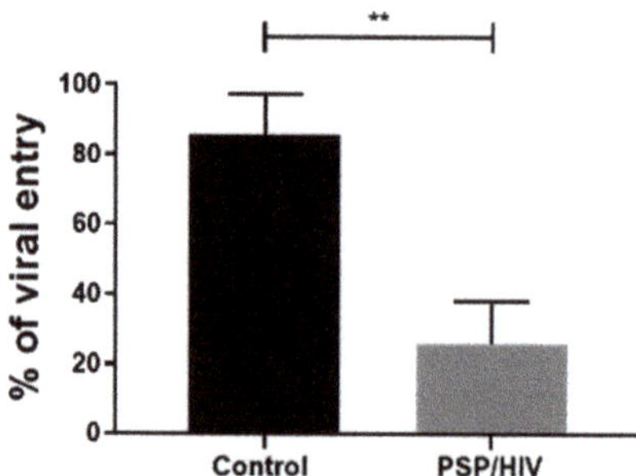

Figure 2. PSP lowers HIV entry in THP1 cells. Percentages of HIV entry in PSP-treated cells analyzed through viral load of HIV p24 antigen and compared with unpaired *t* test. Data are represented as mean ± SEM. Statistically significant difference (**), *p* < 0.01 is shown, *n* = 3.

3.3. Inhibition of PKR and IRE1α Are Associated with an Increase in HIV-1 Entry

Our data indicate that PSP has the potential to restrict HIV-1 entry as well as to induce the differential regulation of cytoskeletal and UPR proteins. To further investigate the signaling complex that influences this restrictive pattern, the potential roles of PKR and the UPR marker IRE1α in HIV-1 entry were studied by subjecting THP1 cells to C16-PKR and 4μ8C-IRE1α pharmaceutical inhibitors. We hypothesized that inhibition of either the activation of PKR or the endoribonuclease activity of IRE1α would affect viral entry into host immune cells. This is based on the overlapping signaling complex with downstream cytoskeletal effectors as seen in the literature research. Therefore, THP1 cells were treated twice with either 56.09 nM of C16 or 221.8 nM of 4μ8C for a period of 5 h prior to PSP treatment and HIV-1 infection. DMSO 0.5% was used as a vehicle control for both blockers. Viral load analysis has revealed no statistical difference in HIV-1 entry for C16, with approximately 89.13% relative to 94.1% for the infected control and 93.96% for the DMSO vehicle group (Figure 3). This correlates to a restrictive percentage of 10.87% for C16-treated PSP-induced cells. Compared to Figure 2, there was a relative increase in viral entry by approximately 63.13% in difference. Moreover, 4μ8C-treated PSP-induced cells showed approximately 69.56% of viral entry (24% restriction relative to the control and vehicle groups). The effects of 4μ8C can be translated to a 19.57% difference in comparison to the C16 blocker, contributing to a significant reduction in viral access (Figure 3). Taking into consideration the data in Figure 2, the inhibition of IRE1α endoribonuclease activity represents an approximate increment difference of 43.56% in viral entry. The results for both drugs show an inversely proportional HIV-1 entry and restriction taking place compared to Figure 2.

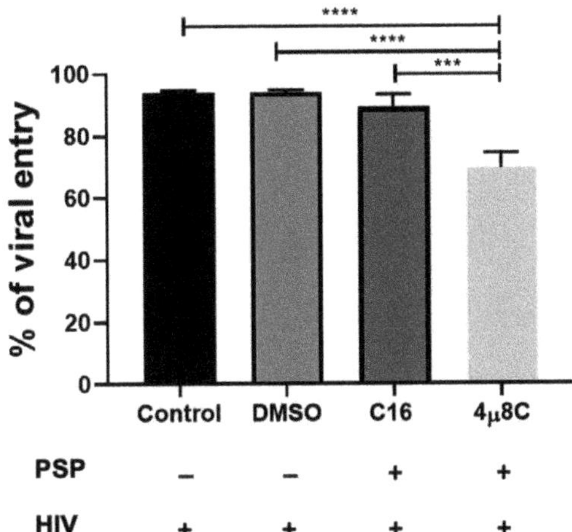

Figure 3. Inhibition of PKR and IRE1α activity facilitates HIV-1 entry. Percentages of HIV entry in PSP-treated cells supplemented with either 56.09 nM of C16-PKR or 221.8 nM of 4μ8C-IRE1α pharmaceutical blockers. Experiments were analyzed through viral loads of the HIV p24 antigen and compared with one-way ANOVA with Tukey multiple comparisons tests. Data are represented as mean ± SEM. Statistically significant difference (***), $p < 0.001$ and (****), $p < 0.0001$ are shown, $n = 3$.

3.4. PSP Induces the Upregulation of Key Cytoskeletal, IFN-IP, and ER Stress Markers

To confirm the regulative patterns obtained through quantitative proteomic analysis results (Tables S1 and S2), we explore the effects of PSP on proteins associated with overlapping cytoskeletal, IFN-IP signaling, and UPR complexes. The selective protein expression levels relating to cofilin-1, gelsolin, PKR, GRP78, and IRE1α were evaluated with Western blot (Figure 4A). Moreover, to grant further insight into PSP's role in its protein regulation, we examined the effects of PSP before (PSP/HIV) and after (HIV/PSP) infection. Immunoblot data showed that there was no statistical difference for the master regulator of

UPR and ER stress marker, GRP78, between PSP-treated cells and the control (Figure 4B). However, HIV infection resulted in an increase in protein expression levels for GRP78 of an average of 1.83-fold. A similar pattern was found in the HIV/PSP group with an increase of 1.60-fold. We observed that PSP lowered GRP78 expression in all instances where treatment was added alone or before infection, with no significance relative to control. The reverse effect was seen for HIV alone or after infection.

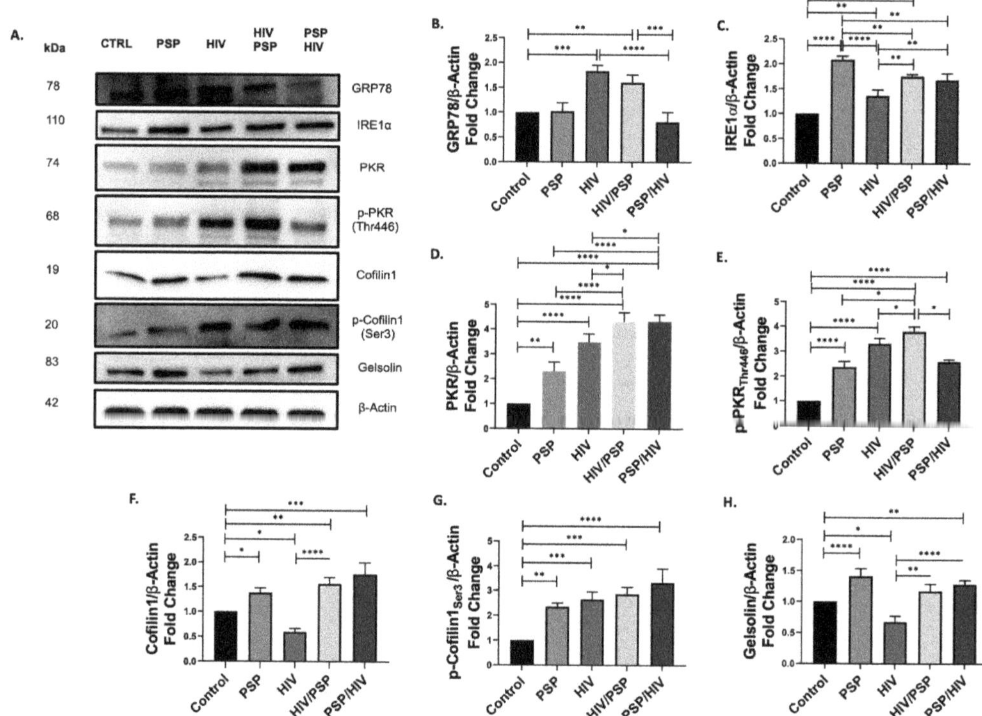

Figure 4. Western blot results related to UPR, IFN, and cytoskeletal biomarkers in PSP-treated THP1 monocytic cells. (**A**) Representative Western blot data for the protein expression levels of UPR: (**B**) GRP78; (**C**) IRE1α; IFN: (**D**) PKR; (**E**) p-PKR; ADF: (**F**) Cofilin1; (**G**) p-Cofilin1; and ABP: (**H**) Gelsolin signaling. β-Actin was used as a loading control and for normalization of data. Images were quantified using the ImageJ software (NIH, version 1.52a) by comparing the integrated density value with the control group. Mean $\pm$ SEM and significant difference (*), $p \leq 0.05$, (**), $p \leq 0.01$, (***), $p \leq 0.001$, (****), $p \leq 0.0001$ are shown and were determined using one-way ANOVA with Tukey multiple comparisons test, $n = 3$. All Western blot images can be found in Figure S2.

The expression patterns for IRE1α were upregulated in all instances for PSP and HIV infection. The highest statistical significance of a 2.08-fold change was seen for PSP alone in comparison to every sample group present (Figure 4C). HIV-1 infection demonstrated the lowest value of a 1.36-fold average relative to the control. In relation to its before and after infection counterparts, IRE1α expression was significantly higher, amounting to 1.75- and 1.68-fold respectively. There was no statistical difference reported relating to HIV/PSP vs. PSP/HIV samples.

Given that PKR has a history of interacting and overlapping with IRE1α and cofilin-1, we included this marker in our Western blot approaches as a marker of interest (Figure 4D). Results demonstrated that PSP has a strong influence on total PKR regulative pattern with a 2.30-fold change vs. the control. Similarly, PKR total expression levels increased in all groups, with the highest statistical significance in PSP treatment before and after infection (4.30- vs. 4.31-fold average respectively). In relation to these two experimental samples, there was no statistical difference seen between each other. Correspondingly, the activated and phosphorylated form of PKR at threonine 446 had a similar role to its total PKR counterpart, favoring upregulation in every group present (Figure 4E). PSP showed significantly higher activation by 3.79-fold expression in HIV/PSP relative to every group present. This data suggests an upregulation for both total and phosphorylated forms of PKR occurring in PSP-treated and infected cells.

ADF cofilin-1 and ABP gelsolin are strong influencers over HIV-1 entry, and it is for these reasons that they were of particular interest to reveal their regulative state in PSP-treated cells. A significant increase in cofilin-1 expression was seen in all instances where PSP was present (Figure 4F). In the present scenario, PSP/HIV had the highest fold value averaging a 1.76-fold difference before infection. Contrary to the previous results for the UPR and IFN-IP markers, there was a significant decrease average of a 0.59-fold change in total cofilin-1 vs. the control for HIV-1 infection. The deactivated/phosphorylated form of cofilin-1 had a similar pattern to the regulation of IRE1α and PKR. These show a tendency towards upregulation in every experimental situation (Figure 4G). Viral infection also demonstrated an increase of a 2.65-fold average of p-cofilin-1 compared to its total counterpart (Figure 4F). The highest reported fold difference of 3.33-fold was revealed before infection took place relative to the control. Lastly, immunoblot data for gelsolin revealed a slight increase in expression levels in all instances where PSP was present with a 1.41-fold difference as the highest in treatment only (Figure 4H). Equivalently to total cofilin-1 expression, viral infection received a significant decrease of 0.67-fold for cytosolic gelsolin. In summary, this data showed increased protein expression levels for key cytoskeletal, UPR, and IFN-IP markers.

3.5. Inhibition of PKR Modulates Cofilin-1 Phosphorylation Expression Patterns

We previously showed that PSP has the capacity to upregulate both total and phosphorylated forms of cofilin-1 in addition to a slight increase in the ABP gelsolin. We also demonstrated that PSP favors upregulation towards the activated and de-activated states of PKR as well as IRE1α. Our viral load data showed that viral entry increases under a PKR-inhibited state. Based on these results, it is possible PKR is involved as an upstream regulator of cofilin-1-mediated phosphorylation, serving as an entryway key factor during HIV infection. Therefore, we aim to assess how cofilin-1 and gelsolin are affected by PKR inhibition. We treated THP1 monocytic cells with 56.09 nM of C16 pharmacological blocker at PKR threonine 443 and subjected the cells to PSP treatment and infection as previously done in our experimental methodologies. Immunoblots showed inverse results to what was observed in PSP-treated cells without any drug intervention. C16 implementation significantly decreased cofilin-1 phosphorylation similar to that of the control group (Figure 5B). PSP treatment before infection suffered the lowest downregulation of 0.54-fold expression vs. control. Interestingly, the total activated form of cofilin-1 resulted in a significant increase in fold expression for every experimental group present (Figure 5C). Taking into consideration the results from Figure 4F, every sample group demonstrated an increase of ± 2-fold with the infection group alone extending this by almost a 5-fold difference. In an opposite manner compared to p-cofilin-1, the highest expression value for its activated counterpart was seen with PSP treatment before infection. This expression amounts to an average of 3.39-fold expression compared to 0.54-fold in phosphorylated cofilin-1. Similar to the data of p-cofilin-1, ABP gelsolin suffered a significant decrease in total protein expression levels for every sample group (Figure 5D). A tendency between these two showed the lowest fold decrease of 0.3 corresponding to PSP treatment before infection. Gelsolin data showed to be directing its focus in an opposite direction to results shown in Figure 4H.

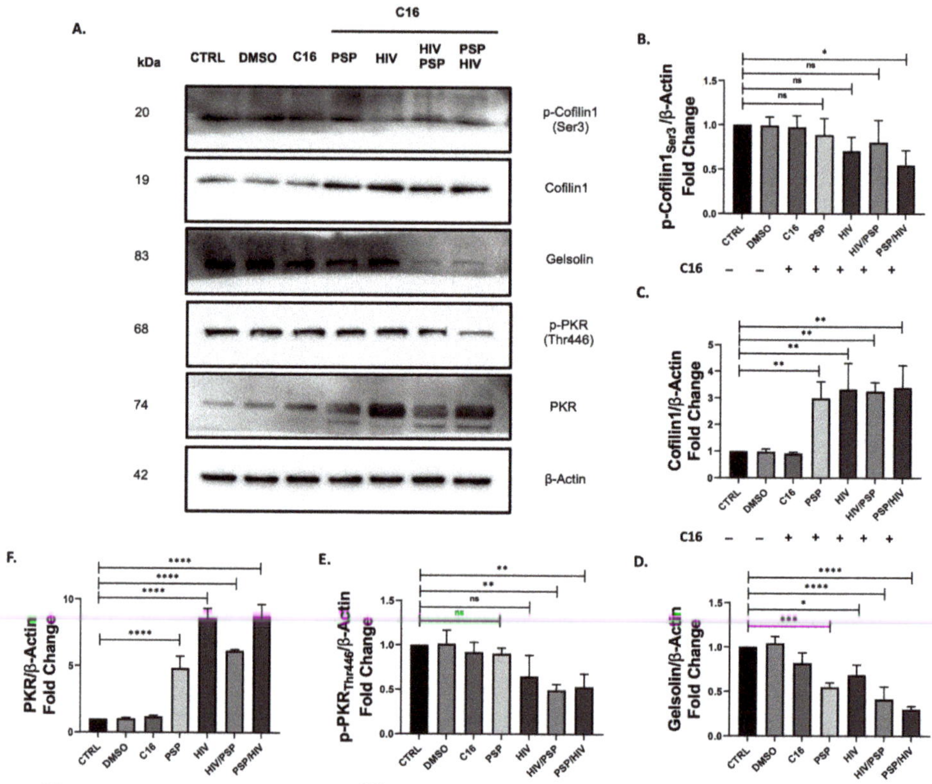

Figure 5. PKR inhibition decreases cofilin-1 phosphorylation and the expression levels relating to gelsolin. The activation of PKR was suppressed after 56.09 nM of C16 inhibitor in THP1 monocytic cells with or without PSP treatment. (**A**) Representative Western blot data for the protein expression levels of cytoskeletal: (**B**) pCofilin1; (**C**) Cofilin1; (**D**) Gelsolin; and IFN-IP: (**E**) pPKR; (**F**) PKR. β-Actin was used as a loading control and for normalization of data. Images were quantified using the ImageJ software (NIH, version 1.52a) by comparing the integrated density value with the control group. Mean ± SEM and significant difference (*), $p \leq 0.05$, (**), $p \leq 0.01$, (***), $p \leq 0.001$, (****), $p \leq 0.0001$ are shown and were determined using one-way ANOVA with Tukey multiple comparisons test, $n = 3$. All Western blot images can be found in Figure S2.

Lastly, we examined the impact of C16-mediated inhibition on the IFN-IP PKR. As expected, phosphorylation of PKR significantly decreases with the implementation of C16 for every situation present (Figure 5E). This downward trend was observed to be directly opposite to PSP-treated cells with no inhibitory conditions from Figure 4E. However, a significant increase in total PKR was seen for every group present with the highest fold difference of 8.77 for PSP treatment before viral infection (Figure 5F). The increments established in total PKR data account for a minimum of 2-folds for all groups with an exception for HIV/PSP which amounts to an increase of ±1.4-fold compared to our data from Figure 4D. These results may imply an accumulation of total protein in the cell. In summary, inhibition of PKR activation significantly decreases cofilin-1 phosphorylation as well as gelsolin protein expression in an inversely proportional manner. Consequently, the reduction of PKR phosphorylation leads to an increase in the total protein counterparts of PKR and cofilin-1.

3.6. Inhibition of Endoribonuclease Activity of IRE1α Is Associated with De-Regulation for Both PKR and Cofilin-1 Phosphorylation

Previous studies have shown that the induction of UPR has had a history of cross-talking with IFNs, specifically PKR [61,63,70]. Additionally, our viral load data have revealed an increase in HIV-1 entry when IRE1α endoribonuclease activity is blocked (Figure 3). Since PKR inhibition led to a significant decrease in p-cofilin-1 and inverse results in its total counterparts (Figure 5), we aim to unveil the potential role of IRE1α in these signaling components. Cells were treated with 221.8 nM of the endoribonuclease blocker 4μ8C for a period of 5 h prior to infection and treatment as previously performed in our methodological approaches. Immunoblot assays have revealed a significant increase in the phosphorylation activity of cofilin-1 in every experimental group (Figure 6B). This regulatory behavior resulted in higher fold expression values than C16 data (Figure 5B) but remained lower in comparison to PSP-treated cells with no drug intervention (Figure 4G). THP1 cells before infection demonstrated a ±50% average fold-change reduction in the midst of 4μ8C compared to no drug treatment. Interestingly, total cofilin-1 protein levels also received an increase in fold change. PSP-treated cells before viral infection obtained the highest fold value of 2.31 on average compared to the control group (Figure 6C). According to what we reported in Figures 4F and 5C, 4μ8C triggered elevated levels of protein expression compared to where no drug was administered but remain lower when cells were exposed to C16. Gelsolin levels were analogous to our results in Figure 5D, favoring a significant downward trend for all groups (Figure 6D). The lowest fold value of 0.47 on average was seen for PSP/HIV samples.

In relation to IFN-IP, we sought to investigate the molecular mechanism behind the regulatory effects of IRE1α and PKR. Our data showed that phosphorylation of PKR had a similar impact on the protein levels of phosphorylated cofilin-1. The expression values of p-PKR endure the inhibitory effects of 4μ8C and in this regard, it remained higher than C16 (Figure 6E). Contrary to what was observed in our data from Figure 4E, this same regulation was inferior where no drug activity happens. The total counterpart PKR would maintain elevated levels (Figure 6F) in contrast to no inhibitor (Figure 4D). However, the same regulatory position did not meet the same standards as its C16 equivalent (Figure 5F), by experiencing overall lesser protein levels across all experimental groups. When compared to the total cytoskeletal marker cofilin-1, both markers' levels increased during 4μ8C inhibition.

To corroborate that the observed results were directly involved with the activity of 4μ8C as well as to validate that our chosen concentration inhibited IRE1α endoribonuclease activity, the UPR marker X-box-binding protein 1 spliced (XBP1s) was included in our Western blot approach. XBP1 is a direct splicing effector downstream of the IRE1α endoribonuclease signaling cascade [71]. A reduction in XBP1s was observed for all experimental situations relative to control, insinuating a 4μ8C-mediated inhibition towards the specific RNase activity of IRE1α (Figure 6G). Taken together, this data is indicating a shift in protein levels and active roles of PKR and cofilin-1. Most importantly, the inhibition of IRE1α is showing a linkage and a closely related pattern between these two markers.

3.7. PSP Regulates the Gene Expressions Associated with Cytoskeleton and IFN-IP Signaling

Since both phosphorylated/total protein expressions of PKR and cofilin-1 are highly overexpressed in PSP-treated cells, we investigated the function of PSP as a regulator of the genes associated with these two signaling pathways. We hypothesized that the active shifting of total and phosphorylated forms of these markers, observed in preceding data, is partially due to the influence of PSP at the gene level. In addition, we decided to incorporate gelsolin in our dataset due to its seemingly tied role in protein regulation with PKR. We performed RT-qPCR analysis for PSP-treated cells under all conditions established in previous experiments. Supplementation of PSP increased the gene expression of cofilin-1 in all treated sample groups with PSP/HIV having the highest fold average of 2.25 (Figure 7A). A slight increase was observed for the infection group alone; however, it proved to have a non-statistical significance compared to the control. Gelsolin has been shown to have a similar gene regulative pattern as cofilin-1, where PSP addition incremented the overall levels of this protein (Figure 7B). The gene expression analysis identified significant upregulation of PKR in all groups present (Figure 7C). Concurrent with our Western blot experiments, RT-qPCR has revealed a close regulatory pattern for all signaling markers under PSP presence. Overall, this data opened a new insightful interpretation for PSP-induced signaling markers. We summarize our findings in Figure 7A–C, as PSP demonstrates control over PKR and gelsolin at the gene level and most importantly, the key HIV-1 entry factor, cofilin-1.

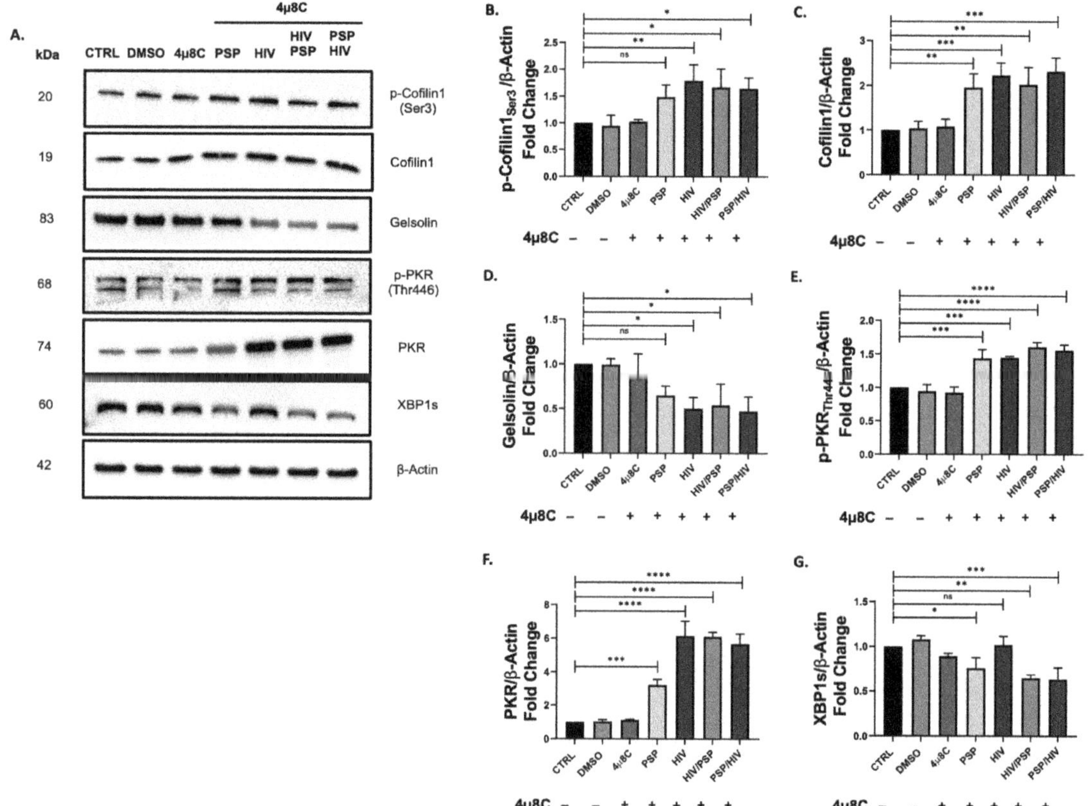

Figure 6. IRE1α modulates downstream phosphorylation activity of PKR and cofilin-1. Endoribonuclease activity of IRE1α was suppressed after 221.8 nM of 4µ8C inhibitor in THP1 monocytic cells with or without PSP treatment. (**A**) Representative Western blot data for the protein expression levels of cytoskeletal: (**B**) pCofilin1; (**C**) Cofilin1; (**D**) Gelsolin; IFN-IP: (**E**) p-PKR; (**F**) PKR and UPR: (**G**) XBP1s. β-Actin was used as a loading control and for normalization of data. Images were quantified using the ImageJ software (NIH, version 1.52a) by comparing the integrated density value with the control group. Mean ± SEM and significant difference (*), $p \leq 0.05$, (**), $p \leq 0.01$, (***), $p \leq 0.001$, (****), $p \leq 0.0001$ are shown and were determined using one-way ANOVA with Tukey multiple comparisons test, $n = 3$. All Western blot images can be found in Figure S2.

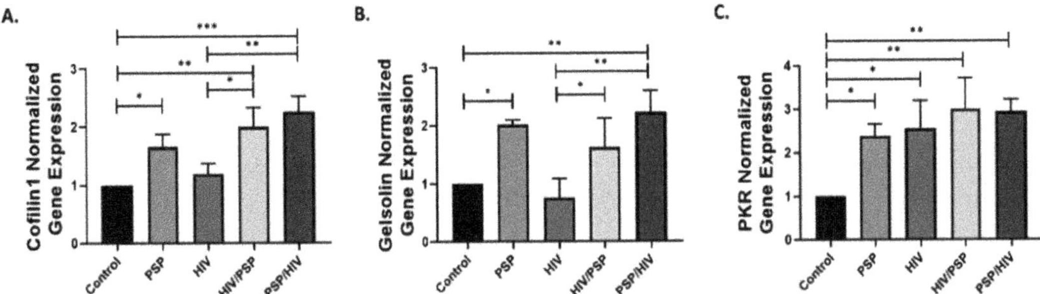

Figure 7. Validation of western blot results using RT-qPCR analysis for cytoskeletal and IFN-IP markers. Relative gene expression analyzed through RT-qPCR approach. The mRNA expression levels for the ADF: (**A**) Cofilin-1; ABP: (**B**) Gelsolin; and IFN-IP: (**C**) PKR are shown. All data were normalized using 18S as a housekeeping gene in response to PSP treatment. Mean $\pm$ SEM and significant difference (*), $p \leq 0.05$, (**), $p \leq 0.01$, (***), $p \leq 0.001$ are shown and were determined using one-way ANOVA, with Tukey multiple comparisons test, $n = 3$.

4. Discussion

Natural products have been wildly used in research as remedies to counteract the effects of diseases, particularly HIV-1. Compounds such as tea polyphenols show great potential at directly inhibiting HIV-associated malignancies [72]. Other organic substances, namely high-mannose oligosaccharides amalgamate, have demonstrated potent antiviral activity by blocking HIV-mediated entry [72,73]. Additional studies have elucidated the roles of carbohydrate-binding agents originating from plant lectins that significantly impact viral infection [74]. Having said this, PSP is a natural and commercially accessible supplement from the mushroom *Coriolus versicolor*. Previously, we published novel findings elucidating the anti-HIV effects of PSP in in vitro HIV-infected THP1 monocytic cells and ex vivo PBMCs. PSP has been demonstrated to possess anti-replicative capabilities against HIV-1 by promoting the upregulation of antiviral chemokines such as RANTES, MIP-1α/β, and SDF-1α. [8]. These chemokines are known to inhibit viral replication and as a result, they serve as immune enhancers [75–77]. Our laboratory has pioneered the immunomodulatory properties of PSP through a TLR4 response, giving way to unexplored signaling pathways that may further explain its immune-boosting factors. Moreover, viral load results have given insightful meaning to the internal antiviral effects that PSP possesses. The average inhibition percentage reported in acutely infected THP1 cells was 61%. This publication served as the basis for proposing this current study since the external role of PSP on HIV-1 entry currently remains unidentified. In addition to its immunological effects, PSP has been incorporated as an anti-cancer medication in clinical approaches with proven efficiency [78]. Due to its natural characteristics, no adverse side effects have been established with the use of PSP. Our research groups have previously performed cytotoxic assays which resulted in no significant signs of toxicity at concentrations higher than the established experimental dose. This essentially confirms the literature research data. Studies have shown that PSP has the innate ability to significantly increase the total count of immune cells without the unfavorable side effects seen in HAART [78,79]. Concurrent with the given history of PSP, other researchers have shown its potentiality with gastral and esophageal cancers due to its characteristics of showing a superior CD4+ and CD8+ cell count among other treatments [80].

The HIV-1 co-receptors CXCR4 and CCR5 signal directly through the cytoskeleton to promote viral entry into host-immune cells (Figure 1). It has been well-stated in the literature that HIV-1 requires manipulation of the actin filaments to promote receptor clustering of CXCR4/CCR5 and increase their spatial orientation and interactions as a pre-entry requisite. However, these receptors alone are not sufficient for infection to take place and pose a challenge for efficient fusion and insertion of the viral capsid. This is mainly due to actin filaments functioning as a physical barrier and blocking HIV-1 entry [16,26–28]. A second factor that comes into play is membrane fusion between the target cell and the virus acting as a means for successful infection in a post-entry step manner [16]. The regulator of these processes can be attributed to cofilin-1 by shifting the actin dynamic complex through the phases of elongation and breakdown. Specifically, this signaling cascade is modulated by the inactivity of phosphorylated cofilin-1 and its activated form respectively [81,82]. It is for these reasons that HIV-1 must overcome this restrictive property for successful infection to occur. Studies have shown that this HIV-induced mechanistic approach has been postulated as a delicate process. Therefore, interference with either form of cofilin-1 leads towards viral restriction [21,22]. In addition, gelsolin has also been associated as a deciding factor for HIV-1 entry, given that it shares the same regulation over cytoskeleton dynamics. Due to this, it is a target of interest in HIV-1 studies.

Given the knowledge of HIV-1 infection, this study investigated the role of PSP in regulating the ADF cofilin-1 required for viral entry through the IFN-IP and UPR signaling. We revealed numerous novel aspects of the external function of PSP as a regulator of viral entry. Among them, we showed that PSP has the potential to hinder HIV-1 entry by an approximately 74% when subjected to its antiviral effects before infection occurs in our in vitro model. The current research also highlights the underlying mechanisms by which PSP induces a shift in the phosphorylation states of PKR and consequently cofilin-1. It has been postulated in the literature research that both PKR and IRE1α can overlap with each other [50]. Inhibitory experiments in our dataset and validated with two pharmaceutical blockers in THP1 cells have confirmed these studies. The present research has demonstrated that PKR and IRE1α seem to have an established correlation under PSP influence. At the same time, inhibition of these biomarkers has led to a significant increase in viral entry for both cases. This suggests that the restrictive nature of PSP is associated with these two molecular signatures.

In this research, we have applied quantitative proteomics analysis to understand the mechanistic pathways that are being influenced by PSP. The current data shows that PSP induces overwhelming deregulation of 111 cytoskeletal as well as 28 proteins pertaining to the UPR. Among these, crucial proteins such as PKR and IRE1α demonstrate favorable up-regulatory roles. Other critical components relating to the small GTPase family specifically required for HIV entry are seen in unfavorable regulatory positions. The indicated GEFS includes RAB3IP, RAB8B, RALGPS2, RASAL2, RASGRP2, ARHGAP 15 and 17. The functions of the referred GEFS are known to signal the exchange of GDP to active GTP, upstream of cofilin-1 activity, and result in cytoskeleton remodeling [83–85]. With significant weight, we highlight the downregulation of SSH3 phosphatase, to which HIV signals through its co-receptor to promote the second phase for capsid fusion [18,19]. This strongly suggests that PSP is signaling toward the regulation of cofilin-1. Additional evidence suggests that PSP exerts control over actin dynamics through the upregulation of tropomyosin and tropomodulin, which directly binds and regulates actin-filaments lengthening [57–59]. Multiple studies have shown that UPR and cytoskeleton signaling overlap with each other and some of these characteristics can be attributed to the upregulation of PDI [40–43]. Its overexpression is mainly associated with disulfide bond formation and protein folding. However, this ER marker is also linked to the direct binding of β-Actin to regulate cytoskeleton reorganization [86]. Taken together, our proteomics data firmly implies that PSP has influence over actin dynamics and UPR signaling. Interestingly, a study has shown the relationship between the cytoskeleton and IFN-IP, particularly with PKR. The referred research demonstrated that PKR could serve as an upstream regulator of cofilin-1. Specifically, PKR can phosphorylate cofilin-1 and render it inactive [54]. Due to our gathered results and documented facts, we were interested in investigating the specific regulatory role of cofilin-1 as well as its correlation with PKR. In particular, PSP-treated cells resulted in an overexpression for both total and phosphorylated forms of cofilin-1 as well as for PKR. Given the literature knowledge that we possess between these two molecules, we proposed that cofilin-1 regulation is being influenced in some way by PKR and essentially can serve as a primary restrictive pathway. To challenge this hypothesis, we directly inhibited the ability for PKR to be activated by either auto-phosphorylation or indirectly stimulated by other factors such as foreign double-stranded RNAs and the endoribonuclease activity of IRE1α. Interestingly, PKR inhibition has led to a significant reduction of phosphorylation activity of cofilin-1 comparable to the control. On the opposite side of the spectrum, there was an increase in total cofilin-1 with a superior expression than PSP-treated cells with no drug intervention. This data suggests that PKR has a strong role in dictating the active state of ADF/cofilin-1 under the influence of PSP.

The ABP gelsolin is known for its abundance and functions in actin capping and severing. To date, this cytoskeletal marker shares three isoforms known as plasma (Isoform 1), cytoplasmic (Isoform 2), and gelsolin-3 [87]. A particular study from Irving et al. has published that PKR is able to counteract viral entry in the innate immune system by directly binding and inhibiting gelsolin [88]. This study raised interest in the interactions between this cytoskeletal marker and PKR since our data reveals both overexpression and restrictive properties of this IFN-IP. Our quantitative proteomic analysis has given insight into the downregulation of gelsolin-3. Since cytoplasmic gelsolin is a primary interest in HIV-entry studies due to its involvement with cytoskeletal events and its role has not been elucidated in PSP before, we decided to incorporate this marker in our experiments to elucidate its role in PSP-induced cells. In contrast to gelsolin-3, PSP seems to have some up-regulatory properties when it comes to its counterpart, showing elevated levels for every experimental condition. When PKR activity was blocked, gelsolin also showed analogous expressions in fold change when compared to p-cofilin-1, suggesting that the functional role of PKR is overlapping with cytoskeletal components. This dataset also serves as an additional validation marker where our chosen concentration for the C16 drug was effective at inhibiting PKR, given that both gelsolin and p-PKR were affected. When taken together, these data heavily suggest that PSP-induced overexpression of PKR is extending its reach by regulating pivotal viral entry components.

It is well known that HIV is recognized by the IFN response and triggers a high production of type I IFNs [89]. This is also observed when THP1 cells were infected, which resulted in a 3.4-fold increase. PKR is expressed at relatively low levels under normal conditions and its transcription begins once type I IFNs are stimulated by viral replication [90]. This study shows that both PSP and HIV-1 independently trigger a high expression of PKR with a greater effect seen when both conditions are combined. When cells were treated with PSP before infection took place, a 4.31-fold average increment of PKR was seen. This group constitutes the same that partook in our viral load analysis. Therefore, since PKR has been postulated before as a restrictive factor for viral entry, our validated results further strengthen these views indicating that PKR plays a major role in hindering viral entry in THP1 cells.

In recent years, studies on the interactions between UPR and PKR have surged with the purpose of promoting an anti-pathogenic environment. Among them, NF-κB signaling and production of type I IFNs have been researched to gain insights into their respective roles such as survivability and immune response [49,91,92]. Interestingly, a study has also shown that IRE1α directly activates PKR in response to Chlamydia trachomatis infection in a TLR4-dependent manner [50]. Having said this, our previously published data stand out by demonstrating the anti-viral effects of PSP through TLR4 and NF-κB. The present research has also demonstrated and validated UPR activity for the first time in PSP-treated cells. Specifically, we saw that PSP highly upregulates the expression of IRE1α by 2 folds while HIV-1 infection seems to minorly lower this response alone or in combination. Given the history of IRE1α in relation to the immune system and PKR, we aim to unveil its specific role in both viral entry and its association with cofilin-1. The gathered data indicates that IRE1α can regulate, to an extent, the phosphorylation patterns of both PKR and cofilin-1. This modulation remains significantly higher than the data seen in the C16-PKR blocker. However, inhibition of IRE1α also led to a decrease in the activity of p-PKR/p-cofilin-1 compared to PSP-treated cells with no drug activity. This renders the effects of this blocker in the middle ground between C16 and PSP non-inhibition, suggesting that PSP-induced IRE1α can serve as an upstream regulator. A peculiar pattern was observed where blockage of IRE1α endoribonuclease activity resulted in elevated levels of PKR/cofilin-1 total counterparts similar to C16. Consequently, gelsolin was also affected by a significant decrease in overall levels comparable to C16 data, suggesting that the overexpression of PKR is once again affecting the regulation of gelsolin. Since we have a clearer insight into the roles of IRE1α in PSP treatment, we were interested if PSP is inducing acute or chronic UPR/ER stress through GRP78. PSP has shown to have no statistical difference in GRP78 compared to the control group. On the other hand, HIV-1 infection significantly increases its overall expression across all groups. In hindsight, while HIV-1 shows superior GRP78 production, an established pattern was observed where PSP actively lowers these levels in all instances where it is present. This indicates that PSP is favoring acute rather than chronic ER stress, by counteracting the apoptotic effects of HIV-1 through GRP78 downregulation. This is evident by the effects of PSP addition before HIV infection which resulted in a 0.81-fold difference compared to 1.83 in the HIV-1 group alone. Additionally, these data are showing concurrency with previous studies as well as our published data, which emphasizes the non-toxic effects of PSP.

We also evaluated the effects of PSP at the gene level due to our hypothesis that accumulation of total PKR and cofilin-1 was due to genetic disturbances. As we expected, PSP induces the upregulation of cofilin-1, gelsolin, and PKR during RT-qPCR analysis. This gives weight to our hypothesis that the overwhelming upregulation of these markers was due to gene influences during inhibitory experiments. In addition, this further explains the reason for both total and phosphorylated forms of PKR/cofilin-1 being in an up-regulatory position. Overall, the corresponding results have led us to propose a mechanistic model that highlights the role of PSP in lowering viral entry through IRE1α and PKR (Figure 8). The overlapping signaling between IRE1α and PKR would allow the immune system to implement the necessary conditions to block viral entry. Consequently, this would restrict HIV-1 at the early entry step due to the overwhelming inactivity of cofilin-1. In general, the study at hand uncovers potential targetable PSP/UPR/PKR pathways that can be implemented in broader research with the aim of future therapeutic approaches.

Future Directions

PSP continues to show anti-HIV capabilities by currently targeting both entry and replicative cycles. Subsequently, to understand its true therapeutic potential, a comprehensive study will be conducted on the adaptive immune response. Specifically, future research will be validated using in vitro/ex vivo models as well as a population consisting of healthy and HIV-infected subjects.

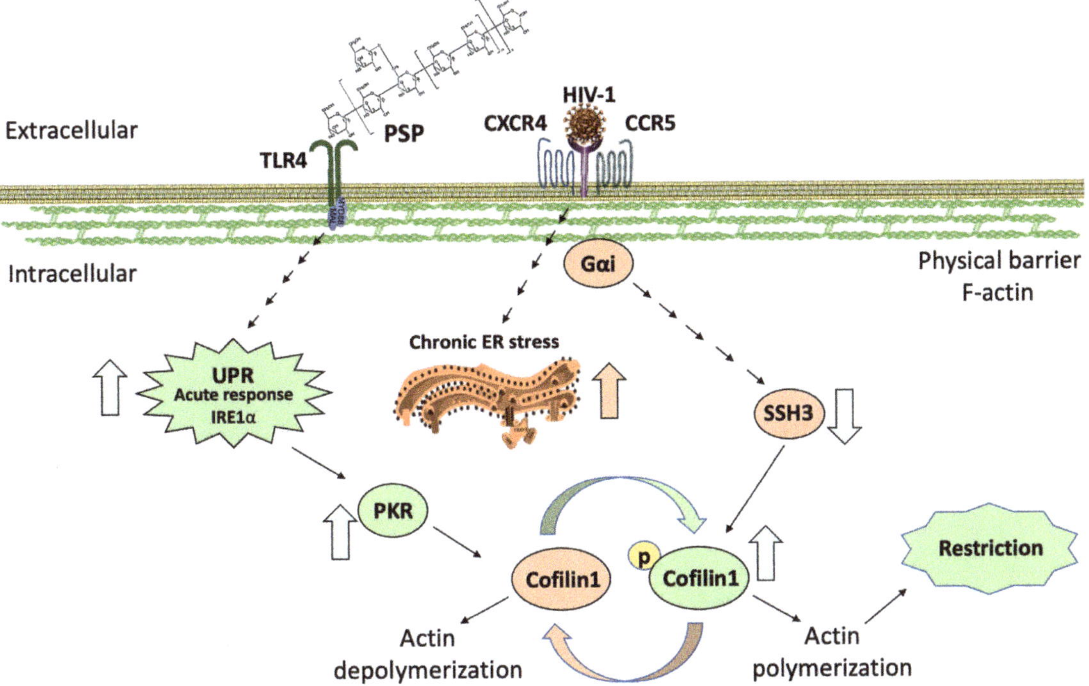

Figure 8. Model depicting PSP signaling through a UPR/IFN-induced pathway. HIV-1 infection results in a chronic ER stress response, while simultaneously dephosphorylating cofilin-1 through SSH3 phosphatase. This results in actin depolymerization for viral entry. Prior to infection, PSP treatment induces actin polymerization via an acute UPR. PKR mediates downstream phosphorylation of IRE1α signals and reverses HIV-induced actin remodeling while infection persists. Legend colors: Green—PSP-mediated signaling and events. Cream- HIV-downstream pathways.

5. Conclusions

Our study demonstrated, for the first time, that PSP hinders HIV-1 entry before infection takes place. Furthermore, we have shown that this restrictive ability is associated with positive regulation of PKR activation as well as IRE1α endoribonuclease activity. Consequently, cofilin-1 phosphorylation is also affected by these two signaling markers and it is reflected in the early entry phase. The present study has provided novel mechanistic insights between the interplay of UPR, IFN-IP, and cytoskeletal events. Taken together, the data provided here suggest that PSP has the potential to be used as a natural alternative to target HIV-1 entry.

Supplementary Materials: The following supporting information can be downloaded at: https://www.mdpi.com/article/10.3390/v15030804/s1, Figure S1: MTT cell viability of C16 and 4μ8C inhibitors. Figure S2: Full western blot images. Table S1: Differentially expressed cytoskeletal-related proteins identified in PSP treated THP1-Blue-CD14 cells. Table S2: Differentially expressed UPR proteins identified in PSP treated THP1-Blue-CD14 cells.

Author Contributions: Conceptualization, E.A.-R. and N.M.B.; methodology, E.A.-R. and M.R.-V.; validation, E.A.-R. and M.R.-V.; formal analysis, E.A.-R. and N.M.B.; investigation, E.A.-R., M.R.-V. and N.M.B.; resources, N.M.B.; data curation, E.A.-R. and M.R.-V.; writing-original draft preparation, E.A.-R.; writing-review and editing, N.M.B.; visualization, E.A.-R. and N.M.B.; supervision, N.M.B.; project administration, N.M.B.; funding acquisition, N.M.B. All authors have read and agreed to the published version of the manuscript.

Funding: This research was funded by The Puerto Rico Science, Technology and Research Trust (PRSTRT), grant number 2022-00158 (N.M.B.); INBRE-PR NIH grant P20GM103475 (N.M.B.) and the Universidad Central del Caribe pre-doctoral grant (E.A.-R.).

Institutional Review Board Statement: Not applicable.

Informed Consent Statement: Not applicable.

Data Availability Statement: The data supporting the results are found within the manuscript and Supplementary Files.

Acknowledgments: The authors wish to express heartfelt appreciations to Angelisa Franceschini, the Borinquen Laboratory and Roche Diagnostics for their support with HIV viral load testing services. We also express our gratitude to Ana M. Espino for donating the THP1-BLUE-CD14 cells and encouragements for this study. We kindly acknowledge The Hispanic Alliance for Clinical and Translational Research (Alliance): supported by the National Institute of General Medical Sciences (NIGMS) National Institutes of Health under the Award Number U54GM133807, for covering the article publishing charges. TMT labeling quantitative proteomics data was provided by the Thermo Fisher Scientific Center for Multiplex Proteomics at Harvard Medical School (http://tcmp.hms.harvard.edu (accessed on 4 February 2018).

Conflicts of Interest: The authors declare no conflict of interest.

References

1. World Health Organization [WHO]. HIV/AIDS. Available online: https://www.who.int/data/gho/data/themes/hiv-aids (accessed on 2 March 2023).
2. UNAIDS. Global HIV & AIDS Statistics. Available online: https://www.unaids.org/en/resources/fact-sheet (accessed on 2 March 2023).
3. McCann, N.C.; Horn, T.H.; Hyle, E.P.; Walensky, R.P. HIV Antiretroviral Therapy Costs in the United States, 2012–2018. *JAMA Intern. Med.* **2020**, *180*, 601–603. [CrossRef]
4. Bingham, A.; Shrestha, R.K.; Khurana, N.; Jacobson, E.U.; Farnham, P.G. Estimated Lifetime HIV–Related Medical Costs in the United States. *Sex. Transm. Dis.* **2021**, *48*, 299–304. [CrossRef] [PubMed]
5. Su, X.-Z.; Miller, L.H. The Discovery of Artemisinin and the Nobel Prize in Physiology or Medicine. *Sci. China Life Sci.* **2015**, *58*, 1175–1179. [CrossRef] [PubMed]
6. Blagodatski, A.; Yatsunskaya, M.; Mikhailova, V.; Tiasto, V.; Kagansky, A.; Katanaev, V.L. Medicinal Mushrooms as an Attractive New Source of Natural Compounds for Future Cancer Therapy. *Oncotarget* **2018**, *9*, 29259. [CrossRef] [PubMed]
7. Elsayed, E.A.; El Enshasy, H.; Wadaan, M.A.M.; Aziz, R. Mushrooms: A Potential Natural Source of Anti-Inflammatory Compounds for Medical Applications. *Mediat. Inflamm.* **2014**, *2014*, 805841. [CrossRef] [PubMed]
8. Rodríguez-Valentín, M.; López, S.; Rivera, M.; Ríos-Olivares, E.; Cubano, L.; Boukli, N.M. Naturally Derived Anti-HIV Polysaccharide Peptide (PSP) Triggers a Toll-like Receptor 4-Dependent Antiviral Immune Response. *J. Immunol. Res.* **2018**, *2018*, 8741698. [CrossRef] [PubMed]
9. Lee, C.; Liu, Q.-H.; Tomkowicz, B.; Yi, Y.; Freedman, B.D.; Collman, R.G. Macrophage Activation through CCR5- and CXCR4-Mediated Gp120-Elicited Signaling Pathways. *J. Leukoc. Biol.* **2003**, *74*, 676–682. [CrossRef] [PubMed]
10. Zhen, A.; Krutzik, S.R.; Levin, B.R.; Kasparian, S.; Zack, J.A.; Kitchen, S.G. CD4 Ligation on Human Blood Monocytes Triggers Macrophage Differentiation and Enhances HIV Infection. *J. Virol.* **2014**, *88*, 9934–9946. [CrossRef]
11. Manches, O.; Frleta, D.; Bhardwaj, N. Dendritic Cells in Progression and Pathology of HIV Infection. *Trends Immunol.* **2014**, *35*, 114–122. [CrossRef]
12. Okoye, A.A.; Picker, L.J. CD4+T-Cell Depletion in HIV Infection: Mechanisms of Immunological Failure. *Immunol. Rev.* **2013**, *254*, 54–64. [CrossRef]
13. Lucera, M.B.; Fleissner, Z.; Tabler, C.O.; Schlatzer, D.M.; Troyer, Z.; Tilton, J.C. HIV Signaling through CD4 and CCR5 Activates Rho Family GTPases That Are Required for Optimal Infection of Primary CD4+ T Cells. *Retrovirology* **2017**, *14*, 4. [CrossRef] [PubMed]
14. Wu, Y.; Yoder, A. Chemokine Coreceptor Signaling in HIV-1 Infection and Pathogenesis. *PLoS Pathog.* **2009**, *5*, e1000520. [CrossRef] [PubMed]
15. Abbas, W.; Herbein, G. Plasma Membrane Signaling in HIV-1 Infection. *Biochim. Biophys. Acta* **2014**, *1838*, 1132–1142. [CrossRef] [PubMed]
16. Lehmann, M.; Nikolic, D.S.; Piguet, V. How HIV-1 Takes Advantage of the Cytoskeleton during Replication and Cell-To-Cell Transmission. *Viruses* **2011**, *3*, 1757–1776. [CrossRef] [PubMed]
17. Yoder, A.; Yu, D.; Dong, L.; Iyer, S.R.; Xu, X.; Kelly, J.; Liu, J.; Wang, W.; Vorster, P.J.; Agulto, L.; et al. HIV Envelope-CXCR4 Signaling Activates Cofilin to Overcome Cortical Actin Restriction in Resting CD4 T Cells. *Cell* **2008**, *134*, 782–792. [CrossRef]

18. Chen, L.; Keppler, O.T.; Schölz, C. Post-Translational Modification-Based Regulation of HIV Replication. *Front. Microbiol.* **2018**, *9*, 2131. [CrossRef]

19. Nishita, M.; Tomizawa, C.; Yamamoto, M.; Horita, Y.; Ohashi, K.; Mizuno, K. Spatial and Temporal Regulation of Cofilin Activity by LIM Kinase and Slingshot Is Critical for Directional Cell Migration. *J. Cell Biol.* **2005**, *171*, 349–359. [CrossRef]

20. Vorster, P.J.; Guo, J.; Yoder, A.; Wang, W.; Zheng, Y.; Xu, X.; Yu, D.; Spear, M.; Wu, Y. LIM Kinase 1 Modulates Cortical Actin and CXCR4 Cycling and Is Activated by HIV-1 to Initiate Viral Infection. *J. Biol. Chem.* **2011**, *286*, 12554–12564. [CrossRef]

21. Yi, F.; Guo, J.; Dabbagh, D.; Spear, M.; He, S.; Kehn-Hall, K.; Fontenot, J.; Yin, Y.; Bibian, M.; Park, C.; et al. Discovery of Novel Small-Molecule Inhibitors of LIM Domain Kinase for Inhibiting HIV-1. *J. Virol.* **2017**, *91*, e02418-16. [CrossRef]

22. Ospina Stella, A.; Turville, S. All-Round Manipulation of the Actin Cytoskeleton by HIV. *Viruses* **2018**, *10*, 63. [CrossRef]

23. Fletcher, D.A.; Mullins, R.D. Cell Mechanics and the Cytoskeleton. *Nature* **2010**, *463*, 485–492. [CrossRef] [PubMed]

24. Parsons, J.T.; Horwitz, A.R.; Schwartz, M.A. Cell Adhesion: Integrating Cytoskeletal Dynamics and Cellular Tension. *Nat. Rev. Mol. Cell Biol.* **2010**, *11*, 633–643. [CrossRef] [PubMed]

25. Seetharaman, S.; Etienne-Manneville, S. Cytoskeletal Crosstalk in Cell Migration. *Trends Cell Biol.* **2020**, *30*, 720–735. [CrossRef] [PubMed]

26. Melikyan, G.B. Common Principles and Intermediates of Viral Protein-Mediated Fusion: The HIV-1 Paradigm. *Retrovirology* **2008**, *5*, 111. [CrossRef]

27. Iyengar, S.; Hildreth, J.E.K.; Schwartz, D.H. Actin-Dependent Receptor Colocalization Required for Human Immunodeficiency Virus Entry into Host Cells. *J. Virol.* **1998**, *72*, 5251–5255. [CrossRef]

28. Fackler, O.T.; Kräusslich, H.-G. Interactions of Human Retroviruses with the Host Cell Cytoskeleton. *Curr. Opin. Microbiol.* **2006**, *9*, 409–415. [CrossRef]

29. Southwick, F.S. Gelsolin and ADF/Cofilin Enhance the Actin Dynamics of Motile Cells. *Proc. Natl. Acad. Sci. USA* **2000**, *97*, 6936–6938. [CrossRef]

30. Kotila, T.; Wioland, H.; Enkavi, G.; Kogan, K.; Vattulainen, I.; Jégou, A.; Romet-Lemonne, G.; Lappalainen, P. Mechanism of Synergistic Actin Filament Pointed End Depolymerization by Cyclase-Associated Protein and Cofilin. *Nat. Commun.* **2019**, *10*, 5320. [CrossRef]

31. García-Expósito, L.; Ziglio, S.; Barroso-González, J.; de Armas-Rillo, L.; Valera, M.-S.; Zipeto, D.; Machado, J.-D.; Valenzuela-Fernández, A. Gelsolin Activity Controls Efficient Early HIV-1 Infection. *Retrovirology* **2013**, *10*, 39. [CrossRef]

32. Piktel, E.; Levental, I.; Durnaś, B.; Janmey, P.A.; Bucki, R. Plasma Gelsolin: Indicator of Inflammation and Its Potential as a Diagnostic Tool and Therapeutic Target. *Int. J. Mol. Sci.* **2018**, *19*, 2516. [CrossRef]

33. Gurel, P.S.; Hatch, A.L.; Higgs, H.N. Connecting the Cytoskeleton to the Endoplasmic Reticulum and Golgi. *Curr. Biol.* **2014**, *24*, R660–R672. [CrossRef] [PubMed]

34. Tsai, F.-C.; Kuo, G.-H.; Chang, S.-W.; Tsai, P.-J. Ca2+Signaling in Cytoskeletal Reorganization, Cell Migration, and Cancer Metastasis. *Biomed Res. Int.* **2015**, *2015*, 409245. [CrossRef]

35. Biwer, L.A.; Isakson, B.E. Endoplasmic Reticulum-Mediated Signalling in Cellular Microdomains. *Acta Physiol.* **2017**, *219*, 162–175. [CrossRef] [PubMed]

36. Schwarz, D.S.; Blower, M.D. The Endoplasmic Reticulum: Structure, Function and Response to Cellular Signaling. *Cell Mol. Life Sci.* **2016**, *73*, 79–94. [CrossRef] [PubMed]

37. Wang, M.; Wey, S.; Zhang, Y.; Ye, R.; Lee, A.S. Role of the Unfolded Protein Response Regulator GRP78/BiP in Development, Cancer, and Neurological Disorders. *Antioxid. Redox. Signal.* **2009**, *11*, 2307–2316. [CrossRef]

38. Read, A.; Schröder, M. The Unfolded Protein Response: An Overview. *Biology* **2021**, *10*, 384. [CrossRef]

39. Hetz, C.; Papa, F.R. The Unfolded Protein Response and Cell Fate Control. *Mol. Cell* **2018**, *69*, 169–181. [CrossRef]

40. van Vliet, A.R.; Agostinis, P. PERK and Filamin a in Actin Cytoskeleton Remodeling at ER-Plasma Membrane Contact Sites. *Mol. Cell Oncol.* **2017**, *4*, e1340105. [CrossRef]

41. Urra, H.; Henriquez, D.R.; Cánovas, J.; Villarroel-Campos, D.; Carreras-Sureda, A.; Pulgar, E.; Molina, E.; Hazari, Y.M.; Limia, C.M.; Alvarez-Rojas, S.; et al. IRE1α Governs Cytoskeleton Remodelling and Cell Migration through a Direct Interaction with Filamin A. *Nat. Cell Biol.* **2018**, *20*, 942–953. [CrossRef]

42. Rainbolt, T.K.; Frydman, J. Dynamics and Clustering of IRE1α during ER Stress. *Proc. Natl. Acad. Sci. USA* **2020**, *117*, 3352–3354. [CrossRef]

43. Dejeans, N.; Pluquet, O.; Lhomond, S.; Grise, F.; Bouchecareilh, M.; Juin, A.; Meynard-Cadars, M.; Bidaud-Meynard, A.; Gentil, C.; Moreau, V.; et al. Autocrine Control of Glioma Cells Adhesion/Migration through IRE1α-Mediated Cleavage of Secreted Protein Acidic Rich in Cysteine (SPARC) MRNA. *J. Cell Sci.* **2012**, *125*, 4278–4287. [CrossRef] [PubMed]

44. van Vliet, A.R.; Sassano, M.L.; Agostinis, P. The Unfolded Protein Response and Membrane Contact Sites: Tethering as a Matter of Life and Death? *Contact* **2018**, *1*, 251525641877051. [CrossRef]

45. Saito, A.; Imaizumi, K. Unfolded Protein Response-Dependent Communication and Contact among Endoplasmic Reticulum, Mitochondria, and Plasma Membrane. *Int. J. Mol. Sci.* **2018**, *19*, 3215. [CrossRef]

46. Ebstein, F.; Poli Harlowe, M.C.; Studencka-Turski, M.; Krüger, E. Contribution of the Unfolded Protein Response (UPR) to the Pathogenesis of Proteasome-Associated Autoinflammatory Syndromes (PRAAS). *Front. Immunol.* **2019**, *10*, 2756. [CrossRef] [PubMed]

47. Kim, T.-H.; Cho, S.-G. Kisspeptin Inhibits Cancer Growth and Metastasis via Activation of EIF2AK2. *Mol. Med. Rep.* **2017**, *16*, 7585–7590. [CrossRef] [PubMed]

48. Inoue, R.; Nishi, H.; Osaka, M.; Yoshida, M.; Nangaku, M. Neutrophil Protein Kinase R Mediates Endothelial Adhesion and Migration by the Promotion of Neutrophil Actin Polymerization. *J. Immunol.* **2022**, *208*, 2173–2183. [CrossRef]
49. Zhu, Z.; Liu, P.; Yuan, L.; Lian, Z.; Hu, D.; Yao, X.; Li, X. Induction of UPR Promotes Interferon Response to Inhibit PRRSV Replication via PKR and NF-KB Pathway. *Front. Microbiol.* **2021**, *12*, 757690. [CrossRef]
50. Webster, S.J.; Ellis, L.; O'Brien, L.M.; Tyrrell, B.; Fitzmaurice, T.J.; Elder, M.J.; Clare, S.; Chee, R.; Gaston, J.S.H.; Goodall, J.C. IRE1α Mediates PKR Activation in Response to Chlamydia Trachomatis Infection. *Microbes Infect.* **2016**, *18*, 472–483. [CrossRef]
51. Lemaire, P.A.; Anderson, E.; Lary, J.; Cole, J.L. Mechanism of PKR Activation by DsRNA. *J. Mol. Biol.* **2008**, *381*, 351–360. [CrossRef]
52. Li, S.; Peters, G.A.; Ding, K.; Zhang, X.; Qin, J.; Sen, G.C. Molecular Basis for PKR Activation by PACT or DsRNA. *Proc. Natl. Acad. Sci. USA* **2006**, *103*, 10005–10010. [CrossRef]
53. Gal-Ben-Ari, S.; Barrera, I.; Ehrlich, M.; Rosenblum, K. PKR: A Kinase to Remember. *Front. Mol. Neurosci.* **2019**, *11*, 480. [CrossRef]
54. Xu, M.; Chen, G.; Wang, S.; Liao, M.; Frank, J.A.; Bower, K.A.; Zhang, Z.; Shi, X.; Luo, J. Double-Stranded RNA-Dependent Protein Kinase Regulates the Motility of Breast Cancer Cells. *PLoS ONE* **2012**, *7*, e47721. [CrossRef]
55. López, S.N.; Rodríguez-Valentín, M.; Rivera, M.; Rodríguez, M.; Babu, M.; Cubano, L.A.; Xiong, H.; Wang, G.; Kucheryavykh, L.; Boukli, N.M. HIV-1 Gp120 Clade B/c Induces a GRP78 Driven Cytoprotective Mechanism in Astrocytoma. *Oncotarget* **2017**, *8*, 68415–68438. [CrossRef] [PubMed]
56. Rivera, M.; Ramos, Y.; Rodríguez-Valentín, M.; López-Acevedo, S.; Cubano, L.A.; Zou, J.; Zhang, Q.; Wang, G.; Boukli, N.M. Targeting Multiple Pro-Apoptotic Signaling Pathways with Curcumin in Prostate Cancer Cells. *PLoS ONE* **2017**, *12*, e0179587. [CrossRef] [PubMed]
57. Ono, S.; Ono, K. Tropomyosin Inhibits ADF/Cofilin-Dependent Actin Filament Dynamics. *J. Cell Biol.* **2002**, *156*, 1065–1076. [CrossRef] [PubMed]
58. Hotulainen, P.; Paunola, E.; Vartiainen, M.K.; Lappalainen, P. Actin-Depolymerizing Factor and Cofilin-1 Play Overlapping Roles in Promoting Rapid F-Actin Depolymerization in Mammalian Nonmuscle Cells. *Mol. Biol. Cell* **2005**, *16*, 649–664. [CrossRef]
59. Dos Remedios, C.G.; Chhabra, D.; Kekic, M.; Dedova, I.V.; Tsubakihara, M.; Berry, D.A.; Nosworthy, N.J. Actin Binding Proteins: Regulation of Cytoskeletal Microfilaments. *Physiol. Rev.* **2003**, *83*, 433–473. [CrossRef]
60. Kimbara, N.; Dohi, N.; Miyamoto, M.; Asai, S.; Okada, H.; Okada, N. Diagnostic Surface Expression of SWAP-70 on HIV-1 Infected T Cells. *Microbiol. Immunol.* **2006**, *50*, 235–242. [CrossRef]
61. Baranov, M.V.; Revelo, N.H.; Verboogen, D.R.J.; ter Beest, M.; van den Bogaart, G. SWAP70 Is a Universal GEF-like Adaptor for Tethering Actin to Phagosomes. *Small GTPases* **2019**, *10*, 311–323. [CrossRef]
62. Yin, W.; Li, W.; Li, Q.; Liu, Y.; Liu, J.; Ren, M.; Ma, Y.; Zhang, Z.; Zhang, X.; Wu, Y.; et al. Real-Time Imaging of Individual Virion-Triggered Cortical Actin Dynamics for Human Immunodeficiency Virus Entry into Resting CD4 T Cells. *Nanoscale* **2020**, *12*, 115–129. [CrossRef]
63. Naghavi, M.H. HIV-1 Capsid Exploitation of the Host Microtubule Cytoskeleton during Early Infection. *Retrovirology* **2021**, *18*, 19. [CrossRef] [PubMed]
64. Sun, G.D.; Kobayashi, T.; Abe, M.; Tada, N.; Adachi, H.; Shiota, A.; Totsuka, Y.; Hino, O. The Endoplasmic Reticulum Stress-Inducible Protein Niban Regulates EIF2α and S6K1/4E-BP1 Phosphorylation. *Biochem. Biophys. Res. Commun.* **2007**, *360*, 181–187. [CrossRef]
65. Olivari, S.; Cali, T.; Salo, K.E.; Paganetti, P.; Ruddock, L.W.; Molinari, M. EDEM1 Regulates ER-Associated Degradation by Accelerating De-Mannosylation of Folding-Defective Polypeptides and by Inhibiting Their Covalent Aggregation. *Biochem. Biophys. Res. Commun.* **2006**, *349*, 1278–1284. [CrossRef] [PubMed]
66. Marcu, M.G.; Doyle, M.; Bertolotti, A.; Ron, D.; Hendershot, L.; Neckers, L. Heat Shock Protein 90 Modulates the Unfolded Protein Response by Stabilizing IRE1α. *Mol. Cell Biol.* **2002**, *22*, 8506–8513. [CrossRef] [PubMed]
67. Chen, X.; Duan, L.-H.; Luo, P.; Hu, G.; Yu, X.; Liu, J.; Lu, H.; Liu, B. FBXO6-Mediated Ubiquitination and Degradation of Ero1L Inhibits Endoplasmic Reticulum Stress-Induced Apoptosis. *Cell. Physiol. Biochem.* **2016**, *39*, 2501–2508. [CrossRef] [PubMed]
68. Liang, Y.; Luo, J.; Yang, N.; Wang, S.; Ye, M.; Pan, G. Activation of the IL-1β/KLF2/HSPH1 Pathway Promotes STAT3 Phosphorylation in Alveolar Macrophages during LPS-Induced Acute Lung Injury. *Biosci. Rep.* **2020**, *40*, BSR20193572. [CrossRef] [PubMed]
69. Clarke, J.P.; Mearow, K.M. Cell Stress Promotes the Association of Phosphorylated HspB1 with F-Actin. *PLoS ONE* **2013**, *8*, e68978. [CrossRef]
70. Lee, E.-S.; Yoon, C.-H.; Kim, Y.-S.; Bae, Y.-S. The Double-Strand RNA-Dependent Protein Kinase PKR Plays a Significant Role in a Sustained ER Stress-Induced Apoptosis. *FEBS Lett.* **2007**, *581*, 4325–4332. [CrossRef]
71. Kishino, A.; Hayashi, K.; Hidai, C.; Masuda, T.; Nomura, Y.; Oshima, T. XBP1-FoxO1 Interaction Regulates ER Stress-Induced Autophagy in Auditory Cells. *Sci. Rep.* **2017**, *7*, 4442. [CrossRef]
72. Fatima, I.; Kanwal, S.; Mahmood, T. Natural Products Mediated Targeting of Virally Infected Cancer. *Dose Response* **2019**, *17*, 155932581881322. [CrossRef]
73. O'Keefe, B.R.; Smee, D.F.; Turpin, J.A.; Saucedo, C.J.; Gustafson, K.R.; Mori, T.; Blakeslee, D.; Buckheit, R.; Boyd, M.R. Potent Anti-Influenza Activity of Cyanovirin-N and Interactions with Viral Hemagglutinin. *Antimicrob. Agents Chemother.* **2003**, *47*, 2518–2525. [CrossRef] [PubMed]

74. Balzarini, J. Carbohydrate-Binding Agents: A Potential Future Cornerstone for the Chemotherapy of Enveloped Viruses? *Antivir. Chem. Chemother.* **2007**, *18*, 1–11. [CrossRef] [PubMed]
75. An, P.; Nelson, G.W.; Wang, L.; Donfield, S.; Goedert, J.J.; Phair, J.; Vlahov, D.; Buchbinder, S.; Farrar, W.L.; Modi, W.; et al. Modulating Influence on HIV/AIDS by Interacting RANTES Gene Variants. *Proc. Natl. Acad. Sci. USA* **2002**, *99*, 10002–10007. [CrossRef]
76. Cocchi, F.; DeVico, A.L.; Yarchoan, R.; Redfield, R.; Cleghorn, F.; Blattner, W.A.; Garzino-Demo, A.; Colombini-Hatch, S.; Margolis, D.; Gallo, R.C. Higher Macrophage Inflammatory Protein (MIP)-1α and MIP-1β Levels from CD8$^+$ T Cells Are Associated with Asymptomatic HIV-1 Infection. *Proc. Natl. Acad. Sci. USA* **2000**, *97*, 13812–13817. [CrossRef]
77. Gianesin, K.; Freguja, R.; Carmona, F.; Zanchetta, M.; Del Bianco, P.; Malacrida, S.; Montagna, M.; Rampon, O.; Giaquinto, C.; De Rossi, A. The Role of Genetic Variants of Stromal Cell-Derived Factor 1 in Pediatric HIV-1 Infection and Disease Progression. *PLoS ONE* **2012**, *7*, e44460. [CrossRef]
78. Tsang, K.W.; Lam, C.L.; Yan, C.; Mak, J.C.; Ooi, G.C.; Ho, J.C.; Lam, B.; Man, R.; Sham, J.S.; Lam, W.K. Coriolus Versicolor Polysaccharide Peptide Slows Progression of Advanced Non-Small Cell Lung Cancer. *Respir. Med.* **2003**, *97*, 618–624. [CrossRef]
79. Tzianabos, A.O. Polysaccharide Immunomodulators as Therapeutic Agents: Structural Aspects and Biologic Function. *Clin. Microbiol. Rev.* **2000**, *13*, 523–533. [CrossRef] [PubMed]
80. Ng, T.B. A Review of Research on the Protein-Bound Polysaccharide (Polysaccharopeptide, PSP) from the Mushroom Coriolus Versicolor (Basidiomycetes: Polyporaceae). *Gen. Pharmacol.* **1998**, *30*, 1–4. [CrossRef]
81. Ohashi, K. Roles of Cofilin in Development and Its Mechanisms of Regulation. *Dev. Growth Differ.* **2015**, *57*, 275–290. [CrossRef]
82. Andrianantoandro, E.; Pollard, T.D. Mechanism of Actin Filament Turnover by Severing and Nucleation at Different Concentrations of ADF/Cofilin. *Mol. Cell* **2006**, *24*, 13–23. [CrossRef]
83. Audoly, G.; Popoff, M.R.; Gluschankof, P. Involvement of a Small GTP Binding Protein in HIV-1 Release. *Retrovirology* **2005**, *2*, 48. [CrossRef] [PubMed]
84. Kjos, I.; Vestre, K.; Guadagno, N.A.; Borg Distefano, M.; Progida, C. Rab and Arf Proteins at the Crossroad between Membrane Transport and Cytoskeleton Dynamics. *Biochim. Biophys. Acta Mol. Cell Res.* **2018**, *1865*, 1397–1409. [CrossRef] [PubMed]
85. Sahai, E.; Olson, M.F.; Marshall, C.J. Cross-Talk between Ras and Rho Signalling Pathways in Transformation Favours Proliferation and Increased Motility. *EMBO J.* **2001**, *20*, 755–766. [CrossRef] [PubMed]
86. Sobierajska, K.; Skurzynski, S.; Stasiak, M.; Kryczka, J.; Cierniewski, C.S.; Swiatkowska, M. Protein Disulfide Isomerase Directly Interacts with β-Actin Cys374 and Regulates Cytoskeleton Reorganization. *J. Biol. Chem.* **2014**, *289*, 5758–5773. [CrossRef] [PubMed]
87. Feldt, J.; Schicht, M.; Garreis, F.; Welss, J.; Schneider, U.W.; Paulsen, F. Structure, Regulation and Related Diseases of the Actin-Binding Protein Gelsolin. *Expert Rev. Mol. Med.* **2019**, *20*, e7. [CrossRef]
88. Irving, A.T.; Wang, D.; Vasilevski, O.; Latchoumanin, O.; Kozer, N.; Clayton, A.H.; Szczepny, A.; Morimoto, H.; Xu, D.; Williams, B.R.G.; et al. Regulation of Actin Dynamics by Protein Kinase R Control of Gelsolin Enforces Basal Innate Immune Defense. *Immunity* **2012**, *36*, 795–806. [CrossRef]
89. Soper, A.; Kimura, I.; Nagaoka, S.; Konno, Y.; Yamamoto, K.; Koyanagi, Y.; Sato, K. Type I Interferon Responses by HIV-1 Infection: Association with Disease Progression and Control. *Front. Immunol.* **2018**, *8*, 1823. [CrossRef]
90. Nakayama, Y.; Plisch, E.H.; Sullivan, J.; Thomas, C.; Czuprynski, C.J.; Williams, B.R.G.; Suresh, M. Role of PKR and Type I IFNs in Viral Control during Primary and Secondary Infection. *PLoS Pathog.* **2010**, *6*, e1000966. [CrossRef]
91. Schmitz, M.; Shaban, M.; Albert, B.; Gökçen, A.; Kracht, M. The Crosstalk of Endoplasmic Reticulum (ER) Stress Pathways with NF-KB: Complex Mechanisms Relevant for Cancer, Inflammation and Infection. *Biomedicines* **2018**, *6*, 58. [CrossRef]
92. Tam, A.B.; Mercado, E.L.; Hoffmann, A.; Niwa, M. ER Stress Activates NF-KB by Integrating Functions of Basal IKK Activity, IRE1 and PERK. *PLoS ONE* **2012**, *7*, e45078. [CrossRef]

viruses

Communication

CXCR4 Recognition by L- and D-Peptides Containing the Full-Length V3 Loop of HIV-1 gp120

Ruohan Zhu [1], Xiaohong Sang [2], Jiao Zhou [2], Qian Meng [1], Lina S. M. Huang [3], Yan Xu [2], Jing An [3,*] and Ziwei Huang [1,2,*]

1　School of Life Sciences, Tsinghua University, Beijing 100084, China
2　Ciechanover Institute of Precision and Regenerative Medicine, School of Medicine, Chinese University of Hong Kong, Shenzhen 518172, China
3　Division of Infectious Diseases and Global Public Heath, Department of Medicine, School of Medicine, University of California at San Diego, 9500 Gilman Drive, La Jolla, CA 92093, USA
*　Correspondence: jan@health.ucsd.edu (J.A.); zwhny@yahoo.com (Z.H.)

Abstract: Human immunodeficiency virus-1 (HIV-1) recognizes one of its principal coreceptors, CXC chemokine receptor 4 (CXCR4), on the host cell via the third variable loop (V3 loop) of HIV-1 envelope glycoprotein gp120 during the viral entry process. Here, the mechanism of the molecular recognition of HIV-1 gp120 V3 loop by coreceptor CXCR4 was probed by synthetic peptides containing the full-length V3 loop. The two ends of the V3 loop were covalently linked by a disulfide bond to form a cyclic peptide with better conformational integrity. In addition, to probe the effect of the changed side-chain conformations of the peptide on CXCR4 recognition, an all-D-amino acid analog of the L-V3 loop peptide was generated. Both of these cyclic L- and D-V3 loop peptides displayed comparable binding recognition to the CXCR4 receptor, but not to another chemokine receptor, CCR5, suggesting their selective interactions with CXCR4. Molecular modeling studies revealed the important roles played by many negative-charged Asp and Glu residues on CXCR4 that probably engaged in favorable electrostatic interactions with the positive-charged Arg residues present in these peptides. These results support the notion that the HIV-1 gp120 V3 loop-CXCR4 interface is flexible for ligands of different chiralities, which might be relevant in terms of the ability of the virus to retain coreceptor recognition despite the mutations at the V3 loop.

Keywords: HIV-1 gp120 V3 loop; chemokine receptor CXCR4; molecular recognition; D-peptide; HIV-1 entry

Citation: Zhu, R.; Sang, X.; Zhou, J.; Meng, Q.; Huang, L.S.M.; Xu, Y.; An, J.; Huang, Z. CXCR4 Recognition by L- and D-Peptides Containing the Full-Length V3 Loop of HIV-1 gp120. *Viruses* **2023**, *15*, 1084. https://doi.org/10.3390/v15051084

Academic Editor: Simone Brogi

Received: 6 April 2023
Revised: 25 April 2023
Accepted: 26 April 2023
Published: 28 April 2023

1. Introduction

CXC chemokine receptor 4 (CXCR4) belongs to class A G-protein-coupled receptors (GPCRs) [1]. It has a unique cognate ligand CXC chemokine ligand 12 (CXCL12), also called stromal-cell-derived factor (SDF-1α) [2,3]. The CXCR4/CXCL12 axis plays important roles in complex biological processes such as inflammatory response, the mobilization and homing of hematopoietic stem cells (HSPCs), angiogenesis, and cell survival [4]. CXCR4 is also shown to be a cell surface receptor of ubiquitin and macrophage migration inhibitory factor (MIF) and can be partially activated upon interaction [5–7].

In 1996, CXCR4 was identified, along with another chemokine receptor CCR5, as the long sought after co-receptor for human immunodeficiency virus type 1 (HIV-1) [2,8–10]. It is thought that the envelope glycoprotein 120 (gp120) of HIV-1 binds the host cell's primary receptor CD4 and undergoes conformation changes which expose its third variable loop (V3 loop) to interact with coreceptor CXCR4 or CCR5, triggering further molecular events that lead to virus–cell membrane fusion and infection [11,12]. The co-receptor usage depends on the cell tropism of an HIV-1 strain, with CXCR4 as the coreceptor for T-cell-line-tropic (T-tropic) HIV-1 and CCR5 for macrophage-tropic (M-tropic) HIV-1 [13]. Usually, viruses using CCR5 as the coreceptor (i.e., R5 viruses) are predominant during the early stage of the infection, whereas viruses using CXCR4 (X4 viruses) are often associated with a late and rapid clinical progression [14,15]. The diverse functions of CXCR4 in physiologies and pathologies as described above have made it a therapeutic target in intensive drug discovery research for HIV and other human diseases [16].

For more than two decades, our laboratories have been interested in the stereochemistry of CXCR4-ligand recognition and its implication in the molecular mechanism of HIV-1 entry via this coreceptor. In 2002, we reported the original finding that CXCR4 could recognize both L- and D-peptides of identical amino acid sequences but with opposite chirality, corresponding to the N-terminus of a viral chemokine vMIP-II, which revealed the flexible stereochemical requirement on the CXCR4-chemokine ligand interface [17]. More recently, we extended this investigation to the viral ligand of CXCR4, HIV-1 gp120, and found that L- and D-peptides derived from fragments of HIV-1 gp120 V3 loop, when combined with another D-peptide derived from vMIP-II to gain higher receptor binding affinity, displayed nanomolar affinity for CXCR4 and strong antagonistic activities [18,19], which suggested that the flexible stereochemistry initially observed for the CXCR4-chemokine peptide interface appeared to hold true for the interaction between CXCR4 and HIV-1 gp120 V3 loop peptide fragments.

Here, in this study, we wanted to further extend the above-described observation with fragments of HIV-1 gp120 V3 loop to the full-length V3 loop. Two L- and D-peptides corresponding to the full-length gp120 V3 loop of HIV-1 89.6 (dual-tropic) strain were chemically synthesized and tested for their interaction with CXCR4 using an anti-CXCR4 antibody competitive binding assay. Both peptides bound CXCR4 with comparable IC_{50} values. The molecular mechanisms of CXCR4's interactions with these two peptides were investigated by molecular dynamics (MD) simulations. These results and their implications for understanding the mechanisms of the entry of HIV-1 via CXCR4 and its evasion of antibody detection are discussed.

2. Results and Discussion

2.1. Design of Cyclic L- and D-Peptides Corresponding to the Full-Length V3 Loop of gp120 of the Dual-Tropic HIV 89.6 Strain

Two cyclic L- and D-peptides containing the amino acid sequence of the entire V3 loop of gp120 of the HIV-1 89.6 strain were chemically synthesized and examined for CXCR4 competitive binding activity (Figure 1). In these peptides, a disulfide bond commonly used for peptide cyclization and conformational restraint was introduced between the two Cys residues with the goal of stabilizing the V3 loop conformation critical for receptor binding. Both peptides displayed competitive binding to CXCR4 with the IC_{50} values of 8.95 μM and 7.8 μM, respectively (Figure 2). The CCR5 competitive binding assay was also performed on these cyclic L- and D-V3 loop peptides. Neither peptides displayed any competitive binding to CCR5 (Figure 3), suggesting that their CXCR4 binding was selective.

2.2. Molecular Modeling of CXCR4 Interactions with Cyclic L- and D-V3 Loop Peptides

To understand the structural basis of CXCR4 recognition by these peptides of opposite chirality, molecular modeling of the docking interactions of the peptides with CXCR4 was conducted using the ROSETTA FlexPepDock procedure. The resulting docking models were further refined with a 100 ns molecular dynamics (MD) simulation. The final models revealed that both L- and D-V3 peptides formed a β-hairpin that inserted into CXCR4 and made contact with nearly all of the seven transmembrane helices (Figure 4). The fragments of the peptides corresponding to the stem and base regions of the native V3 loop interacted with the extracellular loops and N-terminus of CXCR4. On the other hand, there were noticeable differences in the CXCR4 docking modes between L- and D-V3 peptides. The L-V3 peptide maintained the anti-parallel β-strand conformation during docking into the receptor, whereas in the D-V3 peptide this anti-parallel β-strand conformation became distorted, probably due to the changed chirality and, consequently, the changed side-chain orientation. In addition, the peptide fragments corresponding to the stem and base regions of the native V3 loop extended to helices V and VI for the L-V3 peptide, and to helices IV and V for the D-V3 peptide.

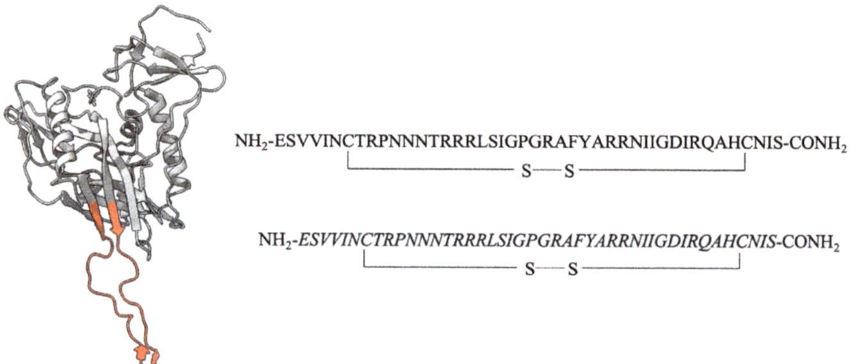

Figure 1. A pair of L- and D-peptides corresponding to the full-length V3 loop of HIV-1 gp120. The V3 loop of the crystal structure (PDB: 2B4C) of HIV-1 gp120 is highlighted in red. The amino acid sequence of the two L- and D-peptides which correspond to the full-length V3 loop of gp120 of the HIV-1 89.6 strain is shown. These two peptides have the same sequence, as shown with the D-peptide of all D-amino acids. These peptides are cyclized through a disulfide bond between the two cysteines as shown.

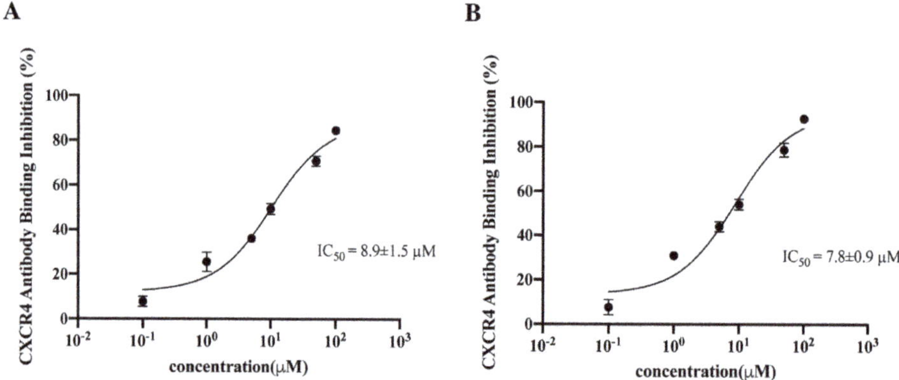

Figure 2. The CXCR4 competitive binding activity of cyclic (**A**) L- and (**B**) D-peptides containing the entire V3 loop sequence of gp120 of the dual-tropic HIV 89.6 strain. The IC_{50} values were determined by the competitive binding assay using an anti-CXCR4 mAb 12G5. The results were from three independent experiments and presented as means ± standard deviation.

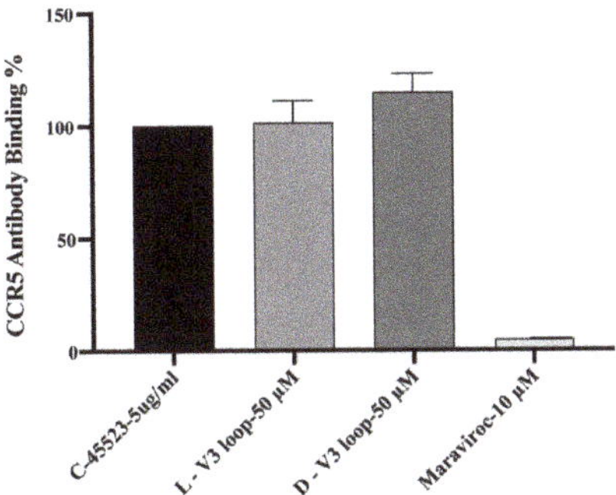

Figure 3. Lack of CCR5 competitive binding of the cyclic L- and D-V3 loop peptides. Maraviroc, a known CCR5 binding drug, was used as the positive control. The results were from three independent experiments.

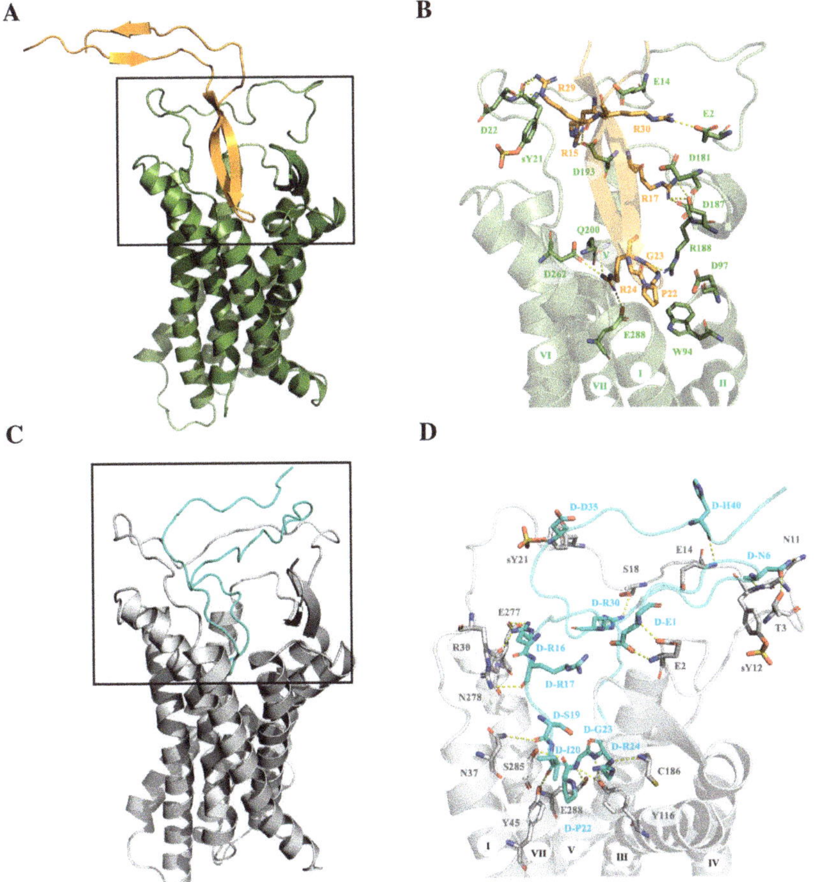

Figure 4. The predicted binding modes of cyclic L-V3 (**A,B**) and D-V3 (**C,D**) loop peptides in CXCR4 crystal structure (PDB: 4RWS). (**A,C**) Cyclic L- and D-V3 loop peptides are shown in orange and cyan cartoon models, respectively. (**B,D**) Key residues in interaction are shown as sticks.

To investigate further the binding recognition mechanism of L- and D-V3 peptides with CXCR4, we employed the PDLD/S-LRA/β method to analyze the contribution of CXCR4 residue to the binding free energy of the peptide ligands (Figure 5). Many negative-charged residues, Asp and Glu, on CXCR4 were seen to contribute to the binding free energy of the receptor and the L- and D-V3 peptides, which seemed to be consistent with the roles of these negative-charged residues in electrostatic interactions with the many positive-charged Arg residues in both peptides. On the other hand, the rank orders of the residue's contribution to the binding free energy differed between these two peptides, which was consistent with the observed differences in their receptor binding modes as described above.

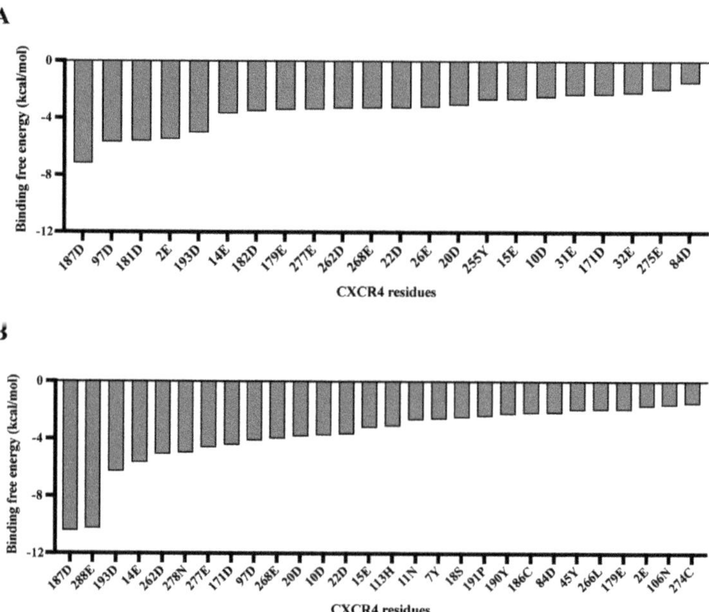

Figure 5. The contribution to ligand–receptor binding free energy of CXCR4 residues for the cyclic L-V3 peptide (**A**) and D-V3 peptide (**B**).

3. Conclusions

In this study, we investigated the stereochemical mechanism of the molecular recognition of HIV-1 by coreceptor CXCR4 by using synthetic peptide probes containing the entire V3 loop of gp120, which is the major site of the virus–coreceptor interaction. To help retain the conformational integrity of the peptide, a disulfide bond that formed between the two ends of the V3 loop was used to cyclize the peptide. In addition, to probe the effect of changed side-chain conformations of the peptide on CXCR4 recognition, an all-D-amino acid analog of the L-V3 loop peptide was generated. Both of these cyclic L- and D-V3 loop peptides displayed comparable binding recognition to the CXCR4 receptor, but not another chemokine receptor, CCR5, suggesting their CXCR4-selective interactions. Further molecular modeling studies revealed the important roles played by many negative-charged Asp and Glu residues on CXCR4 that probably engaged in favorable electrostatic interactions with the positive-charged Arg residues present in these peptides. These results are in line with our recently published data on other V3 loop fragment peptides [19]. Together, they provide further support to the notion that the seemingly flexible V3 loop-CXCR4 interface might be implicated in the viral entry and immune evasion mechanisms. Specifically, with its well-known highly mutable gp120 V3 loop, HIV-1 may evade antibody recognition, which is conformation-dependent, through conformational changes caused by residual mutations reminiscent of the conformational changes caused by residual chirality changes (i.e., L- to D-amino acid residue conversion). At the same time, because of the flexible CXCR4 ligand binding surface able to accommodate residual chirality or conformational changes, HIV-1 can still retain its ability to recognize an important host cell receptor such as CXCR4 for entry and infection. Such insights may have practical implications on the rational design of effective antiviral agents. For example, peptides containing D-amino acids are generally known to be more enzymatically stable than those of L-amino acids. The accommodation of D-amino acids by the CXCR4 ligand binding surface revealed in this study provides a basis for the development of additional D-peptide analogs of high stability for antiviral application.

3.1. Experimental Procedures

3.1.1. Peptide Synthesis

The peptides were synthesized through an Fmoc [N-(9-fluorenyl) methoxycarbonyl] chemistry solid-phase peptide synthesis approach. Fmoc-Rink Amide AM Resin (loading: 0.272 mmol/g, GL) was used as solid support. Before starting synthesis, resins were swelled in DCM (dichloromethane) for 20 min. Nα-Fmoc protecting group was removed with a DMF (N, N-Dimethylformamide) solution of 20% piperidine. 4eq Fmoc-amino acid materials were used in the coupling reaction for 60 min per single reaction. The coupling reagent was 3.8 eq HCTU (O-(6-Chloro-1-hydrocibenzotriazol-1-yl)-1,1,3,3-tetramethyluronium hexafluorophosphate) and 8eq DIEA (diisopropylethylamine) dissolved in DMF. When peptide elongating finished, the peptide was cleavage from resins with cleavage mixture which consisted of TFA (trifluoroacetic acid), phenol, H_2O, and TIS (Triisopropylsilane) (88:2:5:5, $v/w/v/v$). Crude peptides were precipitated from lysate with ice-cold diethyl ether, centrifuged, and lyophilized. The crude peptides were purified by HPLC (high-performance liquid chromatography) using a preparative C18 column (water, 5 μm, 10 × 150 mm) and a linear gradient elution with acetonitrile containing 0.25% TFA. Production peaks were detected at 220 nm and 254 nm UV. Cyclization of peptides was performed by gentle stirring at 4 °C and oxidizing in air overnight; analytical HPLC (C18 column, Shimadzu, 5 μm, 4.6 × 150 mm) monitored the process of the reaction with 220 nm and 254 nm UV. The cyclic peptides with an intramolecular disulfide bond were desalted and purified by preparative HPLC again as mentioned above. The fractions containing target product were collected and lyophilized. Finally, analytical HPLC and high-resolution ESI-MS with positive mode were used to characterize purity and molecular weight, respectively, these results were shown in Supplementary Materials (Figures S1–S4).

3.1.2. Competitive Binding Assay

To evaluate the co-receptor binding affinity of the peptides of interest, competitive binding assay was performed as we described before [18,20,21]. Stable CXCR4-CHO and CCR5 cell lines were constructed and cultured in DMEM medium (10% FBS, 100 IU penicillin, 0.1 mg/mL streptomycin, and 200 μg/mL geneticin). Cells were collected and seeded into a 96-well plate with 5×10^5 cells/well with 100 μL FACS buffer (PBS with 0.5% BSA and 0.05% NaN_3). The monoclonal antibodies used in competitive binding assay were 12G5 (final concentration 250 ng/mL, Sigma-Aldrich, MO, USA) for CXCR4 and C-45523 for CCR5 (final concentration 4 μg/mL, R&D systems, USA). The test compounds with various concentrations were added into each well and then antibody was added. After incubation for 40 min at 4 °C, cells were washed with FACS buffer and centrifugated. The secondary IgG-FITC (goat anti-mouse IgG, Sigma-Aldrich, MO, USA) antibody was added into each well and incubated for 30 min at 4 °C, then washed twice with FACS buffer by centrifugation. Fluorescence intensity ($485_{EX}/535_{EM}$) was measured by using spectrophotometric microplate reader (PerkinElmer EnVision™, Waltham, MA, USA). The experimental data were collected from at least three independent experiments and analyzed by Graphpad Prism 8 and presented as means ± standard deviation.

3.1.3. Peptide-CXCR4 Docking

The template structure of CXCR4 was obtained from the crystal structure (PDB:4RWS), removing chain B and mutating ^{187}Cys to Asp by PyMOL. Missing residues of CXCR4 crystal structure were modeled with ChimeraX [22–24]. Structures of V3 were built in the web server of SWISS-MODEL [25–27]. D-V3 was inverted from the V3 we built before with PyMOL and then relaxed with ROSETTA. The initial complex structures for docking input were modelled based gp120-CD4-CCR5 cryo-EM structure with PyMOL [28]. The initial complex structures were relaxed by ROSETTA3 Relax protocol, and the lowest score out of the outputs was used in next step [29,30]. In order to remove internal clashes that were outside of the docking interface, the input PDB file was prepacked with side chains of each monomer before refinement with the prepack mode of FlexPepDock protocol [31]. Next, the prepacked complex structure was used as input to perform the refinement mode of FlexPepDock with 3000 independent FlexPepDock simulations (producing 3000 output PDB files). To effectively sample the conformational space, a preliminary round of centroid mode optimization was performed before refinement. Finally, the 30 lowest score (reweighted term) outputs were selected and analyzed by clustering [32].

3.1.4. MD Simulation

The complex structures of V3-CXCR4 and D-V3-CXCR4 that were generated from peptide–protein docking were then employed in MD simulation. The bilayer membrane systems were built by CHARMM-GUI input generator [33]. The peptide–protein complexes were embedded in POPC lipid bilayer. The CHARMM36m force field parameters were used in all MD runs. The size of the rectangular water box was adjusted to the structure and filled with 0.15 M NaCl to neutralize the charges of systems. The CHARMM-GUI outputs files were used with default parameters for minimization and pre-equilibrium with Gromacs [34]. MD simulations of each system were performed using a constant number, pressure, and temperature (NPT) ensemble at 310 K for 100 ns in 2 fs intervals. The coordinate files of these two complexes after 100 ns MD simulations were provided in Supplementary Materials.

3.1.5. Binding Free Energy Calculation

MOLARIS-XG 9.15 software was used to calculate the binding free energy. The peptide–protein complex used in the binding free energy calculation was selected from the last frame of MD simulations described above. The binding free energy was calculated using semi-macroscopic protein dipoles Langevin dipoles-linear response approximation/β (PDLD/S-LRA/β) [35]. The relaxation was performed using the standard MOLARIS surface constrained all-atom solvent (SCAAS) boundary conditions and the local reaction field (LRF) long-range treatment. The prot_prot_bind module in MOLARIS-XG 9.15 was used to conduct the analysis. The peptide of the complex was regarded as group_a, and the residues within 60 Å of the peptides were taken as group_b.

Supplementary Materials: The following supporting information can be downloaded at: https://www.mdpi.com/article/10.3390/v15051084/s1, Figure S1: RP-HPLC analyzed the purity of synthesized L-V3 peptide, Figure S2: ESI-MS determined the molecular weight of synthesized of L-V3 peptide, Figure S3: RP-HPLC analyzed the purity of synthesized D-V3 peptide, Figure S4: ESI-MS determined the molecular weight of synthesized of D-V3 peptide. File S1: PDB file of L-V3 peptide bound with CXCR4 after 100 ns MD simulation; File S2: PDB file of D-V3 peptide bound with CXCR4 after 100 ns MD simulation.

Author Contributions: R.Z. performed peptide design and synthesis and wrote the manuscript. X.S. performed the biological assays and analyzed the biological data. R.Z. and J.Z. performed molecular modeling. Q.M., L.S.M.H. and Y.X. contributed to the interpretation and discussion of the results. J.A. and Z.H. supervised the study and revised the manuscript. All authors have read and agreed to the published version of the manuscript.

Funding: Tsinghua-Peking Joint Center for Life Sciences (to Z.H.); NIH grant (GM057761 to J.A.).

Institutional Review Board Statement: Not applicable.

Informed Consent Statement: Not applicable.

Data Availability Statement: Data is contained within the article.

Acknowledgments: This work was supported by Tsinghua-Peking Joint Center for Life Sciences (to Z.H.) and the NIH grant (GM057761 to J.A.).

Conflicts of Interest: The authors declare that they have no conflicts of interest with the contents of the article. During the work, Z.H. held appointments at Tsinghua University, Chinese University of Hong Kong, Shenzhen, and University of California at San Diego.

References

1. Choi, W.T.; Yang, Y.L.; Xu, Y.; An, J. Targeting Chemokine Receptor CXCR4 for Treatment of HIV-1 Infection, Tumor Progression, and Metastasis. *Curr. Top. Med. Chem.* **2014**, *14*, 1574–1589. [CrossRef] [PubMed]
2. Bleul, C.C.; Farzan, M.; Choe, H.; Parolin, C.; ClarkLewis, I.; Sodroski, J.; Springer, T.A. The lymphocyte chemoattractant SDF-1 is a ligand for LESTR/fusin and blocks HIV-1 entry. *Nature* **1996**, *382*, 829–833. [CrossRef]
3. Oberlin, E.; Amara, A.; Bachelerie, F.; Bessia, C.; Virelizier, J.L.; Arenzana-Seisdedos, F.; Schwartz, O.; Heard, J.M.; Clark-Lewis, I.; Legler, D.F.; et al. The CXC chemokine SDF-1 is the ligand for LESTR/fusin and prevents infection by T-cell-line-adapted HIV-1. *Nature* **1996**, *382*, 833–835. [CrossRef] [PubMed]
4. Mazo, I.B.; Massberg, S.; von Andrian, U.H. Hematopoietic stem and progenitor cell trafficking. *Trends Immunol.* **2011**, *32*, 493–503. [CrossRef]
5. Bernhagen, J.; Krohn, R.; Lue, H.; Gregory, J.L.; Zernecke, A.; Koenen, R.R.; Dewor, M.; Georgiev, I.; Schober, A.; Leng, L.; et al. MIF is a noncognate ligand of CXC chemokine receptors in inflammatory and atherogenic cell recruitment. *Nat. Med.* **2007**, *13*, 587–596. [CrossRef]
6. Saini, V.; Marchese, A.; Majetschak, M. CXC chemokine receptor 4 is a cell surface receptor for extracellular ubiquitin. *J. Biol. Chem.* **2010**, *285*, 15566–15576. [CrossRef]
7. Saini, V.; Marchese, A.; Tang, W.J.; Majetschak, M. Structural determinants of ubiquitin-CXC chemokine receptor 4 interaction. *J. Biol. Chem.* **2011**, *286*, 44145–44152. [CrossRef]
8. Feng, Y.; Broder, C.C.; Kennedy, P.E.; Berger, E.A. HIV-1 entry cofactor: Functional cDNA cloning of a seven-transmembrane, G protein-coupled receptor. *Science* **1996**, *272*, 872–877. [CrossRef] [PubMed]
9. Alkhatib, G.; Combadiere, C.; Broder, C.C.; Feng, Y.; Kennedy, P.E.; Murphy, P.M.; Berger, E.A. CC CKR5: A RANTES, MIP-1alpha, MIP-1beta receptor as a fusion cofactor for macrophage-tropic HIV-1. *Science* **1996**, *272*, 1955–1958. [CrossRef]
10. Trkola, A.; Dragic, T.; Arthos, J.; Binley, J.M.; Olson, W.C.; Allaway, G.P.; Cheng-Mayer, C.; Robinson, J.; Maddon, P.J.; Moore, J.P. CD4-dependent, antibody-sensitive interactions between HIV-1 and its co-receptor CCR-5. *Nature* **1996**, *384*, 184–187. [CrossRef]
11. Maddon, P.J.; Dalgleish, A.G.; McDougal, J.S.; Clapham, P.R.; Weiss, R.A.; Axel, R. The T4 gene encodes the AIDS virus receptor and is expressed in the immune system and the brain. *Cell* **1986**, *47*, 333–348. [CrossRef]
12. Wilen, C.B.; Tilton, J.C.; Doms, R.W. HIV: Cell binding and entry. *Cold Spring Harb. Perspect. Med.* **2012**, *2*, a006866. [CrossRef] [PubMed]
13. Berger, E.A.; Doms, R.W.; Fenyo, E.M.; Korber, B.T.; Littman, D.R.; Moore, J.P.; Sattentau, Q.J.; Schuitemaker, H.; Sodroski, J.; Weiss, R.A. A new classification for HIV-1. *Nature* **1998**, *391*, 240. [CrossRef] [PubMed]
14. Iwamoto, A.; Hosoya, N.; Kawana-Tachikawa, A. HIV-1 tropism. *Protein Cell* **2010**, *1*, 510–513. [CrossRef]
15. Connor, R.I.; Sheridan, K.E.; Ceradini, D.; Choe, S.; Landau, N.R. Change in coreceptor use correlates with disease progression in HIV-1–infected individuals. *J. Exp. Med.* **1997**, *185*, 621–628. [CrossRef]
16. Choi, W.T.; Duggineni, S.; Xu, Y.; Huang, Z.; An, J. Drug discovery research targeting the CXC chemokine receptor 4 (CXCR4). *J. Med. Chem.* **2012**, *55*, 977–994. [CrossRef]
17. Zhou, N.M.; Luo, Z.W.; Luo, J.S.; Fan, X.J.; Cayabyab, M.; Hiraoka, M.; Liu, D.X.; Han, X.B.; Pesavento, J.; Dong, C.Z.; et al. Exploring the stereochemistry of CXCR4-peptide recognition and inhibiting HIV-1 entry with D-peptides derived from chemokines. *J. Biol. Chem.* **2002**, *277*, 17476–17485. [CrossRef] [PubMed]
18. Zhang, C.; Huang, L.S.; Zhu, R.; Meng, Q.; Zhu, S.; Xu, Y.; Zhang, H.; Fang, X.; Zhang, X.; Zhou, J.; et al. High affinity CXCR4 inhibitors generated by linking low affinity peptides. *Eur. J. Med. Chem.* **2019**, *172*, 174–185. [CrossRef]
19. Zhu, R.; Meng, Q.; Zhang, H.; Zhang, G.; Huang, L.S.M.; Xu, Y.; Schooley, R.T.; An, J.; Huang, Z. HIV-1 gp120-CXCR4 recognition probed with synthetic nanomolar affinity D-peptides containing fragments of gp120 V3 loop. *Eur. J. Med. Chem.* **2022**, *224*, 114797. [CrossRef]
20. Xu, Y.; Duggineni, S.; Espitia, S.; Richman, D.D.; An, J.; Huang, Z. A synthetic bivalent ligand of CXCR4 inhibits HIV infection. *Biochem. Biophys. Res. Commun.* **2013**, *435*, 646–650. [CrossRef]
21. Yang, Y.; Gao, M.; Zhang, Q.; Zhang, C.; Yang, X.; Huang, Z.; An, J. Design, synthesis, and biological characterization of novel PEG-linked dimeric modulators for CXCR4. *Bioorg. Med. Chem.* **2016**, *24*, 5393–5399. [CrossRef]
22. Pettersen, E.F.; Goddard, T.D.; Huang, C.C.; Meng, E.C.; Couch, G.S.; Croll, T.I.; Morris, J.H.; Ferrin, T.E. UCSF ChimeraX: Structure visualization for researchers, educators, and developers. *Protein Sci.* **2021**, *30*, 70–82. [CrossRef]

23. Goddard, T.D.; Huang, C.C.; Meng, E.C.; Pettersen, E.F.; Couch, G.S.; Morris, J.H.; Ferrin, T.E. UCSF ChimeraX: Meeting modern challenges in visualization and analysis. *Protein Sci.* **2018**, *27*, 14–25. [CrossRef] [PubMed]
24. Webb, B.; Sali, A. Comparative Protein Structure Modeling Using MODELLER. *Curr. Protoc. Bioinform.* **2016**, *54*, 5.6.1–5.6.37. [CrossRef]
25. Waterhouse, A.; Bertoni, M.; Bienert, S.; Studer, G.; Tauriello, G.; Gumienny, R.; Heer, F.T.; de Beer, T.A.P.; Rempfer, C.; Bordoli, L.; et al. SWISS-MODEL: Homology modelling of protein structures and complexes. *Nucleic Acids Res.* **2018**, *46*, W296–W303. [CrossRef] [PubMed]
26. Bienert, S.; Waterhouse, A.; de Beer, T.A.; Tauriello, G.; Studer, G.; Bordoli, L.; Schwede, T. The SWISS-MODEL Repository-new features and functionality. *Nucleic Acids Res.* **2017**, *45*, D313–D319. [CrossRef]
27. Guex, N.; Peitsch, M.C.; Schwede, T. Automated comparative protein structure modeling with SWISS-MODEL and Swiss-PdbViewer: A historical perspective. *Electrophoresis* **2009**, *30* (Suppl. 1), S162–S173. [CrossRef]
28. Shaik, M.M.; Peng, H.; Lu, J.; Rits-Volloch, S.; Xu, C.; Liao, M.; Chen, B. Structural basis of coreceptor recognition by HIV-1 envelope spike. *Nature* **2019**, *565*, 318–323. [CrossRef]
29. Das, R.; Baker, D. Macromolecular Modeling with Rosetta. *Annu. Rev. Biochem.* **2008**, *77*, 363–382. [CrossRef] [PubMed]
30. Conway, P.; Tyka, M.D.; DiMaio, F.; Konerding, D.E.; Baker, D. Relaxation of backbone bond geometry improves protein energy landscape modeling. *Protein Sci.* **2014**, *23*, 47–55. [CrossRef]
31. Raveh, B.; London, N.; Schueler-Furman, O. Sub-angstrom modeling of complexes between flexible peptides and globular proteins. *Proteins* **2010**, *78*, 2029–2040. [CrossRef] [PubMed]
32. Li, S.C.; Ng, Y.K. Calibur: A tool for clustering large numbers of protein decoys. *BMC Bioinform.* **2010**, *11*, 25. [CrossRef]
33. Jo, S.; Kim, T.; Iyer, V.G.; Im, W. CHARMM-GUI: A web-based graphical user interface for CHARMM. *J. Comput. Chem.* **2008**, *29*, 1859–1865. [CrossRef] [PubMed]
34. Van Der Spoel, D.; Lindahl, E.; Hess, B.; Groenhof, G.; Mark, A.E.; Berendsen, H.J. GROMACS: Fast, flexible, and free. *J. Comput. Chem.* **2005**, *26*, 1701–1718. [CrossRef]
35. Singh, N.; Warshel, A. Absolute binding free energy calculations: On the accuracy of computational scoring of protein-ligand interactions. *Proteins* **2010**, *78*, 1705–1723. [CrossRef] [PubMed]

Article

Non-Nucleoside Inhibitors Decrease Foot-and-Mouth Disease Virus Replication by Blocking the Viral 3D^{pol}

Sirin Theerawatanasirikul [1], Ploypailin Semkum [2], Varanya Lueangaramkul [2], Penpitcha Chankeeree [2], Nattarat Thangthamniyom [2] and Porntippa Lekcharoensuk [2,3,*]

1   Department of Anatomy, Faculty of Veterinary Medicine, Kasetsart University, Bangkok 10900, Thailand
2   Department of Microbiology and Immunology, Faculty of Veterinary Medicine, Kasetsart University, Bangkok 10900, Thailand
3   Center for Advanced Studies in Agriculture and Food, Kasetsart University Institute for Advanced Studies, Kasetsart University, Bangkok 10900, Thailand
*   Correspondence: fvetptn@ku.ac.th; Tel.: +66-29428436

Abstract: Foot-and-mouth disease virus (FMDV), an economically important pathogen of cloven-hoofed livestock, is a positive-sense, single-stranded RNA virus classified in the *Picornaviridae* family. RNA-dependent RNA polymerase (RdRp) of RNA viruses is highly conserved. Compounds that bind to the RdRp active site can block viral replication. Herein, we combined double virtual screenings and cell-based antiviral approaches to screen and identify potential inhibitors targeting FMDV RdRp (3D^{pol}). From 5596 compounds, the blind- followed by focus-docking filtered 21 candidates fitting in the 3D^{pol} active sites. Using the BHK-21 cell-based assay, we found that four compounds—NSC217697 (quinoline), NSC670283 (spiro compound), NSC292567 (nigericin), and NSC65850—demonstrated dose-dependent antiviral actions in vitro with the EC50 ranging from 0.78 to 3.49 μM. These compounds could significantly block FMDV 3D^{pol} activity in the cell-based 3D^{pol} inhibition assay with small IC50 values ranging from 0.8 nM to 0.22 μM without an effect on FMDV's main protease, 3C^{pro}. The 3D^{pol} inhibition activities of the compounds were consistent with the decreased viral load and negative-stranded RNA production in a dose-dependent manner. Conclusively, we have identified potential FMDV 3D^{pol} inhibitors that bound within the enzyme active sites and blocked viral replication. These compounds might be beneficial for FMDV or other picornavirus treatment.

Keywords: foot-and-mouth disease virus (FMDV); RNA-dependent RNA polymerase (RdRp); 3D polymerase (3D^{pol}); small molecule; virtual screening; antiviral cell-based assay; antiviral agent

Citation: Theerawatanasirikul, S.; Semkum, P.; Lueangaramkul, V.; Chankeeree, P.; Thangthamniyom, N.; Lekcharoensuk, P. Non-Nucleoside Inhibitors Decrease Foot-and-Mouth Disease Virus Replication by Blocking the Viral 3D^{pol}. *Viruses* **2023**, *15*, 124. https://doi.org/10.3390/v15010124

Academic Editor: Simone Brogi

Received: 15 November 2022
Revised: 26 December 2022
Accepted: 28 December 2022
Published: 30 December 2022

1. Introduction

Foot-and-mouth disease (FMD) is a highly contagious disease caused by the foot-and-mouth disease virus (FMDV). This disease is economically important and affects domestic livestock with cloven hooves worldwide [1]. FMD control relies on two major strategies: culling of infected animals and vaccination [2]. FMDV is a member of the *Picornaviridae* family, which consists of both animal and human pathogens, such as enterovirus, rhinovirus, and poliovirus. The FMDV genome is single-stranded RNA with positive polarity. Viral RNA synthesis during replication and transcription is achieved by RNA-dependent RNA polymerase (RdRp) encoded by 3D, so-called 3D^{pol} [3]. The 3D^{pol} of picornaviruses including FMDV and the RdRp of other RNA viruses share a common characteristic [4,5]. The three-dimensional structure of RdRp resembles a cupped right-handed configuration composed of three subdomains, the so-called thumb, fingers, and palm. The finger subdomain has extensions, termed fingertips, that bridge between the finger and the thumb subdomains to maintain the RdRp active site rearrangement and form the NTP entry site [6]. Therefore, the topology of these subdomains is essential for RdRp's catalytic activity during negative-stranded RNA synthesis by holding the RNA template in place, engaging NTPs, and promoting polymerization. The central region of the 3D^{pol} contains six conserved motifs labelled A, B, C, D, E, and F, which are located mostly in the palm subdomain and form the template-binding channel with the 3D^{Pol} active site [6]. Inhibition of 3D^{Pol} activity can prevent viral genome replication. Therefore, because of its critical role in viral replication, 3D^{Pol} is an attractive target for antiviral drug development.

As FMDV is highly contagious, broad-spectrum antiviral drugs as well as other novel natural and synthesized compounds might support the outbreak control by reducing spillover of the spreading virus into the environment. For example, the nucleoside analogue ribavirin (1-β-d-ribofuranosyl-1,2,4-triazole-3-carboxamide) is an antiviral drug licensed for the treatment of human and animal viral infections. However, ribavirin resistance caused by mutations within the RdRp coding sequences have been reported in many RNA viruses. For instance, poliovirus (PV) has substitution mutations, G64S and L420A, in 3D^{Pol} while M296I was found in the FMDV 3D^{Pol} [7,8]. Furthermore, a human hepatitis C virus (HCV) escape mutant contained the mutation Y415 in the NS5B gene [9]. M296 of FMDV 3D^{Pol} is in the NTP binding site. Thus, alteration at this position may affect nucleotide recognition by the 3D^{Pol} [10]. Previously, nucleoside analogues including favipiravir (T-705), T-1105 (3-oxo-3,4-dihydro-2-pyrazinecarboxamide derivative), and T-1106 were tested for their ability to inhibit FMDV infection in vitro and in vivo [11]. T-1105 demonstrated greater potency than T-705 and T-1106 to inhibit FMDV replication [11,12]. In an experimental FMDV infection, a high dose of T-1105 at 200 mg/kg twice daily for seven consecutive days was required to prevent FMD clinical signs and reduce viral shedding [12]. In a guinea pig model, the prophylactic effect of T-1105 was achieved at a high dose of 400 mg/kg/day twice daily for five consecutive days to give a similar protective level as the full-dose vaccination [13]. Thus far, no antiviral agent has been approved for the treatment or prevention of FMDV infection.

Among antiviral agents, nucleoside-based inhibitors (NIs) are likely to exhibit off-target effects and reduction in their potency in clinical cases [14]. Small molecules that target surface cavities or allosteric site of enzymes, so-called non-nucleoside inhibitors (NNIs), can be either natural or non-natural compounds [15]. Their structures are more flexible and thus can fit in the nucleotide-binding pocket better than the NIs [16]. Several NNI compounds have been reported to inhibit the RdRp activities of RNA viruses such as GPC-N114 in picornaviruses [14], lycorine in MERS-CoV [17], and NITD29 in Zika virus [18]. In addition to RdRp, 3C proteases (3C^{Pro}) are another attractive target for antiviral drug design as it is present and conserved across several positive-sense, single-stranded RNA viruses with picornavirus-like 3C^{Pro} supercluster. It has been shown previously that three dipeptidyl-based synthetic molecules with different interactive chemical groups including an aldehyde (GC373), a bisulfite adduct (GC376), and an α-ketoamide (GC375) possessed broad-spectrum, potent antiviral effects [19]. These compounds effectively inhibited protease (3C^{Pro} or 3CL^{Pro}) activities and replication of viruses from the families *Caliciviridae*, *Coronaviridae*, and *Picornaviridae*, as determined using FRET protease assay and cell-based assays, respectively.

In this study, we screened small molecules that targeted FMDV 3D^{Pol} by integrating computer-aided virtual screening and a cell-based antiviral assay to filter for potential antiviral compounds. We further determined the inhibitory activity of the selected compounds on FMDV 3D^{Pol} using a recently established FMDV minigenome with the green fluorescence reporter protein in the cell-based 3D^{Pol} inhibition assay [20]. Taken together, we have identified four 3D^{Pol} inhibitors and demonstrated the mechanism by which the small molecules could inhibit viral replication in FMDV-infected BHK-21 cells.

2. Materials and Methods

2.1. Virtual Screening of Small Molecules

The crystal structure of FMDV 3D^{Pol} (FMDV RdRp) deposited in the PDB under a code 1wne.pdb was retrieved (https://www.rcsb.org/structure/1WNE). Upon the 3D structure's preparation for the virtual screening, we found that four amino acids (F34, A68, E144, and K148) on the protein structures differed from the deduced amino acid sequence of FMDV O189. Then, homology modeling based on the 3D^{pol} crystal structure (PDB: 1wne.pdb) was performed on the SWISS-MODEL server (https://swissmodel.expasy.org/, accessed on 19 April 2021 [20]) to prepare the complete protein structure of O189 3D^{pol} previously used for the plasmid construction [20]. The structure quality was evaluated using MolProbity [21], Q-MEAN [22], and Ramachanadran plotting [23], as described elsewhere [20]. The FMDV 3D^{pol} model was prepared for the downstream process with the removal of ligands and water molecules and addition of hydrogen atoms. Molecular docking was achieved using the PyRx 0.9.8 virtual screening tool [24].

The compound libraries provided by the Developmental Therapeutics Program (DTP) Open Repository of the National Cancer Institute (NCI) were queried for a subset of the NCI repository collection comprising the NCI Diversity Set III, VI and V, and MechDiv3 libraries containing 5596 freely available models of compounds. Each subset of the compound models was assigned for docking using the Open Babel software toolbox [25], including Gasteiger partial charges, addition of hydrogen atoms, optimization of hydrogen bonds, and removal of atomic clashes, to generate the pdbqt input format.

Virtual screening was performed using AutoDock Vina [26] embedded in the PyRx 0.9.8 virtual screening tool [24]. In the first virtual screening, the small molecules retrieved from the model libraries were blind-docked onto the 3D^{pol} structure, which yielded a set of ligands predicted to bind the 3D^{pol} active sites or the conserved amino acid residues important for 3D^{pol}'s function. Initially, the grid box was set at 65:70:65 (x, y, z dimensions in Angstroms), and the box was centered at 19.34:31.20:23.25 (x, y, z). The virtual screening outputs were presented as the predicted free energy in kcal/mol and the exhaustiveness value for the docking was set to 20. Subsequently, the top-ranked complexes were retrieved for the latter focus-docking. In the focus-docking, the top-ranked compounds with a binding energy lower than −8.0 kcal/mol were included, and the dockings were centered at specific sites or amino acid residues within the FMDV 3D^{pol}. Three sites on the palm subdomain of FMDV 3D^{pol} involved in the nucleotide binding and polymerizations [6] were selected for the focus-docking. According to the 3D structure of FMDV 3D^{pol}, site 1 is composed of D240, Y241, and D245 of Motif A in the loop β8-α9; site 2 comprises M296, S298, G299, S301, and N307 of Motif B in the loop β9-α11; and site 3 contains Y336, D338, and D339 of Motif C in the loop β10-β11. The grid box of 25:25:25 was assigned to these specific sites and centered at 15.50:26.22:15.00 (x, y, z). The promising docking results were harvested for protein–ligand visualization using Discovery Studio Visualizer, version 2021 (BIOVIA, Dassault Systèmes, San Diego, CA, USA) and UCSF Chimera, version 1.16 (UCSF, San Francisco, CA, USA).

2.2. Sources and Preparations of Chemicals

The freely available compounds were provided by the Developmental Therapeutics Program (DTP) Open Repository of the National Cancer Institute (NCI). Ribavirin (Sigma-Aldrich, St. Louis, MO, USA) and rupintrivir (Sigma-Aldrich, St. Louis, MO, US) were used as RdRp and non-RdRp inhibitor controls. All compounds were prepared as 10-mM stock solutions in 100% dimethyl sulfoxide (DMSO) (Sigma-Aldrich, St. Louis, MO, USA), and stored at −20 °C for the following in vitro cell-based experiments.

2.3. Cells and Viruses

Baby Hamster Kidney (BHK−21) cells and HEK-293T cells were obtained from American Type Culture Collection (ATCC®, Manassas, VA, USA). The cells were maintained in a complete medium containing Minimum Essential Medium (MEM, Invitrogen™, Carlsbad, CA, USA), 10% fetal bovine serum (FBS, Invitrogen™, Carlsbad, CA, USA), 2 mM L-glutamine (Invitrogen™, Carlsbad, CA, USA), and 1×Antibiotic-Antimycotic (Invitrogen™, Carlsbad, CA, USA) at 37 °C with 5% CO_2. FMDV serotype A (NP05) was propagated in BHK21 cells at 37 °C with 5% CO_2 for 24 h. The virus stock with a titer of 1×10^9 TCID50/mL was stored at −80 °C in single-use aliquots. All works involving live FMDV were performed at biosafety level-2 with an enhanced facility.

2.4. Cytotoxicity Assay

BHK-21 cells were seeded at 1.8×10^4 cells per well into 96-well plates (Corning Incorporated., Corning, NY, USA) and incubated at 37 °C with 5% CO_2 overnight. The spent media was removed, and the cells were washed twice with $1\times$ PBS. The compounds were serially diluted in serum-free MEM media containing DMSO at a final concentration of $\leq 0.1\%$. The diluted compounds were incubated with the cells or virus and the cells were further incubated at 37 °C for an additional 24 h. Two independent experiments were performed for all biological assays. Cytotoxicity was determined by measuring cell viability using MTS solution provided in the CellTiter 96® Aqueous One Solution Cell Proliferation Assay (Promega, Madison, WI, USA) according to the manufacturer's instructions. The absorbance of the solution in the experiment plates was measured at a wavelength of 490 nm using a multi-mode reader (Synergy H1 Hybrid Multi-Mode Reader, BioTek®, Winooski, VT, USA). Cell viability was calculated using the following formula (1), where OD_{treate} denotes the absorbance of the uninfected BHK-21 cell wells containing cells, media, and compounds; $OD_{cell\ control}$ denotes the absorbance of the cell control wells containing cells and media; and OD_{dmso} denotes the absorbance of the vehicle control wells containing cells, media, and 0.1% DMSO.

$$[OD_{treat} - OD_{cell\ control}]/[OD_{dmso} - OD_{cell\ control}] \times 100 \tag{1}$$

2.5. Antiviral Activity Assay

The compounds that targeted FMDV $3D^{Pol}$ in the virtual screening and could inhibit viral replication were further investigated for the effects on the other processes of viral infection as mentioned in the previous study [27]. Initially, we investigated the effects of compounds on viral entry including binding and penetration (pre-viral-entry experiment). The compound dilutions and FMDV at 10 TCID50 per wells were co-adsorbed onto the overnight-grown BHK-21 cells in a 96-well plate (Corning Incorporated., Corning, NY, USA) at 37 °C for 2 h. All virus–compound mixtures were removed, and the cells were washed twice with $1\times$ PBS. Subsequently, fresh media was added to the cells which were further cultured at 37 °C for an additional 22 h. Secondly, in the post-viral-entry experiment, the BHK-21 cells in the 96-well plate were incubated with 10 TCID50 of FMDV per well at 37 °C for 2 h. Then, the cells were washed twice with $1\times$ PBS, before treatment with the same compound dilutions that were used in the cytotoxicity assay at 37 °C for an additional 22 h to inhibit viral replication.

2.6. Immunoperoxidase Monolayer Assay (IPMA)

Immunoperoxidase monolayer assays (IPMA) were conducted as previously described [27,28]. Briefly, the BHK-21 cells, including no infection with and without compound and FMDV infection with and without compound, were fixed with cold methanol at room temperature for 20 min and washed with PBS plus 0.1% Tween 20 ($1\times$ PBST, Sigma Aldrich®, St. Louis, MO, USA). The cells were then incubated with single-chain variable fragment with Fc fusion protein (scFv-Fc) specific to 3ABC of FMDV [29] at 37 °C for 1 h for viral detection. After primary antibody incubation, the cells were washed with $1\times$ PBST, and subsequently incubated with the protein G, HRP conjugate (dilution 1:1000, EMD Millipore corporation, Temecula, CA, USA) at 37 °C for 1 h. To visualize the antigen–antibody complex, the cells were stained with DAB substrate (DAKO, Santa Clara, CA, USA), and the dark-brown color of the infected FMDV cells was observed using a phase-contrast inverted microscope (Olympus IX73, Tokyo, Japan). The cell images were analyzed using CellProfiler image analysis v.4.2.0 [29]. The resulting data were used to calculate the half-maximal effective concentration (EC50). The EC50 value represented the compound concentration at which the virus infection was reduced by 50% compared to the DMSO control (FMDV infection with 0.1% DMSO). To analyze the data, the DMSO control was set at 100% infection, and the EC50 value of each compound was calculated using GraphPad Software version 9.4.1 (Prism, San Diego, CA, USA). The Z' factor value was analyzed using the following formula (2) to assess the assay performances both within a plate and across plates.

$$\text{Z' factor} = 1 - (3 \times \text{SD of cell control} + 3 \times \text{SD of the virus control}) / (\text{mean cell control signal} - \text{mean virus control signal}) \quad (2)$$

A Z' factor between 0.5 and 1 indicates an acceptable range.

2.7. Real-Time RT-PCR Assay

The whole viral RNA and negative-strand specific RNA of FMDV were detected and quantitated using real-time RT-PCR after treatment with the compounds. The BHK-21 cells were seeded at 1.8×10^5 cells into each well of 24-well plates (Corning Incorporated., Corning, NY, USA) and cultured overnight. The cells were incubated with 10 TCID50 of FMDV for 2 h, and the inoculum was removed after viral adsorption. Then, the FMDV-infected cells were treated with serially diluted compounds at 37 °C as mentioned above. At 24 h post viral infection, intra- and extracellular viral RNA was isolated using Direct-zol^{TM} RNA MiniPrep (Zymo Research Corporation, Tustin, CA, USA) following the manufacturer's instruction. The RNA was quantified using a NanoDrop^{TM} 2000c Spectrophotometer (Thermo Fisher Scientific, Waltham, MA, USA) and used as the templates for cDNA synthesis. The first-strand cDNAs derived from the whole viral RNA were generated using Random hexamers (Invitrogen^{TM}, Carlsbad, CA, USA) for whole viral RNA quantification. On the other hand, a primer specific to the negative-stranded RNA within the 3D^{Pol} coding sequence (5′-AAGGGTTGATTGTTGACA-3′) was employed to amplify the first-strand cDNA for the negative-stranded RNA quantification [30]. Both cDNA syntheses were performed with the enzyme SuperScript III Reverse Transcriptase (Invitrogen^{TM}, Carlsbad, CA, USA) according to the manufacturer's instructions.

The cDNA templates were subjected to downstream DNA quantification with qPCR using iTaq Universal SYBR Green Supermix (Bio-Rad Laboratories, Hercules, CA, USA). The primer sets for the DNA amplifications are listed in Table 1. Briefly, 5 µL mixture containing 0.5 µL of each target-specific forward and reverse primer was mixed with 2 µL cDNA and nuclease-free ddH$_2$O up to a final volume of 10 µL. The qPCR amplifications were conducted in a two-step method: first, denaturation at 95 °C for 30 sec, followed by 40 cycles of denaturation at 95 °C for 5 sec and annealing/extension at 55 °C for 30 sec. A melting curve was analyzed from 65 °C to 95 °C with 0.5 °C increments using a CFX96 touch Real-Time PCR detection system (Bio-Rad Laboratories, Hercules, CA, USA).

Table 1. Primers used in this study for amplification of FMDV cDNAs.

Primers	Sequence (5′–3)	cDNA Derived from	Tm (°C)	References
FMDV-5′UTRF	CTGTTGCTTCGTAGCGGAGC	5′UTR	66.3	[20]
FMDV-5′UTRR	TCGCGTGTTACCTCGGGGTACC	5′UTR	66.3	[20]
FMDV-3DF	TAGAGCAGTAGATGTTG	Negative strand	58	In this study
FMDV-3DR	ATGAACATCATGTTTGAGG	Negative strand	59	In this study

Viral RNA expression was determined using the absolute quantification method [27]. A standard curve of DNA derived from the whole viral RNA was generated from the ten-fold serially diluted plasmid containing FMDV 5′UTR from 10^{-2} to 10^{-7} plasmid molecules/µL. Copy numbers of the DNA were calculated based on the standard curve. The negative-stranded RNA is normally produced by 3D^{Pol} during viral replication. The negative-stranded RNA-derived DNA was quantitated based on the delta Ct values (cycle threshold) by subtracting the Ct values of the virus samples (FMDV-infected BHK-21 cells with compound treatment) from the Ct values of the virus control (FMDV-infected BHK-21 cells without compound treatment). The data were normalized using the following Equation (3).

$$\text{Normalized delta Ct} = 1 - (\text{delta Ct of sample}) / (\text{Ct of dmso}) \quad (3)$$

The normalized data from three independent replications are presented as the means $\pm$ SD using GraphPad Prism version 9.4.1 (Prism, San Diego, CA, USA).

2.8. Cell-Based 3D^{pol} Inhibition Assay

Inhibitory effects of the compounds on FMDV 3D^{pol} were examined using plasmid pKLS3_GFP, an FMDV minigenome expressing GFP, and the two helper plasmids pCAGGS_T7 and pCAGGS_P3 [20]. pKLS3_GFP contains the enhanced green fluorescent protein (GFP) gene inserted between FMDV O189 5' and 3'UTRs required for FMDV transcription and translation, while pCAGGS_T7 and pCAGGS_P3 are mammalian protein expression plasmids containing T7 RNA polymerase and the FMDV P3 region essential for efficient generation of FMDV RNA, respectively.

The three plasmids were transfected onto the BHK-21 cells as described previously [20]. Briefly, BHK-21 cells were seeded at 1×10^4 cells/well into 96-well plates and incubated at 37 °C overnight. On the transfection day, 40 ng of pKLS3_GFP, 120 ng of pCAGGS_T7, and 40 ng of pCAGGS_P3 were mixed with 0.6 µL of Fugene® HD (Promega, Madison, WI, USA) in Opti-MEM™ I Reduced-Serum Medium (Thermo Fisher Scientific, Waltham, MA, USA) to make a final volume of 50 µL per well. Then, the spent media was removed and replaced with the transfection mixture and incubated at 37 °C for 4 h. Subsequently, the transfection mixture was removed and replaced with the serially diluted compounds in Opti-MEM™ I Reduced-Serum Medium (Gibco™ Thermo Fisher Scientific Inc., Waltham, MA, USA). In the vehicle control well, 0.1% DMSO was added in the wells instead of the compounds and it also served as the transfection positive control. pKLS3_GFP without helper plasmid transfection was also included as the plasmid control. The cells were incubated at 37 °C with 5% CO_2 and the fluorescent reporter expression signals were observed at 24–48 h post transfection using a phase-contrast inverted microscope (Olympus IX73, Tokyo, Japan).

The levels of FMDV 3D^{pol} activity corresponded to the numbers and intensity of the bright green fluorescent signal in the positive cells. The background was adjusted for contrast and brightness only. The fluorescent intensities of the images were measured, and the background was subtracted from each image using CellProfiler image analysis v.4.2.0 [29]. The half-maximal inhibitory concentration (IC50) is the concentration at which the 3D^{pol} activity was reduced by 50% compared to that of the DMSO control. For the analysis, the signal of the DMSO control was set at 100% and the IC50 was calculated using GraphPad Software version 9.4.1 (Prism, San Diego, CA, USA).

2.9. Cell-Based FMDV 3C^{pro} Inhibition Assay

3C^{pro} is the main protease crucial for FMDV biology and is an attractive antiviral target. Therefore, we were interested to know whether these compounds could inhibit the 3C^{pro} activity. We determined the effects of the compounds on the main protease using a cell-based FMDV 3C^{pro} inhibition assay as described previously [27,31]. Briefly, HEK-293T cells were grown in 96-well plates at 1×10^3 cells per well. The cells were maintained in Opti-MEM I Reduced-Serum Medium (Gibco™ Thermo Fisher Scientific Inc., Waltham, MA, USA). On the next day, the cells were transfected with plasmids pG5Luc (Promega, Madison, WI, USA) and pBV_3ABCD expressing intact 3C^{pro} or pBV_mu3ABCD expressing inactive 3C^{pro} [32]. The 3ABCD and mu3ABCD genes were inserted between the Gal4-binding domain and the VP16-activation domain of plasmid pBV (Promega, Madison, WI, USA). pG5Luc is a reporter plasmid containing the GAL4 binding site upstream of VP16 and the firefly luciferase reporter gene sequences followed by the downstream *Renilla* luciferase gene. The total 10 µL of co-transfection mixtures comprising 0.1 µg pBV_3ABCD or pBV_mu3ABCD, 0.1 µg pG5Luc, and 0.6 µL Fugene® HD (Promega, Madison, WI, USA) were incubated with HEK-293 T cells at 37 °C for 2 h before adding the diluted compounds. At 16 h post transfection, the transfected cells were lysed with 20 µL passive lysis buffer (Promega, Madison, WI, USA) followed by firefly and *Renilla* luminescence signal determination using the Dual-Glo Luciferase Assay System (Promega, Madison,

WI, USA) in a Synergy H1 Hybrid Multi-Mode Microplate Reader (BioTek, Winooski, VT, USA). The intact 3C^{pro} expressed from pBV_3ABCD would separate the Gal4-binding domain from the VP16-activation domain, resulting in no expression of the firefly luciferase signal. When the compounds could suppress the 3C^{pro}, the Gal4-binding domain and the VP16-activation domain were in close proximity. Thus, GAL4 bound to the Gal4-binding domain and drove the firefly and *Renilla* luciferase expressions. The data were recorded as an inverse correlation of the firefly/*Renilla* luminescent (Fluc/Rluc) signal ratio from compound-treated wells compared to those obtained from wells transfected with pBV_mu3ABCD (3C^{pro} negative control) and pBV16 (empty plasmid control).

3. Results

3.1. Virtual Screening of Small Molecules

We screened compounds from the model libraries to identify novel inhibitors specific to FMDV 3D^{Pol} (Figure 1). In the initial in silico virtual screening of 5596 NCI compounds, the binding energy ranged from −1.5 to −10.1 kcal/mol. We then used a cut-off of −8.0 kcal/mol to select the 722 compounds that occupied the enzyme active sites of FMDV 3D^{Pol} for further focus-docking. The focus-docking was narrowed down to the specific amino acid residues within the three active sites of the FMDV 3D^{Pol} structure: site 1: D240, Y241, and D245 of Motif A; site 2: M296, S298, G299, S301, and N307 of Motif B; and site 3: Y336, D338, and D339 of Motif C. Upon the blind- and focus-virtual screenings, the 21 top-ranked compounds were selected for the following experimental validation. The binding affinities of the 21 compounds are listed in Figure 1c.

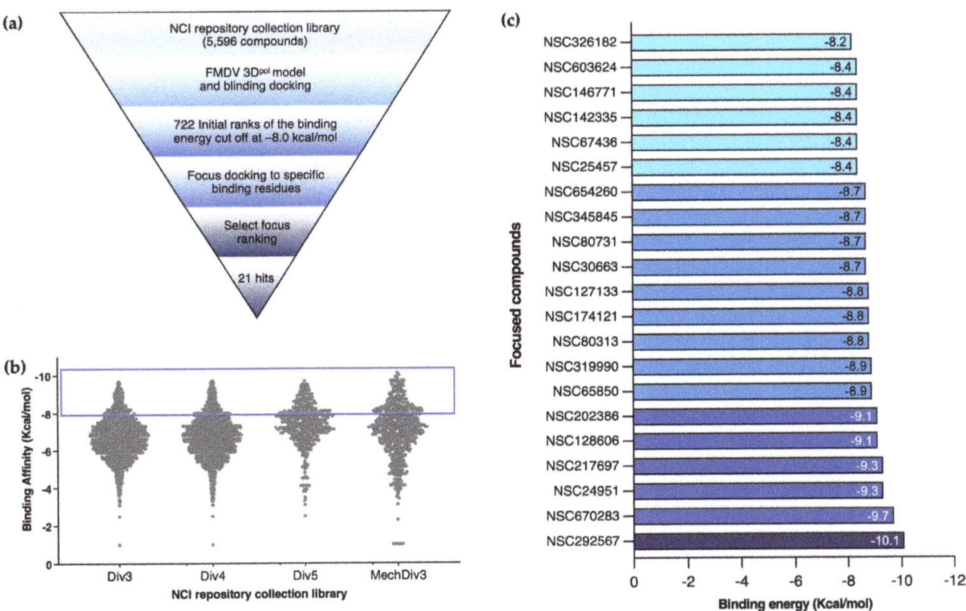

Figure 1. Virtual screening approach in this study. Schematic diagram of virtual screening starting from the initial docking to focus-docking (**a**). The predicted binding affinities of all small molecules, from which the cut-off was set at −8.0 kcal/mol to filter for the second molecular screening (**b**). The twenty-one top-ranking focal compounds chosen for further cell-based experiments (**c**).

3.2. Dose–Response Analysis on the Cell-Based Antiviral Activity of Small Molecules

First, we tested the cytotoxicity and antiviral activities of the 21 focal compounds in BHK-21 cells. The cytotoxicity profile of the small compounds in BHK-21 cells was determined by measuring the MTS formazan product of living cells. The four compounds that were not toxic to the BHK-21 cells and had CC50 values in the mid to high micromolar range (49.91 $\pm$ 1.70 µM to >100 µM) were tested in a dose–response manner. Therefore, the highest non-toxic concentration of each compound was used in the subsequent antiviral activity assay. Among the 21 compounds, NSC217697, NSC670283, NSC292567, and NSC65850 demonstrated antiviral inhibition at the maximum dose of non-toxic concentration under the post-viral infection condition using 10 TCID50 of FMDV and observing at 24 hpi.

To validate these findings in a dose-dependent manner, we investigated the antiviral properties of the four compounds in FMDV-infected BHK-21 cells. In the pre-viral-entry experiment, unabsorbed FMDV were removed, and the cells were washed with PBS before adding the fresh media. The four compounds also showed viral inhibition during co-absorption with the virus on the cell monolayer for 2 h (in Figure S1). High concentrations of the compounds were required to exert the viral-suppression effects compared to the same compound examined in the post-viral-entry assay. We assumed that the compounds still had sufficient potency to reduce FMDV after being removed and washed away.

We next determined whether the candidate compounds had antiviral activity during post-viral infection by infecting BHK-21 cells with FMDV at 10 TCID50, followed by incubating with serial concentrations of each compound. The results showed that the numbers of positive infected cells were markedly decreased with the increased compound concentrations in a dose-dependent manner as shown in Figure 2. The vehicle control (0.1% DMSO) did not influence viral production in the infected cells. NSC217697 and NSC292567 exhibited great antiviral inhibition of 0.78 $\pm$ 0.10 and 0.42 $\pm$ 0.08 µM, respectively. NSC670283 and NSC65850 demonstrated good inhibitory effects with the EC50 of 3.38 $\pm$ 1.02 and 2.73 $\pm$ 0.48 µM, respectively (Table 2). These four compounds showed antiviral activities greater than ribavirin (the broad-spectrum anti-RdRp nucleoside analogue) and had antiviral potency comparable to rupintrivir (the anti-3C^{pro} peptidomimetic drug of human rhinovirus) as shown in Figure 2. In addition, the off-target compounds, which did not bind to the 3D^{pol} active site, or bound with the higher binding affinities (>-8.0 kcal/mol), were also examined for cytotoxicity using MTS and antiviral activities in the cell-based assay with immunostaining by IPMA. The results showed that the random or non-target compounds did not inhibit viral replication in both pre- and post-viral entry as shown in Figure S2.

Furthermore, the EC50 and EC90 values, which are concentrations of compounds that could reduce 50% and 90% of the viral infectivity, respectively, were determined in the cell-based assay and IPMA. Both values were calculated using non-linear regression to determine the SI (selective index) by CC50/EC50. The SIs of the four compounds on suppression of FMDV infection in BHK-21 cells are shown in Table 2. NSC217697 and NSC292567 showed SI values >100.

3.3. Functional Interference on FMDV 3D^{pol}, but Not 3C^{pro}, by the Candidate Compounds

We further investigated the mechanism by which the compounds inhibited FMDV 3D^{pol} activity with a cell-based 3D^{pol} inhibition assay exploiting pKLS3_GFP, an FMDV minigenome with the GFP gene. pKLS3_GFP contains essential elements of FMDV replication and the GFP reporter gene while one of the helper plasmids provides functional FMDV 3D^{pol}. The numbers of GFP-positive cells and their intensity reflect the 3D^{pol} activity. Antiviral replication of the compounds on FMDV 3D^{pol} was measured as the decreased GFP intensity and positive cell numbers in the compound treatment well related to those in the vehicle control well (Figure 3a). The results showed that NSC670283 and NSC292567 could inhibit FMDV 3D^{pol} activity in nanomolar concentrations with the IC50 equal to 71.00 $\pm$ 0.04 nM and 0.8 $\pm$ 0.10 nM, respectively, while NSC217697 and NSC65850 also demonstrated good inhibitions at low micromolar concentrations of 0.22 $\pm$ 0.13 µM and 0.13 $\pm$ 0.10 µM, respectively (Figure 3a and 3b). All candidate compounds showed inhibitory effects greater than ribavirin (IC50 = 2.68 $\pm$ 0.37 µM). We also sought to know the inhibitory effects of the compounds on other viral proteins, because 3C^{pro} plays a major role in picornavirus polyprotein maturation, essential in the viral life cycle. We further investigated this function by testing the compounds using the cell-based FMDV 3C^{pro} inhibition assay. The results demonstrated that none of the candidate compounds showed an effective inhibitory effect on FMDV 3C^{pro}. Therefore, these data highlighted the anti-3D^{pol} activity of the four candidate compounds (Figure 3).

Table 2. The cell-based antiviral activity against FMDV infection and cytotoxicity of compounds in BHK-21 cells.

Compounds	Cytotoxicity CC50 (μM)	Antiviral Activity		SI (CC50/EC50)
		EC50 (μM)	EC90 (μM)	
NSC217697 6,6′-methylene bis(2,2,4-trimethyl-1,2-dihydroquinoline) PubChem ID 99342	>100	0.78 ± 0.10	9.52 ± 0.18	>128.20
NSC670283 2,2′-spirobi [3,6,7,8-tetrahydro-1H-cyclopenta[g]naphthalene]-5,5′-dione PubChem ID 382634	>100	3.38 ± 1.02	30.45 ± 0.25	>28.65
NSC292567 Pandavir/Nigericin PubChem ID 4490	49.91 ± 1.70	0.42 ± 0.08	3.85 ± 0.34	>118.83
NSC65850 sodium;5-[[4-[4-[[2,4-diamino-3-[[4-(carboxymethoxy)phenyl]diazenyl]-5-methylphenyl]diazenyl]phenyl]phenyl]diazenyl]-2-hydroxybenzoic acid PubChem ID 135422251	>100	2.73 ± 0.48	7.19 ± 0.10	>36.63
Ribavirin PubChem ID 37542	>100	42.89 ± 0.42	393.60 ± 2.56	>2.33
Rupintrivir PubChem ID 6440352	>100	2.02 ± 0.30	12.05 ± 1.6	>49.50

3.4. Inhibition of FMDV Replication by Small Compounds

We further examined the effects of the four compounds on the viral replication process by measuring levels of viral loads and negative-stranded RNA synthesis by FMDV 3D^{pol} using RT-qPCR (Figure 4). Firstly, we investigated the negative effects of the compounds on the negative-stranded RNA synthesis at 0.1, 1, 5, 10, 25, and 50 μM (Figure 4). The results showed that all four compounds could reduce the levels of negative-stranded RNA in a dose-dependent manner. Secondly, viral copy numbers derived from FMDV-infected BHK-21 cells treated with serial concentrations of the four compounds were determined to reveal their inhibition activity on viral loads. The results showed that all compounds had inhibition activities over 99% at 5 μM (Figure 4). Among the four compounds, NSC217697 appeared to be the most potent inhibitor with an inhibition ratio > 99% at 1 μM. The results showed that the compounds could actively suppress both viral load and negative-RNA strand production of 3D^{pol}. These data confirmed that the four candidate compounds were potent FMDV 3D^{pol} inhibitors by inhibiting the viral RNA synthesis function of the enzyme.

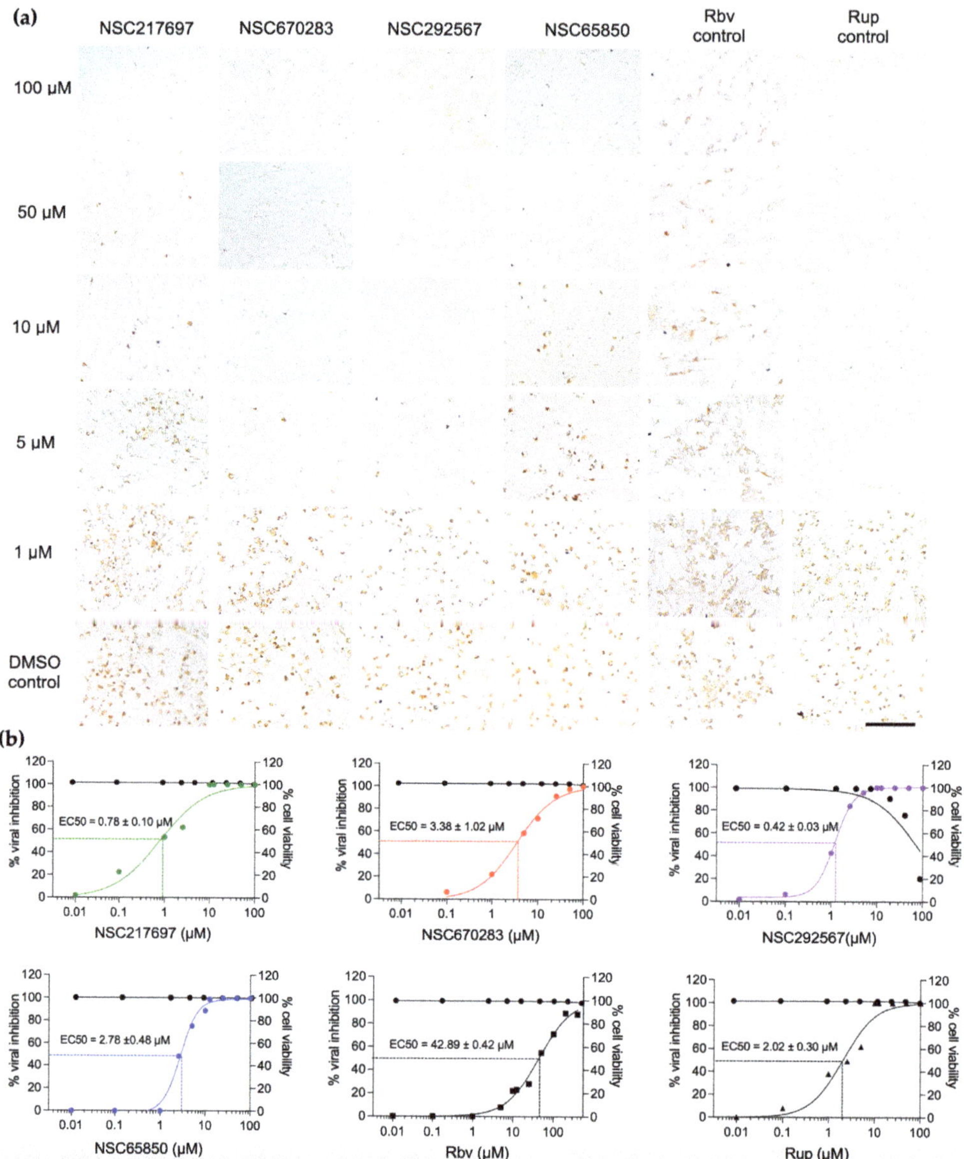

Figure 2. Antiviral activities of the selected compounds against FMDV in cell culture. Four small compounds were evaluated in a dose–response manner, and ribavirin (Rbv) and rupintrivir (Rup) were used as drug controls. The FMDV-infected cells with or without compound treatment were determined using IPMA in which viral antigens in the positive cells were stained with DAB and turned brown (**a**). Graphs of cytotoxicity and antiviral activities of the compounds demonstrating percentage of viable cells (black lines) and percentage of viral inhibition (color lines) at different concentrations of the compounds (**b**). The scale bar is 100 μm.

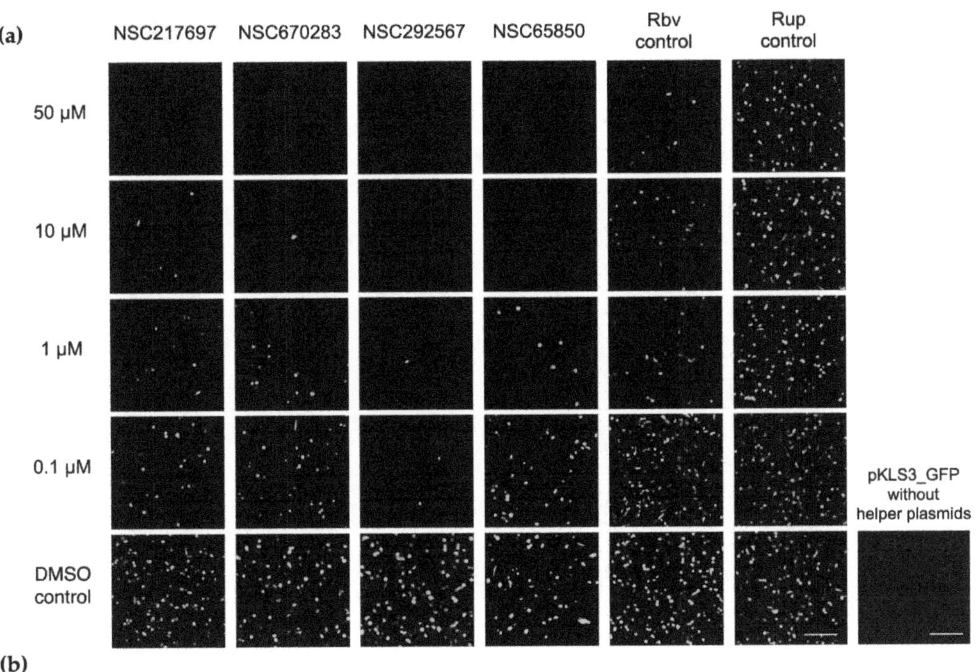

(b)

Assay	IC50 (nM to µM)					
	NSC217697	NSC670283	NSC292567	NSC65850	Rup control	Rbv control
Anti-3D^{pol}	0.22 ± 0.13 µM	71.00 ± 0.04 nM	0.8 ± 0.10 nM	0.13 ± 0.10 µM	Not inhibit	2.68 ± 0.37 µM
Anti-3C^{pro}	Not inhibit	Not inhibit	Not inhibit	Not inhibit	11.06 ± 1.04 µM	Not inhibit

Figure 3. Cell-based 3D^{pol} inhibition assay based on FMDV minigenome expressing green fluorescent protein. 293T cells were transfected with pKLS3_GFP and two helper plasmids, and then treated with the selected compounds in a dose-dependent manner. Ribavirin (Rbv) and rupintrivir (Rup) were used as 3D^{pol} positive and negative drug controls, respectively, and 0.1% DMSO served as a vehicle control. No fluorescent signal was observed in the transfection well containing pKLS3_GFP in the absence of the two helper plasmids (plasmid control). The bars are 100 µM (**a**). The table shows 50% inhibitory concentration (IC50) values of the four compounds, ribavirin (Rbv), and rupintrivir (Rup) (**b**). Please note that ribavirin is the RdRp inhibitor and rupintrivir is a 3C^{pro} inhibitor.

3.5. The Predicted Interaction of the Small Compounds and the Catalytic Active Sites of FMDV 3D^{pol}

The focused molecular docking showed that the four compounds occupied the 3D^{pol} active sites and they inhibited the 3D^{pol} function in the cell-based assay. We further investigated the protein–ligand interactions of the compounds docked to these sites within the 3D^{pol} palm and finger subdomains. The protein–ligand interactions are shown in Figure 5. All compounds properly located to site 1 of the loop β8-α9 and site 3 of the loop β10-β11 rather than site 2. The binding showed that NSC217697 formed a hydrogen bond with Y241 and van der Waals bond with D245 of the loop β8-α9 and reacted to D338 with π-anion. Moreover, this compound could bind to the NTP binding residues (R168, K172, and R179) of the finger subdomain with alkyl and van der Waals interactions (Figure 5a). As the catalytic aspartates and NTP binding site play an important role in RNA–3D^{pol} interaction, NSC217697 was predicted to bind and inhibit the interaction between RNA and 3D^{pol} molecule. NSC670283 is composed of fused bicyclic rings with cyclic ketone. D338 of motif C reacted to the bicyclic rings with π-anion interaction. Moreover, Y336 also shared the binding to cyclic ketone with π-sigma bond. The L386, K387, and R388 of motif

E that are crucial for interacting with the RNA formed π-alkyl/alkyl and π-cation bonds to the bicyclic rings and H-bonds to the ketone group (Figure 5b). When the ionophore antibiotic bound to FMDV 3D^{Pol}, NSC292567 formed three hydrogen bonds with D245, S304, and A116. The Y336 (motif C) of the loop β10-β11 reacted to the compounds by alkyl interaction, and N307 (motif B) formed van der Waals interactions with NSC292567 (Figure 5c). NSC65850 occupied all three sites and formed van der Waals bonds with D240 and Y241 (site 1), amide-π stacked interactions with S298 (site 2), and hydrogen bonds with D338 (site 3) (Figure 5d). These predicted interactions showed that the candidate compounds could structurally interfere with the RNA polymerase activity of FMDV 3D^{Pol}.

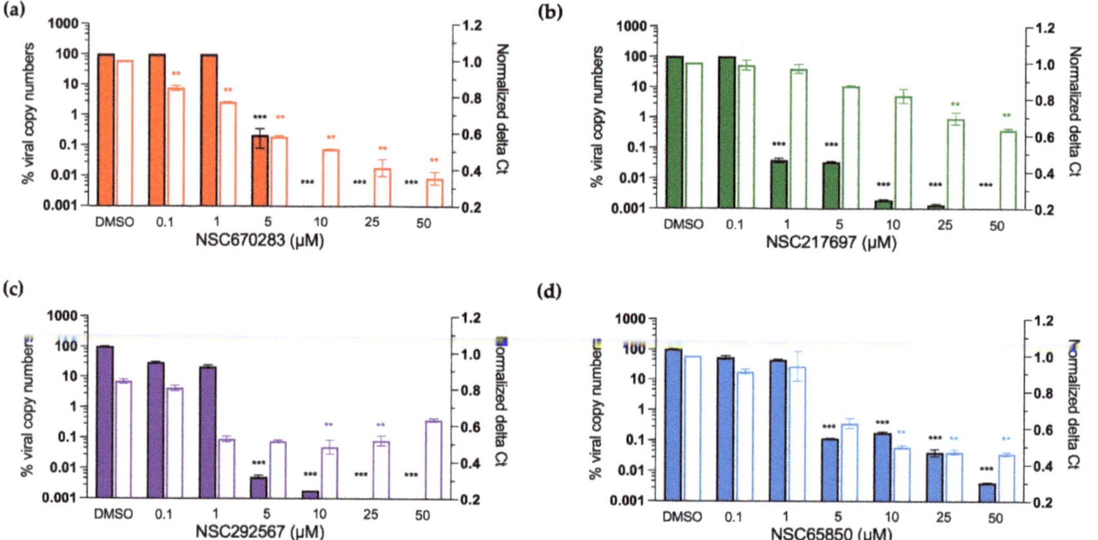

Figure 4. Percent viral copy numbers and the delta Ct values of negative-stranded RNA of FMDV-infected BHK-21 cells with or without compound treatments. The infected cells were treated with serial concentrations of the selected compounds to demonstrate the dose-dependent effect. The selected compounds included NSC670283 (**a**; red), NSC217697 (**b**; green), NSC292567 (**c**; purple), and NSC65850 (**d**; blue). The bar graphs were plotted between differential drug concentrations (*X*-axis), % viral copy number (left *Y*-axis), and the normalized delta Ct values of negative-stranded RNA (right *Y*-axis). The solid-color bars represent % viral copy numbers and the empty bars are the normalized delta Cts. The asterisks denote the significance levels with ** meaning a *p* value of 0.01 and *** a *p* value of 0.001.

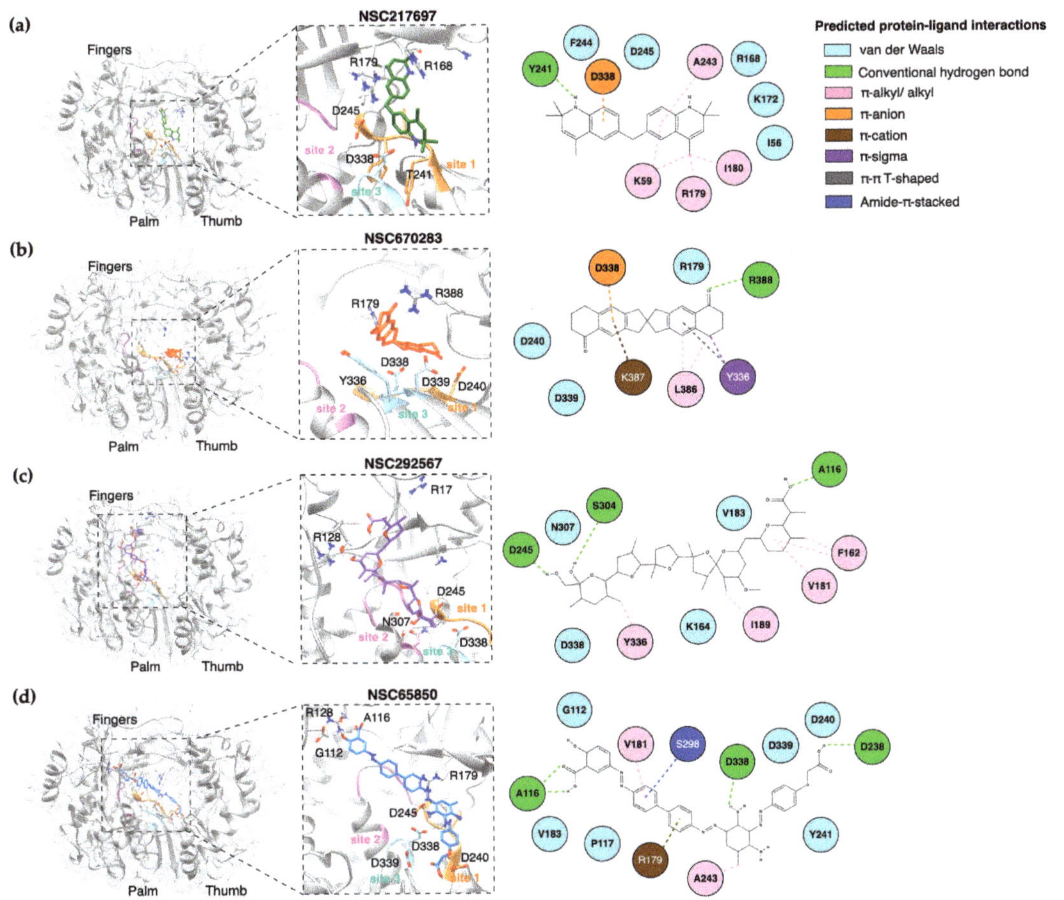

Figure 5. The predicted 3D (left panel) and 2D structures (right panel) of the small molecule–FMDV 3D^{Pol} interactions. Interactions between the residues within sites 1–3 of FMDV 3D^{Pol} and small compounds NSC217697 (**a**), NSC670283 (**b**), NSC292567 (**c**), and NSC65850 (**d**) are demonstrated.

4. Discussion

RNA-dependent RNA polymerase (RdRp) is the enzyme essential for replication and transcription processes of RNA viruses including FMDV and other picornaviruses. Therefore, RdRp is an attractive target for novel inhibitors against RNA viruses. In the case of FMDV, the crystal structures of RdRp (3D^{Pol}) have been resolved, and are similar to the right-handed architecture with three conserved subdomains—palm, finger, and thumb. This characteristic is common among picornaviruses [5,6]. Computational drug discovery has proven to be a guide for screening new drug candidates, and this method can directly provide scientific evidence for further in vitro and in vivo testing. In this study, we optimized the FMDV 3D^{Pol} model for small-molecule screening processes. Initially, we performed the blind virtual screening of small molecules to filter the 3D^{Pol} binding compounds for the second-round focused docking. The focused molecular docking specified the compounds that fit the active sites and involved functional areas of FMDV 3D^{Pol}. We demonstrated that four out of 21 compounds obtained from the double virtual screening inhibited FMDV at the post-infection stage using a cell-based antiviral assay. Moreover, these candidates showed good inhibition specific to FMDV 3D^{Pol} activity, but not 3C^{Pro}. 3D^{Pol} is one of the most important enzymes for viral replication processes and it functions after viral infection. In our study, ribavirin could inhibit FMDV 3D^{Pol} as it is a well-known broad-spectrum anti-RNA virus and the complex of FMDV RdRp, RNA template, and ribavirin triphosphate has been solved and demonstrated [10]. In addition, a ribavirin-resistant mutant, which contained M296I on the β9-α11 loop adjacent to the active site, induced misincorporation of guanosine monophosphate into the RNA chain leading to an error catastrophe [8,10,33]. Therefore, it is suitable to be used as a positive RdRp inhibitor.

We have shown that the four compounds, NSC217697, NSC670283, NSC292567 and NSC65850, interacted with the active sites of FMDV 3D^{Pol} to inhibit the enzyme function, and thus markedly reduced synthesis of viral RNA. NSC217697 (6,6′-methylene bis(2,2,4-trimethyl-1,2-dihydroquinoline)) is a quinoline, which contains fused rings of heterocyclic compounds. We found that this compound could inhibit FMDV replication after infection. Moreover, NSC217697 could inhibit 3D^{Pol} with a micromolar concentration, but had no inhibitory effect on FMDV 3C^{Pro}. This compound was predicted to bind the NTP binding residues (R168, K172, and R179) in the finger subdomain, which possibly masked an initiation site in the complex conformation. Recently, quinoline and quinazoline derivatives have been shown to inhibit SARS-CoV-2 RdRp [34]. We assumed that NSC217697 could be an allosteric inhibitor of FMDV 3D^{Pol}.

NSC670283 (2,2′-spirobi [3,6,7,8-tetrahydro-1H-cyclopenta[g]naphthalene]-5,5′-dione) is a spiro compound that contains a fused unit of 1,2,3,6,7,8-hexahydro-5H-cyclopenta[*b*]na phthalen-5-one at position 2. This spiro compound has been reported to predictably interact with hepatitis C virus E2 envelope glycoprotein (HCV E2). It could bind HCV E2 protein as detected by surface plasmon resonance [35]. According to a viral specificity test, NSC670283 also inhibited RD114 glycoprotein (RD114pp) of endogenous feline retrovirus with IC50 of 3 μM in Huh-7 cells. However, antiviral activity against HCV replication in cell culture was not determined in the previous study. Indeed, we have shown that NSC670283 reacted to specific amino acid residues within Motifs C and E and the catalytic aspartic residue in the palm subdomain of FMDV 3D^{Pol}. This compound could inhibit FMDV replication as well as FMDV 3D^{Pol} activity with micromolar IC50 and EC50, suggesting a potential broad-spectrum antiviral.

Among the four compounds, NSC292567 (nigericin or pandavir) demonstrated the highest binding affinity. This compound is grouped with the ionophore antibiotics, which have inhibitory activities against drug-resistant strains of Gram-positive bacteria and coccidian protozoa. It has been used to treat coccidiosis in poultry [36]. During the global COVID-19 crisis, the ionophore antibiotics were a group of interesting drugs that were repurposed for antiviral drug development. For example, monesin, a monovalent cation/proton antiporter, was found to inhibit MERS-CoV [37] and SARS-CoV-2 [38]. Salinomymin could inhibit influenza A and B viruses by blocking endosomal acidification and interfering with viral matrix protein 2 [39]. In addition, nigericin is an ionophore that affects lipid-soluble molecules. It increases the permeability of an ion across a biological membrane as well as the lipid bilayer of the vesicular transport system that functions in cellular trafficking of protein macromolecules. Picornavirus replication requires maturation of these membranous vesicles in the vesicular compartment during plus- and minus-strand RNA syntheses. Disruption of this membranous vesicle by ionophores thus inhibits viral replication [40]. Nigericin demonstrated a moderate inhibition of SARS-CoV-2 [38]. In this study, nigericin was ranked in the first place of our virtual screening with a binding affinity of −10.1 kcal/mol, and it had a good inhibitory effect on FMDV replication with an EC50 value of 0.42 μM. Moreover, nigericin at 0.8 nM reduced FMDV 3D^{Pol} activity by 50%, and the anti-3D^{Pol} action of this compound was confirmed by the reduction of negative-stranded RNA synthesis. In our predicted model study, nigericin could occupy the finger and palm subdomains of 3D^{Pol}. These results suggest that the ionophore antibiotics could be potential antivirals against FMDV infection by directly suppressing RNA synthesis of 3D^{Pol}.

NSC65850 (5-((4′-((2,4-diamino-3-((4-(carboxymethoxy)phenyl)diazenyl)-5-methylph enyl)diazenyl)[1,1′-biphenyl]-4-yl)diazenyl)-2-hydroxybenzoic acid) is a compound in the

NCI diversity set III. Thus far, its antimicrobial effect has not been reported in the available databases. Therefore, we further investigated the physicochemical properties of this compound. The predicted model based on the SwissAMDE analysis [41] (accessed on 1 August 2022) demonstrated that this compound possessed a poor solubility (Log $S = -11.83$), Log $P_{o/w}$ of 7.17, low skin permeation (log $K_p = -5.14$ cm/s), and low GI absorption. However, its synthetic accessibility score was 4.25, which is moderate for chemical modification in medicinal chemistry to improve its physicochemical property for druggable agents. Nonetheless, our experiments in the cell-based assay showed that this compound could inhibit $3D^{Pol}$, but not $3C^{Pro}$ of FMDV.

5. Conclusions

In conclusion, we have identified novel small-molecule inhibitors of foot-and-mouth disease virus 3D polymerase with virtual screening and further evaluated their actions in both virus-infected cells and a direct target to the polymerase protein. We provided evidence that NSC217697 (quinoline compound), NSC670283 (spiro compound), NSC292567 (ionophore antibiotic), and NSC65850 inhibited FMDV replication by binding to $3D^{Pol}$ and suppressing viral RNA synthesis. Thus, these compounds can be considered as repurposing antiviral agents or as chemical scaffolds for FMDV and other picornavirus drug development.

Supplementary Materials: The following supporting information can be downloaded at: https://www.mdpi.com/article/10.3390/v15010124/s1, Figure S1: Pre-viral entry effect of the selected compounds (colored lines) and their cytotoxicity (black line). The inhibitory effects of the compounds on viral infection were presented by color line graphs plotted between % viral inhibition (left Y-axis) and various concentrations of each compound: NSC217697 (a; green), NSC670283 (b; red), NSC292567 (c; purple), and NSC65850 (d; blue). The EC50 value of each compound was also shown. The cytotoxicity of each compound was depicted as a black line graph by plotting differential doses of each compound (X-axis) against % cell viability (right Y-axis); Figure S2: Non-target compounds from the blind- and focus- molecular dockings. The compounds with the binding affinities higher than the cutoff of -8.0 kcal/mol or those that did not bind in the $3D^{Pol}$ active site were examined for cytotoxicity using MTS (a). The interactions between the non-target compounds (green pointing with yellow arrow) and FMDV $3D^{Pol}$ are demonstrated (b, left panel). The antiviral activities of the compounds at 50 µM were evaluated in the cell-based pre- and post-viral-entry assays and IPMA (b, right panel). The scale bar is 100 µm.

Author Contributions: Conceptualization, P.L. and S.T.; methodology, S.T., P.S., V.L., N.T. and P.C.; Computer analysis, S.T.; resources, P.L.; writing—original draft preparation, S.T.; writing—review and editing, P.L. and S.T.; visualization, S.T.; supervision, P.L.; project administration, P.L.; funding acquisition, P.L. All authors have read and agreed to the published version of the manuscript.

Funding: This research was funded by the National Research Council of Thailand (NRCT), grant number RTA6280011; and Kasetsart University Research and Development Institute (KURDI), grant number FF(KU)17.64.

Institutional Review Board Statement: Not applicable.

Informed Consent Statement: Not applicable.

Data Availability Statement: No new data were created or analyzed in this study. Data are available upon request.

Acknowledgments: We thank the Developmental Therapeutics Program (DTP) Open Repository of the National Cancer Institute (NCI) for providing the chemical libraries and compounds. We greatly thank Pongrama Ramasoota at Mahidol University, Thailand, for providing the scFv-Fc for our work.

Conflicts of Interest: The authors declare no conflict of interest.

References

1. Jamal, S.M.; Belsham, G.J. Foot-and-mouth disease: Past, present and future. *Vet. Res.* **2013**, *44*, 116. [CrossRef] [PubMed]
2. Belsham, G.J. Towards improvements in foot-and-mouth disease vaccine performance. *Acta Vet. Scand.* **2020**, *62*, 20. [CrossRef]

3.	Venkataraman, S.; Prasad, B.; Selvarajan, R. RNA dependent RNA polymerases: Insights from structure, function and evolution. *Viruses* **2018**, *10*, 76. [CrossRef]

4.	Ferrerorta, C.; arias, A.; escarmis, C.; Verdaguer, N. A comparison of viral RNA-dependent RNA polymerases. *Curr. Opin. Struct. Biol.* **2006**, *16*, 27–34. [CrossRef]

5.	Ferrer-Orta, C.; de la Higuera, I.; Caridi, F.; Sánchez-Aparicio, M.T.; Moreno, E.; Perales, C.; Singh, K.; Sarafianos, S.G.; Sobrino, F.; Domingo, E.; et al. Multifunctionality of a picornavirus polymerase domain: Nuclear localization signal and nucleotide recognition. *J. Virol.* **2015**, *89*, 6848–6859. [CrossRef]

6.	Ferrer-Orta, C.; Arias, A.; Perez-Luque, R.; Escarmís, C.; Domingo, E.; Verdaguer, N. Structure of foot-and-mouth disease virus RNA-dependent RNA polymerase and its complex with a template-primer RNA. *J. Biol. Chem.* **2004**, *279*, 47212–47221. [CrossRef]

7.	Kempf, B.J.; Peersen, O.B.; Barton, D.J. Poliovirus polymerase Leu420 facilitates RNA recombination and ribavirin resistance. *J. Virol.* **2016**, *90*, 8410–8421. [CrossRef]

8.	Sierra, M.; Airaksinen, A.; González-López, C.; Agudo, R.; Arias, A.; Domingo, E. Foot-and-mouth disease virus mutant with decreased sensitivity to ribavirin: Implications for error catastrophe. *J. Virol.* **2007**, *81*, 2012–2024. [CrossRef] [PubMed]

9.	Young, K.-C. Identification of a ribavirin-resistant NS5B mutation of hepatitis C virus during ribavirin monotherapy. *Hepatology* **2003**, *38*, 869–878. [CrossRef] [PubMed]

10.	Agudo, R.; Ferrer-Orta, C.; Arias, A.; de la Higuera, I.; Perales, C.; Pérez-Luque, R.; Verdaguer, N.; Domingo, E. A Multi-step process of viral adaptation to a mutagenic nucleoside analogue by modulation of transition types leads to extinction-escape. *PLoS Pathog.* **2010**, *6*, e1001072. [CrossRef]

11.	Sakamoto, K.; Ohashi, S.; Yamazoe, R.; Takahashi, K.; Furuta, Y. *The Inhibition of FMD Virus Excretion from the Infected Pigs by an Antiviral Agent T-1105*; Report of the European Commission for the Control of Foot-and-mouth disease; Appendix 64; FAO: Paphos, Cyprus, 2006; pp. 418–423.

12.	Ohashi, S.; Sakamoto, K.; Fukai, K.; Morioka, K.; Yamazoe, R.; Takahashi, K.; Furuta, Y. An Antiviral Agent, T-1105 Prevents Virus Excretion from Pigs Infected with Porcinophilic Foot-And-Mouth Disease Virus. In *The Global Control of FMD—Tools, Ideas and Ideals, 2008*; Session of the Research Group of the Standing Technical Committee of the European Commission for the Control of Foot-and-Mouth Disease; Appendix 70; Food and Agricultural Organization of the United Nations: Rome, Italy, 2008; pp. 393–398.

13.	De Vleeschauwer, A.R.; Lefebvre, D.J.; Willems, T.; Paul, G.; Billiet, A.; Murao, L.E.; Neyts, J.; Goris, N.; De Clercq, K. A Refined guinea pig model of foot-and-mouth disease virus infection for assessing the efficacy of antiviral compounds. *Transbound. Emerg. Dis.* **2016**, *63*, e205–e212. [CrossRef]

14.	van der Linden, L.; Vives-Adrián, L.; Selisko, B.; Ferrer-Orta, C.; Liu, X.; Lanke, K.; Ulferts, R.; De Palma, A.M.; Tanchis, F.; Goris, N.; et al. The RNA template channel of the RNA-dependent RNA polymerase as a target for development of antiviral therapy of multiple genera within a virus family. *PLoS Pathog.* **2015**, *11*, e1004733. [CrossRef]

15.	Tian, L.; Qiang, T.; Liang, C.; Ren, X.; Jia, M.; Zhang, J.; Li, J.; Wan, M.; YuWen, X.; Li, H.; et al. RNA-dependent RNA polymerase (RdRp) inhibitors: The current landscape and repurposing for the COVID-19 pandemic. *Eur. J. Med. Chem.* **2021**, *213*, 113201. [CrossRef] [PubMed]

16.	Hardy, J.A.; Wells, J.A. Searching for new allosteric sites in enzymes. *Curr. Opin. Struct. Biol.* **2004**, *14*, 706–715. [CrossRef]

17.	Jin, Y.H.; Min, J.S.; Jeon, S.; Lee, J.; Kim, S.; Park, T.; Park, D.; Jang, M.S.; Park, C.M.; Song, J.H.; et al. Lycorine, a non-nucleoside RNA dependent RNA polymerase inhibitor, as potential treatment for emerging coronavirus infections. *Phytomedicine* **2021**, *86*, 153440. [CrossRef]

18.	Gharbi-Ayachi, A.; Santhanakrishnan, S.; Wong, Y.H.; Chan, K.W.K.; Tan, S.T.; Bates, R.W.; Vasudevan, S.G.; El Sahili, A.; Lescar, J. Non-nucleoside Inhibitors of Zika virus RNA-dependent RNA polymerase. *J. Virol.* **2020**, *94*. [CrossRef]

19.	Kim, Y.; Lovell, S.; Tiew, K.-C.; Mandadapu, S.R.; Alliston, K.R.; Battaile, K.P.; Groutas, W.C.; Chang, K.-O. Broad-spectrum antivirals against 3C or 3C-Like proteases of picornaviruses, noroviruses, and coronaviruses. *J. Virol.* **2012**, *86*, 11754–11762. [CrossRef] [PubMed]

20.	Semkum, P.; Kaewborisuth, C.; Thangthamniyom, N.; Theerawatanasirikul, S.; Lekcharoensuk, C.; Hansoongnern, P.; Ramasoota, P.; Lekcharoensuk, P. A novel plasmid DNA-based foot and mouth disease virus minigenome for intracytoplasmic mRNA production. *Viruses* **2021**, *13*, 1047. [CrossRef] [PubMed]

21.	Chen, V.B.; Arendall, W.B.; Headd, J.J.; Keedy, D.A.; Immormino, R.M.; Kapral, G.J.; Murray, L.W.; Richardson, J.S.; Richardson, D.C. MolProbity: All-atom structure validation for macromolecular crystallography. *Acta Crystallogr. Sect. D Biol. Crystallogr.* **2010**, *66*, 12–21. [CrossRef] [PubMed]

22.	Benkert, P.; Biasini, M.; Schwede, T. Toward the estimation of the absolute quality of individual protein structure models. *Bioinformatics* **2011**, *27*, 343–350. [CrossRef] [PubMed]

23.	Ramachandran, P.; Varoquaux, G. Mayavi: 3D visualization of scientific data. *Comput. Sci. Eng.* **2011**, *13*, 40–51. [CrossRef]

24.	Dallakyan, S.; Olson, A.J. Small-molecule library screening by docking with PyRx. *Methods Mol. Biol.* **2015**, *1263*, 243–250. [PubMed]

25.	O'Boyle, N.M.; Banck, M.; James, C.A.; Morley, C.; Vandermeersch, T.; Hutchison, G.R. Open Babel: An open chemical toolbox. *J. Cheminform.* **2011**, *3*, 33. [CrossRef]

26.	Trott, O.; Olson, A.J. AutoDock Vina: Improving the speed and accuracy of docking with a new scoring function, efficient optimization, and multithreading. *J. Comput. Chem.* **2009**, *31*, 455–461. [CrossRef] [PubMed]

27. Theerawatanasirikul, S.; Thangthamniyom, N.; Kuo, C.J.; Semkum, P.; Phecharat, N.; Chankeeree, P.; Lekcharoensuk, P. Natural phytochemicals, luteolin and isoginkgetin, inhibit 3C protease and infection of FMDV, in silico and in vitro. *Viruses* **2021**, *13*, 2118. [CrossRef] [PubMed]

28. Lekcharoensuk, P.; Wiriyarat, W.; Petcharat, N.; Lekcharoensuk, C.; Auewarakul, P.; Richt, J.A. Cloned cDNA of A/swine/Iowa/15/1930 internal genes as a candidate backbone for reverse genetics vaccine against influenza A viruses. *Vaccine* **2012**, *30*, 1453–1459. [CrossRef]

29. Stirling, D.R.; Carpenter, A.E.; Cimini, B.A. CellProfiler Analyst 3.0: Accessible data exploration and machine learning for image analysis. *Bioinformatics* **2021**, *37*, 3992–3994. [CrossRef] [PubMed]

30. Gu, C.; Zheng, C.; Shi, L.; Zhang, Q.; Li, Y.; Lu, B.; Xiong, Y.; Qu, S.; Shao, J.; Chang, H. Plus- and minus-stranded foot-and-mouth disease virus RNA quantified simultaneously using a novel real-time RT-PCR. *Virus Genes* **2007**, *34*, 289–298. [CrossRef]

31. Van Der Linden, L.; Ulferts, R.; Nabuurs, S.B.; Kusov, Y.; Liu, H.; George, S.; Lacroix, C.; Goris, N.; Lefebvre, D.; Lanke, K.H.W.; et al. Application of a cell-based protease assay for testing inhibitors of picornavirus 3C proteases. *Antiviral Res.* **2014**, *103*, 17–24. [CrossRef]

32. Srisombundit, V.; Tungthumniyom, N.; Linchongsubongkoch, W.; Lekcharoensuk, C.; Sariya, L.; Ramasoota, P.; Lekcharoensuk, P. Development of an inactivated 3CPro-3ABC (mu3ABC) ELISA to differentiate cattle infected with foot and mouth disease virus from vaccinated cattle. *J. Virol. Methods* **2013**, *188*, 161–167. [CrossRef] [PubMed]

33. Arias, A.; Arnold, J.J.; Sierra, M.; Smidansky, E.D.; Domingo, E.; Cameron, C.E. Determinants of RNA-dependent RNA polymerase (in)fidelity revealed by kinetic analysis of the polymerase encoded by a foot-and-mouth disease virus mutant with reduced sensitivity to ribavirin. *J. Virol.* **2008**, *82*, 12346–12355. [CrossRef] [PubMed]

34. Zhao, J.; Zhang, Y.; Wang, M.; Liu, Q.; Lei, X.; Wu, M.; Guo, S.; Yi, D.; Li, Q.; Ma, L.; et al. Quinoline and quinazoline derivatives inhibit viral RNA synthesis by SARS-CoV-2 RdRp. *ACS Infect. Dis.* **2021**, *7*, 1535–1544. [CrossRef] [PubMed]

35. Patent: Al-Olaby, R.; Azzazy, H.; Balhorn, R. Ligands that Target Hepatitis C Virus E2 Protein. U.S. Patent 20160361311A1, 15 December 2016.

36. Chapman, H.D.; Jeffers, T.K.; Williams, R.B. Forty years of monensin for the control of coccidiosis in poultry. *Poult. Sci.* **2010**, *89*, 1788–1801. [CrossRef] [PubMed]

37. Dyall, J.; Coleman, C.M.; Hart, B.J.; Venkataraman, T.; Holbrook, M.R.; Kindrachuk, J.; Johnson, R.F.; Olinger, G.G.J.; Jahrling, P.B.; Laidlaw, M.; et al. Repurposing of clinically developed drugs for treatment of Middle East respiratory syndrome coronavirus infection. *Antimicrob. Agents Chemother.* **2014**, *58*, 4885–4893. [CrossRef] [PubMed]

38. Svenningsen, E.B.; Thyrsted, J.; Blay-Cadanet, J.; Liu, H.; Lin, S.; Moyano-Villameriel, J.; Olagnier, D.; Idorn, M.; Paludan, S.R.; Holm, C.K.; et al. Ionophore antibiotic X-206 is a potent inhibitor of SARS-CoV-2 infection in vitro. *Antiviral Res.* **2021**, *185*, 104988. [CrossRef]

39. Jang, Y.; Shin, J.S.; Yoon, Y.-S.; Go, Y.Y.; Lee, H.W.; Kwon, O.S.; Park, S.; Park, M.-S.; Kim, M. Salinomycin inhibits influenza virus infection by disrupting endosomal acidification and viral matrix protein 2 function. *J. Virol.* **2018**, *92*, 1–20. [CrossRef]

40. Irurzun, A.; Perez, L.; Carrasco, L. Involvement of membrane traffic in the replication of poliovirus genomes: Effects of brefeldin A. *Virology* **1992**, *191*, 166–175. [CrossRef]

41. Sander, T.; Freyss, J.; von Korff, M.; Rufener, C. DataWarrior: An open-source program for chemistry aware data visualization and analysis. *J. Chem. Inf. Model.* **2015**, *55*, 460–473. [CrossRef]

Article

In Vitro and Pre-Clinical Evaluation of Locally Isolated Phages, vB_Pae_SMP1 and vB_Pae_SMP5, Formulated as Hydrogels against Carbapenem-Resistant *Pseudomonas aeruginosa*

Samar S. Mabrouk [1], Ghada R. Abdellatif [1], Ahmed S. Abu Zaid [2], Ramy K. Aziz [3,4] and Khaled M. Aboshanab [2,*]

[1] Department of Microbiology, Faculty of Pharmacy, Ahram Canadian University (ACU), 6th October City, Giza 12566, Egypt
[2] Department of Microbiology & Immunology, Faculty of Pharmacy, Ain Shams University, Cairo 11566, Egypt
[3] Department of Microbiology and Immunology, Faculty of Pharmacy, Cairo University, Cairo 11562, Egypt
[4] Department of Microbiology and Immunology, Children's Cancer Hospital Egypt 57357, Cairo 11617, Egypt
* Correspondence: aboshanab2012@pharma.asu.edu.eg

Abstract: The inadequate therapeutic opportunities associated with carbapenem-resistant *Pseudomonas aeruginosa* (CRPA) clinical isolates impose a search for innovative strategies. Therefore, our study aimed to characterize and evaluate two locally isolated phages formulated in a hydrogel, both in vitro and in vivo, against CRPA clinical isolates. The two phages were characterized by genomic, microscopic, phenotypic characterization, genomic analysis, in vitro and in vivo analysis in a *Pseudomonas aeruginosa*-infected skin thermal injury rat model. The two siphoviruses belong to class Caudovirectes and were named vB_Pae_SMP1 and vB_Pae_SMP5. Each phage had an icosahedral head of 60 ± 5 nm and a flexible, non-contractile tail of 170 ± 5 nm long, while vB_Pae_SMP5 had an additional base plate containing a 35 nm fiber observed at the end of the tail. The hydrogel was prepared by mixing 5% w/v carboxymethylcellulose (CMC) into the CRPA propagated phage lysate containing phage titer 10^8 PFU/mL, pH of 7.7, and a spreadability coefficient of 25. The groups were treated with either Phage vB_Pae_SMP1, vB_Pae_SMP5, or a two-phage cocktail hydrogel cellular subepidermal granulation tissues with abundant records of fibroblastic activity and mixed inflammatory cell infiltrates and showed 17.2%, 25.8%, and 22.2% records of dermal mature collagen fibers, respectively. In conclusion, phage vB_Pae_SMP1 or vB_Pae_SMP5, or the two-phage cocktails formulated as hydrogels, were able to manage the infection of CRPA in burn wounds, and promoted healing at the injury site, as evidenced by the histopathological examination, as well as a decrease in animal mortality rate. Therefore, these phage formulae can be considered promising for clinical investigation in humans for the management of CRPA-associated skin infections.

Keywords: carbapenem-resistant; *Pseudomonas aeruginosa*; bacteriophage; hydrogel; thermal injury model; histopathology

check for **updates**

Citation: Mabrouk, S.S.; Abdellatif, G.R.; Abu Zaid, A.S.; Aziz, R.K.; Aboshanab, K.M. In Vitro and Pre-Clinical Evaluation of Locally Isolated Phages, vB_Pae_SMP1 and vB_Pae_SMP5, Formulated as Hydrogels against Carbapenem-Resistant *Pseudomonas aeruginosa. Viruses* **2022**, *14*, 2760. https://doi.org/10.3390/v14122760

Academic Editor: Simone Brogi

Received: 10 November 2022
Accepted: 8 December 2022
Published: 11 December 2022

Publisher's Note: MDPI stays neutral with regard to jurisdictional claims in published maps and institutional affiliations.

1. Introduction

Gram-negative bacterial infections present significant treatment challenges in clinical settings, imposing limited therapeutic options [1–3]. Although many Gram-negative bacteria are clinically significant, *Pseudomonas (P.) aeruginosa* is one of the most common healthcare-associated pathogens and is considered a major threat by the Centers for Disease Control [3]. Approximately 51,000 healthcare-associated infections (HAIs) were caused by the opportunistic pathogen *P. aeruginosa* each year in the United States (USA) from 2011 to 2014 [3]. *P. aeruginosa* ranked third among Gram-negative causes of selected HAIs reported to the National Healthcare Safety Network (NHSN) [1–3].

P. aeruginosa infections are frequently associated with pneumonia, bloodstream, urinary tract, and surgical site infections, as well as significant morbidity and mortality rates.

They are often difficult to treat due to their intrinsic resistance to many commonly used antimicrobial drugs. Accordingly, carbapenems have emerged and have been widely used as crucial antimicrobial agents for the clinical treatment of severe *P. aeruginosa* infections. As a result, an increasing problem of carbapenem resistance has been reported, considering that carbapenem antibiotics are thought to be the final line of defense against extremely severe multidrug-resistant infections [3–7].

Carbapenem-resistant *P. aeruginosa* (CRPA) can contribute to an increase in mortality, prolonged hospital stays, and other issues, including rising medical expenditures [8]. In rare situations, the only antibiotic that is still effective is colistin. However, colistin's therapeutic usage has been constrained by both nephrotoxicity and neurotoxicity [9], and that colistin-resistant strains have also been reported [9–11]. With treatment failures emerging in tandem with rising antibiotic resistance worldwide, interest in new treatment options against CRPA infections has been evoked. One of the most popular alternative treatment options in studies is bacteriophage therapy [12,13].

Phage therapy is known as the direct application of lytic phages to a patient, targeting a lysing bacterial pathogen causing a clinically relevant infection [14]. Lytic bacteriophages are used to treat infections such as upper respiratory tract, abscesses, burns, and wound infections [15]. In addition, bacteriophages offer several benefits, including the ability to combat bacterial biofilms and protect the natural microbiota. They are also non-toxic, inexpensive, and easy to obtain, and hence, they are one of the most promising alternative options in the treatment of such infections [16].

Although several studies have been carried out to test the lytic activity of various bacteriophages either alone as phage lysates or as cocktails against clinically relevant pathogens including *P. aeruginosa*, they are mostly limited to in vitro examination, and only a few studies have recently been conducted in vivo [17–21]. In addition, till now no studies were conducted for pharmaceutical preparation and in vivo evaluation of suitable topical preparations containing these biologically active phage lysates. Therefore, in this study, two newly isolated bacteriophages with lytic activity have been evaluated both in vitro and preclinically, either as lysates or as hydrogel formulae, against a CRPA clinical isolate that was recovered from severely burned infected patient and showed extensively resistant (XDR) phenotype.

2. Materials and Methods

2.1. Clinical Bacterial Isolates: Collection, Identification, and Antimicrobial Susceptibility Testing

Three *P. aeruginosa* clinical isolates were recovered from discharged unidentified wound exudates of patients who had been admitted to the El-Demerdash Tertiary Care Hospital, Cairo, Egypt. Based on the hospital records, the respective samples were collected from unidentified patients suffering from severe burn infections as a routine checkup for culture and sensitivity. The protocol of this study was approved by the Faculty of Pharmacy Ain Shams University Research Ethics Committee (Number, ACUC-FP-ASU RHDIRB2020110301 REC# 41 in September 2021). Using Bergey's manual of determinative bacteriology, isolates were identified macroscopically, microscopically, and biochemically [22]. In addition, bacterial identification was confirmed using the VITEK2 automated system [23]. The isolates were also assessed for their pattern of susceptibility to amikacin (AK), aztreonam (AT), ciprofloxacin (CIP), levofloxacin (LEV), imipenem (IMP), and meropenem (MRP) using the Kirby-Bauer disk diffusion according to CLSI guidelines [24]. The isolates exhibiting the multidrug-resistant (MDR) and XDR phenotypes were defined using previously reported international standard criteria [25]. The XDR isolates that showed a resistance pattern to any of the carbapenems tested were potentially identified as carbapenem-resistant and were chosen to test their MIC against imipenem using the CLSI broth microdilution method according to CLSI guidelines, 2021 [24].

2.2. Phenotypic and Genotypic Detection of Carbapenemase Producers (CPs)

The carbapenemase producer isolates were detected using different phenotypic tests, including the modified carbapenem inactivation method (mCIM), combined disk test (CDT), and blue-Carba test (BCT). The genomic DNA from the phenotypically confirmed XDR CP isolates was isolated using the genomic DNA purification kit (Thermo Fisher Scientific, Massachusetts, USA) and was utilized as a template for PCR to test for the five major carbapenemase genes, including bla_{KPC}, gene coding for *Klebsiella pneumoniae* carbapenemases (KPC); bla_{NDM}, a gene coded for New Delhi metallo-β-lactamase (NDM); imipenem-resistant *Pseudomonas*-type carbapenemases (IMP); bla_{VIM}, a gene coded for Verona integron-encoded metallo-β-lactamase (VIM); and bla_{OXA-48}, oxacillinase (OXA-48-like) types as previously described [26].

2.3. Bacteriophage Recovery from Sewage Samples

2.3.1. Isolation of *P. aeruginosa*-Specific Bacteriophages

The samples were collected from the sewage sources of the specialized Ain Shams University Hospitals, in Cairo, Egypt. The sewage sources of the respective specimens were selected based on the probability of incorporating *P. aeruginosa* and, consequently, bacteriophages specific against it [27]. All samples were kept at 4 °C until processing and were handled based on their apparent clarity [19]. Samples that were visibly clear were used as-is, while Wetted cotton and filter paper were used to thoroughly filter turbid samples [19]. For further processing, only the clear supernatant was preserved. For the isolation of bacteriophages, CRPA isolates served as the bacterial hosts. A loopful of each bacterial isolate cultivated on a nutrient agar plate was inoculated into tryptic soy broth (TSB) and incubated for 6–7 h in a 37 °C, 180 rpm shaking water bath. The suspension of each bacterial isolate was utilized when heavily turbid, with an optical density identical to a bacterial count of 10^9 CFU/mL [28]. For isolation of the phage, a double-strength broth (TSB) was prepared, with 50 µL of 1 M Ca and 25 µL of 0.5 M Mg added in particular [29]. A suspension containing the bacterial host inoculum, the environmental sample, and the isolation medium with a proportion of 1:1:10, respectively, was incubated overnight at 28 °C and 180 rpm [30]. The following day, the co-culture was centrifuged for 20 min at 6000 rpm. The supernatant was agitated vigorously for five minutes after the addition of chloroform with a ratio of 1:10 [19]. The suspensions were allowed to separate for 4–6 h at 4 °C, and the supernatant that formed on top of a plug-like sediment was collected and re-centrifuged under the same circumstances. The collected lysates were stored at 4 °C [30].

2.3.2. Screening for Lytic Activity against CRPA in the Acquired Lysates

A spot test was used to qualitatively screen for anti-CRPA bacteriophages in the fresh lysates. The procedure was carried out in accordance with Adams' original description of the method [31]. The presence of phages active against CRPA was indicated by the presence of clear inhibition spots [31]. The quantitative plaque assay was next performed on the lysates that had positive spot test results using the standard double agar overlay (DAO) method [32]. Plaques were examined and counted the following day. The following equation was used to calculate the phage titer [33]:

Phage titer in plaque-forming unit per ml (PFU/mL) = number of plaques/volume of lysate infected × dilution factor

2.3.3. Phage Propagation

The same isolation procedure was repeated three times, but each time an aliquot of the obtained crude phage lysate was used instead of the starting sewage sample. Essentially, propagation was done frequently to maintain a sizable stock of high-titer phage suspensions [19,34].

2.4. Characterization of the Isolated Bacteriophages Showing Lytic Activities against CRPA

2.4.1. Host Range

The host range was determined as previously reported [35]. The selected lysates showing positive spot test results were tested for lytic activity against the remaining three CRPA isolates and some other clinically relevant pathogens, including three *Klebsiella pneumoniae*, six *Acinetobacter baumannii*, and two *Escherichia coli* clinical isolates that were previously collected and identified in our previous study [26].

2.4.2. Morphology of the Isolated Bacteriophages Showing Lytic Activities against CRPA

For microscopical examination, a concentrated phage suspension of each of the chosen lysates was centrifuged twice at 10,000 rpm for 25 min and was then filtered through a sterile syringe filter (0.22 mm). After that, 20-mL samples were prepared as instructed by Kalatzis et al. and examined via a transmission electron microscope (TEM; version JEOL_JEM_1400 Electron Microscope Nieuw-Vennep, Tokyo, Japan) performed at Cairo University Research Park, Faculty of Agriculture, Cairo, Egypt.

2.4.3. Molecular Analysis of the Bacteriophage

The bacteriophage lysate was sequenced in an Illumina MiSeq instrument (Illumina, La Jolla, CA, USA), and the library was prepared by the Nextera XT DNA Library preparation kit (San Diego, CA, USA). The sequence reads were uploaded into the PATRIC BRC [36] website (now renamed to BV-BRC, URL: https://www.bv-brc.org/, accessed on 23 October 2022) and analyzed through the metagenomics binning pipeline, which uses the BV-BRC database for extraction and annotation of both bacterial and viral genomes from sequence reads [37]. Briefly, reads are assembled by the most optimal assembly tool on the BV-BRC server (either MetaSPAdes or MEGAHIT). The produced assembled contigs are further annotated on PATRIC by the default RAST algorithm [38]. The phage genome was reannotated on the RAST server (https://rast.nmpdr.org, accessed on 23 October 2022) by the RASTtk pipeline, following the customized phage annotation pipeline [39]. The annotation of every protein-coding gene was further checked and confirmed by BLASTP searches against the NCBI NR database, filtered for viruses [40] then by PFAM searches in the case of hypothetical proteins. Curated annotations was generated from the consensus of RAST, PATRIC, and BLAST hits. The creation of the circular image and comparison with other reported similar plasmids were performed using the BLAST Ring Image Generator (BRIG) tool v0.95 (https://sourceforge.net/projects/brig/, accessed on 25 October 2022) [41].

2.5. Formulation of Bacteriophage-Carboxymethyl Cellulose (CMC) Hydrogel

The hydrogel was prepared by mixing 5% w/v CMC (El Nasr Pharmaceutical Chemicals Co. (ADWIC), Cairo, Egypt) in the CRPA propagated phage lysate containing phage titer 10^8 PFU/mL (tested hydrogel). To ensure the formation of a homogeneous hydrogel, CMC was added in portions by sprinkling CMC powder with constant stirring [18]. Another 5% w/v CMC in sterile distilled water was formulated as a negative control (control hydrogel). The hydrogel was prepared with a pH of 7.7 and a spreadability coefficient of 25 measured, as previously reported [42].

2.6. In Vitro Anti-CRPA Activity of the Tested Hydrogels

The anti-CRPA activity of the tested hydrogels was evaluated with the cup-plate method using Mueller–Hinton agar plates. The CRPA isolate was seeded into a sterilized growth medium, which was then transferred aseptically into Mueller-Hinton agar plates to form a double-layer plate [43]. After complete solidification, a sterile cork-borer with a 5 mm diameter was used to make the cups. After that, the prepared cups were filled with the same volume (about 200 µL) of each of the propagated phage lysates alone (positive control), tested hydrogels and control hydrogel (negative-control) separately, and then incubated at 37 °C for 24 h. The antimicrobial activity was determined by the presence or absence of a zone of inhibition around the cups [43].

2.7. Preclinical Evaluation of the Formulated Tested Hydrogels

2.7.1. Laboratory Animals

Throughout the experiment, female Wistar rats weighing approximately 110–120 g were used. All animals were housed in open cages and fed an antibiotic-free diet consisting mainly of 20% protein, 6.5% ash, 5% fiber, and 3.5% fat, with free access to water. They were kept on an alternating 12 h light-dark cycle and a constant temperature of 25 °C adjusted by air conditioning. Animals were maintained in accordance with the Care and Use of Laboratory Animals recommendations and ARRIVE guidelines (https://arriveguidelines.org) (accessed on 23 October 2022) after the study was approved by the Faculty of Pharmacy Ain Shams University Ethics Committee Number, ACUC-FP-ASU RHDIRB2020110301 REC# 41.

2.7.2. Thermal Injury Model

The thermal injury model was performed according to Sakr et al. [18] with a minor modification using a rectangular metal bar (2×2 cm and 1 mm thickness) to induce burns in the backs. The examined animal groups were divided into six control and three test groups, five rats each, as follows:

Group I: Control, Burned, non-infected, untreated.
Group II: Control, burned, infected, untreated.
Group III: Control, Burned, infected, treated with vehicle (control hydrogel).
Group IV: Burned, infected, treated with tested hydrogel-1 (phage vB_Pae_SMP1)
Group V: Burned, infected, treated with tested hydrogel-2 (phage P5).
Group VI: Burned, infected, treated with phage cocktail hydrogel (Phage vB_Pae_SMP1 + phage P5).
Group VII: Positive Control, burned, infected, treated with Silver sulfadiazine 1% (Silvirburn®, MUP Co., Cairo, Egypt).
Group VIII: Positive Control, burned, infected, treated with Collagenase 0.6 IU (Iruxol ®, Abbott Co., Wiesbaden, Germany).
Group IX: Normal Control, intact, non-infected, untreated.

2.7.3. Treatment

The first application of the tested formulae to the burned skin was at 2 h post-infection. On the infected burn, a weight of approximately 1 g of hydrogel was applied topically. The administration of treatment was applied twice daily for 14 days. The survival rate of animals was recorded three days post-infection. Dead animals were eliminated from the groups and were only considered when calculating mortality rates. Before the surviving animals were sacrificed, blood was aseptically withdrawn from the retro-orbital plexus after lidocaine (4%) local anesthesia [44] and analyzed for bacterial counts, as detailed below. The animals were euthanized by cervical dislocation, and the dorsal skin at the site of the wound was also removed immediately after the animals were sacrificed for histopathological examination [18].

2.7.4. Histopathological Examination

For histopathological examination, dorsal skin samples were flushed and fixed in 10% neutral-buffered formalin for 72 h, dehydrated in serial ascending grades of ethanol, cleared in xylene, and then infiltrated with a synthetic paraplast tissue embedding medium [45]. Using a rotatory microtome, 5 μn thick tissue sections were made at the middle zones of various wound samples to show the different skin layers and then stained with hematoxylin and eosin using standard procedures. In a blind manner, tissues were examined for histopathological changes and, additionally, the content of collagen fibers was quantitatively analyzed by Assistant Professor Dr. Mohamed Abdelrazik, Cytology and Histology Department, Veterinary Medicine, Cairo University, using Masson's trichrome stain. In this process, 6 non-overlapping fields were arbitrarily chosen and scanned from the dermal layers of each sample to determine the relative area percentage of collagen fibers in the

Masson's trichrome-stained sections in accordance with Almukainzi et al. [46]. All standard procedures for sample fixation and staining were carried out as previously reported [47].

2.8. Statistical Analysis

To calculate *p*-values and standard deviation, data were analyzed using a one-way ANOVA test using GraphPad Instat-3 software (Graph Pad Software Inc., San Diego, CA, USA). Results were displayed as corresponding average values ± Standard deviation.

3. Results

3.1. Identification, Antimicrobial Susceptibility, Phenotypic and Genotypic Analysis of the Recovered P. aeruginosa Isolates

The three isolates were identified as *P. aeruginosa* PA1, PA2, and PA3, respectively. As shown in Table 1, the susceptibility pattern demonstrated that all the isolates exhibited resistance to all the tested antimicrobial agents, including imipenem and meropenem; therefore, they were categorized as CRPA isolates. The MIC of the three CRPA isolates against imipenem, as well as the phenotypic tests and the detected carbapenemase genes, are shown in Table 1.

Table 1. Summary for susceptibility pattern, MIC, phenotypic and genotypic detection of CPs *P. aeruginosa* isolates.

Isolate Code	Susceptibility Pattern						MIC of IMP (µg/mL)	Phenotypic Tests			Carbapenemase Genes
	AK	AT	CIP	LEV	IMP	MER		CDT	mCIM	BCT	
CRPA1	R	R	R	R	R	R	>1024	−	−	+	*bla*KPC
CRPA2	R	R	R	R	R	R	>512	−	+	+	*bla*OXA-48
CRPA3	R	R	R	R	R	R	32	−	+	+	*bla*OXA-48

Ak, amikacin; AT, aztreonam; CIP, ciprofloxacin; LEV, levofloxacin, IMP, imipenem; MER, meropenem; MIC, minimum inhibitory concentration; blaKPC, gene coded for Klebsiella pneumoniae carbapenemase (group A beta-lactamase); blaOXA-48, gene coded for oxacillinase (group D Beta-lactamase), modified carbapenem inactivation method (mCIM), combined disk test (CDT), and blue-Carba test (BCT).

3.2. Recovery of Bacteriophages and Screening for the Activity against CRPA

Screening 15 sewage samples showed that only two samples (namely, vB_Pae_SMP1 and P5) had a positive spot test against CRPA and, therefore, were chosen for further study. Plaque assay results showed that all lysates had relatively reproducible, high initial titers (>10^8 PFU/mL) (Figure S1). The resulting plaques were clear, circular with regular edges, small (2–5 mm), and encircled by halos (Figure S2).

3.3. Characterization of the Isolated Bacteriophages Showing Lytic Activities against CRPA

3.3.1. Host Range

The vB_Pae_SMP1 lysate was relatively active against CRPA2 and CRPA3 isolates, and the vB_Pae_SMP5 lysate was active against CRPA1 and CRPA2 isolates (Table 2). The two lysates were chosen with isolate CRPA2 for further studies (Figure S3), while none of the two phage lysates showed lytic activity against any of the remaining 11 clinically relevant Gram-negative pathogens.

Table 2. Host range of the lytic properties of phages vB_Pae_SMP1 and vB_Pae_SMP5.

Isolate Code	Microorganism	Spot Test	
		Lysate vB_Pae_SMP1	Lysate vB_Pae_SMP5
CRPA1	*P. aeruginosa*	−	+
CRPA2	*P. aeruginosa*	+	+
CRPA3	*P. aeruginosa*	+	−

3.3.2. Morphology of the Bacteriophages vB_Pae_SMP1 and vB_Pae_SMP5 as Presented by TEM

Phage vB_Pae_SMP1 and vB_Pae_SMP5 each has an icosahedral head of 60 ± 5 nm and a flexible non-contractile tail of 170 ± 5 nm long, while vB_Pae_SMP5 has an additional base plate containing a 35 nm fiber observed at the end of the tail (Figure 1a,b). By matching these observations to the data on the "Viral Zone" website along with the guidelines provided by the International Committee on Virus Taxonomy (ICTV), it could be suggested that both phages vB_Pae_SMP1 and vB_Pae_SMP5 fall into the siphoviral morphotype (formerly family Siphoviridae) of the class Caudoviricetes (formerly order Caudovirales); however, Phage vB_Pae_SMP5 has features of members of the genus *Septimatrevirus* (currently a genus under Caudoviricetes, with no assigned family yet).

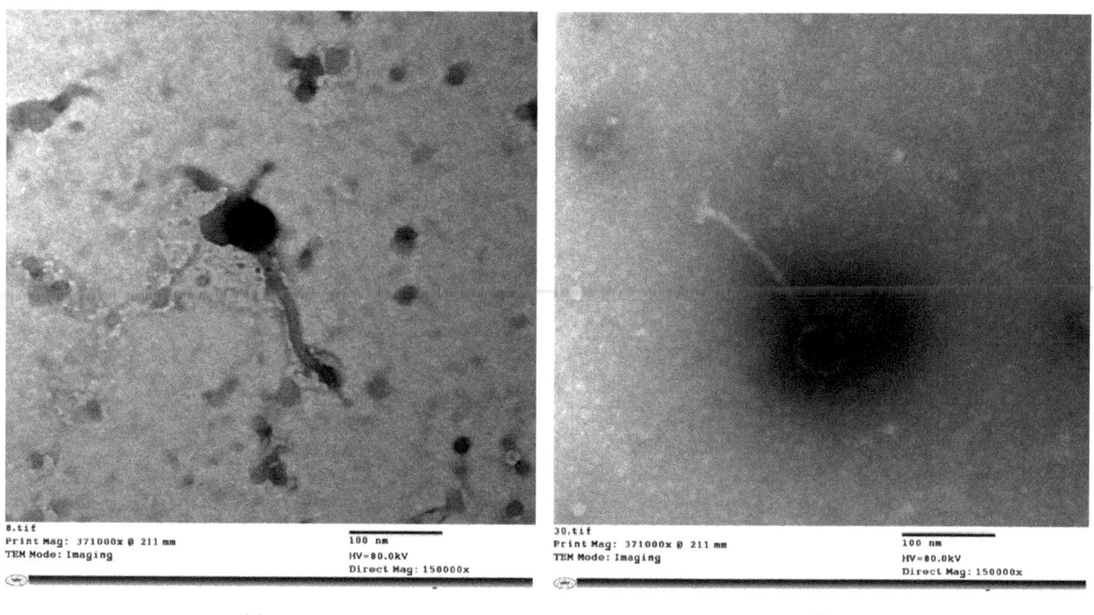

(**a**)   (**b**)

Figure 1. Electron micrograph of bacteriophages; (**a**) vB_Pae_SMP1, 60 ± 5 nm icosahedral head, flexible non-contractile tail 170 ± 5 nm long, (**b**) vB_Pae_SMP5, with an additional base plate containing 35 nm fiber at the end of the tail.

3.3.3. Phage Lysate Sequencing Results a Complete Genome of Phage vB_Pae_SMP5

High-throughput sequencing of the lysate of phage vB_Pae_SMP5 using the Illumina MiSeq instrument yielded a consensus sequence of 43,070 bp (in one full contig) with 57 protein-coding genes. The full description of putative functions of the resulting open reading frames (ORFs) are shown in Table S1. BLASTN [40] alignment showed that the taxonomy of this phage was Viruses; Duplodnaviria; Heunggongvirae; Uroviricota; Caudoviricetes; Septimatrevirus with an alignment score >200 and 98.0% identity. The genomic and phenotypic characteristics of phage vB_Pae_SMP5 are displayed in Table 3. The genomic analysis of phage vB_Pae_SMP1 is undergoing. The circular genome map of vB_Pae_SMP5 is depicted in Figure 2. The genomic sequence of phage vB_Pae_SMP5 has been deposited in the NIH-funded BV-BRC database, and the BV-BRC accession is 2731619.92 (https://www.bv-brc.org/view/Genome/2731619.92; accessed on 7 December 2022).

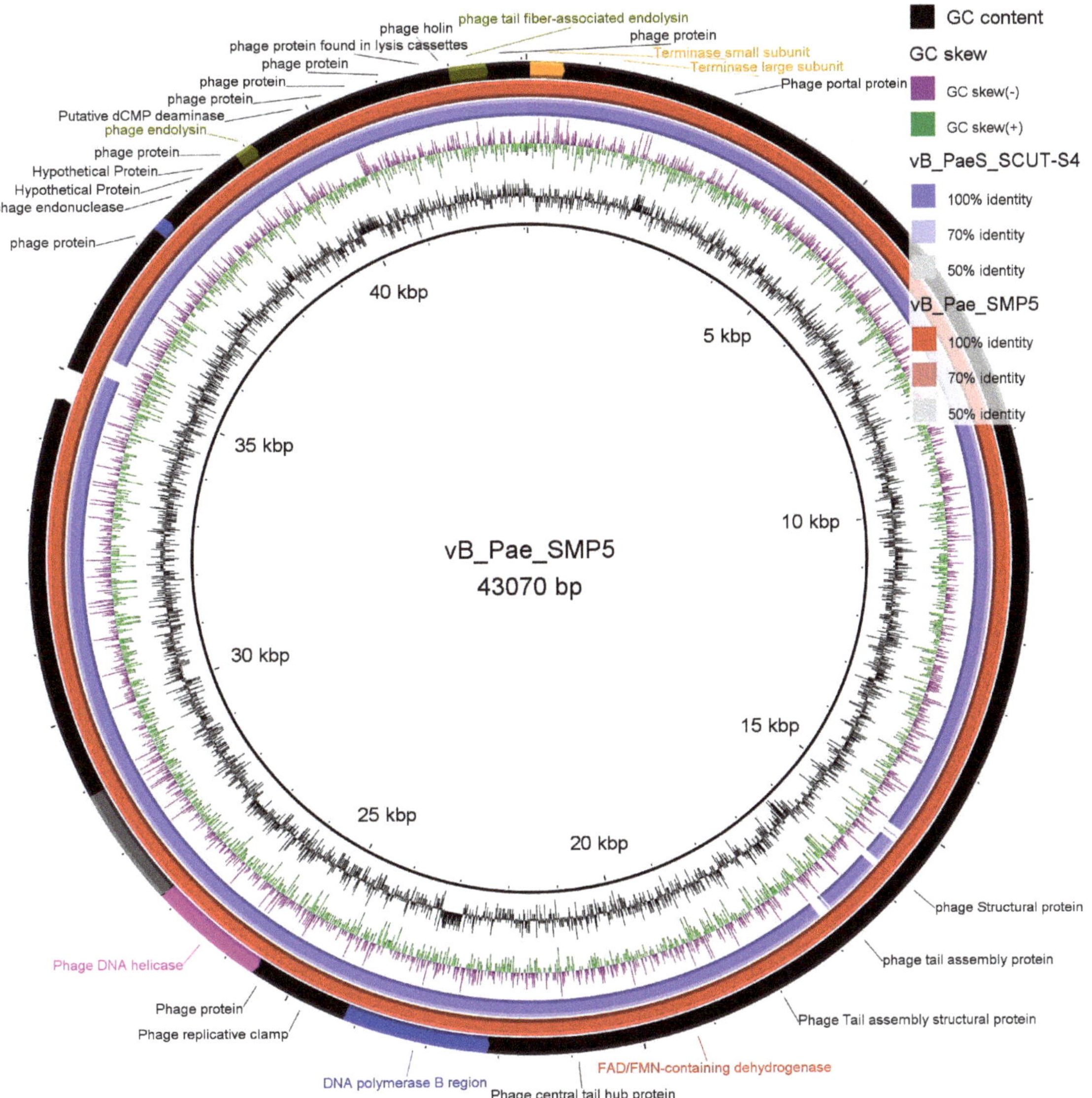

Figure 2. Circular genome map of vB_Pae_SMP5 (Red ring). The BV-BRC accession is 2731619.92. The color coding of genes indicates the functional categories of putative proteins: phage and hypothetical proteins (Black), terminase protein (orange); phage helicase (fuchsia); regulatory protein (purple); phage endolysin (olive); Phage polymerase (blue); exonuclease (Gray); blue ring is the reference phage *Pseudomonas* phage vB_PaeS_SCUT-S4 (NCBI nucleotide accession code MK165658.1). The creation of the circular image and comparison with other reported similar plasmids were performed using the BLAST Ring Image Generator (BRIG) tool v0.95 (https://sourceforge.net/projects/brig/, accessed on 25 October 2022).

Table 3. Genomic and phenotypic characterization of the phage vB_Pae_SMP5.

Parameters	Phage vB_Pae_SMP5
Molecular type	Genomic DNA
Genomic size (bp)	43070 bp
Proteins/ORFs	57 (46 coded by + frames and 11 coded by - frames)
Isolation Source	sewage
Host	XDR *Pseudomonas aeruginosa* clinical isolates CRPA1 and CRPA2
Class	Caudoviricetes
Family	Currently unassigned (formerly Siphoviridae, suggested Septimatreviridae)
Genus	*Septimatrevirus*
Species	unclassified

3.4. In Vitro Anti-CRPA Activity of the Tested Hydrogels

Inhibition zones were observed around the cups containing either phage vB_Pae_SMP1 or vB_Pae_SMP5 lysate (positive control) and around the cups containing either the tested hydrogel of phage vB_Pae_SMP1 or phage vB_Pae_SMP5 (Figure S4), while no inhibition zone was detected around the cups containing the control hydrogel (negative control).

3.5. In Vivo Anti-CRPA Activity of the Formulated Tested Hydrogels

3.5.1. Survival Rate

The survival rate of the tested rats for each of the examined groups, calculated until day 14, is tabulated in Table 4.

Table 4. Percentage of Survival rate of the examined rats.

Group	Description	Rats Survival %
I	Normal Control, intact, non-infected, untreated	100
II	Control, Burned, non-infected, untreated	80
III	Control, burned, infected, untreated	40
IV	Control, Burned, infected, treated with control hydrogel	60
V	Burned, infected, treated with Phage 1 hydrogel	100
VI	Burned, infected, treated with phage 5 hydrogel	100
VII	burned, infected, treated with phage cocktail hydrogel (vB_Pae_SMP1 + vB_Pae_SMP5)	100
VIII	Positive Control, burned, infected, treated with Silverburn®	100
IX	Positive Control, burned, infected, treated with Iruxol®	100

3.5.2. Histopathological Examination

Microscopical examination for different skin samples demonstrated the following: group I revealed normal histological structures of different skin layers, including an apparent intact thin epidermal layer with intact covering epithelium, as well as an intact dermal layer (Figure 3A,B). It also showed normally distributed collagen fibers (Figure 3A) up to 32.1% of the mean area percentage of the dermal layer content, as shown in Masson's trichrome-stained tissue sections of all the samples (Figures 4 and 3A,B). The group II samples (burned, non-infected, untreated) showed a wide area of wound gap, covered with scabs from a necrotic tissue depress with a significant loss and necrosis of the underlying dermal layer, replaced with newly formed granulation tissue with abundant inflammatory cell infiltrates, fibroblastic proliferation, as well as many congested subcutaneous blood vessels (BVs) (Figure 3C,D). This group also displayed minimal records of mature collagen fibers (Figure 4B) up to 10.38% of the mean area percentage of the dermal layer content (Figure 5). Group III (burned, infected, untreated) demonstrated almost the same records as the group II samples, with even more severe records of mixed inflammatory cell infiltrates in dermal and subcutaneous tissue, as well as focal dermal hemorrhagic patches (Figure 3E,F). This group also showed more minimal records of mature collagen fibers (Figure 4C), up to 8.75% of the mean area percentage of the dermal layer content (Figure 5). Group IV (burned, infected, treated with control hydrogel) demonstrated almost the same records as group II (Figure 3G,H), also with minimal records of mature collagen fibers (Figure 4D) up to 10.22% of the mean area percentage of the dermal layer content (Figure 5). Group V (burned, infected, treated with Phage 1 hydrogel) showed persistent records of ulcerated wound gap with epidermal loss and necrotic tissue depression. However, highly cellular subepidermal granulation tissues were observed with abundant records of fibroblastic activity, and mixed inflammatory cell infiltrates (Figure 3I,J). The group also showed mildly higher records of dermal mature collagen fibers (Figure 4E) up to 17.2% of the mean area percentage of the dermal layer content (almost twofold more when compared to Group III Control samples) (Figure 5).

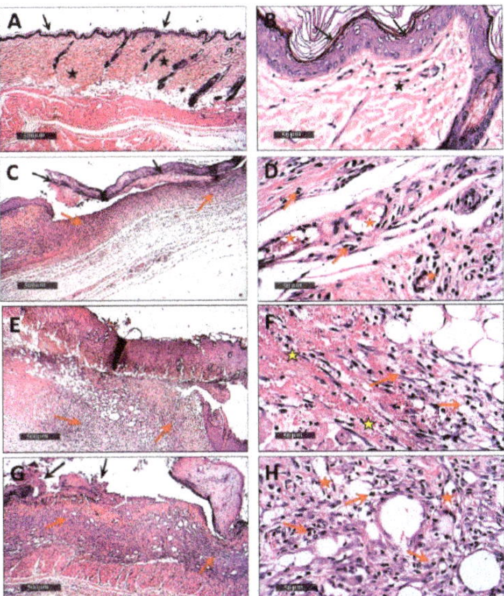
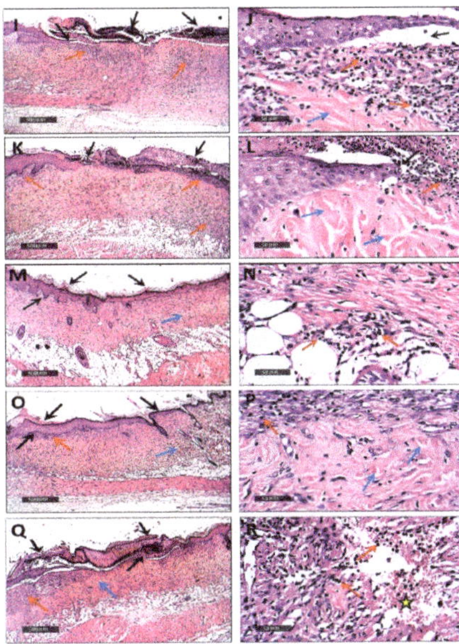

Figure 3. Demonstrating the light microscopic histopathological features of skin layers and wound healing process in different groups; (**A,B**): group I, (**C,D**): group II, (**E,F**): group III, (**G,H**): group IV, (**I,J**): group V, (**K,L**): group VI, (**M,N**): group VII (**O,P**): group VIII, (**Q,R**): group IX. Group codes are shown in Table 4. (H&E stain). Black arrow: Epidermal layer & wound gap. Star: intact skin dermis. Red arrow: Inflammatory cell infiltrates. Red star: congested subcutaneous blood vessels. Blue arrow: Dermal mature collagen fibers. Yellow star: dermal hemorrhagic patches.

The group VI samples (burned, infected, treated with phage 5 hydrogel) showed persistent records of an ulcerated wound gap with epidermal loss and a necrotic tissue depression with moderate persistent records of inflammatory cell infiltrates in deep dermal and subcutaneous layers (Figure 3K,L). However, significant acceleration in dermal mature collagen fiber formation (Figure 4F) up to 25.8% of the mean area percentage of the dermal layer content was shown (almost threefold more when compared to the Group III Control samples) (Figure 5). The group VII samples (burned, infected, treated with phage cocktail hydrogel (vB_Pae_SMP1 + vB_Pae_SMP5) showed a significantly accelerated wound gap closure with a complete re-epithelialization with a thick hyperkeratotic epidermal layer. In addition, there was a significant reduction in inflammatory cell infiltrates with a higher fibroblastic activity (Figure 3M,N) and a moderate maturation of dermal collagen fibers (Figure 4G) up to 22.2% of the mean area percentage of the dermal layer content (almost 2.5-fold higher when compared to Group III Control samples) (Figure 5) just like the positive control group VIII.

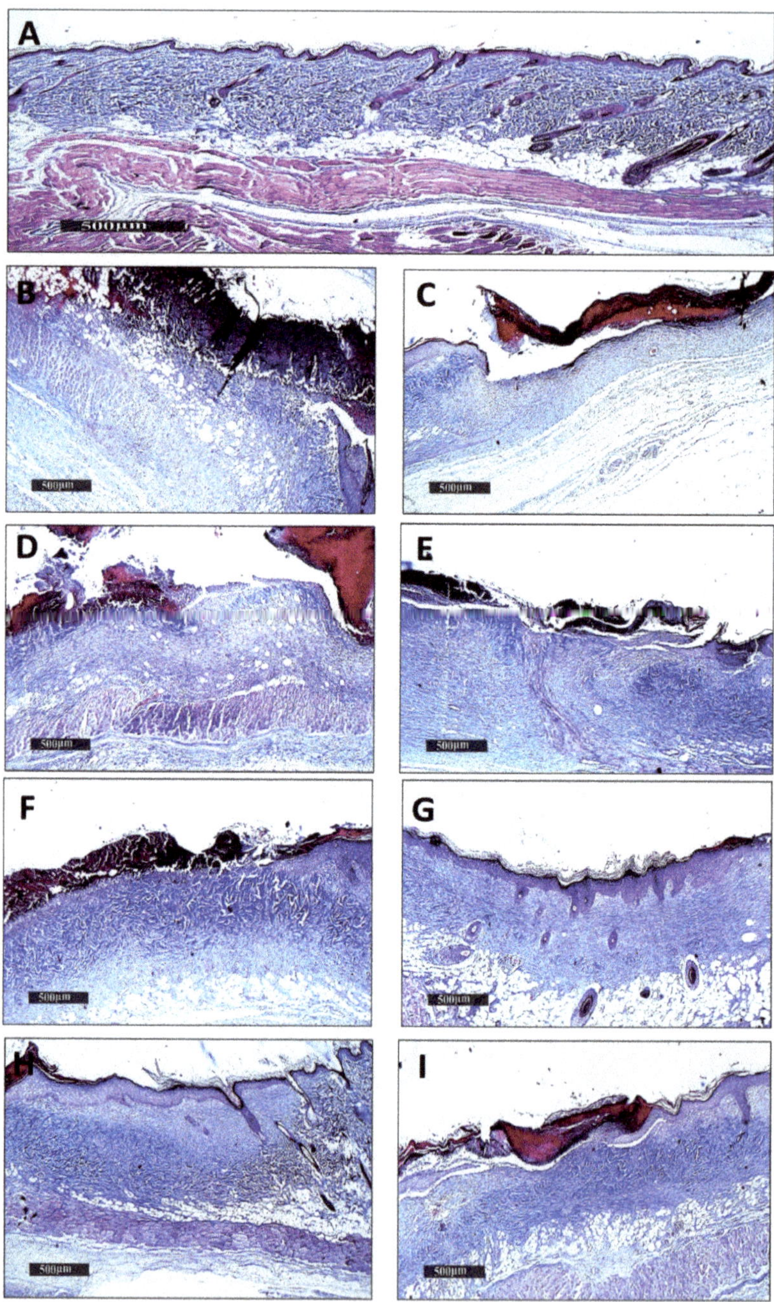

Figure 4. Dermal collagen fibers of different experimental groups; (**A**), dermal collagen fibers of normal health tissue group I; (**B**), group II, (**C**), group III, (**D**): group IV, (**E**), group V, (**F**), group VI; (**G**), group VII; (**H**), group VIII; (**I**), group IX, Group codes are shown in Table 4. (Masson's Trichome stain).

However, the group VIII positive control samples (burned, infected, treated with Silverburn®) showed almost the same records as the Group VII samples (Figure 3O,P) as well as a moderate maturation in dermal collagen fibers (Figure 4H) up to 22% of the mean area percentage of the dermal layer content (Figure 5). However, group IX, the positive control (burned, infected, treated with Iruxol®), showed persistent records of a narrow ulcerated wound gap covered with a scab of a necrotic tissue depress with occasional records of subepidermal hemorrhagic patches and a significant reduction in inflammatory cell infiltrates in dermal and subcutaneous layers (Figure 3Q,R) with a moderate maturation in dermal collagen fibers (Figure 4I) up to 20.6% of the mean area percentage of the dermal layer content (Figure 5).

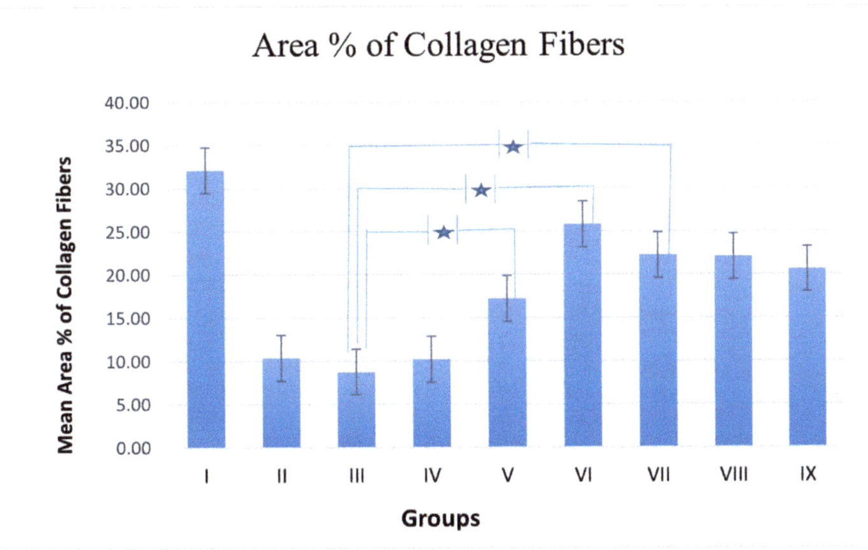

Figure 5. Mean area % of collagen fibers, data presented as mean positive or negative SE, *n* = 6. Star: significant difference.

4. Discussion

CRPA has emerged as a serious pathogen on a global scale due to its limited treatment options and, therefore, is placed in the priority 1-critical category of the World Health Organization's global priority pathogens for research and development of novel antimicrobials [48]. Accordingly, this situation has encouraged a reconsideration of bacteriophage therapy as a possible promising treatment option. In fact, several studies have reported a sporadic rise in the use of antibacterial phage therapy over the course of the past century [49–51]. Additionally, there are some advantages that make phage therapy stand out, including the high host specificity, specific reproduction at the site of infection, the eligibility of single or rare administration, and the benefit of being effective against other pan-drug resistant bacteria [52].

Despite studies demonstrating the clinical effectiveness of phages, further research is still needed in this field due to the information gap regarding the in vivo studies. As a result, our study aimed to test the *lytic activity* of two novel isolated bacteriophages formulated as hydrogels to combat CRPA-infected burns using an appropriate skin animal model. This was accomplished by isolating three clinical *P. aeruginosa* from a third-degree burned patient who is experiencing severe complications and rejection in responding to several antimicrobial agents employed in treatment. The antibiogram analysis showed that the respective isolates had a high level of resistance to several antimicrobial agents, including carbapenems, in addition to the MIC of all isolates revealing that they are all carbapenem-resistant and hence classified as CR-XDR isolates [25]. Therefore, these isolates can undoubtedly cause life-threatening conditions for patients as their colonization of a burn wound frequently results in a disseminated infection and occasionally septic shock and mortality. Accordingly, early disclosure of the widespread dissemination of CRPA with devastating effects within clinical settings is vital for the prompt implementation of infection control measures [53].

CRPA, a perturbing carbapenem-resistant pathogen, was selected as the bacterial host for the purpose of isolating phages active against carbapenem-resistant organisms [12,13,53]. Three CRPA isolates were collected from different clinical specimens. They were resistant to all the antibiotics tested, including imipenem and meropenem. The isolation of the bacteriophage was performed by incubating a freshly grown bacterial host with an environmental sample. Then the bacterial cells were removed, and the suspension was purified using centrifugation and chloroform treatment. For this purpose, 15 environmental samples were collected from a hospital sewage source. Only two lysates continued to produce positive results, suggesting sewage was an outstanding source for phage isolation against *P. aeruginosa* [54,55].

Over the course of time, sewage was believed to be an important reservoir of bacteriophages infecting *P. aeruginosa* [54]. The plaque assay offers information on a specific lysate's purity, as well as the count of viruses [32,56]. A single pure virus is present if it only manifests as one type of plaque with similar morphology, size, and shape, while the presence of multiple viruses is indicated if the plaque assay results in a variety of plaque forms with different characteristics. Additionally, the numbers for identical plaques are inserted into a mathematical formula that roughly predicts the count of viruses that were present in the original lysate in plaque forming units per ml (PFU/mL) [32,57]. The two phage suspensions had initial titers that were relatively high and reproducible (>10^8 PFU/mL). Plaque morphology may reveal the phage type. Lytic (virulent) phages typically have clear, transparent plaques, whereas phages with a lysogenic ability (temperate phages) have opaque, turbid ones [58]. Additionally, some phages produce plaques encircled by halos. Halos are attributed to the diffusion of bacteriophage depolymerase enzymes, which are primarily active against bacterial cell walls biofilms [59]. The vB_Pae_SMP1 and vB_Pae_SMP5 plaques were clear, circular, and had regular whole margins. They were also small (2–5 mm) in diameter, which indicated they were lytic [60]. Additionally, the plaques with halos surrounding them were seen to increase in size with increasing the incubation time or after being kept at room temperature for several days, suggesting an anti-biofilm activity. However, this requires additional testing against organisms that produce biofilms [61].

Normally, candidates for phage therapy are chosen using strict criteria, such as being obligate lytic, having a wide host range, and naturally having high stability [62]. As a result, it was important to investigate some of the phages' primary characteristics. A bacteriophage's host range is restricted to the genus, species, and strains of bacteria that it can infect [57]. This is undoubtedly essential when using a phage in therapy. As shown in the results, both vB_Pae_SMP1 and vB_Pae_SMP5 lysates were active against two out of the three tested XDR *P. aeruginosa* clinical isolates; however, none of the two lysates showed *lytic activity* against any of the remaining 11 clinically relevant Gram-negative pathogens, which suggested their restricted host range and specificity toward *P. aeruginosa* isolates. Therefore, the two lysates were chosen with isolate PA2 for further studies. A phage's host range should be constrained to a single species to prevent it from attacking bacteria other than the disease-causing one, protecting the host's microbiome. Therefore, a phage with a restricted host range is preferred in terms of species. However, a phage that infects many, if not all, strains within this species is advantageous. It suggests that it is suitable for an empirical application, just like broad-spectrum antibiotics [57]. *Furthermore*, the traditional phage isolation technique, which assigns just one host on which phages are supposed to grow, may contribute to their generally restricted host range [57,63,64].

The morphology of phage particles is primarily used for their typical classification. However, the discrepancies among numerous bacteriophage with similar morphology but divergent genomic content has led to recent proposals to cancel the traditional classification of having one bacteriophage order Caudovirales with three major families [65]. Currently, the class Caudoviricetes comprises tailed phage families, most famously the former families Myoviridae, Siphoviridae, and Podoviridae representing the different tailed morphotypes. Phages belonging to these morphotypes share icosahedral capsids with double-stranded (ds) DNA, but their tail shapes vary [66]. Transmission electron microscopy (TEM) examination is the most significant technique for phage visualization [67]. vB_Pae_SMP1 and vB_Pae_SMP5 phage suspensions were introduced for TEM [28]. The morphological features of the vB_Pae_SMP1 and vB_Pae_SMP5 virus particles demonstrated that they seemed to be tailed; hence, they are proposed to be members to class Caudoviricetes. They have an icosahedral head and lack connectors at the head-to-tail junction. However, vB_Pae_SMP5 had an extra base plate with a 35 nm fiber noticed at the end of the tail. These properties excluded the possibility of belonging to the former family Myoviridae [67]. They were marked with a long, flexible non-contractile tail (~170 nm) and an icosahedral head, ~60 nm in diameter. It was suggested that the vB_Pae_SMP1 most likely belong to the siphoviruses based on the morphological similarity to the descriptions presented on the Viral Zone website ("Siphoviridae ViralZone") and in accordance with the guidelines compiled by the International Committee on the Taxonomy of Viruses (ICTV)—ninth report [68]. Phage vB_Pae_SMP5 was genotypically confirmed to belong to the class Caudoviricetes, genus *Septimatrevirus*, which is one of the newly established genera with no assigned family yet, and we propose that it can be considered family Septimatreviridae. Genomic analysis of vB_Pae_SMP1 is undergoing.

For testing the in vitro effect of the isolated phages formulated as hydrogels on a CRPA-infected burn, CMC was chosen for our hydrogel preparation. Cellulose is a widely available biopolymer with distinct properties, most importantly being that it is very water-soluble and forms superabsorbent hydrogels with excellent mechanical and viscoelastic properties. Cross-linked CMC-based hydrogels have recently been researched as potential dermal drug delivery systems for bacteriophages and antibiotics because of these features [69,70].

The anti-CRPA activity of the tested hydrogel was evaluated using the cup-plate method. The results showed the formation of inhibition zones around both the positive control cups containing either phage vB_Pae_SMP1 or vB_Pae_SMP5 lysates and around the cups containing either the tested hydrogel of phage vB_Pae_SMP1 or vB_Pae_SMP5 while the negative control cups containing the CMC- hydrogel alone exhibited no inhibition zones indicating the benefit of CMC hydrogel as a vehicle for delivering the bacteriophages vB_Pae_SMP1 and vB_Pae_SMP5 in a sustained manner [71].

To evaluate the ability of the bacteriophages formulated as hydrogels to alleviate the pathogenicity of such CRPA isolate (PA2), a thermal injury model was designed in rats infected with the CRPA isolate (PA2). Numerous studies in the literature applying superficial skin infection in mice/rats have been reported [18,72,73]. The prepared bacteriophage-hydrogel was applied to the infected burn following the establishment of the infection. The calculated survival rates for the treated groups were compared to those of other control groups. Histopathological examination and the percentage of dermal collagen fibers of the burned infected rat skin were determined as well [18]. Our study's findings showed that the tested hydrogel containing the bacteriophages increased the survival rate, as all the animals given this tested hydrogel survived. These findings come in accordance with previous studies, which reported a higher survival rate after phage treatment compared to the control [74–76]. This could be attributed to the beneficial topical use of the phages in reducing the number of bacteria in the wound and reducing their dissemination at the end of phage treatment, as previously reported [77]. Additionally, the results revealed that rats treated with control hydrogel (the vehicle) without the bacteriophages had higher survival rates than the untreated rats. This could be due to the moisturizing properties of the hydrogel and its function as a physical barrier that shields the wound, as was mentioned in a previous study that discussed the benefits of hydrogels in wound care [78].

The histological examination of various wounded groups showed different degrees of tissue damage and healing processes. To ensure the reliability of the results and relate the healing activity entirely to bacteriophage, an additional control group (group IV), aside from the untreated group, was designed for our study. In this control group (Group IV), a hydrogel made solely from the vehicle without bacteriophage was topically applied to the infected burn, and the results were evaluated. In a similar manner, group II, which had non-infected burns, was utilized to assess the damage produced by XDR *P. aeruginosa* infection in the other groups relative to this group. Moreover, another control group (group III) was designed in which XDR *P. aeruginosa* infection was induced to the burn and left untreated. This group was established to investigate the virulence and the properties of *P. aeruginosa* pathogenicity, assess its persistent inflammation that delays healing and boosts antimicrobial tolerance, and the efficacy of topical bacteriophages in the burn-wound infection model, as previously reported [18,79]. Two groups were designed as positive controls to be used as references for comparing antibacterial activity and wound healing properties. In one positive control group (group VIII), the infected burn was treated with silver sulfadiazine, which is well-known for its antibacterial activity. Silver sulfadiazine is the topical treatment of choice in severe burns and is used almost globally today in preference to compounds such as silver nitrate and mafenide acetate [80]. In the other positive control group (group IX), the infected burn was treated with collagenase due to its wound-healing properties.

Collagenase, a Clostridium histolyticum-derived enzyme debriding agent, is utilized in clinical practice to treat infected and surgical wounds [81]. Previous research has found that collagenase cleans the necrotic tissue of the wound using an enzymatic technique, accelerates the formation of granulation tissue and subsequent re-epithelialization, and increases the activation of fibroblasts, myofibroblasts, and collagen in rat dermal wounds [80]. Therefore, silver sulfadiazine and collagenase were used in our study as positive controls for the treatment of rat skin wounds, as reported in several studies [80,82,83]. Our results revealed that topical application of bacteriophage hydrogel in groups (V, VI, VII) showed a significant difference ($p < 0.05$) in suppressing local wound infection and promoting skin regeneration, as well as higher records and significant acceleration of dermal mature collagen fiber formation compared to the control group III (burned, infected, untreated) indicating wound healing process. Thus, our results confirmed that phages could be promising in preventing wound-associated infections with *P. aeruginosa* [84]. Our findings were in accordance with the study conducted by Mendes et al., which proved an epithelial gap reduction in wounds treated by phages infected by *P. aeruginosa* and, therefore, was proof that phages can enhance wound healing as well as combat the infection [85].

Several studies have reported that a topically administered bacteriophage treatment may be effective in resolving and healing chronic wound infections in animal models [86–88]. Our findings revealed that the group treated with the phage cocktail (Group VII) showed more significant accelerated wound gap closure with complete re-epithelialization. The treatments also resulted in a significant reduction ($p < 0.05$) in inflammatory cell infiltrates with higher fibroblastic activity when compared to the other single phage-treated groups (Group V and Group VI). This could suggest the synergistic activity of using the phage cocktail compared to using a single phage alone. Phage cocktails have the potential to target a wider variety of strain-specific microorganisms, and their use is considered a crucial strategy for limiting the development of bacterial resistance [84]. In fact, Pinto and his colleagues reported that after just a few hours of monophage therapy, several resistant bacterial strains had emerged. When compared to monophage therapy, phage cocktails accelerate the rate of bacterial death and decrease the number of phage-resistant bacterial mutants [89]. Additionally, a study performed on mice revealed that phage cocktails led to a significant decrease in wound bioburden, faster wound closure, and greater wound contraction. Furthermore, in vitro stability studies and in vivo phage titer determination have shown a correlation between better phage persistence at the wound site and liposomal entrapment of the phage cocktail [90].

5. Conclusions

In this, study two lytic phages of the class Caudoviricetes, named VB_Pae_SMP1 and VB_Pae_SMP5, were isolated from sewage samples, purified and charcaterized. The lytic activities of the respective phages either each alone or as cocktail were evlauted both in vitro and in vivo using XDR *Pseudomonas aeruginosa*-infected skin thermal injury rat model. The phae VB_Pae_SMP5 was genotypically confirmed to beloged the class Caudoviricetes; genus Septimatrevirus. Each of the phage lysate or the two-phage cocktail formulated as a hydrogel for topical treatment has the power to manage the infection of XDR CRPA in burn wounds and decrease the mortality rate of the tested rats. Additionally, it encourages healing at the injury site, as evidenced by the histopathological examination where restored epithelium and activated fibroblasts were noticed. The two phages formulated hydrogels improved wound healing as well as mortality rates of the tested animals. Therefore, each of the phage lysate or the two-phage cocktail formulated as a hydrogel is a promising topical formula for future clinical use in burn wounds caused by severe CRPA infections. However, further clinical studies should be done to ensure suitability of the respective phage hydrogels for clinical application in humans.

Supplementary Materials: The following supporting information can be downloaded at: https://www.mdpi.com/article/10.3390/v14122760/s1, Figure S1: Plaque assay results of vB_Pae_SMP5 at different dilutions; Figure S2: Plaque morphology as presented by phages vB_Pae_SMP1 and vB_Pae_SMP5. Plaques are clear, regularly circular, small in size (diameter of 2–5 mm), and halos are shown around the plaques; Figure S3: Spot test of phages vB_Pae_SMP1 and vB_Pae_SMP5 against *Pseudomonas aeruginosa* isolate showing clear spots and proving their lytic abilities; Figure S4: In vitro anti-CRPA activity of the tested hydrogels. Phage vB_Pae_SMP1 and vB_Pae_SMP5: Positive control. vB_Pae_SMP1 +CMC, vB_Pae_SMP5 +CMC: Tested hydrogels, C: negative control; Table S1: Genomic analysis (resulted contigs and putative functions of the resulted open reading frames) of the phage vB_Pae_SMP5.

Author Contributions: Conceptualization, S.S.M., G.R.A., A.S.A.Z., R.K.A. and K.M.A. methodology, S.S.M., G.R.A., A.S.A.Z., R.K.A. and K.M.A.; writing—original draft preparation, S.S.M., G.R.A. and A.S.A.Z. writing—review and editing, G.R.A., R.K.A. and K.M.A.; supervision, G.R.A., A.S.A.Z. and K.M.A. All authors have read and agreed to the published version of the manuscript.

Funding: This research received no external funding.

Institutional Review Board Statement: The protocol of this study was approved by the Faculty of Pharmacy Ain Shams University Research Ethics Committee (Number, ACUC-FP-ASU RHDIRB2020110301 REC# 41 in September 2021).

Data Availability Statement: The data supporting reported results are found in the manuscript and supplementary file. The genomic sequence of phage vB_Pae_SMP5 has been deposited in the NIH-funded BV-BRC database, and the genome will be publicly released upon manuscript publication. The BV-BRC accession is 2731619.92, and its link is https://www.bv-brc.org/view/Genome/2731619.92. (accessed on 7 December 2022).

Acknowledgments: The authors acknowledge the Microbiology and Immunology Department, Faculty of Pharmacy, Ahram Canadian University (ACU), and Ain Shams University (ASU), Egypt, for providing all the required facilities and support for the accomplishment of this work.

Conflicts of Interest: The authors declare no conflict of interest.

References

1. Diekema, D.J.; Hsueh, P.-R.; Mendes, R.E.; Pfaller, M.A.; Rolston, K.V.; Sader, H.S.; Jones, R.N. The Microbiology of Bloodstream Infection: 20-Year Trends from the SENTRY Antimicrobial Surveillance Program. *Antimicrob. Agents Chemother.* **2019**, *63*, e00355-19. [CrossRef] [PubMed]
2. Weiner, L.M.; Webb, A.K.; Limbago, B.; Dudeck, M.A.; Patel, J.; Kallen, A.J.; Edwards, J.R.; Sievert, D.M. Antimicrobial-Resistant Pathogens Associated with Healthcare-Associated Infections: Summary of Data Reported to the National Healthcare Safety Network at the Centers for Disease Control and Prevention, 2011–2014. *Infect. Control Hosp. Epidemiol.* **2016**, *37*, 1288–1301. [CrossRef] [PubMed]

3. Kadri, S.S. Key Takeaways from the US CDC's 2019 Antibiotic Resistance Threats Report for Frontline Providers. *Crit. Care Med.* **2020**, *48*, 939–945. [PubMed]
4. Motbainor, H.; Bereded, F.; Mulu, W. Multi-Drug Resistance of Blood Stream, Urinary Tract and Surgical Site Nosocomial Infections of *Acinetobacter baumannii* and *Pseudomonas aeruginosa* among Patients Hospitalized at Felegehiwot Referral Hospital, Northwest Ethiopia: A Cross-Sectional Study. *BMC Infect. Dis.* **2020**, *20*, 92. [CrossRef] [PubMed]
5. Nimer, N.A. Nosocomial Infection and Antibiotic-Resistant Threat in the Middle East. *Infect. Drug Resist.* **2022**, *15*, 631. [CrossRef] [PubMed]
6. Garg, A.; Garg, J.; Kumar, S.; Bhattacharya, A.; Agarwal, S.; Upadhyay, G.C. Molecular Epidemiology & Therapeutic Options of Carbapenem-Resistant Gram-Negative Bacteria. *Indian J. Med. Res.* **2019**, *149*, 285.
7. Paul, M.; Carrara, E.; Retamar, P.; Tängdén, T.; Bitterman, R.; Bonomo, R.A.; De Waele, J.; Daikos, G.L.; Akova, M.; Harbarth, S. European Society of Clinical Microbiology and Infectious Diseases (ESCMID) Guidelines for the Treatment of Infections Caused by Multidrug-Resistant Gram-Negative Bacilli (Endorsed by European Society of Intensive Care Medicine). *Clin. Microbiol. Infect.* **2022**, *28*, 521–547. [CrossRef]
8. Zhen, X.; Stålsby Lundborg, C.; Sun, X.; Gu, S.; Dong, H. Clinical and Economic Burden of Carbapenem-Resistant Infection or Colonization Caused by Klebsiella Pneumoniae, *Pseudomonas aeruginosa*, *Acinetobacter baumannii*: A Multicenter Study in China. *Antibiotics* **2020**, *9*, 514. [CrossRef]
9. Azimi, L.; Lari, A.R. Colistin-Resistant *Pseudomonas aeruginosa* Clinical Strains with Defective Biofilm Formation. *GMS Hyg. Infect. Control* **2019**, *14*, 1–6. [CrossRef]
10. Erol, H.B.; Kaskatepe, B. Comparison of the in Vitro Efficacy of Commercial Bacteriophage Cocktails and Isolated Bacteriophage vB_Pa01 against Carbapenem Resistant Nosocomial *Pseudomonas aeruginosa*. *J. Res. Pharm.* **2021**, *25*, 407–414.
11. Jahangiri, A.; Neshani, A.; Mirhosseini, S.A.; Ghazvini, K.; Zare, H.; Sedighian, H. Synergistic Effect of Two Antimicrobial Peptides, Nisin and P10 with Conventional Antibiotics against Extensively Drug-Resistant *Acinetobacter baumannii* and Colistin-Resistant *Pseudomonas aeruginosa* Isolates. *Microb. Pathog.* **2021**, *150*, 104700. [PubMed]
12. Broncano-Lavado, A.; Santamaría-Corral, G.; Esteban, J.; García-Quintanilla, M. Advances in Bacteriophage Therapy against Relevant Multidrug-Resistant Pathogens. *Antibiotics* **2021**, *10*, 672. [CrossRef] [PubMed]
13. Chen, P.; Liu, Z.; Tan, X.; Wang, H.; Liang, Y.; Kong, Y.; Sun, W.; Sun, L.; Ma, Y.; Lu, H. Bacteriophage Therapy for Empyema Caused by Carbapenem-Resistant *Pseudomonas aeruginosa*. *Biosci. Trends* **2022**, *16*, 158–162. [PubMed]
14. Fathima, B.; Archer, A.C. Bacteriophage Therapy: Recent Developments and Applications of a Renaissant Weapon. *Res. Microbiol.* **2021**, *172*, 103863.
15. Erol, H.B.; Kaskatepe, B. Isolation of newly isolated vb_k1 bacteriophage and investigation of susceptibility on esbl positive *Klebsiella* spp. strains. *J. Fac. Pharm. Ankara Univ.* **2021**, *45*, 515–523. [CrossRef]
16. Kalelkar, P.P.; Riddick, M.; García, A.J. Biomaterial-Based Antimicrobial Therapies for the Treatment of Bacterial Infections. *Nat. Rev. Mater.* **2022**, *7*, 39–54.
17. Rivera, D.; Moreno-Switt, A.I.; Denes, T.G.; Hudson, L.K.; Peters, T.L.; Samir, R.; Aziz, R.K.; Noben, J.-P.; Wagemans, J.; Dueñas, F. Novel Salmonella Phage, vB_Sen_STGO 35-1, Characterization and Evaluation in Chicken Meat. *Microorganisms* **2022**, *10*, 606.
18. Sakr, M.M.; Elkhatib, W.F.; Aboshanab, K.M.; Mantawy, E.M.; Yassien, M.A.; Hassouna, N.A. In Vivo Evaluation of a Recombinant N-Acylhomoserine Lactonase Formulated in a Hydrogel Using a Murine Model Infected with MDR *Pseudomonas aeruginosa* Clinical Isolate, CCASUP2. *AMB Express* **2021**, *11*, 109. [CrossRef]
19. Abd-Allah, I.M.; El-Housseiny, G.S.; Alshahrani, M.Y.; El-Masry, S.S.; Aboshanab, K.M.; Hassouna, N.A. An Anti-MRSA Phage From Raw Fish Rinse: Stability Evaluation and Production Optimization. *Front. Cell. Infect. Microbiol.* **2022**, *12*, 904531. [CrossRef]
20. Camens, S.; Liu, S.; Hon, K.; Bouras, G.S.; Psaltis, A.J.; Wormald, P.-J.; Vreugde, S. Preclinical Development of a Bacteriophage Cocktail for Treating Multidrug Resistant *Pseudomonas aeruginosa* Infections. *Microorganisms* **2021**, *9*, 2001. [CrossRef]
21. Kim, H.Y.; Chang, R.Y.K.; Morales, S.; Chan, H.K. Bacteriophage-Delivering Hydrogels: Current Progress in Combating Antibiotic Resistant Bacterial Infection. *Antibiotics* **2021**, *10*, 130. [CrossRef] [PubMed]
22. Holt, J.G.; Krieg, N.R.; Sneath, P.H.A. *Bergey's Manual of Determinative Bacteriology*, 9th ed.; Lippincott Williams and Wilkins: Baltimore, Maryland, 1994; Available online: https://www.scirp.org/(S(i43dyn45teexjx455qlt3d2q))/reference/ReferencesPapers.aspx?ReferenceID=42336 (accessed on 7 December 2022).
23. Shetty, N.; Hill, G.; Ridgway, G.L. The Vitek Analyser for Routine Bacterial Identification and Susceptibility Testing: Protocols, Problems, and Pitfalls. *J. Clin. Pathol.* **1998**, *51*, 316–323. [CrossRef] [PubMed]
24. Weinstein, M.P. *Performance Standards for Antimicrobial Susceptibility Testing*; Clinical and Laboratory Standards Institute: Wayne, PA, USA, 2021; ISBN 1684401046.
25. Magiorakos, A.; Srinivasan, A.; Carey, R.B.; Carmeli, Y.; Falagas, M.E.; Giske, C.G.; Harbarth, S.; Hindler, J.F.; Kahlmeter, G.; Olsson-Liljequist, B. Multidrug-resistant, Extensively Drug-resistant and Pandrug-resistant Bacteria: An International Expert Proposal for Interim Standard Definitions for Acquired Resistance. *Clin. Microbiol. Infect.* **2012**, *18*, 268–281. [CrossRef]
26. Mabrouk, S.S.; Abdellatif, G.R.; El-Ansary, M.R.; Aboshanab, K.M.; Ragab, Y.M. Carbapenemase Producers Among Extensive Drug-Resistant Gram-Negative Pathogens Recovered from Febrile Neutrophilic Patients in Egypt. *Infect. Drug Resist.* **2020**, *13*, 3113. [CrossRef] [PubMed]
27. Wommack, K.E.; Williamson, K.E.; Helton, R.R.; Bench, S.R.; Winget, D.M. Methods for the Isolation of Viruses from Environmental Samples. In *Bacteriophages*; Springer: Berlin/Heidelberg, Germany, 2009; pp. 3–14.

28. Ackermann, H.-W. Basic Phage Electron Microscopy. In *Bacteriophages*; Springer: Berlin/Heidelberg, Germany, 2009; pp. 113–126.
29. El-Dougdoug, N.; Nasr-Eldin, M.; Azzam, M.; Mohamed, A.; Hazaa, M. Improving Wastewater Treatment Using Dried Banana Leaves and Bacteriophage Cocktail. *Egypt. J. Bot.* **2020**, *60*, 199–212. [CrossRef]
30. Hussein, Y.S.; El-Masry, S.S.; Faiesal, A.A.; El-Dougdoug, K.A.; Othman, B.A. Physico-chemical properties of some listeria phages. *Arab Univ. J. Agric. Sci.* **2019**, *27*, 175–183. [CrossRef]
31. Adams, M.H. *Bacteriophages*; Citeseer: 1959; Wiley Interscience: New York, NY, USA; Available online: https://www.scirp.org/ (S(351jmbntvnsjt1aadkposzje))/reference/ReferencesPapers.aspx?ReferenceID=1855130 (accessed on 7 December 2022).
32. Anderson, B.; Rashid, M.H.; Carter, C.; Pasternack, G.; Rajanna, C.; Revazishvili, T.; Dean, T.; Senecal, A.; Sulakvelidze, A. Enumeration of Bacteriophage Particles: Comparative Analysis of the Traditional Plaque Assay and Real-Time QPCR-and Nanosight-Based Assays. *Bacteriophage* **2011**, *1*, 86–93. [CrossRef]
33. Kropinski, A.M. *Bacteriophages, Methods and Protocols, Volume 1: Isolation, Characterization, and Interactions*; Clokie, M.R.J., Kropinski, A.M., Eds.; Part of the book series: Methods in Molecular Biology; Humana Totowa: New York, NY, USA, 2009; Volume 501. [CrossRef]
34. Carlson, K. *Working with Bacteriophages: Common Techniques and Methodological Approaches*; CRC Press: Boca Raton, FL, USA, 2005; Volume 1.
35. Kutter, E. Phage Host Range and Efficiency of Plating. In *Bacteriophages*; Springer: Berlin/Heidelberg, Germany, 2009; pp. 141–149.
36. Davis, J.J.; Wattam, A.R.; Aziz, R.K.; Brettin, T.; Butler, R.; Butler, R.M.; Chlenski, P.; Conrad, N.; Dickerman, A.; Dietrich, E.M. The PATRIC Bioinformatics Resource Center: Expanding Data and Analysis Capabilities. *Nucleic Acids Res.* **2020**, *48*, D606–D612. [CrossRef]
37. Parrello, B.; Butler, R.; Chlenski, P.; Pusch, G.D.; Overbeek, R. Supervised Extraction of Near-Complete Genomes from Metagenomic Samples: A New Service in PATRIC. *PLoS ONE* **2021**, *16*, e0250092. [CrossRef]
38. Aziz, R.K.; Bartels, D.; Best, A.A.; DeJongh, M.; Disz, T.; Edwards, R.A.; Formsma, K.; Gerdes, S.; Glass, E.M.; Kubal, M. The RAST Server: Rapid Annotations Using Subsystems Technology. *BMC Genom.* **2008**, *9*, 75. [CrossRef]
39. McNair, K.; Aziz, R.K.; Pusch, G.D.; Overbeek, R.; Dutilh, B.E.; Edwards, R. Phage Genome Annotation Using the RAST Pipeline. In *Bacteriophages*; Springer: Berlin/Heidelberg, Germany, 2018; pp. 231–238.
40. Altschul, S.F.; Gish, W.; Miller, W.; Myers, E.W.; Lipman, D.J. Basic local alignment search tool. *J. Mol. Biol.* **1990**, *215*, 403–410. [CrossRef] [PubMed]
41. Alikhan, N.-F.; Petty, N.K.; Ben Zakour, N.L.; Beatson, S.A. BLAST Ring Image Generator (BRIG): Simple Prokaryote Genome Comparisons. *BMC Genom.* **2011**, *12*, 402. [CrossRef] [PubMed]
42. Khullar, R.; Saini, S.; Seth, N.; Rana, A.C. Emulgels: A Surrogate Approach for Topically Used Hydrophobic Drugs. *Int. J. Pharm. Bio. Sci.* **2011**, *1*, 117–128.
43. Singh, B.; Sharma, S.; Dhiman, A. Design of Antibiotic Containing Hydrogel Wound Dressings: Biomedical Properties and Histological Study of Wound Healing. *Int. J. Pharm.* **2013**, *457*, 82–91. [CrossRef] [PubMed]
44. Moustafa, P.E.; Abdelkader, N.F.; El Awdan, S.A.; El-Shabrawy, O.A.; Zaki, H.F. Liraglutide Ameliorated Peripheral Neuropathy in Diabetic Rats: Involvement of Oxidative Stress, Inflammation and Extracellular Matrix Remodeling. *J. Neurochem.* **2018**, *146*, 173–185. [CrossRef]
45. Werner, M.; Chott, A.; Fabiano, A.; Battifora, H. Effect of Formalin Tissue Fixation and Processing on Immunohistochemistry. *Am. J. Surg. Pathol.* **2000**, *24*, 1016–1019. [CrossRef]
46. Almukainzi, M.; El-Masry, T.A.; Negm, W.A.; Elekhnawy, E.; Saleh, A.; Sayed, A.E.; Khattab, M.A.; Abdelkader, D.H. Gentiopicroside PLGA Nanospheres: Fabrication, in Vitro Characterization, Antimicrobial Action, and in Vivo Effect for Enhancing Wound Healing in Diabetic Rats. *Int. J. Nanomed.* **2022**, *17*, 1203. [CrossRef]
47. Culling, C.F.A. *Handbook of Histopathological and Histochemical Techniques: Including Museum Techniques*; Butterworth-Heinemann: Oxford, UK, 2013; ISBN 1483164799.
48. Shrivastava, S.R.; Shrivastava, P.S.; Ramasamy, J. World Health Organization Releases Global Priority List of Antibiotic-Resistant Bacteria to Guide Research, Discovery, and Development of New Antibiotics. *J. Med. Soc.* **2018**, *32*, 76. [CrossRef]
49. Abedon, S.T.; García, P.; Mullany, P.; Aminov, R. Phage Therapy: Past, Present and Future. *Front. Microbiol.* **2017**, *8*, 981. [CrossRef]
50. Jennes, S.; Merabishvili, M.; Soentjens, P.; Pang, K.W.; Rose, T.; Keersebilck, E.; Soete, O.; François, P.-M.; Teodorescu, S.; Verween, G. Use of Bacteriophages in the Treatment of Colistin-Only-Sensitive *Pseudomonas aeruginosa* Septicaemia in a Patient with Acute Kidney Injury—A Case Report. *Crit. Care* **2017**, *21*, 129. [CrossRef]
51. Nir-Paz, R.; Gelman, D.; Khouri, A.; Sisson, B.M.; Fackler, J.; Alkalay-Oren, S.; Khalifa, L.; Rimon, A.; Yerushalmy, O.; Bader, R. Successful Treatment of Antibiotic-Resistant, Poly-Microbial Bone Infection with Bacteriophages and Antibiotics Combination. *Clin. Infect. Dis.* **2019**, *69*, 2015–2018. [PubMed]
52. Domingo-Calap, P.; Delgado-Martínez, J. Bacteriophages: Protagonists of a Post-Antibiotic Era. *Antibiotics* **2018**, *7*, 66. [CrossRef] [PubMed]
53. Laurie, C.D.; Hogan, B.K.; Murray, C.K.; Loo, F.L.; Hospenthal, D.R.; Cancio, L.C.; Kim, S.H.; Renz, E.M.; Barillo, D.; Holcomb, J.B. Contribution of Bacterial and Viral Infections to Attributable Mortality in Patients with Severe Burns: An Autopsy Series. *Burns* **2010**, *36*, 773–779.
54. Chen, F.; Cheng, X.; Li, J.; Yuan, X.; Huang, X.; Lian, M.; Li, W.; Huang, T.; Xie, Y.; Liu, J. Novel Lytic Phages Protect Cells and Mice against *Pseudomonas aeruginosa* Infection. *J. Virol.* **2021**, *95*, e01832-20. [CrossRef]

55. Aghaee, B.L.; Mirzaei, M.K.; Alikhani, M.Y.; Mojtahedi, A. Sewage and Sewage-Contaminated Environments Are the Most Prominent Sources to Isolate Phages against *Pseudomonas aeruginosa*. *BMC Microbiol.* **2021**, *21*, 132. [CrossRef]
56. Bhat, T.; Cao, A.; Yin, J. Virus-like Particles: Measures and Biological Functions. *Viruses* **2022**, *14*, 383.
57. Hyman, P. Phages for Phage Therapy: Isolation, Characterization, and Host Range Breadth. *Pharmaceuticals* **2019**, *12*, 35. [CrossRef]
58. Park, D.-W.; Lim, G.; Lee, Y.; Park, J.-H. Characteristics of Lytic Phage vB_EcoM-ECP26 and Reduction of Shiga-Toxin Producing *Escherichia coli* on Produce Romaine. *Appl. Biol. Chem.* **2020**, *63*, 19.
59. Vukotic, G.; Obradovic, M.; Novovic, K.; Di Luca, M.; Jovcic, B.; Fira, D.; Neve, H.; Kojic, M.; McAuliffe, O. Characterization, Antibiofilm, and Depolymerizing Activity of Two Phages Active on Carbapenem-Resistant *Acinetobacter baumannii*. *Front. Med.* **2020**, *7*, 426. [CrossRef]
60. Duc, H.M.; Son, H.M.; Ngan, P.H.; Sato, J.; Masuda, Y.; Honjoh, K.; Miyamoto, T. Isolation and Application of Bacteriophages Alone or in Combination with Nisin against Planktonic and Biofilm Cells of *Staphylococcus aureus*. *Appl. Microbiol. Biotechnol.* **2020**, *104*, 5145–5158. [CrossRef]
61. Dakheel, K.H.; Rahim, R.A.; Neela, V.K.; Al-Obaidi, J.R.; Hun, T.G.; Isa, M.N.M.; Yusoff, K. Genomic Analyses of Two Novel Biofilm-Degrading Methicillin-Resistant Staphylococcus Aureus Phages. *BMC Microbiol.* **2019**, *19*, 114. [CrossRef] [PubMed]
62. Klumpp, J.; Loessner, M.J. Listeria Phages: Genomes, Evolution, and Application. *Bacteriophage* **2013**, *3*, e26861. [PubMed]
63. Ross, A.; Ward, S.; Hyman, P. More Is Better: Selecting for Broad Host Range Bacteriophages. *Front. Microbiol.* **2016**, *7*, 1352. [PubMed]
64. de Jonge, P.A.; Nobrega, F.L.; Brouns, S.J.J.; Dutilh, B.E. Molecular and Evolutionary Determinants of Bacteriophage Host Range. *Trends Microbiol.* **2019**, *27*, 51–63.
65. Turner, D.; Kropinski, A.M.; Adriaenssens, E.M. A Roadmap for Genome-Based Phage Taxonomy. *Viruses* **2021**, *13*, 506.
66. Othman, B.A.; Askora, A.; Abo-Senna, A.S.M. Isolation and Characterization of a Siphoviridae Phage Infecting *Bacillus megaterium* from a Heavily Trafficked Holy Site in Saudi Arabia. *Folia Microbiol.* **2015**, *60*, 289–295.
67. Ackermann, H.-W. Bacteriophage Electron Microscopy. *Adv. Virus Res.* **2012**, *82*, 1–32.
68. King, A.M.Q.; Lefkowitz, E.; Adams, M.J.; Carstens, E.B. *Virus Taxonomy: Ninth Report of the International Committee on Taxonomy of Viruses*; Elsevier: Amsterdam, The Netherlands, 2011; Volume 9. ISBN 0123846854
69. Sezer, S.; Şahin, İ.; Öztürk, K.; Şanko, V.; Koçer, Z.; Sezer, Ü.A. Cellulose-Based Hydrogels as Biomaterials. In *Cellulose-Based Superabsorbent Hydrogels*; Springer: Berlin/Heidelberg, Germany, 2019; pp. 1177–1203.
70. Alibolandi, M.; Bagheri, E.; Mohammadi, M.; Sameiyan, E.; Ramezani, M. Biopolymer-Based Hydrogel Wound Dressing. In *Modeling and Control of Drug Delivery Systems*; Elsevier: Amsterdam, The Netherlands, 2021; pp. 227–251.
71. Li, Z.; He, C.; Yuan, B.; Dong, X.; Chen, X. Injectable Polysaccharide Hydrogels as Biocompatible Platforms for Localized and Sustained Delivery of Antibiotics for Preventing Local Infections. *Macromol. Biosci.* **2017**, *17*, 1600347.
72. El-Gayar, M.H.; Ishak, R.A.H.; Esmat, A.; Aboulwafa, M.M.; Aboshanab, K.M. Evaluation of Lyophilized Royal Jelly and Garlic Extract Emulgels Using a Murine Model Infected with Methicillin-Resistant *Staphylococcus aureus*. *AMB Express* **2022**, *12*, 37. [CrossRef]
73. Masson-Meyers, D.S.; Andrade, T.A.M.; Caetano, G.F.; Guimaraes, F.R.; Leite, M.N.; Leite, S.N.; Frade, M.A.C. Experimental Models and Methods for Cutaneous Wound Healing Assessment. *Int. J. Exp. Pathol.* **2020**, *101*, 21–37.
74. Pires, D.P.; Vilas Boas, D.; Sillankorva, S.; Azeredo, J. Phage Therapy: A Step Forward in the Treatment of *Pseudomonas aeruginosa* Infections. *J. Virol.* **2015**, *89*, 7449–7456. [PubMed]
75. McVay, C.S.; Velásquez, M.; Fralick, J.A. Phage Therapy of *Pseudomonas aeruginosa* Infection in a Mouse Burn Wound Model. *Antimicrob. Agents Chemother.* **2007**, *51*, 1934–1938. [CrossRef] [PubMed]
76. Moghadam, M.T.; Khoshbayan, A.; Chegini, Z.; Farahani, I.; Shariati, A. Bacteriophages, a New Therapeutic Solution for Inhibiting Multidrug-Resistant Bacteria Causing Wound Infection: Lesson from Animal Models and Clinical Trials. *Drug Des. Devel. Ther.* **2020**, *14*, 1867.
77. Jault, P.; Leclerc, T.; Jennes, S.; Pirnay, J.P.; Que, Y.-A.; Resch, G.; Rousseau, A.F.; Ravat, F.; Carsin, H.; Le Floch, R. Efficacy and Tolerability of a Cocktail of Bacteriophages to Treat Burn Wounds Infected by *Pseudomonas aeruginosa* (PhagoBurn): A Randomised, Controlled, Double-Blind Phase 1/2 Trial. *Lancet Infect. Dis.* **2019**, *19*, 35–45. [PubMed]
78. Koehler, J.; Brandl, F.P.; Goepferich, A.M. Hydrogel Wound Dressings for Bioactive Treatment of Acute and Chronic Wounds. *Eur. Polym. J.* **2018**, *100*, 1–11. [CrossRef]
79. Maura, D.; Bandyopadhaya, A.; Rahme, L.G. Animal Models for *Pseudomonas aeruginosa* Quorum Sensing Studies. In *Quorum Sensing*; Springer: Berlin/Heidelberg, Germany, 2018; pp. 227–241.
80. Durmus, A.S.; Han, M.C.; Yaman, I. Comperative Evaluation of Collagenase and Silver Sulfadiazine on Burned Wound Healing in Rats. *Firat Univ. Saglik Bilim. Vet. Derg.* **2009**, *23*, 135–139.
81. Mekkes, J.R.; Zeegelaar, J.E.; Westerhof, W. Quantitative and Objective Evaluation of Wound Debriding Properties of Collagenase and Fibrinolysin/Desoxyribonuclease in a Necrotic Ulcer Animal Model. *Arch. Dermatol. Res.* **1998**, *290*, 152–157.
82. Khan, M.A.; Hussain, Z.; Ali, S.; Qamar, Z.; Imran, M.; Hafeez, F.Y. Fabrication of Electrospun Probiotic Functionalized Nanocomposite Scaffolds for Infection Control and Dermal Burn Healing in a Mice Model. *ACS Biomater. Sci. Eng.* **2019**, *5*, 6109–6116. [CrossRef]
83. Pereira Beserra, F.; Sérgio Gushiken, L.F.; Vieira, A.J.; Augusto Bérgamo, D.; Luísa Bérgamo, P.; Oliveira de Souza, M.; Alberto Hussni, C.; Kiomi Takahira, R.; Henrique Nóbrega, R.; Monteiro Martinez, E.R. From Inflammation to Cutaneous Repair: Topical

Application of Lupeol Improves Skin Wound Healing in Rats by Modulating the Cytokine Levels, NF-KB, Ki-67, Growth Factor Expression, and Distribution of Collagen Fibers. *Int. J. Mol. Sci.* **2020**, *21*, 4952. [CrossRef]

84. Steele, A.; Stacey, H.J.; De Soir, S.; Jones, J.D. The Safety and Efficacy of Phage Therapy for Superficial Bacterial Infections: A Systematic Review. *Antibiotics* **2020**, *9*, 754.

85. Mendes, J.J.; Leandro, C.; Corte-Real, S.; Barbosa, R.; Cavaco-Silva, P.; Melo-Cristino, J.; Górski, A.; Garcia, M. Wound Healing Potential of Topical Bacteriophage Therapy on Diabetic Cutaneous Wounds. *Wound Repair Regen.* **2013**, *21*, 595–603. [PubMed]

86. Gupta, P.; Singh, H.S.; Shukla, V.K.; Nath, G.; Bhartiya, S.K. Bacteriophage Therapy of Chronic Nonhealing Wound: Clinical Study. *Int. J. Low. Extrem. Wounds* **2019**, *18*, 171–175. [PubMed]

87. Khalid, F.; Siddique, A.B.; Nawaz, Z.; Shafique, M.; Zahoor, M.A.; Nisar, M.A.; Rasool, M.H. Efficacy of Bacteriophage against Multidrug Resistant *Pseudomonas aeruginosa* Isolates. *Southeast Asian J. Trop. Med. Public Health* **2017**, *48*, 1056–1062.

88. Raz, A.; Serrano, A.; Hernandez, A.; Euler, C.W.; Fischetti, V.A. Isolation of Phage Lysins That Effectively Kill *Pseudomonas aeruginosa* in Mouse Models of Lung and Skin Infection. *Antimicrob. Agents Chemother.* **2019**, *63*, e00024-19. [PubMed]

89. Pinto, A.M.; Cerqueira, M.A.; Bañobre-Lópes, M.; Pastrana, L.M.; Sillankorva, S. Bacteriophages for Chronic Wound Treatment: From Traditional to Novel Delivery Systems. *Viruses* **2020**, *12*, 235. [PubMed]

90. Chhibber, S.; Kaur, J.; Kaur, S. Liposome Entrapment of Bacteriophages Improves Wound Healing in a Diabetic Mouse MRSA Infection. *Front. Microbiol.* **2018**, *9*, 561.

Article

Recent Information on Pan-Genotypic Direct-Acting Antiviral Agents for HCV in Chronic Kidney Disease

Fabrizio Fabrizi [1,*], Federica Tripodi [1], Roberta Cerutti [1], Luca Nardelli [1], Carlo M. Alfieri [1,2], Maria F. Donato [3] and Giuseppe Castellano [1,2]

[1] Division of Nephrology, Dialysis, and Kidney Transplant, Foundation IRCCS Cà Granda Ospedale Maggiore Policlinico, 20122 Milano, Italy
[2] Department of Clinical Sciences and Community Health, University School of Medicine, 20122 Milano, Italy
[3] Division of Gastroenterology and Hepatology, Foundation IRCCS Cà Granda Ospedale Maggiore Policlinico, 20122 Milano, Italy
* Correspondence: fabrizio.fabrizi@policlinico.mi.it; Tel.:+39-2-55034553; Fax: +39-2-55034550

Abstract: Background: Hepatitis C virus (HCV) is still common in patients with chronic kidney disease. It has been recently discovered that chronic HCV is a risk factor for increased incidence of CKD in the adult general population. According to a systematic review with a meta-analysis of clinical studies, pooling results of longitudinal studies (n = 2,299,134 unique patients) demonstrated an association between positive anti-HCV serologic status and increased incidence of CKD; the summary estimate for adjusted HR across the surveys was 1.54 (95% CI, 1.26; 1.87), ($p < 0.0001$). The introduction of direct-acting antiviral drugs (DAAs) has caused a paradigm shift in the management of HCV infection; recent guidelines recommend pan-genotypic drugs (i.e., drugs effective on all HCV genotypes) as the first-choice therapy for HCV, and these promise to be effective and safe even in the context of chronic kidney disease. Aim: The purpose of this narrative review is to show the most important data on pan-genotypic DAAs in advanced CKD (CKD stage 4/5). Methods: We recruited studies by electronic databases and grey literature. Numerous key-words ('Hepatitis C' AND 'Chronic kidney disease' AND 'Pan-genotypic agents', among others) were adopted. Results: The most important pan-genotypic combinations for HCV in advanced CKD are glecaprevir/pibrentasvir (GLE/PIB) and sofosbuvir/velpatasvir (SOF/VEL). Two clinical trials (EXPEDITION-4 and EXPEDITION-5) and some 'real-world' studies (n = 6) reported that GLE/PIB combinations in CKD stage 4/5 gave SVR12 rates ranging between 86 and 99%. We retrieved clinical trials (n = 1) and 'real life' studies (n = 6) showing the performance of SOF/VEL; according to our pooled analysis, the summary estimate of SVR rate was 100% in studies adopting SOF/VEL antiviral combinations. The drop-out rate (due to AEs) in patients on SOF/VEL ranged between 0 and 4.8%. Conclusions: Pan-genotypic combinations, such as GLE/PIB and SOF/VEL, appear effective and safe for HCV in advanced CKD, even if a limited number of studies with small sample sizes currently exist on this issue. Studies are under way to assess whether successful antiviral therapy with DAAs will translate into better survival in patients with advanced CKD.

Keywords: dialysis; end-stage kidney disease; hepatitis C virus; sofosbuvir; sustained viral response

Citation: Fabrizi, F.; Tripodi, F.; Cerutti, R.; Nardelli, L.; Alfieri, C.M.; Donato, M.F.; Castellano, G. Recent Information on Pan-Genotypic Direct-Acting Antiviral Agents for HCV in Chronic Kidney Disease. *Viruses* 2022, 14, 2570. https://doi.org/10.3390/v14112570

Academic Editor: Simone Brogi

Received: 8 October 2022
Accepted: 14 November 2022
Published: 20 November 2022

Publisher's Note: MDPI stays neutral with regard to jurisdictional claims in published maps and institutional affiliations.

1. Introduction

HCV was identified in 1989; since then, HCV has emerged as a major public health problem [1]. According to Polaris models, there was a global prevalence of HCV RNA-positive patients of 0.7% (95% UI, 0.7%; 0.9%), corresponding to 56.8 million (95% UI, 55.2; 67.8) infections on January 1, 2020 [2]. In 2020, an estimated 641,1000 (95% CI, 623,000; 765,000) individuals started antiviral treatment. Chronic HCV infection can lead to liver cirrhosis, hepatic failure, and hepatocellular carcinoma in patients with or without kidney disease. After identification of HCV, a high frequency of serum anti-HCV antibodies and HCV ribonucleic acids (RNA) has been found in patients with chronic kidney disease [1].

The relationship between chronic kidney disease and HCV infection is bi-directional; HCV infection is both a cause and consequence of chronic kidney disease [1]. According to a systematic review with a meta-analysis of clinical studies, pooling results of longitudinal studies (n = 15 studies, n = 2,299,134 unique patients) demonstrated an association between positive anti-HCV serologic status and increased incidence of CKD; the summary estimate for adjusted HR across the surveys was 1.54 (95% CI, 1.26; 1.87) (p < 0.0001). However, between-study heterogeneity was observed (Q value by chi-squared test 500.3, p < 0.0001) [3].

The introduction of direct-acting antiviral agents (DAAs) has created a paradigm shift in the management of infection by HCV [1,4]. More recently, WHO recommended the adoption of pan-genotypic DAAs in order to support the campaign to eliminate HCV infection all over the world [4]. Genotype-specific DAAs have a limited antiviral spectrum of activity and frequently need ribavirin, which is commonly associated with important side-effects. Pan-genotypic regimens offer oral administration, eliminate the need for genotype testing, and provide favourable efficacy and tolerability. There is poor evidence in the medical literature concerning tolerability and effectiveness of pan-genotypic DAAs in end-stage renal disease [1]. The aim of this narrative review is to provide information on the efficacy and safety of pan-genotypic regimens of DAAs for patients with advanced CKD.

2. Materials and Methods

2.1. Information Sources and Search Strategy

Studies were identified by searching electronic databases and sources of grey literature. The literature search was applied to PubMed MEDLINE, EMBASE, and Google Scholar. The following key words were adopted: ('Hepatitis C Virus' OR 'HCV' OR 'Hepatitis C') AND ('Chronic Kidney Disease' OR 'End-stage kidney disease' OR 'Renal Insufficiency' OR 'Renal failure' or 'Renal impairment') AND ('Direct-acting antiviral agents' OR 'DAAs' OR 'Pan-genotypic agents' OR 'Sofosbuvir' OR 'Velpatasvir').

2.2. Statistical Methods

We performed pooled quantitative summary estimates of the sustained viral response (SVR) and discontinuation rates of antiviral therapies (pan-genotypic DAAs for HCV in advanced CKD) across individual studies using the inverse-variance method. A random-effects meta-analysis was conducted [5,6]. Outcomes were analysed on an intention-to-treat basis; all patients enrolled in these studies were included for the calculation of the response rate, whereas patients without an end-point were categorized as failures.

Observational studies that compared the all-cause mortality in anti-HCV positive patients who received antiviral therapy compared with those who did not were also retrieved. The aRR was obtained in each study by multivariate analysis to find the impact of antiviral therapy *per se* on death rate, irrespective of the role of covariates. Pooled RRs and their 95% CIs were estimated by the weighted inverse of their variance. Heterogeneity was evaluated using Der Simonian and Laird's Q test and quantified by calculating the proportion of the total variance attributable to between-study variance (Ri). We computed both fixed- and random-effects models but used the latter in case of large heterogeneity. We adopted HEpiMA, a novel software program that carries out a complete study of heterogeneity of study effects [6]. Novel and useful estimators of heterogeneity, such as R_i and CV$_B$ were used. Heterogeneity was considered substantial if Ri was $\geq$ 0.75. A two-sided p-value of <0.05 was considered statistically significant.

3. Results

3.1. Epidemiology

Abundant information exists on the epidemiology of hepatitis C virus infection in patients with end-stage renal disease. In 2012–2015, anti-HCV antibody prevalence among prevalent HD patients in the DOPPS was 9.9% overall (21 countries all over the world); it ranged from 4.1% in Belgium to 20.1% across the Gulf Cooperation Council Countries

(GCCC; Bahrain, Kuwait, Oman, Qatar, Saudi Arabia, and United Arab Emirates), whereas >8% of patients receiving HD in China, Italy, Japan, Russia, and Spain were anti-HCV positive [7].

Another survey was conducted among Medicare beneficiaries with HCV undergoing haemodialysis in the United States (2005–2016). A total of 291,663 patients on haemodialysis were enrolled. The prevalence of HCV in patients on haemodialysis was greater than in individuals not on HD (4.2 vs. <1%) [8].

Evidence on the prevalence and incidence rates of HCV among patients on long-term dialysis in the emerging world was not satisfactory—several studies with small sample sizes have been published, and recorded prevalence rates of up to around 80% [9–12]. Some systematic reviews have been made on this point; Harfouche et al. collected data from 289 studies ($n = 106, 463$ unique patients) and found a regional pooled mean estimate of 29.2% (95% CI, 25.6%; 32.8%) for HCV antibody prevalence among patients on long-term haemodialysis in the Middle East and North Africa (MENA) [10].

3.2. Natural History of HCV Infection

It is difficult to make a detailed evaluation of the natural history of HCV infection in patients with chronic kidney disease, particularly those on maintenance dialysis. Various reasons explain this—the natural history of HCV usually spans decades in patients with intact kidneys, whereas dialysis patients have limited life expectancies. In fact, patients with chronic kidney disease have higher morbidity and mortality than the general population, due to aging and comorbidities. HCV infection is frequently asymptomatic, with an apparently indolent course even in patients with advanced chronic kidney disease. Aminotransferase values are lower in patients on maintenance dialysis; thus, it is not easy to recognise the occurrence of liver disease on the grounds of biochemical abnormalities. A total of 506 patients undergoing regular dialysis in northern Italy (Lombardy) were tested by anti-HCV ELISA and PCR assays for the detection of anti-HCV antibodies and HCV RNA in serum, respectively. Serum transaminase values were significantly greater in HCV RNA-positive than HCV RNA-negative patients, 19.3 ± 1.6 vs. 15.7 ± 1.6 ($p = 0.008$) and 22.8 ± 1.7 vs. 16.1 ± 1.7 ($p = 0.0001$). According to logistic regression analysis, detectable HCV RNA in serum was a strong predictor of raised AST ($p = 0.0001$) and ALT ($p = 0.000001$) values [13].

The current availability of direct-acting antiviral agents, which are considered to be safe and have high efficacy, precludes the implementation of large observational studies and prolonged follow-up to assess the course of chronic HCV infection in end-stage kidney disease. It has been stated that survival in most patients with stage 1 and 2 CKD is not different from that observed in the general population with intact kidneys. Survival in patients with CKD stages 3–5 is lower than that observed in the general population, and some information has been recently accumulated on the link between positive anti-HCV serologic status and the death rate in the dialysis population. Death can be considered a reliable endpoint in the context of observational studies evaluating the course of HCV over time in patients with intact kidneys or end-stage kidney disease, and some clinical studies have carried out such analyses. We recently conducted a systematic review with a meta-analysis of observational studies ($n = 23$ studies; $n = 574,081$ patients on long-term dialysis). We found that positive anti-HCV serologic status was an independent and significant risk factor for death in the dialysis population. The overall estimate for adjusted mortality (all-cause death risk) with HCV was 1.26 (95% CI, 1.18; 1.34) ($p < 0.0001$) [14]. We performed stratified analyses to assess the causes of the increased death risk. The summary estimate for adjusted mortality (liver disease-related mortality) was 5.05 (95% CI, 2.53; 10.0) ($p < 0.0001$). The overall estimate for cardiovascular death risk was 1.18 (95% CI, 1.085; 1.29) ($p < 0.0001$). Using meta-regression, we observed that the relationship between positive anti-HCV serologic status and all-cause death risk was more evident in surveys with a larger size ($p < 0.0001$), higher proportion of diabetics ($p = 0.0005$), and HCV-positive patients ($p = 0.001$) [14].

The latest report on this issue was published a few months ago by Ko and coworkers. After adjusting for potential confounding factors, the multivariate Cox regression resulted in a significant association between serum HCV RNA-positive status and death rate in a cohort of haemodialysis patients who started chronic HD after 2002 (AHR, 1.48; 95% CI, 1.13; 1.93, p = 0.005). The study group in such a report was large (n = 1,437 patients on maintenance HD in Hiroshima, Japan) [15].

3.3. Antiviral Therapy of HCV and Its Purpose (Pan-Genotypic Regimens)

The purpose of antiviral therapy of HCV was the achievement of SVR12. SVR12 is the elimination of HCV RNA from serum which persists at least 12 weeks after completing antiviral therapy [16]. Antiviral treatment was indicated for patients showing anti-HCV antibody and detectable HCV RNA in serum. Solid evidence in the medical literature exists, suggesting that the achievement of SVR12 ('the cure') was associated with better survival and quality of life in patients with intact kidneys [17]. Patients with chronic kidney disease, HCV/HIV co-infection, and HCV/HBV co-infection who have had previous unsuccessful DAA regimens were defined as "special populations", where the antiviral approach was biased by low efficacy and safety (due to high rate of co-infections and comorbidities). It appears now that DAAs have abolished the notion of 'special populations' and pan-genotypic regimens promise to be effective and safe, even in this context. Pan-genotypic agents need only minimal monitoring and this encourages a test-and-treat approach as the focus of HCV management moves toward global elimination (with simplified protocols).

3.4. Natural History of HCV, HD Population, and Antivirals

The evidence in the medical literature supporting the antiviral treatment of HCV in patients receiving long-term dialysis is poor. Prior to the advent of DAAs, IFN-based regimens were the standard of care for the antiviral treatment of HCV. Clinicians had been reluctant to adopt IFN-based regimens for HCV in patients on maintenance dialysis, due to limited efficacy and low tolerability of IFN-based regimens in the haemodialysis setting. The effectiveness and tolerability of combined antiviral therapy (pegylated interferon plus ribavirin) in patients on regular dialysis had been addressed in a meta-analysis of clinical studies. We identified eleven clinical studies (n = 287 unique patients) and the summary estimate for SVR and drop-out rate was 0.60 (95% CI, 0.47; 0.71) and 0.18 (95% CI, 0.08; 0.35), respectively. The most common source of drop-out was anaemia (11/46 = 23%) [18]. The limited efficacy and tolerability of combined antiviral therapy in the dialysis population meant that only a minority of dialysis patients were 'cured' with such approach.

To date, a few studies with small sample sizes have recorded higher survival rates in haemodialysis patients who received antiviral therapy compared with HCV-infected patients on HD who did not receive it. Goodkin and colleagues [19] evaluated the Dialysis Outcomes and Practice Patterns Study; 49,762 patients on long-term haemodialysis in 12 countries were enrolled (1996–2011), and survival and other clinical parameters were reviewed over a median 1.4 year per study phase. There were 4,735 (9.5%) patients with HCV infection and 4589 (96.9%) of them had a history of a prescription of antivirals. In the group of HCV-infected patients with an overlapping propensity for antiviral treatment, there was no difference regarding the death rate among patients who received antiviral treatment who died and those who did not (Table 1). The adjusted mortality risk (aHR for mortality) was 0.47 (95% CI, 0.17; 1.26, NS).

Table 1. The impact of antiviral therapy upon survival in HCV-positive patients who received antiviral therapy: univariate analysis.

Study	Reference Year	Death Rate (Treated)	Death Rate (Untreated)	p	Country
Goodkin D, *et al.*	2013	4/42 (9.5%)	638/3,307 (21%)	NS	USA
Hsu Y, *et al.*	2015	7/134 (5.2%)	581/2,097 (27.7%)	0.0001	Taiwan
Soderholm J, *et al.*	2018	11/45 (24%)	124/223 (56%)	0.0001	Sweden
Chen Y, *et al.*	2019	61/482 (12.7%)	648/1,928 (33.6%)	0.0001	Taiwan
Perez de Josè A, *et al.*	2021	13/124 (10.5%)	10/15 (67%)	0.0001	Spain

Soderholm and coworkers [20] gave antiviral treatment to 45 of 268 patients on long-term haemodialysis with chronic HCV. Antiviral treatment was associated with a favourable outcome; the death rate was lower during the study period for treated patients than for untreated patients ($p = 0.0001$) (Table 1). According to their multivariate analysis, kidney transplant (aOR, 2.97, 95% CI, 1.64; 5.37, $p = 0.0001$), acute kidney failure before renal replacement therapy (aOR, 2.518, 95% CI, 1.39; 4.54, $p = 0.002$), and antiviral treatment (aOR, 3.54, 95% CI, 1.63; 7.79, $p = 0.001$) were independent predictors of improved survival in patients on maintenance haemodialysis with chronic HCV. Age at HD initiation was linked to lower survival (aOR, 0.968, 95% CI, 0.94; 0.98, $p = 0.004$).

Hsu and colleagues [21] worked on the National Health Insurance program in Taiwan and investigated whether interferon-based treatment was associated with improved survival in ESRD with HCV infection. In their cohort of HCV-infected patients ($n = 2231$), 134 (6.01%) patients received interferon and 2097 (93.9%) did not; the mean follow-up duration was 3.22 years (Table 1). The aHR for mortality was 3.91 (95% CI, 0.54; 28.1) in the untreated HCV cohort. In the subset of patients with HCV and without cirrhosis, patients who did not receive IFN-based therapy had a greater risk of death in comparison with the treated group (aHR, 6.31, 95% CI, 1.57; 25.4). In the subgroup of HCV-infected patients with cirrhosis and/or liver cancer, no differences in the risk of death occurred between those who received IFN or not (NS).

Chen and coworkers [22] made a large nationwide retrospective cohort study ($n = 93,894$ Taiwanese adults diagnosed with stage 1–5 CKD and without HBV infection) and used propensity score-matched and competing risk analyses to evaluate the long-term patient survival (death rate) of anti-HCV therapy, especially interferon-based therapy, in CKD patients. They observed that the treated cohort had a 29% (95% CI, 0.54; 0.92) ($p = 0.011$) decrease in death compared with the untreated cohort.

The last report on this topic was published by Perez de Josè and coworkers [23]. They retrieved a large cohort of patients with HCV-related mixed cryoglobulinemia. At baseline, mean serum creatinine and GFR were 1.4 mg/dL and 56 mL/min, respectively, and mean proteinuria was 2.1 gr/day. Overall, 100 patients underwent antiviral therapy with DAAs, 24 with IFN plus ribavirin, and 15 remained untreated. The death rate was greater in those patients who did not receive antiviral therapy (Table 1) compared with patients who underwent therapy for HCV; patients treated with DAAs had reduced mortality, aHR, 0.12 (95% CI, 0.04; 0.40, $p < 0.001$).

All the studies reported in Table 1 performed multivariate analysis in order to assess a significant and independent relationship between death rate and antiviral therapy. On the grounds of our meta-analysis of observational studies (random-effects model) [5,6], the pooled adjusted RR was 0.66 (95% CI, 0.52; 0.84, $p < 0.001$) (Figure 1). The results obtained with the fixed-and the random-effects models were similar (Figure 1); however, there was some heterogeneity ($R_i = 0.82$, CV_between 1.07). Further study is needed to make more definitive conclusions.

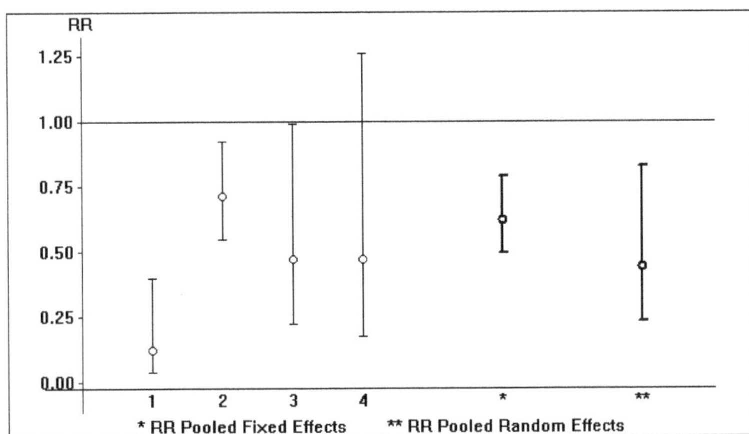

Figure 1. The impact of antiviral therapy on death rate in patients with advanced CKD: pooled adjusted RR according to fixed- and random-effects models.

3.5. Pan-Genotypic DAAs (Sofosbuvir)

Sofosbuvir was approved in 2013 and is now the backbone of the most commonly used DAA regimens. SOF is an oral nucleoside analogue and potent inhibitor of the NS5B RNA-dependent RNA polymerase. Upon oral administration, SOF is metabolized (at liver level) to 2′-deoxy-2′-alpha-fluoro-beta-C-methyluridine-5′-monophophate, which undergoes conversion to the active triphosphate form (GS-461203). SOF acts as a HCV RNA chain terminator by inhibiting NS5B RNA-dependent RNA polymerase, which is essential for the replication of the HCV RNA viral genome (Figure 2). The dephosphorylation of GS-461203 produces an inactive metabolite (GS-331007) that undergoes large clearance by the kidneys. The administration of a single full-dose of SOF reported a greater plasma AUC in individuals with CKD stage 5 (1.33-fold) and CKD stage 4 (2.73-fold) than in patients with an estimated GFR > 80 mL/min/1.73 m^2. In addition, the administration of a single full dose of SOF revealed that the plasma AUC of GS-331007 were 5.6-fold and 6.83-fold higher in patients with CKD stages 4 and 5, respectively, than those with intact kidneys [24]. A pharmacokinetic study with SOF at a dose of 400 mg per day or 400 mg thrice weekly for 12–24 weeks in patients undergoing long-term haemodialysis did not result in SOF accumulation between HD sessions or throughout the treatment course [25]. At the beginning, SOF use had not been recommended in patients with end-stage renal disease due to the fear of accumulation of SOF or its active metabolites. Since then, several studies reported satisfactory efficacy and tolerability regarding SOF use in advanced CKD. In November 2019, the US FDA amended the package inserts for sofosbuvir-containing regimens to allow use in patients with renal disease, including those with CKD stage 4 and 5 and those on dialysis.

1) Sofosbuvir- activation (in hepatocytes) to triphosphate GS-461203 by hydrolysis of the carboxylate ester (by enzymes cathepsin A or carboxylesterase 1), followed by cleaving of the phosphoramidate by the enzyme histidine triad nucleotide-binding protein 1 (HINT1), and subsequent repeated phosphorylation

2) Dephosphorylation of GS-461203 resulting in GS-331007 (inactive metabolite)

3) GS-331007 clearance by kidneys (80%), feces (14%), or respiration (2.5%)

4) Pharmacokinetic study (SOF, 400 mg/day or 400 mg x3/weekly for 12-24 weeks) in patients on long-term HD. no accumulation of plasma levels of SOF or GS-331007 with either regimen between HD procedures or throughout the treatment course

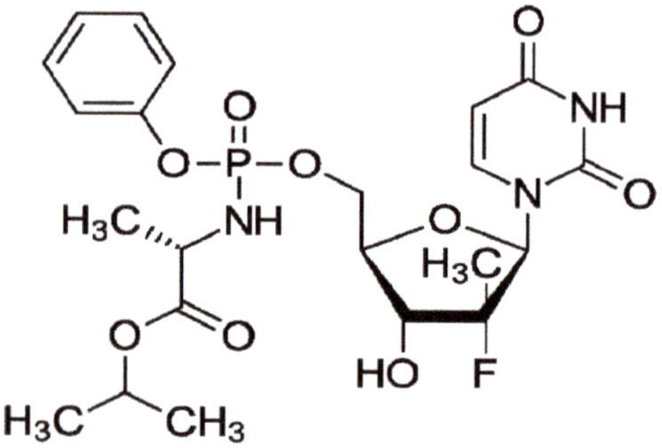

Figure 2. Sofosbuvir: metabolism in advanced kidney impairment and structure.

A systematic review of the medical literature with a meta-analysis of clinical studies has been recently published with the aim to assess the effectiveness and tolerability of SOF-based regimens in patients with advanced CKD (CKD stage 4 and 5, including patients receiving long-term haemodialysis). The primary end-point was the frequency of sustained viral response (as a measure of efficacy); the secondary outcomes were the rates of SAEs and drop-out due to AEs (as measures of tolerability). We identified 30 clinical studies ($n = 1537$ unique patients). The overall frequency of SVR12 was 0.99 (95% CI, 0.97; 1.0, $I^2 = 99.8\%$); the pooled frequency of SAEs was 0.09 (95% CI, 0.05; 0.13, $I^2 = 84.3\%$). Some ($n = 6$; 69 unique patients) clinical studies reported eGFR values at the beginning and end of antiviral therapy, and no consistent changes were recorded [26]. The conclusion of the authors was that SOF-based regimens appear safe and effective even in patients with CKD stage 4 and 5. Serum creatinine levels should be carefully monitored during therapy with SOF in the CKD population [26].

3.6. Pan-Genotypic DAAs (Glecaprevir/Pibrentasvir)

EXPEDITION-4 was a phase III multi-center open-label trial to assess the efficacy and safety of antiviral therapy for HCV (combined therapy of the NS3/4A protease inhibitor glecaprevir and the NS5A inhibitor pibrentasvir) for 12 weeks in adults with HCV infection and HCV genotype 1, 2, 3, 4, 5, or 6 and compensated liver disease (with or without cirrhosis) [27]. All patients in the study group had chronic kidney disease stages 4/5 (dialysis-dependent or not). The primary endpoint was the sustained viral response, 12 weeks after the end of treatment. There were 104 patients in the study group, 52% had genotype 1 infection, 16% had genotype 2 infection, 11% had genotype 3 infection, 19% had genotype 4 infection, and 2% had genotype 5 and 6 infections. The frequency of SVR was 98% (102/104) (95% CI, 95%; 100%), according to ITT analysis. Two patients failed to achieve SVR12 because of early discontinuation and loss to follow-up. SAEs were noted in 24% (25/104) of patients; none of the SAEs were considered by the study investigators to be drug related. Four patients discontinued the study due to adverse events, three of them had SVR. Some (20%, 21/104) patients complained of itching. EXPEDITION-5 is a phase III trial aimed to assess efficacy and safety of glecaprevir/pibrentasvir by oral route (300/120 mg daily, consisting of three tablets of 100/40 mg each), once a day for 8 to 16 weeks. A total of 101 patients with CKD stages 3b–5 were recruited. Fifty-five per cent of patients had HCV genotype 1 infection, 27% had genotype 2 infection, 15% had genotype 3 infection, and 4% had genotype 4; there were no patients with HCV genotype 5 or 6 infections. The SVR12 rate was 97% (98/101, 95% CI, 91.6; 99) by ITT. Serious AEs were reported in 12% of patients; none were related to the study drug [28].

Some 'real-life' studies have also been published in patients with advanced CKD and showed SVR12 rates similar to those observed in EXPEDITION-4 and EXPEDITION-5 trials (Table 2) [29–34]. The most common side-effect was pruritus (range, 0–61%), but it was mild in most patients. The frequency of AEs resulting in discontinuation of therapy was extremely low (Table 2). The most frequent causes of discontinuation of therapy due to AEs were pruritus ($n = 3$) and raised serum creatinine ($n = 2$). Other reasons were cerebral infarction ($n = 1$), fungal peritonitis ($n = 1$), patient preference ($n = 1$), and cardiomyopathy ($n = 1$).

Table 2. Pan-genotypic agents (GLE/PIB) for HCV in advanced CKD: real-life studies.

Study	Reference Year	SVR Rate	AEs Resulting in Drug Discontinuation	Country	Study Design
Suda G, *et al.*	2019	26/27 (96.3%)	2 (7.4%)	Japan	Prospective
Atsukawa M, *et al.*	2019	140/141 (99.3%)	3 (7.2%)	Japan	Prospective
Yen H, *et al.*	2020	42/44 (95.5%)	1 (2.3%)	Taiwan	Retrospective
Yap D, *et al.*	2020	18/21 (85.7%)	1 (4%)	Hong Kong/Taiwan	Prospective
Liu C, *et al.*	2020	107/108 (99%)	2 (3%)	Taiwan	Retrospective
Stein K, *et al.*	2022	29/33 (87.9%)	0%	Germany	Prospective

3.7. Pan-Genotypic DAAs (Sofosbuvir/Velpatasvir)

Borgia et al. performed a phase II single-arm study to treat 59 patients with genotype 1–6 HCV infection on long-term haemodialysis ($n = 54$) or peritoneal dialysis ($n = 5$) [35]. These patients received SOF/VEL (400/100 mg) once daily for 12 weeks. HCV-infected patients who were treatment-naïve or treatment-experienced with compensated cirrhosis or without cirrhosis were included. Out of 59 patients, 56 achieved SVR12 (95% CI, 86–99%). There were three patients who did not achieve SVR12: two had virological relapse (at post-treatment 4) and one of them prematurely discontinued antiviral treatment. Another patient died from suicide after obtaining SVR12. Most patients experienced an adverse event (80%), the majority of which were mild or moderate in severity. No patients prematurely discontinued SOF/VEL due to AEs.

Some real-world studies [36–42] and meta-analyses [43] have been published regarding SOF-VEL for HCV in ESRD (Tables 3 and 4). The SVR12 rates ranged between 89 and 100% in patients with advanced CKD, according to ITT analysis. Our pooled analysis of clinical studies showed that the summary estimate of SVR rates after pan-genotypic antiviral therapy (SOF/VEL) in ESRD was 1.0 (95% CI, 1.0; 1.0) (Table 5). The most common adverse event was nausea (reported in five studies), followed by headache (in three reports). Some patients ($n = 7$) showed early discontinuation of therapy due to AEs, and the causes were weakness ($n = 1$), anaemia ($n = 1$), dizziness ($n = 1$), and gastrointestinal disorders ($n = 4$).

Table 3. Studies on SOF/VEL in advanced CKD: characteristics of study patients (and viral response).

Study	Reference Year	Study Size	SVR Rate	Age, years	Males, n	Country
Borgia S, *et al.*	2019	59	56/59 (94.9%)	60 (33; 91)	35 (59%)	Canada
Gohel K, *et al.*	2020	3	3/3 (100%)	46.5	NA	India
Mostafi M, *et al.*	2020	44	44/44 (100%)	43.7 ± 12	NA	Bangladesh
Gaur N, *et al.*	2020	31	30/31 (96.8%)	39.8 ± 10.8	24 (77.5%)	India
Yu M, *et al.*	2021	105	94/105 (89.5%)	66.2 ± 10	54 (51.4%)	Taiwan
Taneja S, *et al.*	2021	51	49/51 (96%)	42.8 ± 14.6	41 (80.4%)	India
Liu C, *et al.*	2022	191	181/191 (94.8%)	65 (23; 95)	104 (54.5%)	Taiwan

Table 4. Studies on SOF/VEL in advanced CKD: characteristics of study patients (and drop-out rate).

Study	HBsAg, n	HCV Genotype 1, n	Cirrhosis, n	Treatment-Naïve, n	Diabetics, n	Drop-Out Rate (Due to AEs), n
Borgia S, *et al.*	NA	27 (45.8%)	17 (28.8%)	46 (77.9%)	19 (32%)	0
Gohel K, *et al.*	NA	NA	0	3 (100%)	NA	0
Mostafi M, *et al.*	0	NA	10 (23%)	44 (100%)	28 (63.6%)	0
Gaur N, *et al.*	6 (19.3%)	21 (67.7%)	3 (9.6%)	31 (100%)	6 (19%)	0
Yu M, *et al.*	8 (7.6%)	46 (43.8%)	37 (35.2%)	NA	65 (61.9%)	5/105 (4.8%)
Taneja S, *et al.*	NA	15 (79%)	10 (19.6%)	43 (84.3%)	NA	0
Liu C, *et al.*	5 (2.6%)	112 (58.6%)	27 (14.1%)	175 (91.6%)	NA	2/191 (1%)

Table 5. Pooled SVR rate after antiviral therapy with pan-genotypic DAAs (SOF/VEL) in advanced CKD. Test for heterogeneity: $Chi^2 = 32.13$, df = 6 ($p < 0.0001$), $I^2 = 81.3\%$. Test for overall effect: $Z = 2716.6$ ($p < 0.0001$).

	Weight (%)	SVR Rate (SE)	SVR Rate (Random-Effects Model) 95% CI	Year
Borgia A, *et al.*	0.02	0.94 (0.0286)	0.95 (0.89; 1.01)	2019
Gohel K, *et al.*	49.95	1 (0.0001)	1.0 (1.00; 1.00)	2020
Mostafi M, *et al.*	49.95	1 (0.0001)	1.0 (1.00; 1.00)	2020
Gaur N, *et al.*	0.01	0.96 (0.0352)	0.96 (0.89; 1.03)	2020
Yu M, *et al.*	0.01	0.89 (0.0305)	0.89 (0.83; 0.95)	2021
Taneja S, *et al.*	0.02	0.96 (0.0270)	0.96 (0.91; 0.97)	2021
Liu C, *et al.*	0.05	0.94 (0.0170)	0.94 (0.91; 0.97)	2022
Total (95% CI)	100.00		1.0 (1.00; 1.00)	

No data have been currently published on the antiviral combination of sofosbuvir/velpatasvir/voxilaprevir (SOF/VEL/VOX) in patients with advanced CKD.

4. Conclusions

Patients with advanced CKD have been defined a 'special population' or 'difficult-to-treat population', where antiviral therapy has historically been unable to generate high efficacy and safety. DAA-based regimens have revolutionized the management of HCV, and recommended DAA regimens can vary in duration, dosing frequency, pill burden, and co-administration of ribavirin. Pan-genotypic drugs allow the simplification of the management of HCV. These drugs reduce the need for pre-treatment testing, and consequently, the time between HCV diagnosis and therapy initiation is shortened. Pan-genotypic agents promise to be safe and effective, even in patients with advanced CKD, and more studies are needed to confirm and expand such pieces of evidence. An important limitation of the present narrative review was the limited number of included studies and the small number of patients overall. Research aimed to assess whether successful antiviral therapy with DAAs will improve survival of patients with advanced CKD is in progress.

Author Contributions: Methodology, G.C.; Formal analysis, F.T. and R.C.; Investigation, C.M.A. and M.F.D.; Data curation, L.N.; Writing–review & editing, F.F. All authors have read and agreed to the published version of the manuscript.

Funding: This research received no external funding.

Institutional Review Board Statement: Not applicable.

Informed Consent Statement: Not applicable.

Data Availability Statement: Not applicable.

Conflicts of Interest: The authors declare no conflict of interest.

Abbreviations

AH	Arterial hypertension
AUC	Area under curve
CI	Confidence intervals
CKD	Chronic kidney disease
CLD	Chronic liver disease
COPD	Chronic obstructive pulmonary disease
CVD	Cardiovascular disease
DAAs	Direct acting antiviral agents
DM	Diabetes mellitus
eGFR	Estimated glomerular filtration rate
ESRD	End-stage renal disease
GLE	Glecaprevir
HBsAg	Hepatitis B virus antigen
HCV	Hepatitis C virus
HCW	Health care worker
HD	Haemodialysis
IFN	Interferon
ITT	Intention-to-treat
KDIGO	Kidney Disease: Improving Global Outcomes
NA	Not available
PD	Peritoneal dialysis
PIB	Pibrentasvir
RRT	Renal replacement therapy
SAE	Severe adverse event
SE	Standard error
SOF	Sofosbuvir
SVR	Sustained virological response
VEL	Velpatasvir
VOX	Voxilaprevir
WHO	World Health Organization

References

1. Fabrizi, F.; Cerutti, R.; Messa, P. An updated view on the antiviral therapy of hepatitis C in chronic kidney disease. *Pathogens* **2021**, *1*, 1381. [CrossRef]
2. Polaris Observatory HCV Collaborators. Global change in hepatitis C virus prevalence and cascade of care between 2015 and 2020: A modelling study. *Lancet Gastroenterol. Hepatol.* **2022**, *7*, 396–415. [CrossRef]
3. Fabrizi, F.; Donato, F.; Messa, P. Association between hepatitis C virus and chronic kidney disease: A systematic review and meta-analysis. *Ann Hepatol.* **2018**, *17*, 364–391. [CrossRef]
4. Cox, A.; El-Sayed, M.; Kao, J.; Lazarus, J.; Lemoine, M.; Lok, A.; Zoulim, F. Progress towards elimination goals for viral hepatitis. *Nat. Rev. Gastroenterol. Hepatol.* **2020**, *17*, 533–542. [CrossRef] [PubMed]
5. Der Simonian, R.; Laird, D. Meta-analysis in clinical trials. *Control. Clin. Trials* **1986**, *7*, 177–188. [CrossRef]
6. Costa Bouzas, J.; Takkouche, B.; Cadarso-Suarez, C.; Spiegelman, D. HEpiMA: Software for the identification of heterogeneity in meta-analysis. *Comput. Meth. Programs Biomed.* **2001**, *64*, 101–107. [CrossRef]
7. Jadoul, M.; Bieber, B.; Martin, P.; Akiba, T.; Nwanko, C.; Arduino, M.; Goodkin, D.; Pisoni, R. Prevalence, incidence, and risk factors for hepatitis C virus infection in haemodialysis patients. *Kidney Int.* **2019**, *95*, 939–947. [CrossRef] [PubMed]
8. Deshpande, R.; Stepanova, M.; Golabi, P.; Brown, K.; Younossi, Z. Prevalence, mortality and healthcare utilization among Medicare beneficiaries with hepatitis C in haemodialysis units. *J. Viral. Hepat.* **2019**, *26*, 1293–1300. [CrossRef]
9. Adane, T.; Getawa, S. The prevalence and associated factors of hepatitis B and C virus in haemodialysis patients in Africa: A systematic review and meta-analysis. *PLoS ONE* **2021**, *16*, e0251570. [CrossRef] [PubMed]
10. Harfouche, M.; Chemaitelly, H.; Mahmud, S.; Chabna, K.; Kouyoumjian, S.; Kanaani, Z.; Abu-Raddad, L. Epidemiology of hepatitis C virus among haemodialysis patients in the Middle East and North Africa: Systematic syntheses, meta-analyses, and meta-regressions. *Epidemiol. Infect.* **2017**, *145*, 3243–3263. [CrossRef] [PubMed]
11. Akhtar, S.; Nasir, J.; Usman, M.; Sarwar, A.; Majeed, R.; Billah, B. The prevalence of hepatitis C virus in haemodialysis in Pakistan: A systematic review and meta-analysis. *PLoS ONE* **2020**, *15*, e0232931. [CrossRef] [PubMed]
12. Chuaypen, N.; Khlaiphuengsin, A.; Prasoppokakorn, T.; Susantitaphong, P.; Prasithsirikul, W.; Avihingsanon, A.; Tangkijvanich, P.; Praditpornsilpa, K. Prevalence and genotype distribution of hepatitis C virus within haemodialysis units in Thailand: Role of HCV core antigen in the assessment of viremia. *BMC Infect. Dis.* **2022**, *22*, 79. [CrossRef]
13. Fabrizi, F.; Lunghi, G.; Andrulli, S.; Pagliari, B.; Mangano, S.; Faranna, P.; Pagano, A.; Locatelli, F. Influence of hepatitis C virus (HCV) viraemia upon serum aminotransferase activity in chronic dialysis patients. *Nephrol. Dial. Transpl.* **1997**, *12*, 1394–1398. [CrossRef]
14. Fabrizi, F.; Dixit, V.; Messa, P. Hepatitis C virus and mortality among patients on dialysis: A systematic review and meta-analysis. *Clin. Res. Hepatol. Gastroenterol.* **2019**, *43*, 244–254. [CrossRef] [PubMed]
15. Ko, K.; Nagashima, S.; Yamamoto, C.; Takahashi, K.; Matsuo, J.; Ohisa, M.; Akita, T.; Matyakubov, J.; Mizeav, U.; Katayama, K.; et al. Eighteen-year follow-up cohort study on hepatitis B and C virus infections related long-term prognosis among haemodialysis patients in Hiroshima. *J. Med. Virol.* **2020**, *92*, 3436–3447. [CrossRef] [PubMed]
16. Liu, C.; Kao, J. Pan-genotypic direct-acting antivirals for patients with hepatitis C virus infection and chronic kidney disease stage 4 or 5. *Hepatol. Int.* **2022**, *16*, 1001–1019. [CrossRef]
17. AASLD-IDSA HCV Guidance Panel. Hepatitis C Guidance: AASLD-IDSA Recommendations for Testing, Managing, and Treating Hepatitis C. Available online: https://www.hcvguidelines.org/uniquepopulations/renalimpairment (accessed on 24 October 2022).
18. Fabrizi, F.; Dixit, V.; Messa, P.; Martin, P. Antiviral therapy (pegylated interferon and ribavirin) of hepatitis C in dialysis patients: Meta-analysis of clinical studies. *J. Viral. Hepat.* **2014**, *21*, 681–689. [CrossRef]
19. Goodkin, D.; Bieber, B.; Gillespie, B.; Robinson, B.; Jadoul, M. Hepatitis C infection is very rarely treated among haemodialysis patients. *Am. J. Nephrol.* **2013**, *38*, 405–412. [CrossRef]
20. Soderholm, J.; Millbourn, C.; Busch, K.; Kovamees, J.; Schvarcz, R.; Lindahl, K.; Bruchfeld, A. Higher risk of renal disease in chronic hepatitis C patients: Antiviral therapy survival benefit in patients on haemodialysis. *J. Hepatol.* **2018**, *68*, 904–911. [CrossRef]
21. Hsu, Y.; Hung, P.; Muo, C.; Tsai, W.; Hsu, C.; Kao, C. Interferon-based treatment of hepatitis C virus infection reduces all-cause mortality in patients with end-stage renal disease. An 8-year nationwide cohort study in Taiwan. *Medicine* **2015**, *94*, e2113. [CrossRef]
22. Chen, Y.; Li, C.; Tsai, S.; Chen, Y. Anti-hepatitis C virus therapy in chronic kidney disease patients improves long-term renal and patient survivals. *World J. Clin. Cases* **2019**, *7*, 1270–1281. [CrossRef] [PubMed]
23. Perez de Josè, A.; Carbayo, J.; Pocurull, A.; Bada-Bosch, T.; Corona, C.; Shabaka, A.; Terrada, N.; Valenzuela, L.; Huerta, A.; Lorente, F.; et al. Direct-acting antiviral therapy improves kidney survival in hepatitis C virus-associated cryoglobulinaemia: The RENALCRYOGLOBULINEMIC study. *Clin. Kidney J.* **2021**, *14*, 586–592. [CrossRef] [PubMed]
24. Lawitz, E.; Marbury, T.; Kirby, B.; Au, N.; Mathias, A.; Stamm, L.; Sajwani, K.; Klein, G.; Gane, E. The effect of renal or hepatic impairment on the pharmacokinetics of GS-9857, a pangenotypic HCV NS3/4A protease inhibitor. *J. Hepatol.* **2016**, *64*, S613. [CrossRef]

25. Desnoyer, A.; Pospai, D.; Lè, M.; Gervais, A.; Heurguè–Berlot, A.; Laradi, A.; Harent, S.; Pinto, A.; Salmon, D.; Hillaire, S.; et al. Pharmacokinetics, safety and efficacy of a full dose sofosbuvir-based regimen given daily in haemodialysis patients with chronic hepatitis C. *J. Hepatol.* **2016**, *65*, 40–47. [CrossRef]

26. Fabrizi, F.; Cerutti, R.; Dixit, V.; Ridruejo, E. Sofosbuvir-based regimens for HCV in stage 4–5 chronic kidney disease. A systematic review with meta-analysis. *Nefrologia* **2021**, *41*, 578–589. [CrossRef]

27. Gane, E.; Lawitz, E.; Pugatch, D.; Papatheodoridis, G.; Brau, N.; Brown, A.; Pol, S.; Leroy, V.; Persico, M.; Moreno, C.; et al. Glecaprevir and pibrentasvir in patients with HCV and severe renal impairment. *N. Engl. J. Med.* **2017**, *377*, 1448–1455. [CrossRef]

28. Lawitz, E.; Flisiak, R.; Abunimeh, M.; SIse, M.; Park, J.; Kaskas, M.; Bruchfeld, A.; Worns, M.; Aglitti, A.; Zamor, P.; et al. Efficacy and safety of glecaprevir/pibrentasvir in renally impaired patients with chronic HCV infection. *Liver Int.* **2020**, *40*, 1032–1041. [CrossRef]

29. Suda, G.; Hasebe, C.; Abe, M.; Kurosaki, M.; Itakura, J.; Izumi, N.; Uchida, Y.; Mochida, S.; Haga, H.; Ueno, Y.; et al. Safety and efficacy of glecaprevir and pibrentasvir in Japanese haemodialysis patients with genotype 2 hepatitis C virus infection. *J. Gastroenterol.* **2019**, *54*, 641–649. [CrossRef]

30. Atsukawa, M.; Tsubota, A.; Toyoda, H.; Takaguchi, K.; Nakamuta, M.; Watanabe, T.; Michitaka, K.; Ikegami, T.; Nozaki, A.; Uojima, H.; et al. The efficacy and safety of glecaprevir plus pibrentasvir in 141 patients with severe renal impairment; a prospective, multicenter study. *Aliment. Pharmacol. Ther.* **2019**, *49*, 1230–1241. [CrossRef]

31. Yen, H.; Su, P.; Zeng, Y.; Liu, I.; Huang, S.; Hsu, Y.; Chen, Y.; Yang, C.; Wu, S.; Chou, K. Glecaprevir-pibrentasvir for chronic hepatitis C: Comparing treatment effect in patients with and without end-stage renal disease in a real-world setting. *PLoS ONE* **2020**, *15*, e02337582. [CrossRef]

32. Yap, D.; Liu, K.; Hsu, Y.; Wong, G.; Tsai, M.; Chen, C.; Hsu, C.; Hui, Y.; Li, M.; Liu, C.; et al. Use of glecaprevir/pibrentasvir in patients with chronic hepatitis C virus infection and severe renal impairment. *Clin. Mol. Hepatol.* **2020**, *26*, 554–561. [CrossRef] [PubMed]

33. Liu, C.; Yang, S.; Peng, C.; Lin, W.; Liu, C.; Su, T.; Tseng, T.; Chen, P.; Chen, D.; Kao, J. Glecaprevir/pibrentasvir for patients with chronic hepatitis C virus infection and severe renal impairment. *J. Viral. Hepat.* **2020**, *27*, 568–575. [CrossRef] [PubMed]

34. Stein, K.; Stoehr, A.; Klinker Hm Teuber, G.; Naumann, U.; Christine, J.; Heyne, R.; Serfert, Y.; Niederau, C.; Zeuzem, S.; Berg, T.; et al. Hepatitis C therapy with grazoprevir/elbasvir and glecaprevir/pibrentasvir in patients with advanced chronic kidney disease: Data from the Germany Hepatitis C-Registry (DHC-R). *Eur. J. Gastroenterol. Hepatol.* **2022**, *34*, 76–83. [CrossRef] [PubMed]

35. Borgia, S.; Dearden, J.; Yoshida, E.; Shafran, S.; Brown, A.; Ben-Ari, Z.; Cramp, M.; Cooper, C.; Foxton, M.; Rodriguez, F.; et al. Sofosbuvir/velpatasvir for 12 weeks in hepatitis C virus-infected patients with end-stage renal disease undergoing dialysis. *J. Hepatol.* **2019**, *71*, 660–665. [CrossRef] [PubMed]

36. Begovac, J.; Krznaric, J.; Bogdanic, N.; Mocibob, L.; Zekan, S. Successful treatment of genotype 3 hepatitis C infection in a non-cirrhotic HIV infected patient on chronic dialysis with the combination of sofobuvir and velpatasvir. *Medicine* **2018**, *97*, e13671. [CrossRef] [PubMed]

37. Gaur, N.; Malhotra, V.; Agrawal, D.; Singh, S.; Beniwal, P.; Sharma, S.; Jhorawat, R.; Rathore, V.; Joshi, H. Sofosbuvir-velpatasvir fixed drug combination for the treatment of chronic hepatitis C infection in patients with end-stage renal disease and kidney transplantation. *J. Clin. Exp. Hepatol.* **2020**, *10*, 189–193. [CrossRef]

38. Taneja, S.; Duseja, A.; Mehta, M.; De, A.; Verma, N.; Premkumar, M.; Dhiman, R.; Singh, V.; Singh, M.; Ratho, R.; et al. Sofosbuvir and velpatasvir combination is safe and effective in treating chronic hepatitis C in end-stage renal disease on maintenance haemodialysis. *Liver. Int.* **2021**, *41*, 705–709. [CrossRef]

39. Liu, C.; Chen, C.; Su, W.; Tseng, K.; Lo, C.; Liu, C.; Chen, J.; Peng, C.; Shih, Y.; Yang, S.; et al. Sofosbuvir/velpatasvir with or without low-dose ribavirin for patients with chronic hepatitis C virus infection and severe renal impairment. *Gut* **2022**, *71*, 176–184. [CrossRef]

40. Yu, M.; Huang, C.; Wei, Y.; Lin, W.; Lin, Y.; Hsu, P.; Hsu, C.; Liu, T.; Lee, J.; Niu, S.; et al. Establishment of an outreach, grouping healthcare system to achieve micro-elimination of HCV for uremic patients in haemodialysis centres (ERASE-C). *Gut* **2020**, *70*, 2349–2358. [CrossRef]

41. Gohel, K.; Borasadia, P. Sosfobuvir-based HCV treatment in maintenance haemodialysis patients: A single-center study. *Transplant. Proc.* **2020**, *52*, 1684–1686. [CrossRef]

42. Mostafi, M.; Jabin, M.; Chowdhury, Z.; Khondoker, M.; Ali, S.; Tamanna, R.; Rezwan, R.; Alomgir, S. The outcome of daclatasvir and low dose sofosbuvir therapy in end-stage renal disease patients with hepatitis C virus infection. *Ukrainian K Nephrol Dial* **2020**, *2*, 3–8. [CrossRef]

43. De, A.; Roy, A.; Verma, N.; Mishra, S.; Premkumar, M.; Taneja, S.; Singh, V.; Duseja, A. Sofosbuvir plus velpatasvir combination for the treatment of chronic hepatitis C in patients with end-stage renal disease on renal replacement therapy: A systematic review and meta-analysis. *Nephrology* **2022**, *27*, 82–89. [CrossRef] [PubMed]

viruses

MDPI

Article

Characterization of Anti-Poliovirus Compounds Isolated from Edible Plants

Minetaro Arita [1,*] and Hiroyuki Fuchino [2,*]

[1] Department of Virology II, National Institute of Infectious Diseases, 4-7-1 Gakuen, Musashimurayama-shi 208-0011, Tokyo, Japan

[2] Research Center for Medicinal Plant Resources, National Institutes of Biomedical Innovation, Health and Nutrition, 1-2 Hachimandai, Tsukuba 305-0843, Ibaraki, Japan

* Correspondence: minetaro@niid.go.jp (M.A.); fuchino@nibiohn.go.jp (H.F.); Tel.: +81-42-561-0771 (M.A.); +81-29-838-0572 (H.F.); Fax: +81-42-561-4729 (M.A.); +81-29-838-0575 (H.F.)

Abstract: Poliovirus (PV) is the causative agent of poliomyelitis and is a target of the global eradication programs of the World Health Organization (WHO). After eradication of type 2 and 3 wild-type PVs, vaccine-derived PV remains a substantial threat against the eradication as well as type 1 wild-type PV. Antivirals could serve as an effective means to suppress the outbreak; however, no anti-PV drugs have been approved at present. Here, we screened for effective anti-PV compounds in a library of edible plant extracts (a total of 6032 extracts). We found anti-PV activity in the extracts of seven different plant species. We isolated chrysophanol and vanicoside B (VCB) as the identities of the anti-PV activities of the extracts of *Rheum rhaponticum* and *Fallopia sachalinensis*, respectively. VCB targeted the host PI4KB/OSBP pathway for its anti-PV activity (EC$_{50}$ = 9.2 μM) with an inhibitory effect on in vitro PI4KB activity (IC$_{50}$ = 5.0 μM). This work offers new insights into the anti-PV activity in edible plants that may serve as potent antivirals for PV infection.

Keywords: virus; picornavirus; enterovirus; antiviral; edible plant; PI4KB

check for
updates

Citation: Arita, M.; Fuchino, H. Characterization of Anti-Poliovirus Compounds Isolated from Edible Plants. *Viruses* **2023**, *15*, 903. https://doi.org/10.3390/v15040903

Academic Editor: Simone Brogi

Received: 27 February 2023
Revised: 29 March 2023
Accepted: 30 March 2023
Published: 31 March 2023

1. Introduction

Poliovirus (PV) is a small non-enveloped virus with a positive-sense single-stranded RNA genome of about 7500 nt belonging to the family *Picornaviridae*, including poliovirus (PV, species *Enterovirus C*) [1]. PV is the causative agent of poliomyelitis, which mainly affects children under 5 years of age, and is a target of global eradication by the World Health Organization (WHO). Through vaccination programs of the Global Polio Eradication Initiative beginning in 1988 with a live oral PV vaccine (OPV) and/or an inactivated PV vaccine (IPV), type 2 and 3 wild-type PVs (WPVs) have been eradicated (declared in 2015 and 2019, respectively), and only Pakistan and Afghanistan remain as endemic countries of type 1 WPV as of 2022. However, circulating vaccine-derived PV (cVDPV) remains a substantial threat against the eradication (724 cases in 2022), especially type 2 cVDPV that emerged after the global cessation of type 2 OPV in 2016 [2], as well as type 1 WPV (30 cases in 2022) [3]. Transmission of PV could be re-established by the importation of the following strains: polio cases by type 1 WPV in Malawi in 2021 and Mozambique in 2022, a case by type 2 cVDPV, and the silent circulation in the United States of America and [4] the United Kingdom in 2022 [5]. To interrupt PV circulation, only an OPV campaign for a potentially susceptible population in the area could serve as the effective mean at present (in case of an outbreak response in Israel, the target population was children under 10 years of age who have received at least one dose of IPV) [6]. In addition to conventional OPV, novel OPV type 2 (nOPV2), which was designed to have more genetic stability than type 2 OPV and to decrease the risk of VDPV [7], is available for the outbreak response under Emergency Use Listing of the WHO authorized in 2020. Currently, nOPV2 has mainly been used in African countries since 2021 (reviewed in [8]).

In the global eradication, antivirals for PV are anticipated to serve as an effective means to suppress cVDPV outbreaks and to treat patients chronically infected with PV [9,10]. As the candidate compounds, direct-acting antivirals targeting viral capsid proteins (inhibitors of viral binding/uncoating steps), proteases (viral 2A and 3C/3CD proteins), helicase (viral 2C protein), and polymerase (viral 3D protein) have been reported (reviewed in [11]). Host-targeting antivirals have also been reported, including inhibitors for host GBF1 (a guanine-nucleotide exchange factor) [12,13], eIF4A (protein synthesis) [14], HSP90 (folding of viral capsid proteins) [15], DHODH (de novo pyrimidine synthesis) [16–18], ribosome (protein synthesis) [19], PI4KB (phosphatidylinositol 4-monophosphate production) [20–27], and OSBP (exchanger of cholesterol and phosphatidylinositol 4-monophosphate) [28–32]. However, there is no antiviral available for PV infection at present.

Target population of PV is mainly for children under five years of age as well as other enteroviruses (EVs) [33,34]; therefore, safety is one of the major challenges for the antiviral development [10]. In a previous study, a highly active anti-EV compound (EC_{50} = 2.0 µM) was isolated from avocado [35], suggesting that edible plants provide a promising source for potent anti-EV compounds. Here, we isolated anti-PV compounds from edible plants, *Rheum rhaponticum* and *Fallopia sachalinensis*, and analyzed the potency and the mechanism of action of their isolated compounds.

2. Materials and Methods

Cells. Cells were cultured as monolayers in Dulbecco's modified Eagle medium (DMEM, FUJIFILM Wako Pure Chemical Corporation, Osaka, JPN, 044-29765) supplemented with 10% foetal calf serum (FCS). RD cells (human rhabdomyosarcoma cells) were used for the titration of PV and evaluation of anti-PV activity of plant extracts. HEK293 cells (human embryonic kidney cells) were used for the production of type 1 PV pseudovirus ($PV1_{pv}$). A *PI4KB*-knockout RD cell line (RD[$\Delta PI4KB$]) was used to evaluate the antiviral effects of plant extracts targeting PI4KB/OSBP-independent viral replication [36].

Viruses. Type 1 PV Sabin 1 strain (PV1[Sabin 1]) (GenBank: AY184219), type 3 PV Sabin 3 strain (PV3[Sabin 3]) (GenBank: AY184221), EV-A71 (Nagoya) (GenBank: AB482183), and EV-D68 (Fermon) (GenBank: AY426531) were used for the screening of plant extracts. Luciferase-encoding Sendai virus (SeV-luc) was a kind gift from Atsushi Kato. $PV1_{pv}$ mutants were produced with a firefly luciferase-encoding type 1 PV Mahoney strain (PV1[Mahoney]) (GenBank: V01149) replicon and the capsid proteins of PV1(Mahoney) [37]. The $PV1_{pv}$ mutants used in this study are as follows: an enviroxime (PI4KB inhibitor)-resistant mutant (with a G5318A [3A-Ala70Thr] mutation)[21], a guanidine hydrochloride (GuaHCl) (viral 2C helicase inhibitor)-resistant mutant (with a U4614A [2C-Phe164Tyr] mutation) [38], a brefeldin A (GBF1 inhibitor)-resistant mutant (with a G4361A [2C-Val80Ile] mutation and a C5190U [3A-Ala27Val] mutation) [39], a rupintrivir (viral 3C protease inhibitor)-resistant mutant (with a G5819A [3C-Gly128Ser] mutation) [40], a disoxaril (viral capsid-binding uncoating inhibitor)-resistant mutant (with an A3059U [VP1-Ile194Phe] mutation)[41], the $\Delta PI4KB$-resistant [−2C] mutant (a PI4KB/OSBP-independent mutant) (with a U3623C [2A-Phe80Leu] mutation, a U3881C [2B-Phe17Leu] mutation, a G3892U [2B-Gln20His] mutation, an A5269U [3A-Glu53Asp] mutation, and an A5270U [3A-Arg54Trp] mutation) [36]. Plasmids for the rupintrivir-resistant mutant and the disoxaril-resistant mutant were constructed in this study as below.

Chemicals. Chrysophanol and vanicoside B (VCB) were obtained from the roots of *R. rhaponticum* and of *F. sachalinensis*, respectively (purity >95% determined by NMR, HPLC). MDL-860 was a kind gift from Angel S. Galabov (purity >99.5%, determined by NMR) [42].

General methods for molecular cloning. *Escherichia coli* strain XL10gold (Agilent Technologies, Inc., Santa Clara, CA, USA) was used for the preparation of plasmids. PCR was performed using KOD Plus DNA polymerase (TOYOBO CO., LTD., Osaka, Japan). DNA sequencing was performed using a BigDye Terminator v3.1 cycle sequencing ready reaction kit (Thermo Fisher Scientific Inc., Waltham, MA, USA) and then analyzed with a 3500xL genetic analyzer (Thermo Fisher Scientific Inc., Waltham, MA, USA).

Plasmids:

Rupintrivir-resistant PV replicon. A resistant mutation to rupintrivir was introduced into a plasmid encoding the cDNA of a PV replicon (pPV-Fluc mc) [43], by PCR with primer set 1.

Primer set 1:

5′-GGATATCTAAATCTCAGTGGGCGCCAAAC-3′

5′-GTTTGGCGCCCACTGAGATTTAGATATCC-3′

Disoxaril-resistant PV capsid expression vector. A resistant mutation to disoxaril was introduced into a plasmid expression vector for the capsid protein of PV1(Mahoney) [37], by PCR with primer set 2.

Primer set 2:

5′-CAGCTCCAGCCCGGTTCTCGGTACCGTATG-3′

5′-CATACGGTACCGAGAACCGGGCTGGAGCTG-3′

Edible plant extract library. A plant extract library was prepared in the Research Center for Medicinal Plant Resources (NIBIOHN), by the methanol extraction of dried and pulverized plant materials, with evaporation, dissolution in DMSO, and filtration. The concentration of the final DMSO solution was adjusted to 40 mg/mL for all extracts. This plant extract library is a collection of extracts from a wide range of wild plants in Japan, which can be widely used not only in drug discovery but also in the life sciences field, such as health food development. All the original plants in the library are annotated with information on whether or not they have been eaten by humans. The presence of food experience in the original plant reflects the safety of the extract for human beings, which is useful information for drug discovery. We selected samples from this library derived from the original plants with food experience and used them in this study.

Screening for anti-PV compounds from edible plant extract library. RD cells (2.0×10^4 cells per well in 50 µL medium) were cultured in 96-well plates, and then infected with PV1(Sabin 1) (2000 50% cell culture infectious dose [$CCID_{50}$]) in the presence of a plant extract (final 0.2 mg/mL) (total 200 µL/well). The cells were incubated at 37 °C, and then observed for CPE at 1, 2, 3, and 7 days post-infection (p.i.).

Measurement of cytotoxicity and anti-PV activity of compounds. For the measurement of cytotoxicity, RD cells (8×10^3 cells per well in 20 µL medium) were cultured at 37 °C in 384-well plates (781,080, Greiner Bio-One, Kremsmünster, Austria), followed by the addition of 20 µL of a compound solution at an indicated final concentration. The cells were incubated at 37 °C for 7 h or 2 days and then the cell viability was measured by using a CellTiter-Glo 2.0 Cell Viability Assay kit (G9241, Promega Corporation, Madison, WI, USA) using a 2030 ARVO X luminometer (PerkinElmer, Waltham, MA, USA). The 50% cytotoxic concentration (CC_{50}) values were determined by a nonlinear regression analysis of the dose–response curves.

For the measurement of anti-PV activity with $PV1_{pv}$, RD cells (8×10^3 cells per well in 20 µL of medium) in 384-well plates (781,080, Greiner Bio-One, Kremsmünster, Austria) were inoculated with 10 µL of $PV1_{pv}$ (800 infectious units [IU]) and 10 µL of compound solution at an indicated final concentration. The cells were incubated at 37 °C for 7 h. Luciferase activity in the infected cells was measured at 7 h p.i. with the Steady-Glo luciferase assay system (Promega Corporation, Madison, WI, USA) using a 2030 ARVO X luminometer (PerkinElmer, Waltham, MA, USA). The 50% effective concentration (EC_{50}) values were determined via a nonlinear regression analysis of the dose–response curves.

For the measurement of anti-PV activity of VCB, RD cells (2.8×10^4 cells per well in 50 µL of medium) were cultured in 96-well plates, and then infected with PV1(Sabin 1) (100 $CCID_{50}$), EV-A71(Nagoya) (1000 $CCID_{50}$), and EV-D68(Fermon) (10^4 $CCID_{50}$) in the presence of VCB at an indicated final concentration (total 200 µL/well). The cells were incubated at 37 °C for 2 days (for PV), 35 °C for 4 days (for EV-A71), or 33 °C for 4 days (for EV-D68). The cells were then fixed and stained with formalin and crystal violet (final concentration of 5% and 0.25%, respectively).

Measurement of inhibitory effect of compounds on in vitro PI4KB activity. The in vitro activity of purified GST-PI4KB (PV5277, Thermo Fisher Scientific Inc., Waltham, MA, USA) was evaluated by using an ADP-Glo Lipid Kinase Systems kit (Promega Corporation, Madison, WI, USA) as previously described. In a total 5.5 μL reaction solution, the PI4KB activity of 32 ng of purified GST-PI4KB (final concentration of 48 nM) with lipid substrates (0.025 mg/mL of phosphatidylinositol and 0.075 mg/mL or phosphatidylserine) and 25 μM of ATP was measured in the presence or the absence of compounds. The net signals of the mock-treated samples were taken as 100% of the PI4KB activity. The 50% inhibitory concentration (IC_{50}) values of the compounds were determined by nonlinear regression analyses of the dose–response curves.

Statistical analysis. The results of the experiments are shown as means with standard deviations. Values of $p < 0.05$ by one-tailed t tests were considered to indicate a significant difference, and were indicated by asterisks (* $p < 0.05$, ** $p < 0.01$, *** $p < 0.001$).

3. Results

3.1. Screening of Edible Plant Extracts for Anti-PV Activity

We screened a total of 6032 edible plant extracts for anti-PV activity in RD cells as that previously performed for the screening for anti-EV-D68 activity [35] (Figure 1). The plant extracts were added to the RD cells (final concentration of 0.2 mg/mL), and then infected with PV1(Sabin 1) at a multiplicity of infection (MOI) of 0.1. The extracts that completely protected the cells from the viral infection after 1 day post-infection (p.i.) were identified as initial hit extracts. We identified 8 hit extracts, which consisted of 7 plant species. In this study, we focused on the identification of the identity of the antiviral effects in *R. rhaponticum* and *F. sachalinensis*, among the hits for availability of the plant materials.

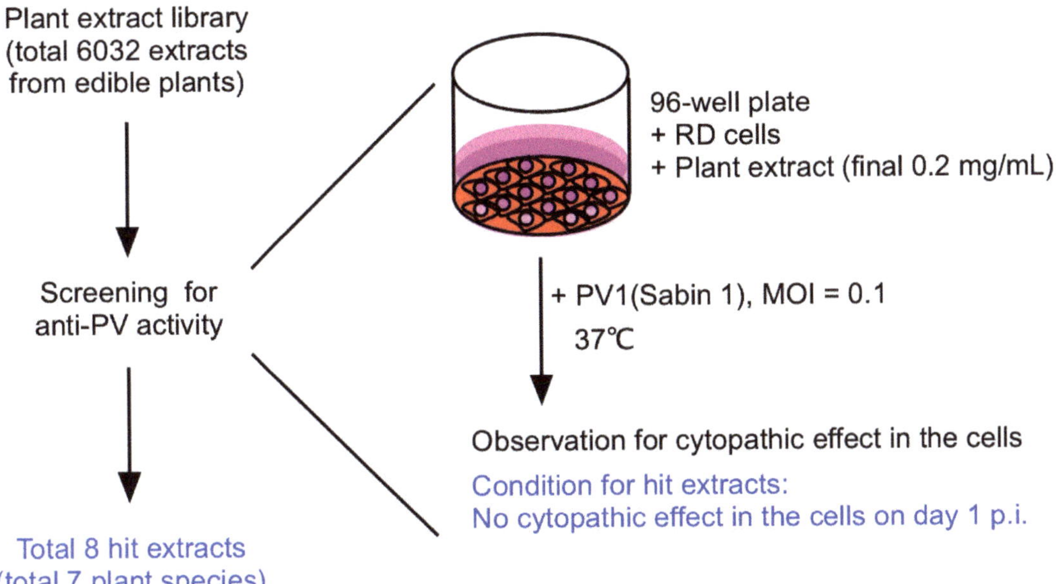

Figure 1. Summary of the screening of anti-PV compound from edible plant extract library.

3.2. Purification and Structure Determination of Anti-PV Compound in R. Rhaponticum

Methanolic extracts of *R. rhaponticum* (*Polygonaceae*) root was partitioned with *n*-hexane, ethyl acetate, *n*-buthanol, and water, successively. The *n*-hexane layer was purified via silica-gel column chromatography, then, HPLC to give chrysophanol (21 mg) [44], and 6-*O*-methylemodin (4.2 mg) [45] (Figure 2). The ethyl acetate layer was separated by silica-gel column chromatography and HPLC (Supplementary data 1), repeatedly, to obtain rhapontigenin (3.7 mg) [46], *trans*-resveratrol (3.1 mg) [47], pulmatin (0.9 mg) [48], 4-methylresveratrol-3-glucopyranoside (6.5 mg) [49], ε-viniferin (4.2 mg) [50], δ-viniferin (11.9 mg) [51], and deoxyrhapontigenin (3.0 mg) [46]. Their chemical structures were determined by NMR and LC/MS (Supplementary data 2). The major active component in the active fractions was revealed to be chrysophanol (Figure 2A). Chrysophanol has been reported as the anti-PV component of an Australian medicinal plant *Dianella longifolia*, and targets the early stage of PV infection [52]. The antiviral effect of chrysophanol was specific to PV; no antiviral effect was observed on the infection of EV-A71 or EV-D68 (Figure S1). The EC_{50} of chrysophanol for type 1 PV pseudovirus ($PV1_{pv}$) infection was 8.0 μM; however, the infection could be suppressed only moderately even at 790 μM (12% of that in mock-treated cells), (Figure 2B). Consistent with a previous report, a disoxaril (viral capsid-binding uncoating inhibitor)-resistant mutant showed substantial resistance to purified chrysophanol (Figure 2B). This suggested that chrysophanol was the major identity for the anti-PV activity of the *Rheum rhaponticum* extract and targeted the PV capsid protein.

3.3. Purification and Structure Determination of Anti-PV Compound in F. Sachalinensis

The methanolic extracts of the *F. sachalinensis* (*Polygonaceae*) root was purified by silica-gel column chromatography with chloroform–methanol as an eluent to give 20 fractions (Figure 3A). Fractions eluted with 50% methanol/chloroform were combined and subjected to HPLC separation to obtain vanicoside B (VCB, 1.1 mg). The chemical structure was determined via comparison with NMR data from the literature [53]

3.4. Characterization of Anti-PV Activity of VCB

To evaluate the potency of VCB as an anti-PV compound, we determined the 50% cytotoxic concentration (CC_{50}) in human RD cells and 50% effective concentration (EC_{50}) for the PV infection of VCB (Figure 3B). The cytotoxicity of VCB was not observed when the cells were treated with 100 μM VCB for 7 h, but the CC_{50} of VCB after 2 days of treatment was 27 μM. The EC_{50} of VCB for $PV1_{pv}$ infection was 9.2 μM. The selectivity index (SI) of VCB for anti-PV activity in RD cells was 2.9, suggesting a low specificity for the anti-PV activity of VCB (e.g., SI of PI4KB inhibitors for the anti-PV activities could be >1000) [25]. VCB protected the RD cells from the infection of PV1(Sabin 1), EV-A71(Nagoya), or EV-D68(Fermon) only at 20 μM, suggesting that the potential therapeutic window of VCB is quite narrow.

3.5. Mechanism of Anti-PV Activity of VCB

To evaluate the specificity of the anti-PV activity for VCB, we analyzed the antiviral activity with a panel of drug-resistant PV mutants in parental (wild-type) RD cells (RD[WT]) and *PI4KB*-knockout RD cells (RD[Δ*PI4KB*]) (Figure 4A). The panel included resistant mutants to the direct-acting antivirals guanidine hydrochloride (GuHCl) (viral 2C helicase inhibitor), rupintrivir (viral 3C protease inhibitor), and disoxaril (viral capsid-binding uncoating inhibitor) and to the host-targeting antivirals brefeldin A (host GBF1 inhibitor), enviroxime (host PI4KB inhibitor), and a PI4KB/OSBP-independent mutant (Δ*PI4KB*-resistant [−2C]). The antiviral effect of VCB on the infection of the $PV1_{pv}$ mutants was analyzed in RD(WT) cells, except for that of the Δ*PI4KB*-resistant (−2C) mutant. The potential antiviral effects on PI4KB/OSBP-independent replication was analyzed in the infection of the Δ*PI4KB*-resistant (−2C) mutant in RD(Δ*PI4KB*) cells. Among the mutants, the enviroxime-resistant mutant and Δ*PI4KB*-resistant (−2C) mutant showed significant resistance to VCB, suggesting that VCB targets the host PI4KB/OSBP pathway in PV replication.

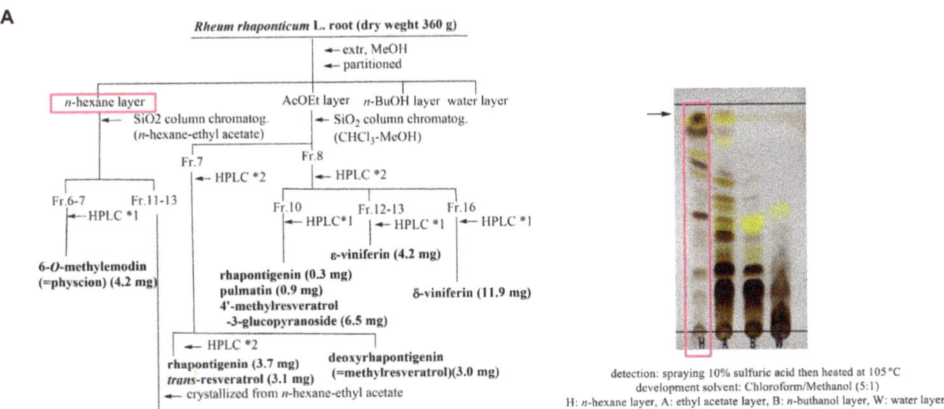

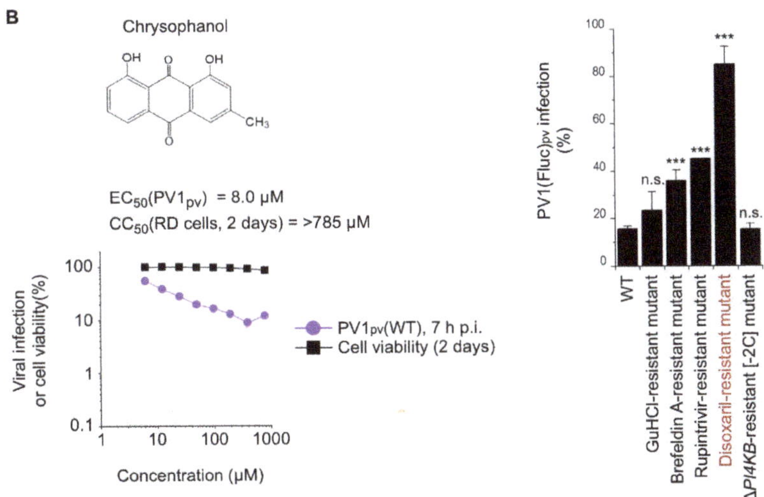

Figure 2. Identification and purification of an anti-PV compound from *R. rhaponticum* extract. (**A**) Procedure of purification of anti-PV compound from *R. rhaponticum* extract. Boxes highlight the fraction containing chrysophanol. Arrow indicates the band of chrysophanol. (**B**) Antiviral effect of chrysophanol on PV infection. (Left) Structure, cytotoxicity, and antiviral effect of chrysophanol. Viral infection and viability of RD cells in the presence of chrysophanol are shown. RD cells were infected with $PV1_{pv}$, then luciferase activity was measured at 7 h p.i. Marked precipitation of chrysophanol was observed above 98 μM. Viral infection or the cell viability in the absence of chrysophanol were taken as 100%. (Right) Inhibitory effect of chrysophanol on the infection of a panel of drug-resistant $PV1_{pv}$ mutants. $PV1_{pv}$ infection at 7 h p.i. in RD(WT) cells in the presence or absence of chrysophanol (790 μM) is shown. $PV1_{pv}$ infection in the absence of chrysophanol is taken as 100%. n.s., not significant. ***, $p < 0.001$.

A

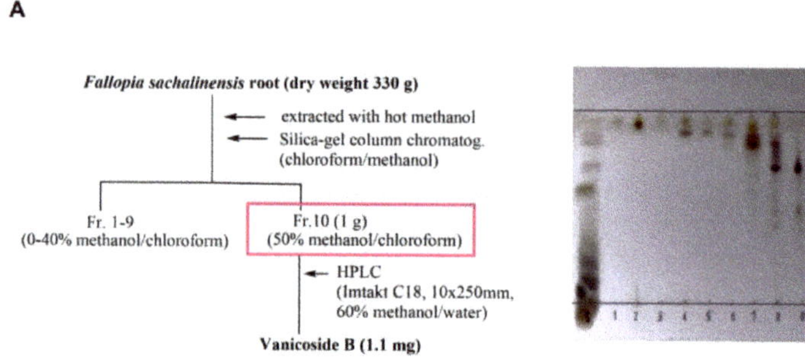

B

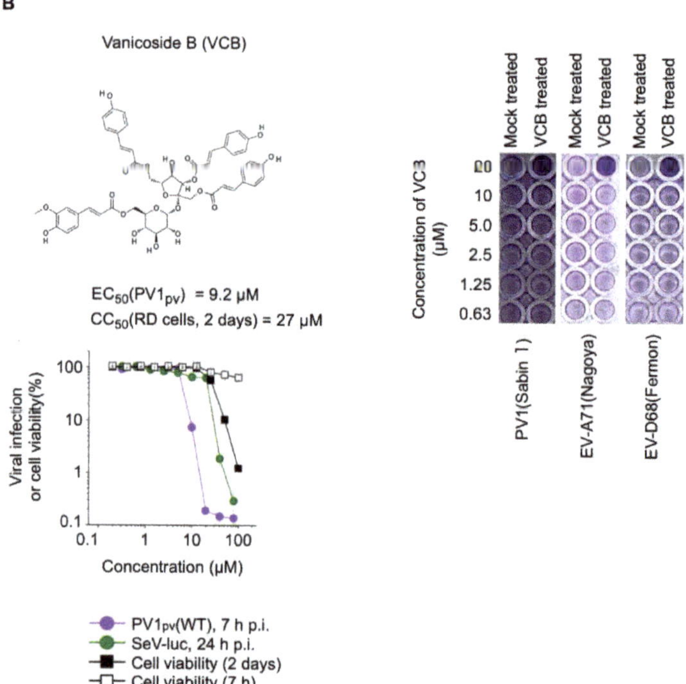

Figure 3. Identification and purification of an anti-PV compound from *F. sachalinensis* extract. (**A**) Procedure of purification of anti-PV compound from *F. sachalinensis* root extract. Boxes highlight the fraction containing vanicoside B. (**B**) Antiviral effect of VCB on PV infection. (Left) Structure of VCB, cytotoxicity, and antiviral activity of VCB. Viral infection and viability of RD cells in the presence of VCB are shown. RD cells were infected with $PV1_{pv}$ or SeV-luc, then luciferase activity was measured at 7 h p.i. (for $PV1_{pv}$) or 24 h p.i. (for SeV-luc). Viral infection or the cell viability in the absence of VCB were taken as 100%. (Right) Antiviral effect of VCB on EV infection. RD cells on 96-well plates were infected with EV at an MOI of 0.005 (for PV), 0.05 (for EV-A71) or 0.5 (for EV-D68) in the absence (mock-treated) or the presence (VCB treated) of VCB at the indicated final concentration. The cells were fixed and stained at 2 days p.i. (for PV) or 4 days p.i. (for EV-A71 and EV-D68). The data are representative of two independent experiments with two to three biological replicates.

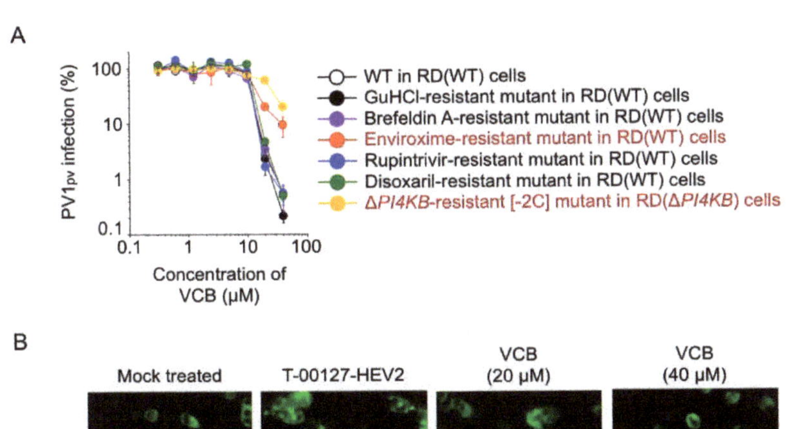

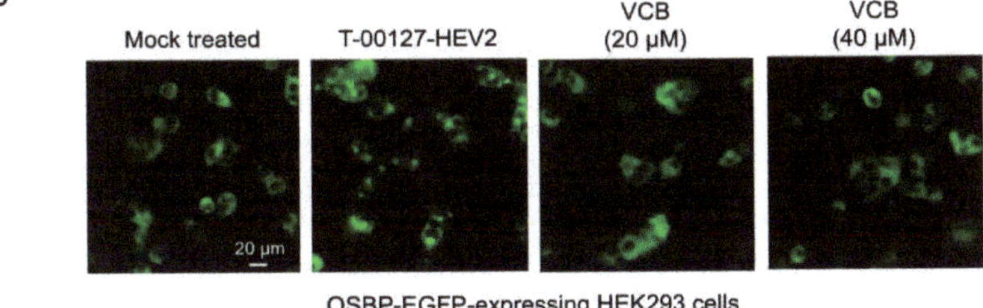

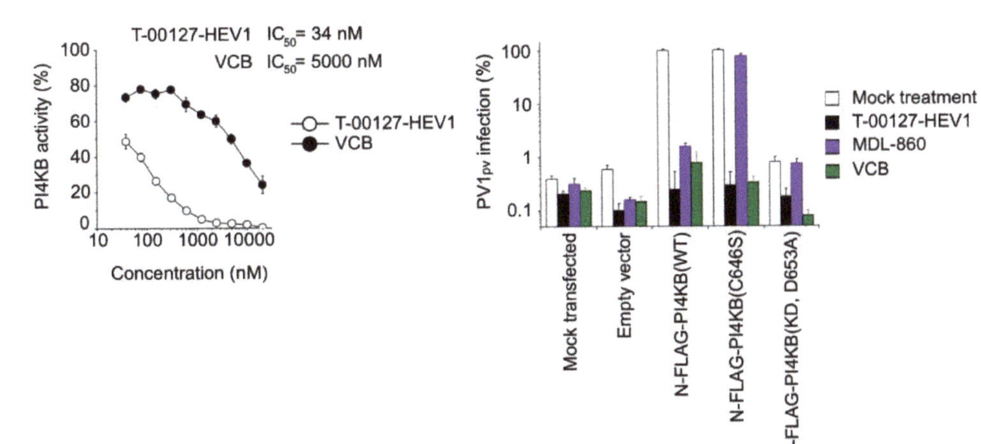

Figure 4. VCB is a novel PI4KB inhibitor. (**A**) Inhibitory effect of VCB on the infection of a panel of drug-resistant $PV1_{pv}$ mutants. $PV1_{pv}$ infection at 7 h p.i. in RD(WT) cells in the presence or absence of VCB is shown, except for the infection of the Δ*PI4KB*-resistant [−2C] mutant, which was analyzed at 17 h p.i. in RD(Δ*PI4KB*) cells. $PV1_{pv}$ infection in the absence of VCB is taken as 100%. (**B**) OSBP-EGFP-expressing HEK293 cells were incubated at 37 °C for 30 min in the presence of the compounds (10 μM T-00127-HEV2, 20 or 40 μM VCB, respectively). Localization of OSBP-EGFP in the cells is shown. (**C**) (Left) Inhibitory effect of VCB on in vitro PI4KB activity. (Right) *Trans*-rescue of PV infection in RD(Δ*PI4KB*) cells by indicated PI4KB variants in the presence of the compounds (20 μM T-00127-HEV1, 40 μM MDL-860, or 20 μM VCB, respectively). $PV1_{pv}$ infection in RD(Δ*PI4KB*) cells transfected with pTK-N-FLAG-PI4KB(WT, C646S, or D653A variants) in the absence of the compound is taken as 100%. The data are representative of two independent experiments with three biological replicates.

To further dissect the target of VCB, we analyzed the effect of VCB on the subcellular localization of host OSBP. OSBP relocalizes to the Golgi in the presence of OSBP inhibitors via the lipid-transfer domain [30,54,55]. HEK293 cells overexpressing C-terminally EGFP-fused OSBP were treated with an OSBP inhibitor T-00127-HEV2 or VCB (Figure 4B). While treatment of the cells with T-00127-HEV2 caused the relocalization of the ectopically expressed OSBP to the Golgi, the treatment with VCB did not affect the subcellular localization, suggesting that OSBP is not the target of VCB. Next, we analyzed the effect of VCB on the PI4KB activity (Figure 4C). VCB showed an inhibitory effect on the in vitro PI4KB activity albeit with low potency (IC_{50} = 5000 nM) compared to a PI4KB inhibitor T-00127-HEV1 (IC_{50} = 34 nM), which possibly targets the ATP-binding site of PI4KB, similar to its analogue [26]. We also analyzed a potential allosteric effect of VCB on the PI4KB activity. VCB inhibited the activity of a PI4KB variant (C646S) as well as T-00127-HEV1, in contrast to MDL-860 that has an allosteric inhibitory effect on PI4KB via a covalent modification of the Cys646 residue [27,56]. These results suggested that PI4KB is the direct target of VCB for the anti-PV activity.

4. Discussion

Several potent anti-EV compounds have been isolated from plants: pachypodol (Ro 09-0179) (PI4KB inhibitor) [57], oxoglaucine (PI4KB inhibitor) [58,59], chrysin (viral 3C protease inhibitor) [60], prunin (viral protein synthesis inhibitor) [61], and avoenin (viral capsid-binding uncoating inhibitor) [35]. However, the availability of these plant-derived compounds for treatment or prophylactic use in PV infection has yet to be established. In the present study, we isolated chrysophanol and vanicoside D (VCD) as anti-PV compounds from an *R. rhaponticum* extract and *F. sachalinensis* extract, respectively.

The petiole of *R. rhaponticum* is edible, called rhubarb, and is used mainly as a jam. On the other hand, the roots of many *Rheum* spp. are considered medicinal and laxative because they contain many anthraquinones. *F. sachalinensis* is a large herbaceous plant whose grass can grow up to 2 m tall. The Ainu tribe of Japan traditionally eats its young stems and sprouts. The rhizome of *F. sachalinensis* has antibacterial, antitussive, and diuretic properties, as well as an improvement in its laxative effects. The Japanese name for this plant ("ooitadori") is derived from the fact that when bruised, its leaves can be applied to the affected area to relief pain. Approximately 360 g of the root of *R. rhaponticum* contained 0.2 g of chrysophanol (about 0.05%w/w). The content of VCB in the roots of *F. sachalinensis* dry was estimated to be about 0.001%w/w, based on the weight of the isolate and its presence in other fractions. Although the roots of these two plant species are not edible parts, other parts have been traditionally consumed. Therefore, a certain degree of safety is considered to be assured.

Chrysophanol is a well-known purgative component of the roots of the *Rheum* spp. and Senna leaf (the leaves of *Cassia angustifolia, C. acutifolia*). It has also been reported as a constituent of *Fallopia japonica*, a closely related plant to *F. sachalinensis*. Chrysophanol is also known as an anti-PV compound [52]. Emodin, a compound structurally related to chrysophanol, has anti-EV activity targeting viral protein synthesis or virion maturation [62,63]. Chrysophanol targeted a site of a PV capsid protein similar to the capsid-binding uncoating inhibitors; however, the inhibitory effect of the chrysophanol on $PV1_{pv}$ infection was rather weak even at a high concentration (about an 8-fold reduction at 790 μM, Figure 2B). The infection cycle of $PV1_{pv}$ includes viral binding, uncoating, and replication, but not virion production (assembly, encapsidation, maturation, and egress) [37]; therefore, chrysophanol may have additional targets after the replication step, such as virion maturation, similar to emodin. The partial resistance of a brefeldin A-resistant mutant and a rupintrivir-resistant mutant against chrysophanol suggest the effects on the replication step that could have functional coupling to virion production [64,65]. Pocapavir (viral capsid-binding uncoating inhibitor) [66,67] and V-7404 (3C protease inhibitor) [68–70] have been considered as candidate antivirals in the polio eradication program [71,72]. The availability of chrysophanol in a broad plant species would allow further evaluation of the potency of the extracts, possibly in combination with other drugs/extracts with different antiviral mechanisms.

VCB was first isolated from nature in 1994 as a protein kinase C inhibitor from *Polygonum pensylvanicum* (*Polygonaceae*) together with vanicoside A [73]. Vanicosides have inhibitory effects on the viral proteases of the human immunodeficiency virus or SARS-CoV-2 [74,75], but its antiviral effects have yet to be evaluated. Unexpectedly, resistant mutants (the enviroxime-resistant mutant and Δ*PI4KB*-resistant [−2C] mutant) suggested that the target of VCB for the anti-PV activity is the host PI4KB/OSBP pathway rather than viral proteases (Figure 3). The inhibitory effect on in vitro PI4KB activity (IC_{50} = 5.0 μM) suggested that VCB is a novel PI4KB inhibitor. PI4KB is a host factor required for the replication of EV identified by Hsu et al. [22]. The subsequent analysis on a group of anti-EV compounds (designated enviroxime-like compounds), which have PI4KB and an unknown factor as the targets for the anti-EV activity [23,29], revealed the host oxysterol-binding protein (OSBP) family I (OSBP and OSBP2/ORP4) as another target of this compound group [30,31]. PI4KB and OSBP form an inseparable functional axis for the formation of a viral replication complex by providing phosphatidylinositol 4-monophosphate (PI4P) for the recruitment of OSBP on viral replication organelles (ROs), and the accumulation of unesterified cholesterol on the ROs by OSBP [76]. This process enhances the cleavage of the viral 3AB protein and development of the RO for the synthesis of viral plus-strand RNA [36,57,77–79]. In addition to the 3AB protein, the viral 2B protein is essential to complement the functional axis [36,65], while the functional role of the 2B protein remains largely unknown. PI4KB inhibitors generally show low cytotoxicity to cultured cells [20,21,23,28,43,80,81]; however, the antiproliferative effect in lymphocytes [80] and lethality in a mouse line [81] raised concerns on the safety in vivo. A recent study revealed a protective effect in PI4KB heterozygous kinase-dead mice against EV infection and the therapeutic potency of a specific PI4KB inhibitor in vivo [82], supporting the potential safety of PI4KB inhibitors in clinical use as opposed to earlier findings. Therefore, the PI4KB inhibitors isolated from edible plants could have a more important role than ever thought.

The limitations of this study include elucidation of the mechanism of the inhibitory effect of VCB on PI4KB activity and its off-target effect against clinical applicability. Most of the identified PI4KB inhibitors target the ATP-binding site of PI4KB [26,83], with MDL-860 as the exception. The inhibitory effect of VCB on in vitro PI4KB activity might suggest the direct interaction with PI4KB (Figure 3C), but the target site remained to be determined. While the specificity to the PI4KB/OSBP pathway in terms of the anti-PV activity was clear, we could not exclude the potential contribution of the off-target effect of VCB (CC_{50} = 27 μM) to the observed anti-PV activity, which had quite a narrow therapeutic window (complete protection of the cells from PV1[Sabin 1] infection at 20 μM).

Supplementary Materials: The following supporting information can be downloaded at: https://www.mdpi.com/article/10.3390/v15040903/s1, Figure S1: Antiviral effect of chrysophanol on EV infection. Supplementary data 1: HPLC chart. Supplementary data 2: NMR spectra.

Author Contributions: Conceptualization, M.A. and H.F.; methodology, M.A. and H.F.; software, M.A. and H.F.; validation, M.A. and H.F.; formal analysis, M.A. and H.F.; investigation, M.A. and H.F.; resources, M.A. and H.F.; data curation, M.A. and H.F.; writing—original draft preparation, M.A. and H.F.; writing—review and editing, M.A. and H.F.; visualization, M.A. and H.F.; supervision, M.A. and H.F.; project administration, M.A. and H.F.; funding acquisition, M.A. and H.F. All authors have read and agreed to the published version of the manuscript.

Funding: This study was in part supported by AMED (Grant number: 22fk0108627j0001), and by JSPS KAKENHI (Grant Number: 22K07107) to M.A and by AMED (Grant number: 22fk0108627h1101) to H.F. The funders had no role in the study design, data collection and analysis, decision to publish, or preparation of the manuscript.

Institutional Review Board Statement: Not applicable.

Informed Consent Statement: Not applicable.

Data Availability Statement: Raw data sets not included in this paper are available from the corresponding authors upon request.

Acknowledgments: We are grateful to Yuzuru Aoi for her excellent technical assistance. We thank Angel S. Galabov for kindly providing MDL-860, Atsushi Kato for luciferase-encoding the Sendai virus, and Masamichi Muramatsu for his kind supports.

Conflicts of Interest: The authors declare no conflict of interest.

List of Abbreviations

$CCID_{50}$	50% cell culture infectious dose
cVDPV	circulating vaccine-derived poliovirus
EC_{50}	50% effective concentration
EV	enterovirus
GuHCl	guanidine hydrochloride
IC_{50}	50% inhibitory concentration
IPV	inactivated PV vaccine
MOI	multiplicity of infection
OPV	oral PV vaccine
OSBP	oxysterol-binding protein
P.i.	post-infection
PI4P	phosphatidylinositol 4-monophosphate
PV	poliovirus
$PV1_{pv}$	type 1 PV pseudovirus
RO	replication organelle
SeV-luc	luciferase-encoding Sendai virus
SI	selectivity index
VCB	vanicoside B

References

1. Kitamura, N.; Semler, B.L.; Rothberg, P.G.; Larsen, G.R.; Adler, C.J.; Dorner, A.J.; Emini, E.A.; Hanecak, R.; Lee, J.J.; van der Werf, S.; et al. Primary structure, gene organization and polypeptide expression of poliovirus RNA. *Nature* **1981**, *291*, 547–553. [CrossRef] [PubMed]
2. Macklin, G.R.; O'Reilly, K.M.; Grassly, N.C.; Edmunds, W.J.; Mach, O.; Santhana Gopala Krishnan, R.; Voorman, A.; Vertefeuille, J.F.; Abdelwahab, J.; Gumede, N.; et al. Evolving epidemiology of poliovirus serotype 2 following withdrawal of the serotype 2 oral poliovirus vaccine. *Science* **2020**, *368*, 401–405. [CrossRef] [PubMed]
3. WHO. Available online: https://polioeradication.org/polio-today/polio-now/this-week/circulating-vaccine-derived-poliovirus/ (accessed on 27 February 2023).
4. Link-Gelles, R.; Lutterloh, E.; Schnabel Ruppert, P.; Backenson, P.B.; St George, K.; Rosenberg, E.S.; Anderson, B.J.; Fuschino, M.; Popowich, M.; Punjabi, C.; et al. Public Health Response to a Case of Paralytic Poliomyelitis in an Unvaccinated Person and Detection of Poliovirus in Wastewater—New York, June-August 2022. *MMWR Morb. Mortal Wkly. Rep.* **2022**, *71*, 1065–1068. [CrossRef] [PubMed]
5. Klapsa, D.; Wilton, T.; Zealand, A.; Bujaki, E.; Saxentoff, E.; Troman, C.; Shaw, A.G.; Tedcastle, A.; Majumdar, M.; Mate, R.; et al. Sustained detection of type 2 poliovirus in London sewage between February and July, 2022, by enhanced environmental surveillance. *Lancet* **2022**, *400*, 1531–1538. [CrossRef]
6. Anis, E.; Kopel, E.; Singer, S.R.; Kaliner, E.; Moerman, L.; Moran-Gilad, J.; Sofer, D.; Manor, Y.; Shulman, L.M.; Mendelson, E.; et al. Insidious reintroduction of wild poliovirus into Israel, 2013. *Eurosurveillance* **2013**, *18*, 20586. [CrossRef]
7. Yeh, M.T.; Bujaki, E.; Dolan, P.T.; Smith, M.; Wahid, R.; Konz, J.; Weiner, A.J.; Bandyopadhyay, A.S.; Van Damme, P.; De Coster, I.; et al. Engineering the Live-Attenuated Polio Vaccine to Prevent Reversion to Virulence. *Cell Host Microbe* **2020**, *27*, 736–751.e8. [CrossRef]
8. Bandyopadhyay, A.S.; Zipursky, S. A novel tool to eradicate an ancient scourge: The novel oral polio vaccine type 2 story. *Lancet Infect. Dis.* **2023**, *23*, e67–e71. [CrossRef]
9. Collett, M.S.; Neyts, J.; Modlin, J.F. A case for developing antiviral drugs against polio. *Antiviral Res.* **2008**, *79*, 179–187. [CrossRef]
10. Committee on Development of a Polio Antiviral and Its Potential Role in Global Poliomyelitis Eradication, National Research Council. *Exploring the Role of Antiviral Drugs in the Eradication of Polio: Workshop Report*; The National Academies Press: Washington, DC, USA, 2006.

11. De Palma, A.M.; Vliegen, I.; De Clercq, E.; Neyts, J. Selective inhibitors of picornavirus replication. *Med. Res. Rev.* **2008**, *28*, 823–884. [CrossRef]
12. Maynell, L.A.; Kirkegaard, K.; Klymkowsky, M.W. Inhibition of poliovirus RNA synthesis by brefeldin A. *J. Virol.* **1992**, *66*, 1985–1994. [CrossRef]
13. Irurzun, A.; Perez, L.; Carrasco, L. Involvement of membrane traffic in the replication of poliovirus genomes: Effects of brefeldin A. *Virology* **1992**, *191*, 166–175. [CrossRef] [PubMed]
14. Bordeleau, M.E.; Mori, A.; Oberer, M.; Lindqvist, L.; Chard, L.S.; Higa, T.; Belsham, G.J.; Wagner, G.; Tanaka, J.; Pelletier, J. Functional characterization of IRESes by an inhibitor of the RNA helicase eIF4A. *Nat. Chem. Biol.* **2006**, *2*, 213–220. [CrossRef] [PubMed]
15. Geller, R.; Vignuzzi, M.; Andino, R.; Frydman, J. Evolutionary constraints on chaperone-mediated folding provide an antiviral approach refractory to development of drug resistance. *Genes Dev.* **2007**, *21*, 195–205. [CrossRef]
16. Hoffmann, H.H.; Kunz, A.; Simon, V.A.; Palese, P.; Shaw, M.L. Broad-spectrum antiviral that interferes with de novo pyrimidine biosynthesis. *Proc. Natl. Acad. Sci. USA* **2011**, *108*, 5777–5782. [CrossRef]
17. Lee, K.; Kim, D.E.; Jang, K.S.; Kim, S.J.; Cho, S.; Kim, C. Gemcitabine, a broad-spectrum antiviral drug, suppresses enterovirus infections through innate immunity induced by the inhibition of pyrimidine biosynthesis and nucleotide depletion. *Oncotarget* **2017**, *8*, 115315–115325. [CrossRef]
18. Zhang, Z.; Yang, E.; Hu, C.; Cheng, H.; Chen, C.Y.; Huang, D.; Wang, R.; Zhao, Y.; Rong, L.; Vignuzzi, M.; et al. Cell-Based High-Throughput Screening Assay Identifies 2′,2′-Difluoro-2′-deoxycytidine Gemcitabine as a Potential Antipoliovirus Agent. *ACS Infect. Dis.* **2017**, *3*, 45–53. [CrossRef]
19. Grollman, A.P. Inhibitors of protein biosynthesis. V. Effects of emetine on protein and nucleic acid biosynthesis in HeLa cells. *J. Biol. Chem.* **1968**, *243*, 4089–4094. [CrossRef] [PubMed]
20. Wikel, J.H.; Paget, C.J.; DeLong, D.C.; Nelson, J.D.; Wu, C.Y.; Paschal, J.W.; Dinner, A.; Templeton, R.J.; Chaney, M.O.; Jones, N.D.; et al. Synthesis of syn and anti isomers of 6-[[(hydroxyimino)phenyl]methyl]-1-[(1-methylethyl)sulfonyl]-1H-benzimidazol-2-amine. Inhibitors of rhinovirus multiplication. *J. Med. Chem.* **1980**, *23*, 368–372. [CrossRef]
21. Heinz, B.A.; Vance, L.M. The antiviral compound enviroxime targets the 3A coding region of rhinovirus and poliovirus. *J. Virol.* **1995**, *69*, 4189–4197. [CrossRef]
22. Hsu, N.Y.; Ilnytska, O.; Belov, G.; Santiana, M.; Chen, Y.H.; Takvorian, P.M.; Pau, C.; van der Schaar, H.; Kaushik-Basu, N.; Balla, T.; et al. Viral reorganization of the secretory pathway generates distinct organelles for RNA replication. *Cell* **2010**, *141*, 799–811. [CrossRef]
23. Arita, M.; Kojima, H.; Nagano, T.; Okabe, T.; Wakita, T.; Shimizu, H. Phosphatidylinositol 4-kinase III beta is a target of enviroxime-like compounds for antipoliovirus activity. *J. Virol.* **2011**, *85*, 2364–2372. [CrossRef] [PubMed]
24. Delang, L.; Paeshuyse, J.; Neyts, J. The role of phosphatidylinositol 4-kinases and phosphatidylinositol 4-phosphate during viral replication. *Biochem. Pharmacol.* **2012**, *84*, 1400–1408. [CrossRef] [PubMed]
25. MacLeod, A.M.; Mitchell, D.R.; Palmer, N.J.; Van de Poel, H.; Conrath, K.; Andrews, M.; Leyssen, P.; Neyts, J. Identification of a series of compounds with potent antiviral activity for the treatment of enterovirus infections. *ACS Med. Chem. Lett.* **2013**, *4*, 585–589. [CrossRef] [PubMed]
26. Mejdrova, I.; Chalupska, D.; Kogler, M.; Sala, M.; Plackova, P.; Baumlova, A.; Hrebabecky, H.; Prochazkova, E.; Dejmek, M.; Guillon, R.; et al. Highly Selective Phosphatidylinositol 4-Kinase IIIbeta Inhibitors and Structural Insight into Their Mode of Action. *J. Med. Chem.* **2015**, *58*, 3767–3793. [CrossRef]
27. Arita, M.; Dobrikov, G.; Purstinger, G.; Galabov, A.S. Allosteric Regulation of Phosphatidylinositol 4-Kinase III Beta by an Antipicornavirus Compound MDL-860. *ACS Infect. Dis.* **2017**, *3*, 585–594. [CrossRef]
28. De Palma, A.M.; Thibaut, H.J.; van der Linden, L.; Lanke, K.; Heggermont, W.; Ireland, S.; Andrews, R.; Arimilli, M.; Altel, T.; De Clercq, E.; et al. Mutations in the non-structural protein 3A confer resistance to the novel enterovirus replication inhibitor TTP-8307. *Antimicrob. Agents Chemother.* **2009**, *53*, 1850–1857. [CrossRef]
29. Arita, M.; Takebe, Y.; Wakita, T.; Shimizu, H. A bifunctional anti-enterovirus compound that inhibits replication and early stage of enterovirus 71 infection. *J. Gen. Virol.* **2010**, *91*, 2734–2744. [CrossRef]
30. Arita, M.; Kojima, H.; Nagano, T.; Okabe, T.; Wakita, T.; Shimizu, H. Oxysterol-binding protein family I is the target of minor enviroxime-like compounds. *J. Virol.* **2013**, *87*, 4252–4260. [CrossRef]
31. Strating, J.R.; van der Linden, L.; Albulescu, L.; Bigay, J.; Arita, M.; Delang, L.; Leyssen, P.; van der Schaar, H.M.; Lanke, K.H.; Thibaut, H.J.; et al. Itraconazole Inhibits Enterovirus Replication by Targeting the Oxysterol-Binding Protein. *Cell Rep.* **2015**, *10*, 600–615. [CrossRef]
32. Albulescu, L.; Bigay, J.; Biswas, B.; Weber-Boyvat, M.; Dorobantu, C.M.; Delang, L.; van der Schaar, H.M.; Jung, Y.S.; Neyts, J.; Olkkonen, V.M.; et al. Uncovering oxysterol-binding protein (OSBP) as a target of the anti-enteroviral compound TTP-8307. *Antiviral Res.* **2017**, *140*, 37–44. [CrossRef]
33. Ho, M.; Chen, E.R.; Hsu, K.H.; Twu, S.J.; Chen, K.T.; Tsai, S.F.; Wang, J.R.; Shih, S.R. An epidemic of enterovirus 71 infection in Taiwan. *N. Engl. J. Med.* **1999**, *341*, 929–935. [CrossRef] [PubMed]
34. Schuffenecker, I.; Mirand, A.; Josset, L.; Henquell, C.; Hecquet, D.; Pilorge, L.; Petitjean-Lecherbonnier, J.; Manoha, C.; Legoff, J.; Deback, C.; et al. Epidemiological and clinical characteristics of patients infected with enterovirus D68, France, July to December 2014. *Eurosurveillance* **2016**, *21*, 30226. [CrossRef] [PubMed]

35. Arita, M.; Fuchino, H.; Kawakami, H.; Ezaki, M.; Kawahara, N. Characterization of a New Antienterovirus D68 Compound Purified from Avocado. *ACS Infect. Dis.* **2020**, *6*, 2291–2300. [CrossRef] [PubMed]
36. Arita, M.; Bigay, J. Poliovirus Evolution toward Independence from the Phosphatidylinositol-4 Kinase III beta/Oxysterol-Binding Protein Family I Pathway. *ACS Infect. Dis.* **2019**, *5*, 962–973. [CrossRef]
37. Arita, M.; Nagata, N.; Sata, T.; Miyamura, T.; Shimizu, H. Quantitative analysis of poliomyelitis-like paralysis in mice induced by a poliovirus replicon. *J. Gen. Virol.* **2006**, *87 Pt 11*, 3317–3327. [CrossRef]
38. Baltera, R.F., Jr.; Tershak, D.R. Guanidine-resistant mutants of poliovirus have distinct mutations in peptide 2C. *J. Virol.* **1989**, *63*, 4441–4444. [CrossRef]
39. Crotty, S.; Saleh, M.C.; Gitlin, L.; Beske, O.; Andino, R. The poliovirus replication machinery can escape inhibition by an antiviral drug that targets a host cell protein. *J. Virol.* **2004**, *78*, 3378–3386. [CrossRef]
40. Tanner, E.J.; Liu, H.M.; Oberste, M.S.; Pallansch, M.; Collett, M.S.; Kirkegaard, K. Dominant drug targets suppress the emergence of antiviral resistance. *eLife* **2014**, *3*, e03830. [CrossRef]
41. Mosser, A.G.; Sgro, J.Y.; Rueckert, R.R. Distribution of drug resistance mutations in type 3 poliovirus identifies three regions involved in uncoating functions. *J. Virol.* **1994**, *68*, 8193–8201. [CrossRef]
42. Purstinger, G.; De Palma, A.M.; Zimmerhofer, G.; Huber, S.; Ladurner, S.; Neyts, J. Synthesis and anti-CVB 3 evaluation of substituted 5-nitro-2-phenoxybenzonitriles. *Bioorg. Med. Chem. Lett.* **2008**, *18*, 5123–5125. [CrossRef]
43. Arita, M.; Wakita, T.; Shimizu, H. Characterization of pharmacologically active compounds that inhibit poliovirus and enterovirus 71 infectivity. *J. Gen. Virol.* **2008**, *89 Pt 10*, 2518–2530. [CrossRef] [PubMed]
44. Meselhy, M.R. Inhibition of LPS-induced NO production by the oleogum resin of *Commiphora wightii* and its constituents. *Phytochemistry* **2003**, *62*, 213–218. [CrossRef] [PubMed]
45. Dias, D.A.; Urban, S. Phytochemical investigation of the Australian lichens *Ramalina glaucescens* and *Xanthoria parietina*. *Nat. Prod. Commun.* **2009**, *4*, 959–964. [CrossRef] [PubMed]
46. Lee, H.S.; Lee, B.W.; Kim, M.R.; Jun, J.G. Syntheses of Resveratrol and its Hydroxylated Derivatives as Radical Scavenger and Tyrosinase Inhibitor. *Bull. Korean Chem. Soc.* **2010**, *31*, 971–975. [CrossRef]
47. Commodari, F.; Khiat, A.; Ibrahimi, S.; Brizius, A.R.; Kalkstein, N. Comparison of the phytoestrogen trans-resveratrol (3,4′,5-trihydroxystilbene) structures from x-ray diffraction and solution NMR. *Magn. Reson. Chem.* **2005**, *43*, 567–572. [CrossRef] [PubMed]
48. Kubo, I.; Murai, Y.; Soediro, I.; Soetarno, S.; Sastrodihardjo, S. Cytotoxic anthraquinones from *Rheum pulmatum*. *Phytochemistry* **1992**, *31*, 1063–1065. [CrossRef]
49. Dai, L.-M.; Tang, J.; Li, H.-L.; Shen, Y.-H.; Peng, C.-Y.; Zhang, W.-D. A new stilbene glycoside from the n-butanol fraction of Veratrum dahuricum. *Chem. Nat. Compd.* **2009**, *45*, 325–329. [CrossRef]
50. Ngoc, T.M.; Hung, T.M.; Thuong, P.T.; Na, M.; Kim, H.; Ha, D.T.; Min, B.-S.; Minh, P.T.H.; Bae, K. Inhibition of Human Low Density Lipoprotein and High Density Lipoprotein Oxidation by Oligostilbenes from Rhubarb. *Biol. Pharm. Bull.* **2008**, *31*, 1809–1812. [CrossRef]
51. Pezet, R.; Perret, C.; Jean-Denis, J.B.; Tabacchi, R.; Gindro, K.; Viret, O. Delta-viniferin, a resveratrol dehydrodimer: One of the major stilbenes synthesized by stressed grapevine leaves. *J. Agric. Food Chem.* **2003**, *51*, 5488–5492. [CrossRef]
52. Semple, S.J.; Pyke, S.M.; Reynolds, G.D.; Flower, R.L. In vitro antiviral activity of the anthraquinone chrysophanic acid against poliovirus. *Antiviral Res.* **2001**, *49*, 169–178. [CrossRef]
53. Kiem, P.V.; Nhiem, N.X.; Cuong, N.X.; Hoa, T.Q.; Huong, H.T.; Huong, L.M.; Minh, C.V.; Kim, Y.H. New phenylpropanoid esters of sucrose from Polygonum hydropiper and their antioxidant activity. *Arch. Pharm. Res.* **2008**, *31*, 1477–1482. [CrossRef] [PubMed]
54. Ridgway, N.D.; Dawson, P.A.; Ho, Y.K.; Brown, M.S.; Goldstein, J.L. Translocation of oxysterol binding protein to Golgi apparatus triggered by ligand binding. *J. Cell Biol.* **1992**, *116*, 307–319. [CrossRef] [PubMed]
55. Kobayashi, J.; Arita, M.; Sakai, S.; Kojima, H.; Senda, M.; Senda, T.; Hanada, K.; Kato, R. Ligand Recognition by the Lipid Transfer Domain of Human OSBP Is Important for Enterovirus Replication. *ACS Infect. Dis.* **2022**, *8*, 1161–1170. [CrossRef] [PubMed]
56. Torney, H.L.; Dulworth, J.K.; Steward, D.L. Antiviral activity and mechanism of action of 2-(3,4-dichlorophenoxy)-5-nitrobenzonitrile (MDL-860). *Antimicrob. Agents Chemother.* **1982**, *22*, 635–638. [CrossRef]
57. Ishitsuka, H.; Ohsawa, C.; Ohiwa, T.; Umeda, I.; Suhara, Y. Antipicornavirus flavone Ro 09-0179. *Antimicrob. Agents Chemother.* **1982**, *22*, 611–616. [CrossRef]
58. Galabov, A.S.; Nikolaeva, L.; Philipov, S. Aporphinoid alkaloid glaucinone: A selective inhibitor of poliovirus replication. *Antiviral. Res.* **1995**, *26* (Suppl. 1), A347. [CrossRef]
59. Nikolaeva-Glomb, L.; Philipov, S.; Galabov, A.S. Oxoglaucin—Enterovirus replication inhibitor: Study on the antiviral spectrum, mode of action and development of resistance (in Bulgarian). *J. Bulg. Acad. Sci.* **2007**, *120*, 22–26.
60. Wang, J.; Zhang, T.; Du, J.; Cui, S.; Yang, F.; Jin, Q. Anti-enterovirus 71 effects of chrysin and its phosphate ester. *PLoS ONE* **2014**, *9*, e89668. [CrossRef]
61. Gunaseelan, S.; Wong, K.Z.; Min, N.; Sun, J.; Ismail, N.; Tan, Y.J.; Lee, R.C.H.; Chu, J.J.H. Prunin suppresses viral IRES activity and is a potential candidate for treating enterovirus A71 infection. *Sci. Transl. Med.* **2019**, *11*, eaar5759. [CrossRef]
62. Zhang, H.M.; Wang, F.; Qiu, Y.; Ye, X.; Hanson, P.; Shen, H.; Yang, D. Emodin inhibits coxsackievirus B3 replication via multiple signalling cascades leading to suppression of translation. *Biochem. J.* **2016**, *473*, 473–485. [CrossRef]

63. Zhong, T.; Zhang, L.Y.; Wang, Z.Y.; Wang, Y.; Song, F.M.; Zhang, Y.H.; Yu, J.H. Rheum emodin inhibits enterovirus 71 viral replication and affects the host cell cycle environment. *Acta Pharmacol. Sin.* **2017**, *38*, 392–401. [CrossRef] [PubMed]

64. Nugent, C.I.; Johnson, K.L.; Sarnow, P.; Kirkegaard, K. Functional coupling between replication and packaging of poliovirus replicon RNA. *J. Virol.* **1999**, *73*, 427–435. [CrossRef] [PubMed]

65. Arita, M. High-Order Epistasis and Functional Coupling of Infection Steps Drive Virus Evolution toward Independence from a Host Pathway. *Microbiol. Spectr.* **2021**, *9*, e0080021. [CrossRef] [PubMed]

66. Buontempo, P.J.; Cox, S.; Wright-Minogue, J.; DeMartino, J.L.; Skelton, A.M.; Ferrari, E.; Albin, R.; Rozhon, E.J.; Girijavallabhan, V.; Modlin, J.F.; et al. SCH 48973: A potent, broad-spectrum, antienterovirus compound. *Antimicrob. Agents Chemother.* **1997**, *41*, 1220–1225. [CrossRef]

67. Oberste, M.S.; Moore, D.; Anderson, B.; Pallansch, M.A.; Pevear, D.C.; Collett, M.S. In vitro antiviral activity of V-073 against polioviruses. *Antimicrob. Agents Chemother.* **2009**, *53*, 4501–4503. [CrossRef]

68. Patick, A.K.; Brothers, M.A.; Maldonado, F.; Binford, S.; Maldonado, O.; Fuhrman, S.; Petersen, A.; Smith, G.J., 3rd; Zalman, L.S.; Burns-Naas, L.A.; et al. In vitro antiviral activity and single-dose pharmacokinetics in humans of a novel, orally bioavailable inhibitor of human rhinovirus 3C protease. *Antimicrob. Agents Chemother.* **2005**, *49*, 2267–2275. [CrossRef]

69. Rhoden, E.; Liu, H.M.; Wang-Chern, S.W.; Oberste, M.S. Anti-poliovirus activity of protease inhibitor AG-7404, and assessment of in vitro activity in combination with antiviral capsid inhibitor compounds. *Antiviral Res.* **2013**, *98*, 186–191. [CrossRef]

70. Kankam, M.K.; Burns, J.M.; Collett, M.S.; Corrado, M.L.; Hincks, J.R. A Phase 1 Study of the Safety, Tolerability, and Pharmacokinetics of Single and Multiple Oral Doses of V-7404 in Healthy Adult Volunteers. *Antimicrob. Agents Chemother.* **2021**, *65*, e0102921. [CrossRef]

71. McKinlay, M.A.; Collett, M.S.; Hincks, J.R.; Oberste, M.S.; Pallansch, M.A.; Okayasu, H.; Sutter, R.W.; Modlin, J.F.; Dowdle, W.R. Progress in the development of poliovirus antiviral agents and their essential role in reducing risks that threaten eradication. *J. Infect. Dis.* **2014**, *210* (Suppl. 1), S447–S453. [CrossRef]

72. WHO. Available online: https://polioeradication.org/tools-and-library/current-research-areas/antivirals/ (accessed on 27 February 2023).

73. Zimmermann, M.L.; Sneden, A.T. Vanicosides A and B, protein kinase C inhibitors from *Polygonum pensylvanicum*. *J. Nat. Prod.* **1994**, *57*, 236–242. [CrossRef]

74. Ahmad, R.; Sahidin, I.; Taher, M.; Low, C.; Noor, N.M.; Sillapachaiyaporn, C.; Chuchawankul, S.; Sarachana, T.; Tencomnao, T.; Iskandar, F.; et al. Polygonumins A, a newly isolated compound from the stem of *Polygonum minus* Huds with potential medicinal activities. *Sci. Rep.* **2018**, *8*, 4202. [CrossRef] [PubMed]

75. Nawrot-Hadzik, I.; Zmudzinski, M.; Matkowski, A.; Preissner, R.; Kesik-Brodacka, M.; Hadzik, J.; Drag, M.; Abel, R. Reynoutria Rhizomes as a Natural Source of SARS-CoV-2 Mpro Inhibitors-Molecular Docking and In Vitro Study. *Pharmaceuticals* **2021**, *14*, 742. [CrossRef] [PubMed]

76. Arita, M. Phosphatidylinositol-4 kinase III beta and oxysterol-binding protein accumulate unesterified cholesterol on poliovirus-induced membrane structure. *Microbiol. Immunol.* **2014**, *58*, 239–256. [CrossRef] [PubMed]

77. Arita, M. Mechanism of Poliovirus Resistance to Host Phosphatidylinositol-4 Kinase III β Inhibitor. *ACS Infect. Dis.* **2016**, *2*, 140–148. [CrossRef] [PubMed]

78. Lyoo, H.; Dorobantu, C.M.; van der Schaar, H.M.; van Kuppeveld, F.J.M. Modulation of proteolytic polyprotein processing by coxsackievirus mutants resistant to inhibitors targeting phosphatidylinositol-4-kinase IIIbeta or oxysterol binding protein. *Antiviral Res.* **2017**, *147*, 86–90. [CrossRef]

79. Melia, C.E.; van der Schaar, H.M.; Lyoo, H.; Limpens, R.; Feng, Q.; Wahedi, M.; Overheul, G.J.; van Rij, R.P.; Snijder, E.J.; Koster, A.J.; et al. Escaping Host Factor PI4KB Inhibition: Enterovirus Genomic RNA Replication in the Absence of Replication Organelles. *Cell Rep.* **2017**, *21*, 587–599. [CrossRef]

80. Lamarche, M.J.; Borawski, J.; Bose, A.; Capacci-Daniel, C.; Colvin, R.; Dennehy, M.; Ding, J.; Dobler, M.; Drumm, J.; Gaither, L.A.; et al. Anti-hepatitis C virus activity and toxicity of type III phosphatidylinositol-4-kinase beta inhibitors. *Antimicrob. Agents Chemother.* **2012**, *56*, 5149–5156. [CrossRef]

81. Spickler, C.; Lippens, J.; Laberge, M.K.; Desmeules, S.; Bellavance, E.; Garneau, M.; Guo, T.; Hucke, O.; Leyssen, P.; Neyts, J.; et al. Phosphatidylinositol 4-Kinase III Beta Is Essential for Replication of Human Rhinovirus and Its Inhibition Causes a Lethal Phenotype In Vivo. *Antimicrob. Agents Chemother.* **2013**, *57*, 3358–3368. [CrossRef]

82. Matsui, T.; Fujita, M.; Ishibashi, Y.; Nomanbhoy, T.; Rosenblum, J.S.; Nagasawa, M. 134. KRP-A218, an Orally Active and Selective PI4KB Inhibitor with Broad-Spectrum Anti-Rhinovirus Activity, Has Potent Therapeutic Antiviral Activity In Vivo. *Open Forum Infect. Dis.* **2021**, *8* (Suppl. 1), S82. [CrossRef]

83. Knight, Z.A.; Gonzalez, B.; Feldman, M.E.; Zunder, E.R.; Goldenberg, D.D.; Williams, O.; Loewith, R.; Stokoe, D.; Balla, A.; Toth, B.; et al. A pharmacological map of the PI3-K family defines a role for p110alpha in insulin signaling. *Cell* **2006**, *125*, 733–747. [CrossRef]

viruses

Review

Structural and Synthetic Aspects of Small Ring Oxa- and Aza-Heterocyclic Ring Systems as Antiviral Activities

Sibasish Manna [1], Koushik Das [1], Sougata Santra [2], Emily V. Nosova [2,3], Grigory V. Zyryanov [2,3] and Sandipan Halder [1,*]

[1] Department of Chemistry, Visvesvaraya National Institute of Technology, Nagpur 440010, India
[2] Department of Organic and Biomolecular Chemistry, Chemical Engineering Institute, Ural Federal University, 19 Mira Street, 620002 Yekaterinburg, Russia; sougatasantra85@gmail.com (S.S.); emilia.nosova@yandex.ru (E.V.N.); g.v.zyrianov@urfu.ru (G.V.Z.)
[3] I. Ya. Postovskiy Institute of Organic Synthesis, Ural Division of the Russian Academy of Sciences, 22 S. Kovalevskoy Street, 620219 Yekaterinburg, Russia
* Correspondence: sandipanhalder@chm.vnit.ac.in

Abstract: Antiviral properties of different oxa- and aza-heterocycles are identified and properly correlated with their structural features and discussed in this review article. The primary objective is to explore the activity of such ring systems as antiviral agents, as well as their synthetic routes and biological significance. Eventually, the *structure–activity relationship* (SAR) of the heterocyclic compounds, along with their salient characteristics are exhibited to build a suitable platform for medicinal chemists and biotechnologists. The synergistic conclusions are extremely important for the introduction of a newer tool for the future drug discovery program.

Keywords: antiviral agents; heterocycles; natural products; *Structure Activity Relationship* (SAR); synthetic methods

Citation: Manna, S.; Das, K.; Santra, S.; Nosova, E.V.; Zyryanov, G.V.; Halder, S. Structural and Synthetic Aspects of Small Ring Oxa- and Aza-Heterocyclic Ring Systems as Antiviral Activities. *Viruses* **2023**, *15*, 1826. https://doi.org/10.3390/v15091826

Academic Editors: Simone Brogi and Daniel DiMaio

Received: 31 March 2023
Revised: 17 August 2023
Accepted: 21 August 2023
Published: 28 August 2023

1. Introduction

The terrible impact of viral diseases has become a severe concern for the whole animal kingdom, including human beings, during the last few decades [1–3]. Several categories of viruses, given their diverse behavior against biological systems, are the major reasons for the incidence of chronic health issues. It is indeed extremely important to address the epidemic nature and the corresponding mortality rate of such diseases in comparison to other fatal infections [4,5]. According to recent reports, the highest number of deaths have been caused by cardiovascular diseases throughout the globe [6], but still, viral infections also have been responsible for millions of human casualties every year (Figure 1). Human immunodeficiency virus (HIV) is considered to be very fatal in nature [7], but it is also very important to note the pervasive nature of other viruses with respect to geographic and economic diversity [8]. As, for example, Rabies disease (originated from domestic dogs) is 100% fatal to humans, but not a pandemic [9]. Disease due to the Ebola virus is a global pandemic with very high average fatality rate (~55%), but it varies depending upon the region [10–12].

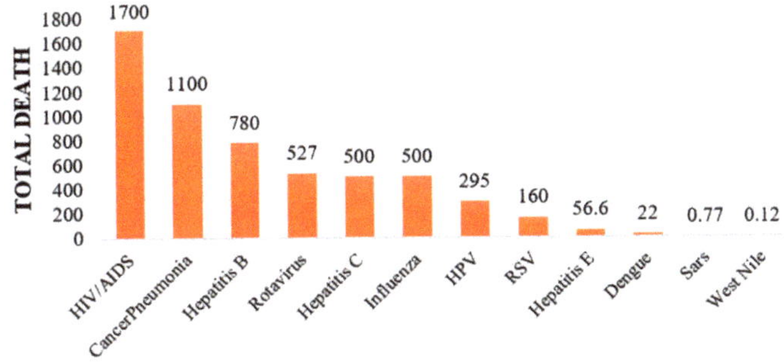

Figure 1. Global mortality rate per year due to different viral infections [13].

In the year of 1918, there was an outbreak of the H1N1 Influenza virus, which is definitely considered to be the world's biggest pandemic, known as the Spanish flu [14,15]. It was exceptionally severe; 50 million people (~2% of the world's population) become infected during 1918–1920 [16,17]. In recent times, the entire world has gone through a catastrophic situation due to the rapid spread of a newly emerged COVID-19, or severe acute respiratory syndrome CoV-2 (SARS-CoV-2) virus [18]. The existence of this highly contagious virus was first reported in December 2019 in Wuhan, China. The World Health Organization (WHO) began immediate special surveillance on this particular issue and effectively framed several protocols throughout the globe. But to the present day, there are very limited remedies available to fight against this COVID-19 [19–23]. The vaccines listed and recommended by WHO are the only solutions to combat these deadly infections [24]. After facing such situations, it is undoubtedly needed to develop libraries of newer drug candidates corresponding to each and every virus, in order to control their fatal outcomes.

In drug discovery research, the role of heterocyclic moieties is extremely significant in order to make stable interactions with the targeted proteins [25,26]. Such ring systems offer suitable coordination with the specific proteins by tailoring the bulk size and pertinent electron density to modulate the efficacy of the drug compounds in the biological environments [27–32]. For these reasons, the synthetic and biological studies of structurally designed heterocyclic moieties are really a crucial measure for the development of future generation drug candidates [33–37]. In this review article, our main focus will be on the structural features, along with the synthetic routes, of oxygen- and nitrogen-containing ring systems (saturated/unsaturated derivatives) present in several natural products, marketed drugs and synthetic analogues having prominent antiviral activities.

2. Overview of the Viral Diseases

Based on its genomic aspects, a virus could be classified as a DNA virus [38] (DNA as a genetic material), which replicates with the help of DNA polymerase (e.g., HSV and HCMV). Secondly, it could be categorized as an RNA virus (RNA as a genetic material), which replicates in the presence of RNA polymerase (e.g., HCV, HBV, RSV and Ebola virus) [39]. Thirdly, it could be categorized as a reverse-transcribing virus [40]; for these viruses, the genome is RNA, but by using a reverse transcriptase enzyme, it is able to form DNA (e.g., HIV) [41].

2.1. Human Immunodeficiency Virus (HIV)

This retrovirus was discovered in the year 1983 [42] and is mainly transmitted through bodily fluids or by bodily contacts of HIV-positive patients. According to the WHO global health survey, there are 38.0 million people living with HIV, and 690,000 people had died by the year 2021 [43]. There is no specific vaccine or medicine available to cure AIDS, but there are some natural and synthetic drug candidates available [44,45]. Dolutegravir, sold under the brand name of Tivicay, was the bestselling anti-HIV drug in 2018 [46]. Some other synthetic drugs like zalcitabine [47], zidovudine [48] and emtricitabine [49] terminate the viral DNA chain by inhibiting the reverse transcriptase.

2.2. Hepatitis C Virus (HCV)

This is a blood-borne virus, discovered in 1989, which generally is transmitted in a similar way to HIV [50]. According to the WHO global health survey, hepatitis C generates chronic diseases like liver cirrhosis and jaundice. Till now there is no specific vaccine or medicine available for the treatment of hepatitis C. There are some marketed drugs available for the general treatment of HCV.

Mavyret, which is the composition of glecaprevir and pibrentasvir, was the best-selling anti-HCV drug in 2018 [51]. Asunaprevir, boceprevir and grazoprevir inhibit the proteolytic activity of HCV NS3/4A protease and show promising anti-HCV properties [52–54].

2.3. Hepatitis B Virus (HBV)

The existence of this virus was confirmed in the year of 1963; it generally is transmitted through sexual contact, blood transfusion or by bodily fluids [55]. According to the WHO global health survey, HBV causes approximately 780,000 deaths every year. The vaccine that corresponds to HBV is available in the market as a hepatitis B surface antigen (HBsAg) [56]; along with that, there are several other drugs that are also available in the market for the treatment of hepatitis B, with the names entecavir, telbivudine and lamivudine [57–59].

2.4. Respiratory Syncytial Virus (RSV)

The World Health Organization (WHO) has reported that RSV causes a significant number of casualties each year, ranging from 66,000 to 199,000 [60]. Moreover, in the year 2005, it was estimated that RSV had infected around 33.8 million children [61,62]. This virus is transmitted through the respiratory systems, by droplets, or from contaminated substances [63].

2.5. Human Cytomegalovirus (HCMV)

HCMV, a member of the herpes virus family, is a prevalent virus that often presents with mild or no symptoms in healthy individuals [64]. Moreover, individuals with weakened immune systems, including newborns, pregnant women and immune compromised individuals can experience severe complications. This infectious virus generally spreads with the help of bodily fluids; it causes serious organ damage, including gastrointestinal problems and colitis [65]. The management of HCMV infection involves the use of antiviral medications which effectively control the virus and alleviate symptoms [66].

2.6. Herpes Simplex Virus (HSV)

Herpes simplex virus (HSV) is a highly contagious virus that causes recurrent infections. The virus is transmitted through direct contact with an infected person's skin, mucous membranes, or bodily fluids [67]. Although there is no cure for HSV, antiviral medications can help to manage symptoms and reduce the frequency and severity of outbreaks. According to the WHO global health survey (2016), over half a billion people worldwide are estimated to have genital herpes caused by herpes simplex virus type 1 (HSV-1) or type 2 (HSV-2) [68].

2.7. Ebola Virus (EBOV)

The existence of the Ebola virus was first understood in the year of 1976 [69]. The primary mode of transmission for the Ebola virus is through direct contact with infected blood, bodily fluids, or tissues [70]. In the year 2018, WHO documented approximately 11,500 deaths globally attributed to Ebola. However, it is important to note that these figures are not fixed and can fluctuate over time and across different regions. The impact of Ebola outbreaks can vary depending on factors such as healthcare infrastructure, access to resources, and public health interventions. Efforts are continuously underway to improve surveillance and response capabilities and institute preventive measures to minimize the spread of the virus and reduce the number of Ebola-related fatalities.

2.8. Severe Acute Respiratory Syndrome CoV-2 (SARS-CoV-2) Virus

Coronaviruses are positive single-stranded RNA viruses that have an enveloped structure and can infect humans and various animals [71]. In 1965, Tyrrell and Bynoe made a significant contribution to the history of human coronaviruses [72]. They discovered that a virus named B814, obtained from the respiratory tract of an adult with a common cold, could be successfully propagated in human embryonic tracheal organ cultures. This discovery paved the way for further research on coronaviruses and their potential impact on human health [73]. These viruses have a spherical shape, with surface projections resembling the solar corona, hence their name "coronaviruses" (derived from the Latin word "corona", meaning "crown"). There are four subfamilies of coronaviruses, including alpha, beta, gamma and delta. The alpha and beta variants likely originate from mammals, predominantly bats, while gamma and delta are associated with pigs and birds. Coronaviruses have genome sizes ranging from 26 to 32 kb. Among the seven subtypes of coronaviruses that can infect humans, beta-coronaviruses can lead to severe illness and fatalities, whereas alpha-coronaviruses typically result in mild or asymptomatic infections [74].

2.9. The Human Papillomavirus (HPV)

The human papillomavirus is a common virus that can infect the body's skin and mucous membranes. It is usually transferred through sexual contact or skin-to-skin contact [75]. According to the WHO report, the consequence of this virus is very much prominent in patients with cervical cancer; the majority of the deaths of such women patients are due to a HPV infection [76].

2.10. Rabies Virus

Rabies virus is a deadly virus that affects the central nervous system of mammals, including humans. It primarily spreads through the saliva of infected animals, usually through a bite. The virus is believed to have originated from bats, but is also found in other animals such as dogs, cats and other wild animals (e.g., fox, raccoon and skunk) [77].

2.11. Zika Virus

The Zika virus is a mosquito-borne virus that was first identified in the Zika Forest of Uganda in 1947 [78]. It remained relatively unknown until a major outbreak occurred in the year 2015 in Brazil, which rapidly spread throughout the Americas and the Caribbean [79].

Most people infected by the Zika virus experience mild symptoms, and the incidence of casualties is also rare, but the viral infection is found to link with birth defects in newborn babies. In most cases, it causes microcephaly, in which the infants are born with abnormally small or underdeveloped brains [80]. The exact number of deaths related to the Zika virus is difficult to determine, as many deaths may have been caused by co-infections or other complications [81].

2.12. The Poliovirus

The poliovirus was first isolated in 1909 by Karl Landsteiner and Erwin Popper; it is a highly contagious virus that primarily affects young children and can lead to paralysis, and even death [82]. It is believed to have originated in ancient times and has been a major public health concern worldwide since the early 20th century [83].

Polio generally is transmitted through contaminated food and water or direct contact with an infected person's saliva [84]. As per the WHO report, the Global Polio Eradication Initiative has led to an impressive 99% reduction in polio infections. In 2020, only 140 polio cases were found worldwide [85].

2.13. West Nile Virus

West Nile virus (WNV) is a virus that generally enters into humans through the bite of infected mosquitoes. The virus originated in Africa, and was first identified in the West Nile district of Uganda in 1937 [86]. The virus can cause a range of symptoms, from mild flu to more severe neurological problems like meningitis and encephalitis [87].

2.14. The Chickenpox Virus

The virus that corresponds to chickenpox is known as varicella–zoster virus (VZV), which is a highly contagious virus that causes a characteristic itchy rash and fever. The virus mostly spreads through respiratory droplets or direct contact with the fluid from the blisters of infected individuals [88].

2.15. The Influenza Virus

The influenza virus is a highly infectious respiratory virus that can cause mild to severe illness, and even death. It is believed that this virus originated in birds and transmitted to humans through close contact with the infected birds or contaminated surfaces [89]. The severity of the illness varies from a mild fever to severe symptoms, or even death. Globally, this annual epidemic leads to approximately three to five million cases of severe illness each year [90].

2.16. Yellow Fever

Yellow fever virus is a Flavivirus which originated in Africa and is transmitted by the *Aedes aegypti* mosquito. The virus can cause a wide range of symptoms, from mild flu to hemorrhagic fever [91]. Yellow fever is endemic in tropical regions of Africa and South America, where it affects thousands of people every year. The majority of infections are asymptomatic, but in severe cases, the virus can cause liver damage, kidney failure and death. According to the WHO report, yellow fever still causes an estimated 200,000 cases and 30,000 deaths per year, mostly in the tropical regions where it is endemic [92].

3. Overview of the Antiviral Drugs

The period of antiviral drugs begins from the year 1959 with the introduction of *Idoxuridine* (5-iodo-2′-deoxyuridine) (Figure 2), the first antiviral drug (Figure 2), by the American pharmacologist William H. Prusoff for the treatment of HSV keratitis in humans [93–95]. This drug was formally approved by FDA in June of 1963; subsequently, different categories of drugs have been discovered and marketed to combat other viral infections [96].

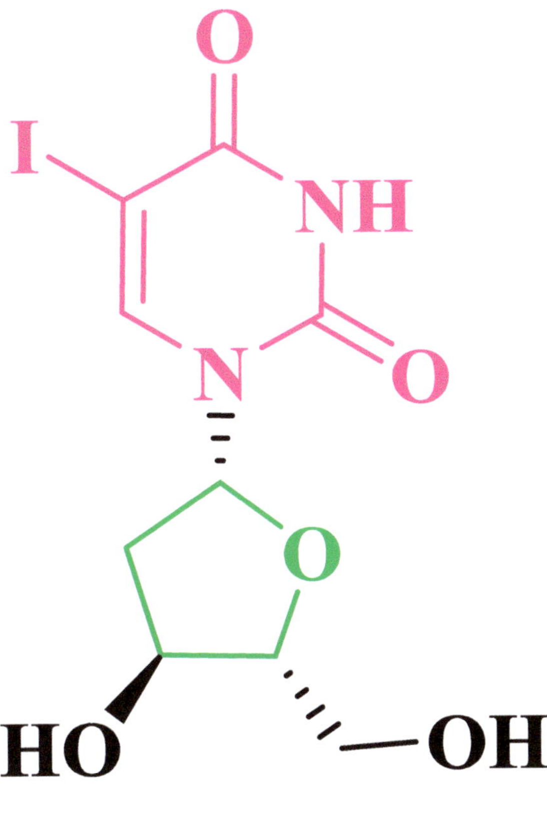

Idoxuridine

Figure 2. First FDA-approved antiviral drug Idoxuridine.

3.1. Representative Antiviral Drug Candidates

Antiviral drugs (Figure 3) target specific enzymes and proteins involved in the viral life cycle, such as RNA-dependent DNA polymerase, RNA-dependent RNA polymerase, proteases and neuraminidases. RNA-dependent DNA polymerase inhibitors are used for retroviral infections like HIV, while RNA-dependent RNA polymerase inhibitors are efficient against RNA viral infections. Protease inhibitors are effective against viruses that require proteases, and neuraminidase inhibitors treat influenza [97].

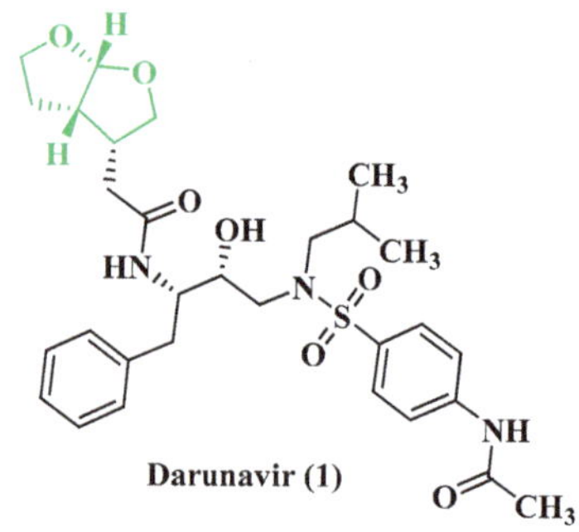

Darunavir (1)

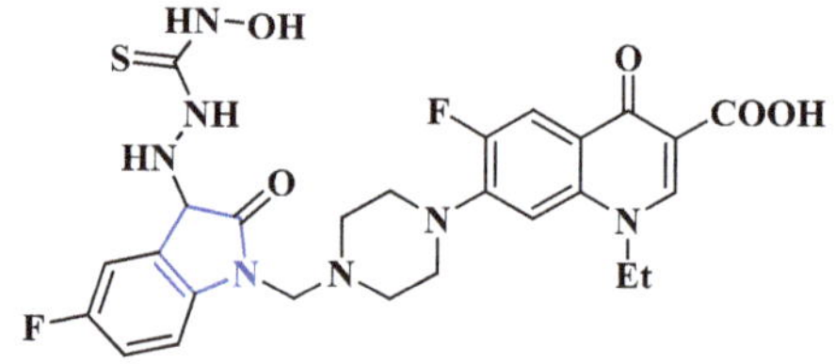

Fluoroquinolone-isatin-thiosemicarbazone (2)

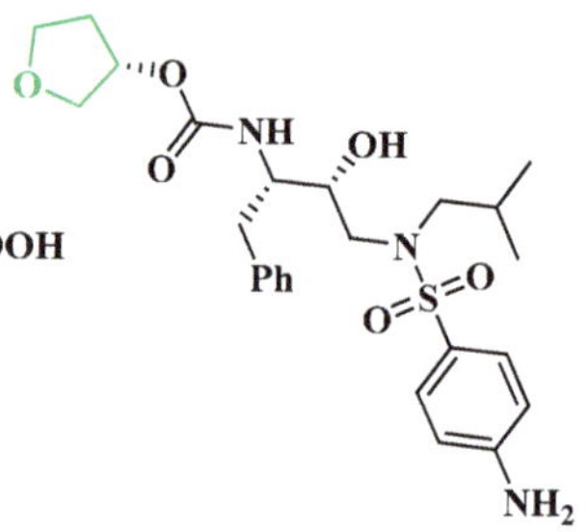

Amprenavir (3)

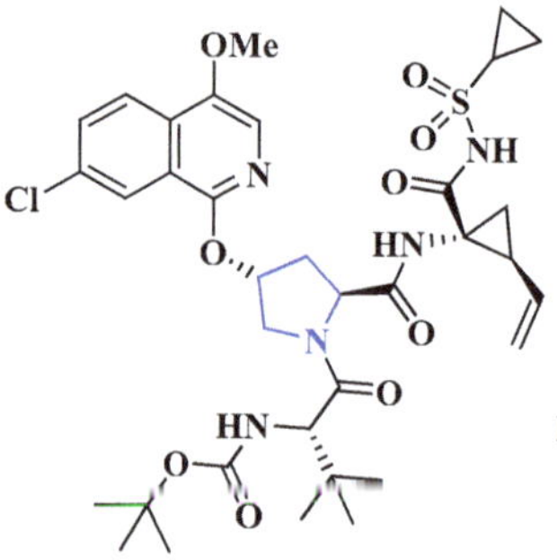

Asunaprevir (4)

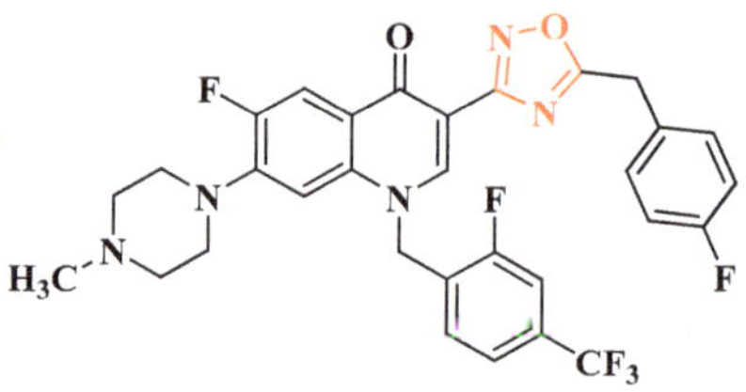

3-(1,2,4-oxadiazole)-quinolone derivatives (5)

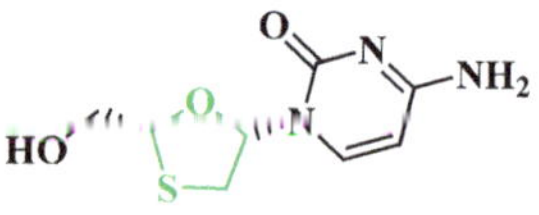

Lamivudine (6)

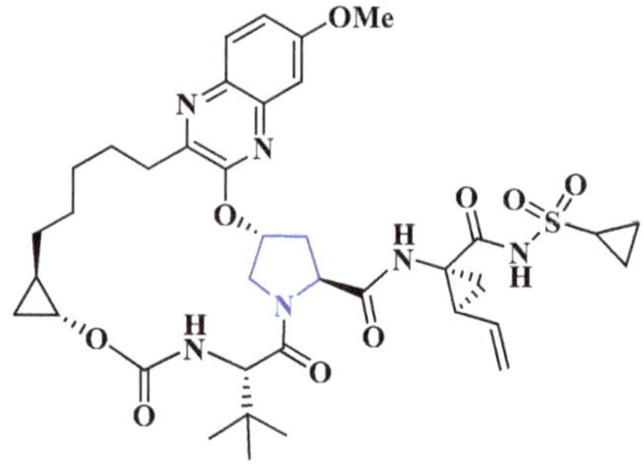

Grazoprevir (7)

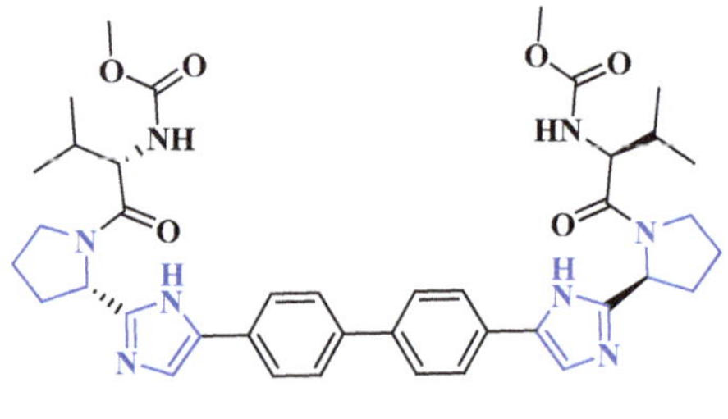

Daclatasvir (8)

Figure 3. *Cont.*

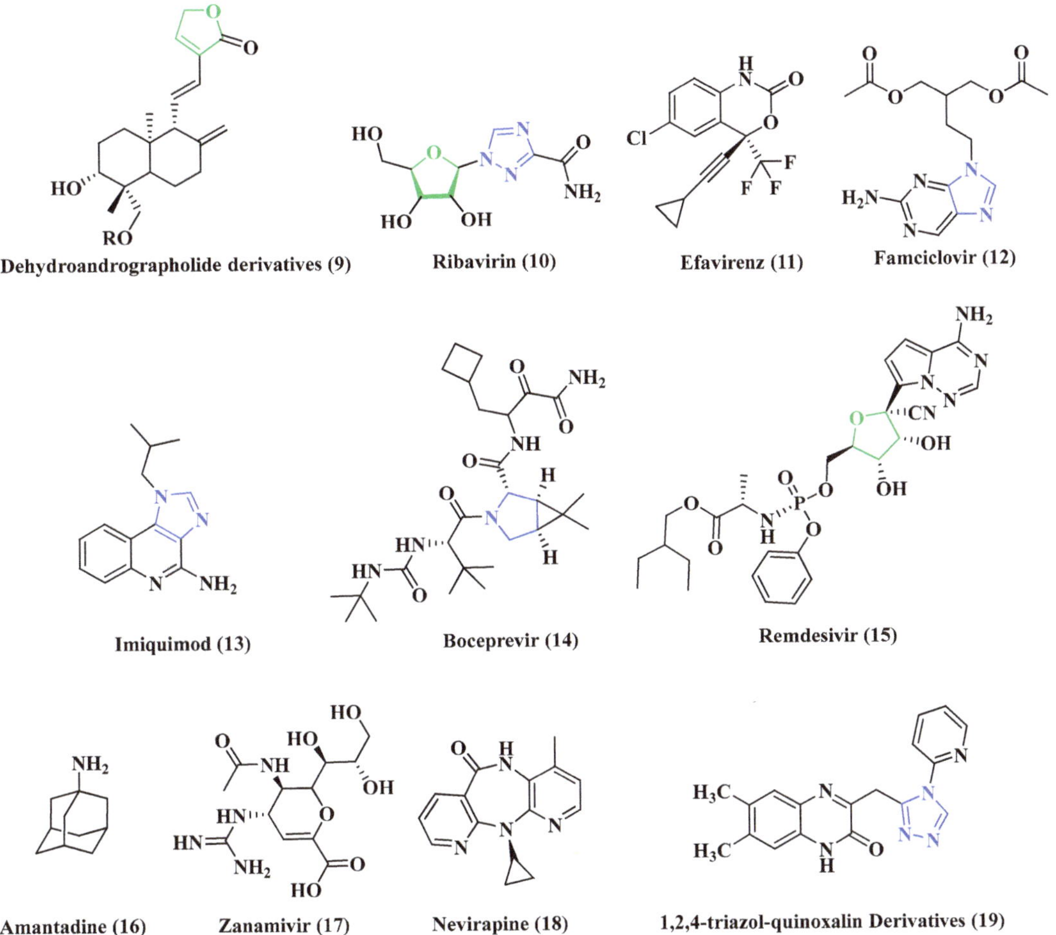

Figure 3. Representative examples of antiviral drug candidates.

3.2. Antiviral Drugs Containing Nucleoside Subunit

These drugs contain nucleobase/substituted nucleobase and a sugar derivative (Figure 4) having prominent antiviral properties. Arabinosyl nucleoside analogues were isolated initially from sponges [98].

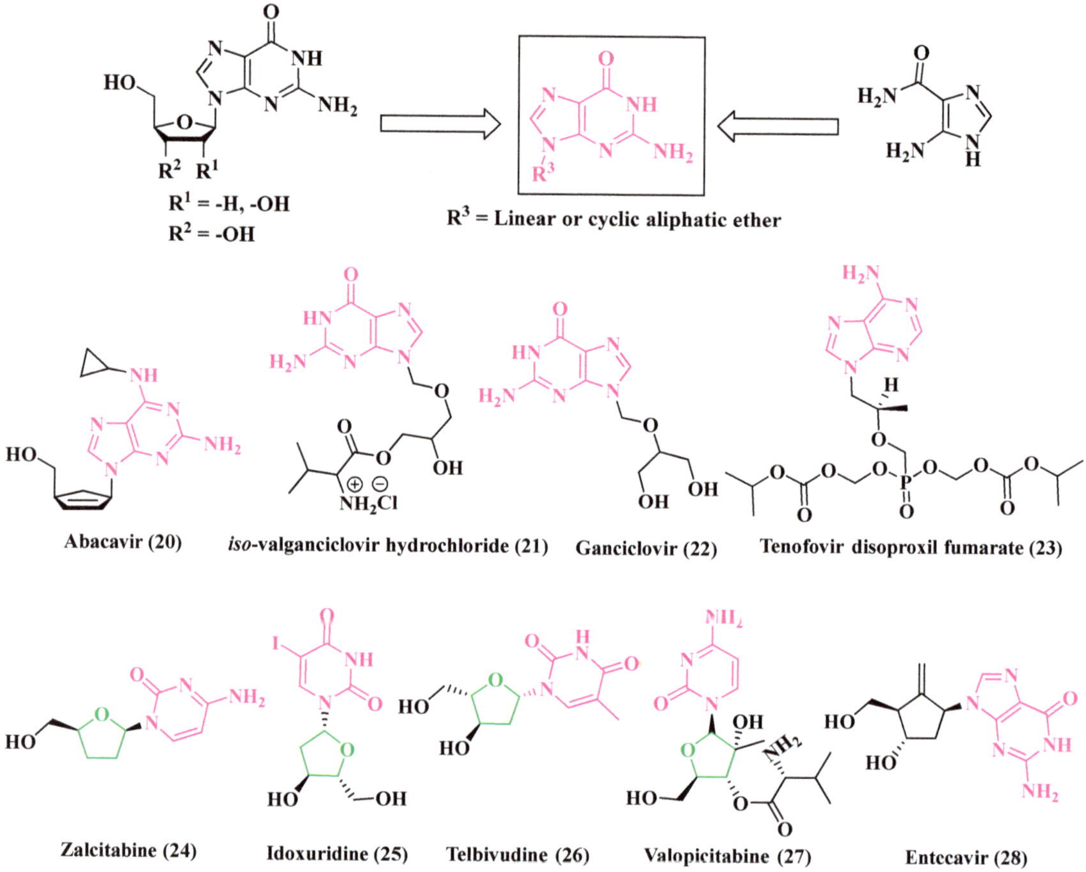

Figure 4. Representative nucleoside drugs.

3.3. Examples of Natural Products with Antiviral Properties

Nature is a continuous source for the provision of different kinds of natural products having excellent biological activities (Figure 5). The alkaloids found in numerous plants and marine algae show important antiviral properties.

- Chalepin (**29**), from a *Ruta angustifolia* species plant, shows a good inhibitory effect against HCV [99].
- Lamellarin α-20 sulfate (**30**) is an alkaloid found in marine Lamellarins [100], and is responsible for inhibiting the integration of HIV-1 replication in its very early stages.
- Lycogarubins (A, B and C) are isolated from fruit bodies of Myxomycetes *Lycogala epidendnrm*, and contain two indole groups connected with dimethyl pyrrole-dicarboxylate, in which Lycogarubin C (**31**) shows activity against HSV [101].
- Silvestrol (**32**), from the bark of the *Aglaia foveolate* type of plants, contains a substituted dioxane and acts as a potent inhibitor of the Ebola virus [102].
- Manassantin B (**33**), extracted from *Saururus chinensis Baill* plants, shows inhibitory properties against the Epstein–Barr virus [103].
- Harmaline (**34**) is an indole alkaloid from *Peganum harmala*, and shows antiviral properties against HSV-1 [104].
- Dehydro-Andrographolide (**35**) and Andrographolide (**36**) are two types of natural diterpenoids that have been extracted from *Andrographis paniculata*. These compounds have demonstrated the ability to inhibit the replication of HBV DNA [105].

- (+)-Dehydrod-iconiferyl alcohol (**37**) that has been isolated from *Swertia patens* shows inhibitory activities on the secretion of HBsAg, with IC$_{50}$ value of 1.94 mM [106].
- Syringaresinol 4''-*O*-β-D-glucopyranoside (**38**), which was extracted from *Swertia chirayita*, exhibited an inhibitory effect on the secretion of HBsAg, with IC$_{50}$ values of 1.49 ± 0.033 mM [107].

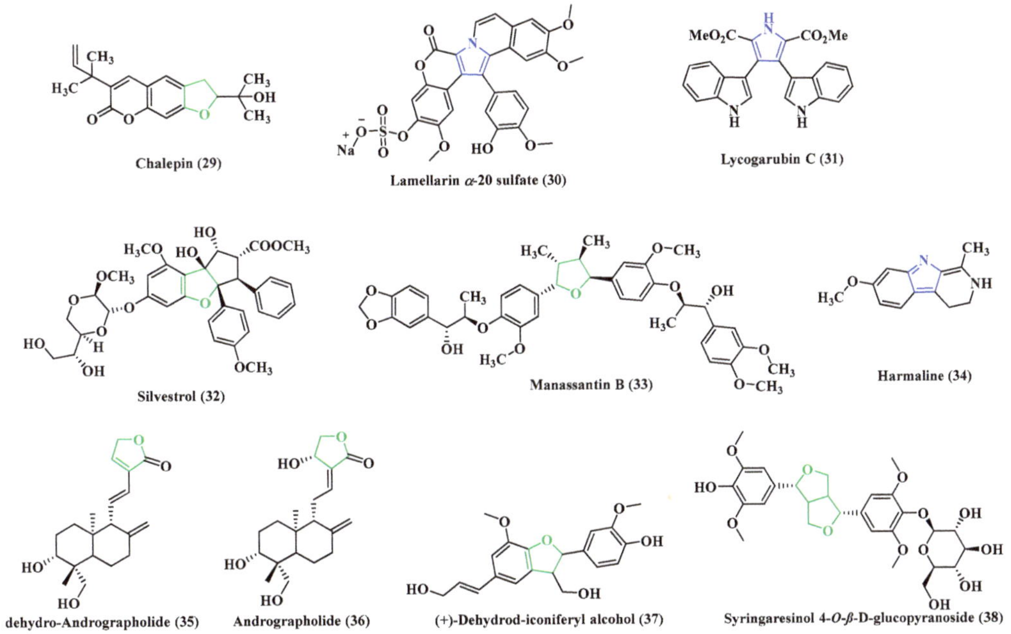

Figure 5. Representative natural products having antiviral activities.

4. Importance of Heterocyclic Ring Systems as Antiviral Agents

Nitrogen- and oxygen-containing ring systems offer significant activities as antiviral candidates [108–111]. These ring systems have either been constructed by linear synthetic steps or could be present from particular starting substrates (like sugar or aza-sugar derivatives). Several research groups have made immense efforts seeking the development of such heterocyclic subunits. Here, the heterocyclic moieties have been screened based on their profound antiviral properties and the corresponding synthetic protocols have been discussed.

5. Synthetic Outlines of Representative Antiviral Drug Candidates

5.1. Anti-HIV Agent

5.1.1. Anti-HIV Agent Darunavir

Kate et al. have demonstrated the synthesis of darunavir (Figure 6), which is considered to be a protease inhibitor [112] and used in low doses for the treatment of HIV. Raltegravir [113] and stavudine [114] are the other available drugs which also show similar properties in this particular domain.

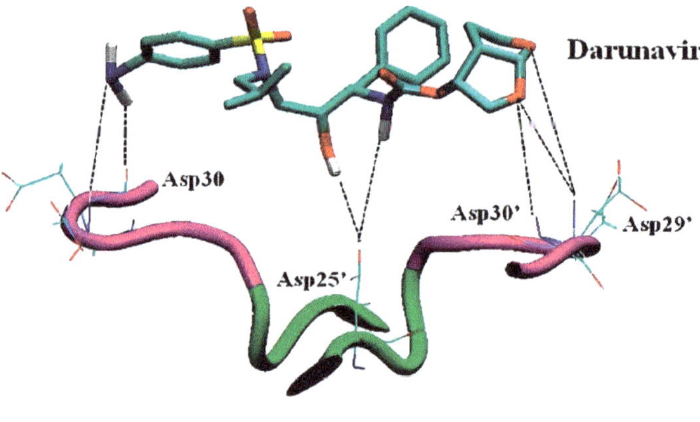

Figure 6. Chemical structure of darunavir (**1**).

Darunavir becomes stabilized inside the cavity of the enzyme by making a hydrogen bonding interaction through the coordination of the hydroxyl group, 4-amino phenyl and tetrahydrofuran ring system with the active site (Asp25′) or near to the active site (Asp29/29′, Asp30/30′) of the amino acid residues as present in HIV-1 PR (Figure 7) [115].

Figure 7. The interaction of darunavir with HIV-1 PR.

Key synthetic steps for Scheme 1 [116]: (a) ring opening of epoxide with primary amine; (b) N-protected 4-aminobenzenesulfonyl halide; (c) Boc deprotection; and (d) nucleophilic addition followed by separation of the product **1**.

Reagents and conditions: (a) A mixture of (2*S*,3*S*)-1,2-Epoxy-3-(Boc-amino)–4-phenylbutane and isobutyl amine was heated at 65–75 °C and (b) N-acetyl sulphanilyl chloride was added at 5–15 °C to the pre-cooled mixture of (1*S*,2*R*)-(1-Benzyl-2-hydroxy-3-(isobutyl-amino) propyl) carbamic acid tert-butyl ester in N, N-dimethylacetamide. Then, triethyl amine was added to the reaction mixture at a temperature below 30 °C and (c) Boc deprotection was done by taking the corresponding carbamic acid tert-butyl ester in isopropyl alcohol at 25–35 °C; after that, aqueous sulphuric acid solution was added to the reaction mixture at 25–35 °C. Then, the amino- N-((2*R*,3*S*)-3-amino-2-hydroxy-4-phenylbutyl)-N-isobutylbenzene sulfonamide sulphate salt was treated with potassium carbonate solution in water and 4-Amino-N-((2*R*,3*S*)-3-amino–2-hydroxy-4-phenylbutyl)-N–isobutylbenzenesulfonamide was obtained above in water. Then, to a stirred mixture of (d) potassium carbonate, isopropyl acetate and water, 4-amino-N-((2*R*,3*S*)-3-amino-2-hydroxy-4-phenylbutyl)-N-isobutyl benzene sulphonamide was added at 25–35 °C, and the reaction mixture was cooled to 15–25 °C; after that, (3*R*,3a*S*,6a*R*)-Hydroxyhexa hydrofuro [2,3-b] furanyl succinimidyl carbonate was added at 15–25 °C. After completion of the reaction, the crude reaction mixture was purified to obtain the final product **1**.

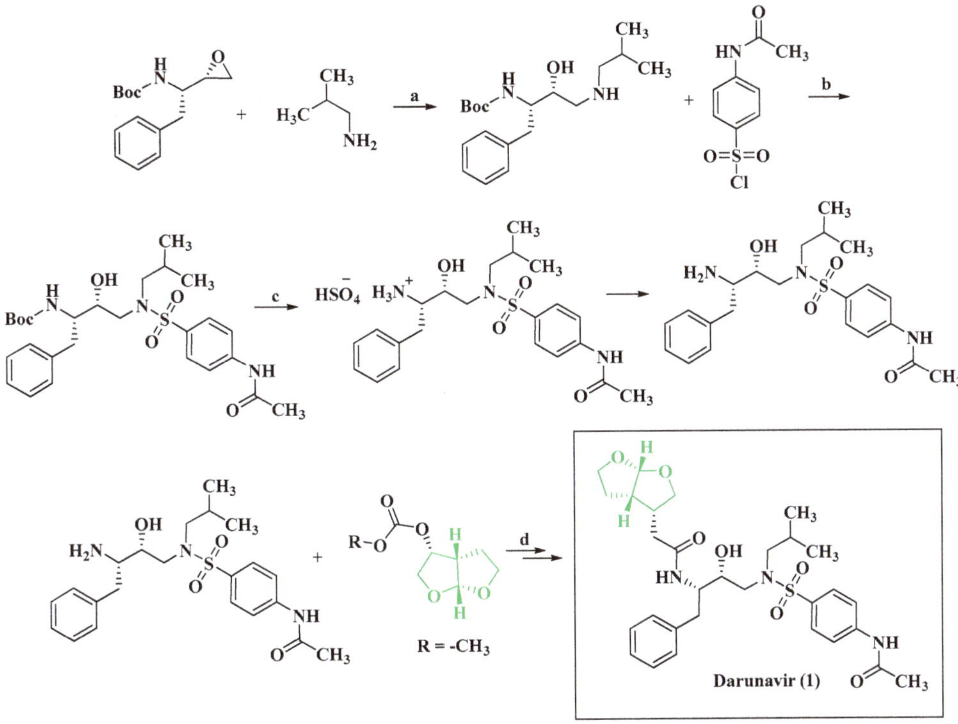

Scheme 1. Synthesis of darunavir.

Structure–Activity Relationship of Darunavir: The structure–activity relationship of darunavir analogues is shown in Table 1, in which the incorporation of thiazole and ethyl phosphonate subunit as $\mathbf{R^1}$, along with benzo[d][1,3]dioxole moiety as $\mathbf{R^2}$, increases its activity significantly [117]. Darunavir and its corresponding synthetic analogues show a distinctive mechanism of action, as characterized by dual functionality. It works as HIV-1 protease inhibitor and also hinders the dimerization process of the HIV-1 protease [118]. Mostly, the Darunavir class of compounds exhibits a binding affinity towards plasma proteins such as alpha-1-acid glycoprotein (AAG or AGP) [119]. The CheckMate™ Mammalian Two-Hybrid System was utilized to establish a dual luciferase assay. This assay was employed to assess the susceptibility of HIV-1$_{\text{LAI}}$ to a variety of drugs and evaluate the cytotoxic effects of the drugs.

A 3-(4,5-dimethylthiazol-2-yl)-2,5-diphenyltetrazolium bromide assay was employed for determining drug susceptibility and cytotoxicity [120]. The chromogenic substrate Lys-Ala-Arg-Val-Nle-paranitro-Phe-Glu-Ala-Nle-amide was used to determine the kinetic parameters [121].

Mechanism of Action of Darunavir: Darunavir interacts with the protease enzyme of HIV-1 to prevent the dimerization and enhance the catalytic activity. As a result, the cleavage of the proteins is disturbed, and ultimately the replication of the virus is stopped, after the application of this drug to HIV-infected cells [122]. Generally, the interaction takes place with the primary chains of Asp-29 and Asp-30 amino acids present in the active site of the protease enzyme.

Table 1. SAR analysis of darunavir.

Compound	Substitution		Activity	
	$-R^1$	$-R^2$	Inhibition Concentration (nM)	Inhibitor Constant (K_i)
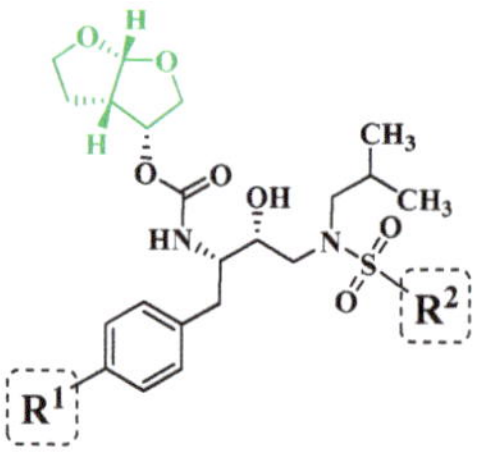	H	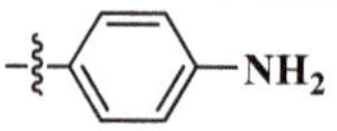 —⟨⟩—NH_2	$IC_{90} = 4.1$	16 pM
	H	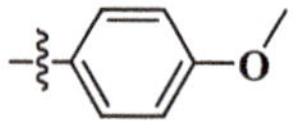 —⟨⟩—O	$IC_{90} = 1.4$	14 pM
	H	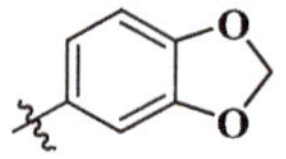 benzodioxole	$IC_{50} = 0.22$	14 pM
	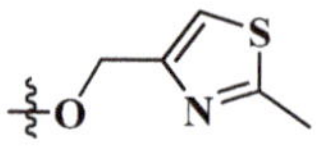 thiazolylmethoxy	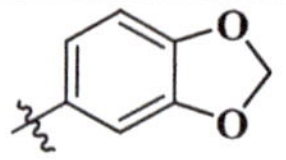 benzodioxole	$IC_{50} = 0.7$	15 fM
	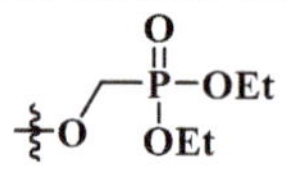 phosphonate (O-P(=O)(OEt)OEt)	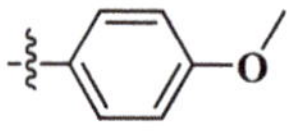 —⟨⟩—O	$IC_{50} = 1$	8 pM

5.1.2. Anti-HIV Agent Fluoroquinolone-Isatin Thiosemicarbazone Hybrids

The molecular-hybrid approach was introduced to understand the synergistic effects of the two compounds in order to generate a new structural entity with superior properties [123] for inhibiting the viral replication. Isatin derivatives have been shown to exhibit antiviral activity against a range of viruses, including HIV-1. Recently, hybrid fluoroquinolone-isatin derivatives have attracted attention due to their promising anti-HIV properties (Figure 8) [124].

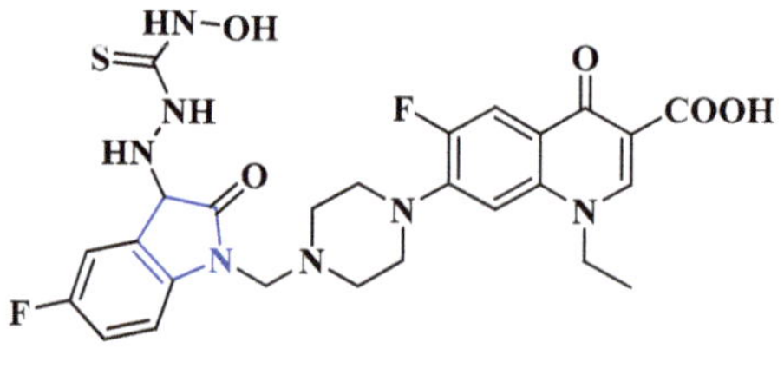

Figure 8. Fluoroquinolone-isatin-thiosemicarbazone hybrid (**2**), a potential anti-HIV agent.

Key synthetic steps for Scheme 2 [125]: (a) A solution of *N*-hydroxylamine in absolute ethanol was added to potassium hydroxide and carbon disulphide, and the mixture was stirred at 0–5 °C to form the corresponding potassium salt of dithiocarbamates; (b) hydrazine hydrate was added to the reaction mixture and stirred at 80 °C; after completion of the reaction, it was cooled to 0 °C to obtain the corresponding thiosemicarbazide; (c) to a hot dispersion of thiosemicarbazide in ethanol was added an equimolar aqueous solution of sodium acetate, and to this solution a further equimolar ethanolic solution of 5-F-isatin was added, and the mixture was stirred while being heated on a hot plate for 4–15 min. The resultant precipitate was filtered off and dried. The product was recrystallized from 95% ethanol; (d) the *N*-Mannich bases were further synthesized by condensing the acidic imino group of isatin derivatives with formaldehyde and secondary amine

(4-ethyl-7-fluoro-1-oxo-6-(piperazin-1-yl)-1,4-dihydronaphthalene-2-carboxylic acid) by irradiating the reaction vessel in a microwave reactor for 3—15 min at 455 W, followed by purification to obtain the corresponding product **2**.

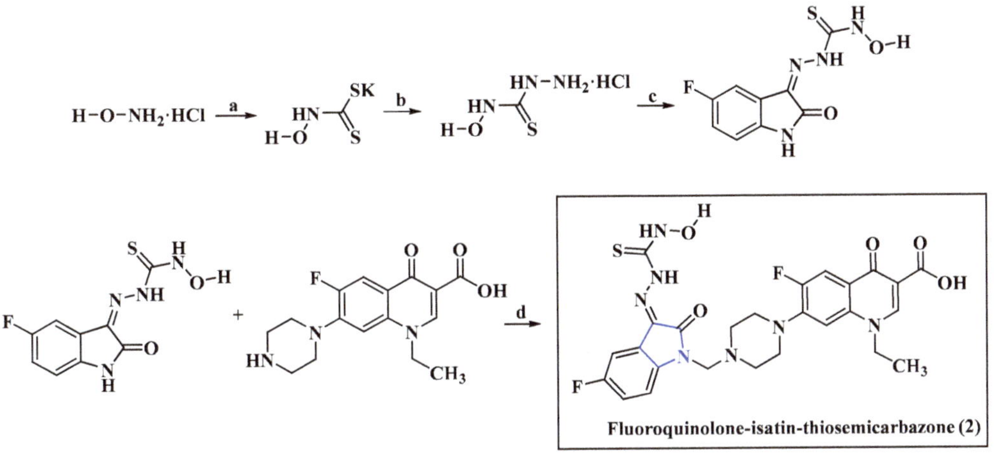

Scheme 2. Synthesis of fluoroquinolone-isatin-thiosemicarbazone hybrid.

Reagents and conditions: (a) CS_2, KOH, C_2H_5OH, 0—5 °C; (b) $NH_2NH_2.H_2O$, 80 °C, conc. HCl; (c) 5-F-Isatin, CH_3COONa; (d) 30% HCHO, 455 W, 3—15 min.

Mechanism of Action of Fluoroquinolone-isatin-thiosemicarbazone: Isatinyl thiosemicarbazone derivatives have been found to exhibit anti-HIV activity by hindering the viral protease enzyme's function. The viral protease enzyme plays a pivotal role in the maturation of the HIV virion by interrupting its activity; as a result, the production of the corresponding infectious virions stops. Isatinyl thiosemicarbazone derivatives interact with the active site of the viral protease enzyme, and hence its activity is stopped [126,127].

5.1.3. Anti-HIV Agent Amprenavir

The drug amprenavir (Figure 9) is primarily used to treat HIV infections; it acts as a protease inhibitor. It binds to the active site of the enzyme and inhibits its activity. It prevents the cleavage of viral polyproteins, which leads to the development of immature non-infectious viruses [128]. Amprenavir's hydroxyl group interacts with Asp25 and Asp25′ residues of the protein at the catalytic site. Along with that, there are stable H-bonding interactions between the hydroxyl group of the drug with the catalytic site of the aspartic acid side chains (Figure 10) [129].

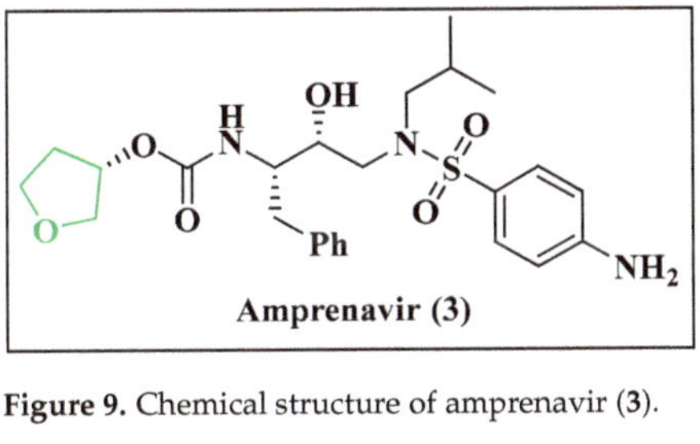

Figure 9. Chemical structure of amprenavir (**3**).

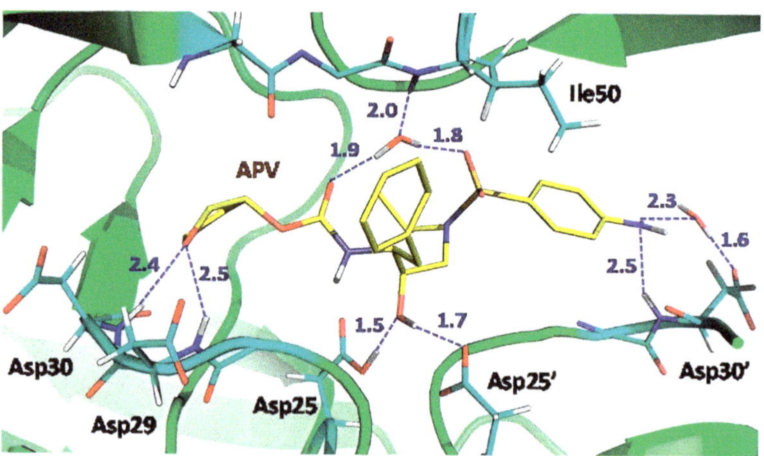

Figure 10. H-bonds and water-mediated interactions formed by amprenavir with the active site.

Key synthetic steps for Scheme 3 [130]: (a) Formation of chalcone by the reaction of 2-phenylacetaldehyde and Ph_3PCHCO_2Et at 90 °C in toluene; (b) reduction of the ester by lithium aluminium hydride and $AlCl_3$ in diethyl ether; (c) chiral epoxidation; (d) epoxide ring opening, followed by (e) epoxide ring closing and further (f) ring opening, and formation of gem-diol derivatives which formed product **3** by reacting with (g) isobutyl amine and (h) N-hydroxysuccinimidyl carbonate of (S)-3-hydroxytetrahydrofuran.

Scheme 3. Synthesis of amprenavir.

Reagents and conditions: (a) Ph_3PCHCO_2Et, PhH, 90 °C; (b) $LiAlH_4$, $AlCl_3$ (30 mol %), Et_2O, 0 °C; (c) mCPBA, CH_2Cl_2, 0 °C; (d) $Ti(O^iPr)_4$, $TMSN_3$, C_6H_6, 70 °C; (e) p-TsCl, Bu_2SnO (2 mol %), Et_3N, DMAP (10 mol %), CH_2Cl_2, 0 °C; (f) K_2CO_3, MeOH, 0 °C; (g) (S,S)-

Co(salen)OAc (0.5 mol %), THF, H_2O (0.5 equiv), 25 °C; (h) (1) [i]BuNH$_2$, [i]PrOH, 50 °C, (2) PPh$_3$, H_2O, THF, 25 °C; (2) N-hydroxysuccinimidyl carbonate of (S)-3-hydroxytetrahydrofuran, Et$_3$N, CH_2Cl_2, 25 °C.

Structure–Activity Relationship of Amprenavir: It shows antiviral properties against the HIV-1 virus in an in vitro study against the C8166 cell line [131]. The efficacy of HIV-1 protease inhibitors was assessed using the fluorescence resonance energy transfer (FRET) technique. A specific protease substrate [Arg-Glu(EDANS)-Ser-Gln-Asn-Tyr-Pro-Ile-Val-Gln-Lys(DABCYL)-Arg], was employed. The determination of the inhibitor's binding dissociation constant (K_i) involved fitting the initial velocity plot against inhibitor concentrations to the Morrison equation through non-linear regression analysis [132]. The EC_{50} values were compared with those of amprenavir, resulting in the conclusion that the biaryl subunit with varying substituents showed lower efficiency. However, compounds that featured substituents of –NH$_2$ for **R^1** and 3-pyridyl or 4-pyridyl for **R^2** exhibited increased solubility and stronger enzyme inhibitory activity at a sub-nanomolar level. These compounds were found to be 2–10 times more active than amprenavir, as described in Table 2 [133].

Table 2. SAR analysis of amprenavir.

Compound	Substitution		Activity	
	$-R^1$	$-R^2$	EC_{50} (nM)	Inhibitor Constant (K_i) (nM)
	$-NH_2$	$-H$	2.93	0.84
	$-OCH_3$	phenyl–F	4.29	5.8
	$-OCH_3$	phenyl–Cl	2.32	6.4
	$-OCH_3$	phenyl–COMe	6.89	5.8
	$-OCH_3$	phenyl–SMe	6.04	7.4
	$-OCH_3$	phenyl–OCF$_3$	3.63	5.2
	$-OCH_3$	phenyl (CH$_3$, OCH$_3$)	5.53	1.8
	$-OCH_3$	phenyl–COOH	4.30	1.9
	$-OCH_3$	phenyl–OH	3.46	4.3
	$-OCH_3$	phenyl–CO$_2$Me	9.2	1.6

Table 2. *Cont.*

Compound	Substitution		Activity	
	$-R^1$	$-R^2$	EC$_{50}$ (nM)	Inhibitor Constant (K$_i$) (nM)
	$-OCH_3$	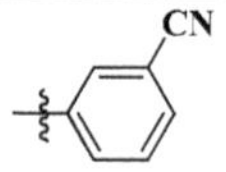 (3-cyanophenyl) CN	4.48	8.1
	$-OCH_3$	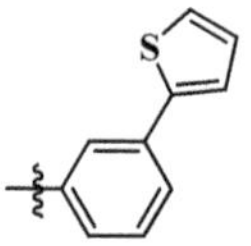 (3-thiophen-2-yl-phenyl) S	3.08	3.5
	$-OCH_3$	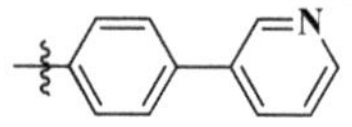 (4-pyridin-3-yl-phenyl) N	2.68	1.6
	$-OCH_3$	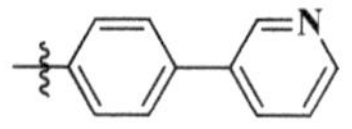 (4-pyridin-3-yl-phenyl) N	2.25	3.8
	$-NH_2$	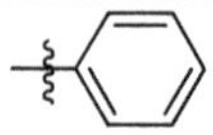 (phenyl)	0.10	1.95
	$-NH_2$	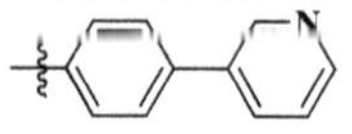 (4-pyridin-3-yl-phenyl) N	0.10	0.33
	$-NH_2$	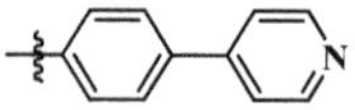 (4-pyridin-4-yl-phenyl) N	2.93	0.086

Mechanism of Action of Amprenavir: Amprenavir binds to the active site of the protease and inhibits the activity of the enzyme. This inhibition prevents the cleavage of the gag-pol polyprotein. Amprenavir competes with the natural substrate of the viral protease enzyme, which is a precursor protein of the viral genome. This competition leads to the formation of a stable complex between the drug and the enzyme, preventing the enzyme from cleaving the protein. As a result, non-infectious viral particles are produced, leading to a reduction in the number of viral particles in the body [134]. In Table 3 other marketed anti-HIV drugs are listed with their mechanism of action, ways of use and side effects.

Table 3. Anti-HIV drugs.

Sl. No.	Drug Name	Drug Target	Mechanism of Action	Ways of Use	Side Effect	Brand Name
1.	Zalcitabine [135]	Reverse transcriptase/RNase H (Human immunodeficiency virus 1)	Inhibiting reverse transcriptase and terminating the viral DNA chain	Oral	Liver failure, inflammation of the pancreas	Hivid
2.	Amprenavir [136]	Human immunodeficiency virus type 1 protease	Blocks the active site of HIV protease to prevent cleavage of the viral precursor proteins	Oral	Diarrhea, kidney failure, weakness	Agenerase
3.	Maraviroc [137]	C-C chemokine receptor type 5	Blocks GP120-CCR5 interaction to inhibit HIV entry	Oral	Liver problems, nausea, allergic reactions	Selzentry
4.	Tenofovir disoproxil Fumarate [138]	Reverse transcriptase/RNaseH (Human immunodeficiency virus 1)	Competes with dATP and inhibits the activity of HIV RT	Oral	Trouble sleeping, dizziness, diarrhea	Viread

Table 3. *Cont.*

Sl. No.	Drug Name	Drug Target	Mechanism of Action	Ways of Use	Side Effect	Brand Name
5.	Emtricitabine [139]	Reverse transcriptase/RNase H (Human immunodeficiency virus 1)	Through nucleoside reverse transcriptase inhibition	Oral	Body aches, ough, diarrhea, fever	Truvada
6.	Cabotegravir [140]	Integrase (Human immunodeficiency virus 1)	By HIV-1 integrase inhibition	Oral	Abnormal dreams, dark urine, difficulty in breathing	Vocabria
7.	Lenacapavir [141]	Gag-Pol polyprotein	Lenacapavir works by blocking the HIV-1 virus' protein shell (the capsid), thereby interfering with multiple essential steps of the viral lifecycle	Oral/subcutaneous injection	Nausea	Sunlenca

Some other synthesized compounds that show activity against HIV are given in Table 4.

Table 4. Synthesized anti-HIV compounds.

Sl. No.	Antiviral Agent	Drug Target	Activity
1.		HIV [142]	Methylthiazole derivatives effectively inhibit HIV-1 entry into cells, demonstrating low micromolar potency (IC_{50} = 1.6 μM) and a favorable selectivity index (SI) in both single-cycle TZM-bl cell assays and multicycle MT-2 cell assays.
2.		HIV-1 and HIV-2 [143]	3″-alkenyl-substituted TSAO derivatives exhibit anti-HIV-1 and anti-HIV-2 activity at subtoxic concentrations against HIV-2 RT involved the RNA-dependent DNA polymerase assay with $EC_{50}(HIV-1_{MT-4})$ = 0.06 ± 0.03 μM and $EC_{50}(HIV-2_{MT-4})$ >10 μM.
3.		HIV [144]	4′-ethynyl-2-fluoro-2′-deoxyadenosine demonstrates exceptional potency in inhibiting HIV-1 replication in phytohemagglutinin-activated peripheral blood mononuclear cells with $EC_{50}(HIV-1_{NL4-3})$ = 50 pM.
4.		HIV-1 [145]	Deoxythreosyl phosphonate nucleosides exhibit potent anti-HIV-1 activity ($EC_{50}(HIV-1)$ = 2.53 μM) by binding effectively to the active site pocket of HIV-1 reverse transcriptase. These compounds show no cytotoxicity, even at high concentrations (CC_{50} > 316 μM). The incorporation kinetics of these compounds into DNA were studied using their diphosphate form as a substrate and HIV-1 reverse transcriptase as the catalyst.
5.		HIV-1 [146]	Flavopiridol, a cyclin-dependent kinase (CDK) inhibitor, blocks HIV-1 Tat transactivation and viral replication by inhibiting P-TEFb kinase activity. The cytotoxicity of Flavopiridol and its analogues was evaluated using an MTT-based cell viability assay, demonstrating their potential as anti-HIV-1 therapeutics with EC_{50} = 7.4 nM.

5.2. Anti-HCV Agent

5.2.1. Anti-HCV Agent Asunaprevir

Asunaprevir (Figure 11) is an orally efficacious NS3 protease inhibitor used for the treatment of hepatitis C virus infection. This tripeptidic acyl sulfonamide is an inhibitor of the enzyme NS3/4A, and is now in phase III clinical trials for the treatment of hepatitis C virus infection. The activity of asunaprevir showed a robust antiviral response in early clinical trials. Suzuki et al. have studied the antiviral activity and toxicological profile of asunaprevir [147]. It inhibits the activity of proteases by binding to the active site. Viral polyproteins cannot be cleaved by this inhibition, which produces undeveloped and non-infectious viral particles (Figure 12) [148].

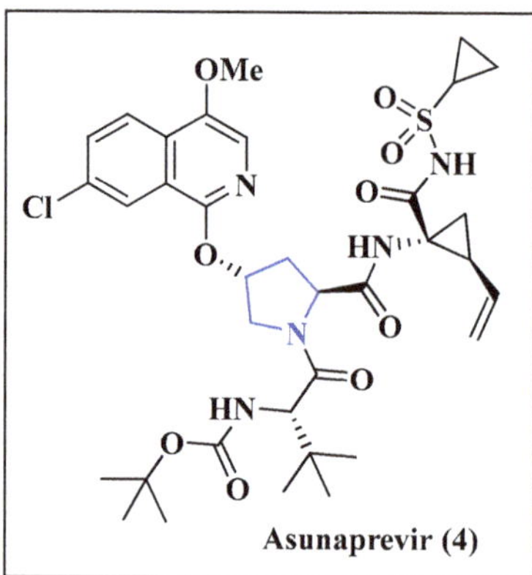

Figure 11. Chemical structure of asunaprevir (**4**).

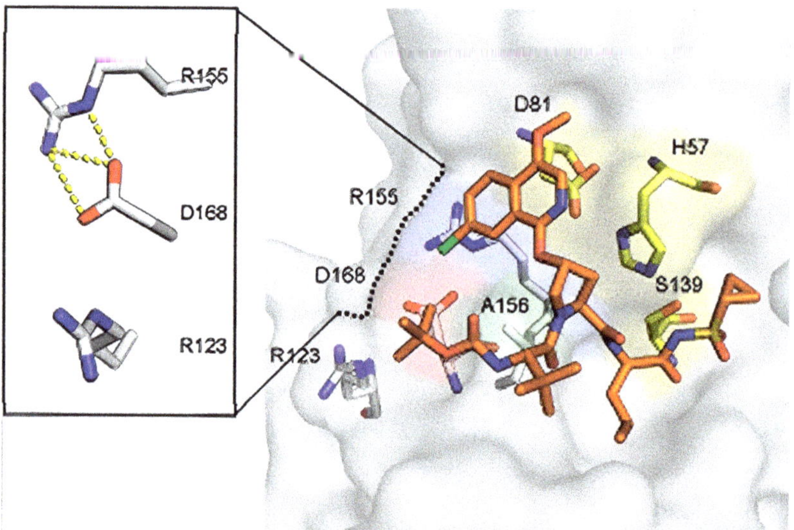

Figure 12. Structure of HCV NS3/4A protease in complex with asunaprevir.

Key synthetic steps for the Scheme 4 [149]: (a,b) (*E*)-3-(4-chlorophenyl)acrylic acid is cyclized, and the derivatization (c) reacts with *N*-Boc-3-(*R*)-hydroxy-L-proline via nucleophilic addition (d) by peptide coupling reaction. The synthesized fragment reacts with (1*R*,2*S*)-1-amino-*N*-(cyclopropylsulfonyl)-2-vinylcyclopropanecarboxamide (TsOH salt), followed by (e) deprotection of proline nitrogen. (f) The final product **4** was obtained by the peptide coupling reaction with *N*-Boc-t-butyl-L-glycine.

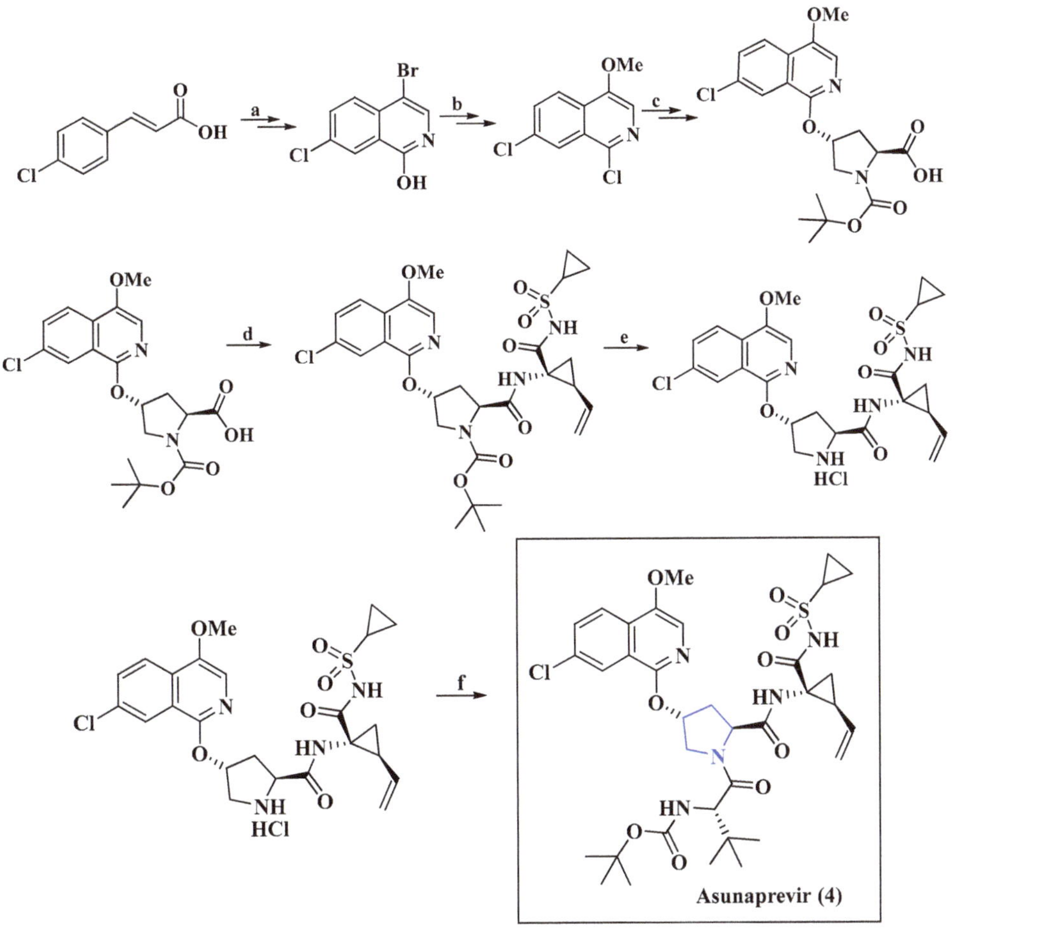

Scheme 4. Synthesis of anti-HCV agent asunaprevir.

Reagents and conditions: (a) (i) DPPA, Et$_3$N, benzene, rt; (ii) Ph$_2$CH$_2$, reflux; (iii) NBS, MeCN, reflux. (b) (i) POCl$_3$, reflux; (ii) (1) n-Bu-Li, THF, $-78\,^\circ$C, (2) (*i*-PrO)$_3$B, $-78\,^\circ$C; (3) 50% H$_2$O$_2$, Na$_2$SO$_3$ $-78\,^\circ$C; (iii) MeOH, MeCN, TMSCHN$_2$, $0\,^\circ$C–rt. (c) *N*-Boc-3-(*R*)-hydroxy-L-proline, t-BuOK, DMSO, $10\,^\circ$C. (d) HATU, Hunig's base *i*-Pr$_2$Net, (1*R*,2*S*)-1-amino-*N*-(cyclopropylsulfonyl)-2-vinylcyclopropanecarboxamide (TsOH salt), rt; (e) HCl (conc.), MeOH, reflux; (f) HATU, Hunig's base, *N*-Boc-t-butyl-L-glycine, DCM, $0\,^\circ$C–rt.

Structure–Activity Relationship of Asunaprevir: The structure-activity relationship studies show that the incorporation of a hydroxyl group at the **R^1** position decreases its antiviral activity (Table 5). The isoquinoline series with methoxy and chlorinated analogues proved to be potent inhibitors of the NS3/4A protease (GT-1a NS3/4A enzyme) which extended to excellent inhibitory activity in the replicon at the **R^2** substituent. The antiviral activity of the drug was evaluated through a two-part study. Initially, a single ascending dose (SAD) study was conducted in patients infected with genotype 1, followed by a subsequent multiple ascending dose (MAD) study [149].

Table 5. SAR analysis of asunaprevir derivatives.

Compound	Substitution				Activity
	$-R^1$	$-R^2$	$-R^3$	$-R^4$	IC$_{50}$ (nM)
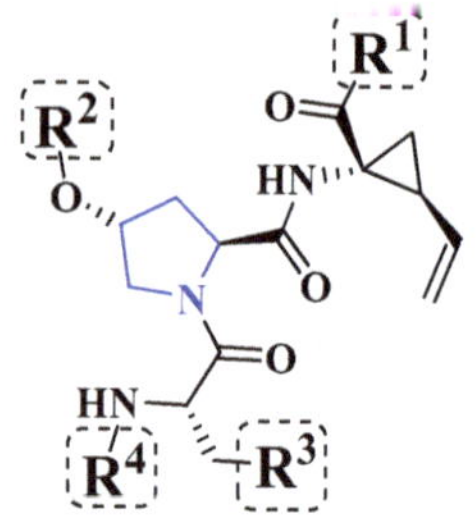	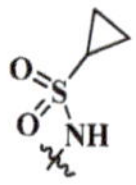	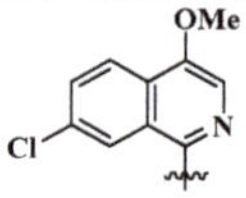		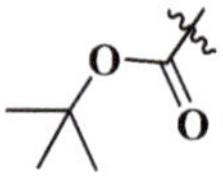	2
	$-$OH	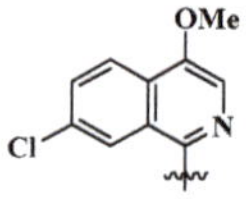		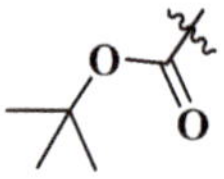	247
	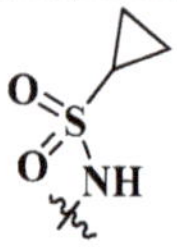	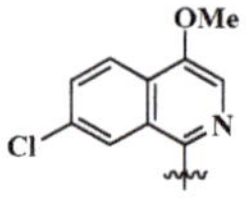		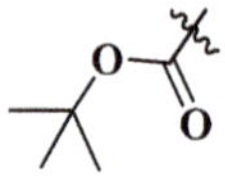	2
	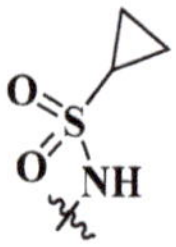	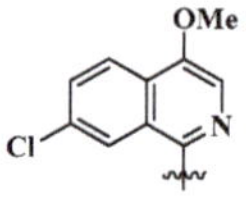		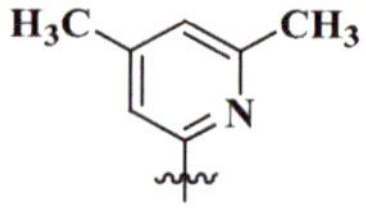	4
	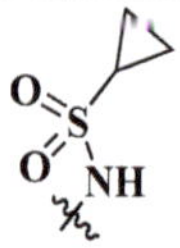	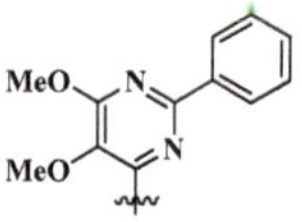		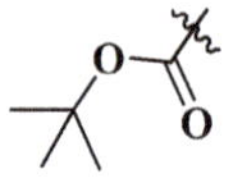	4
	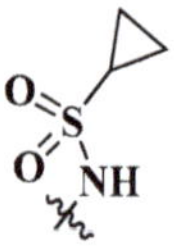	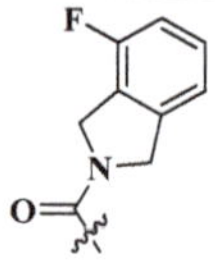		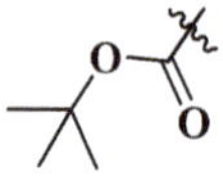	2
	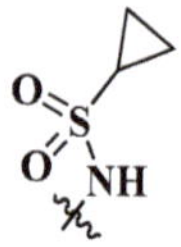	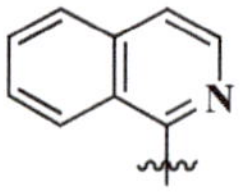		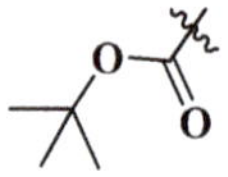	7
	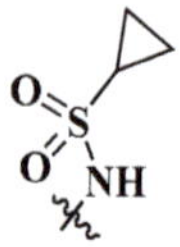	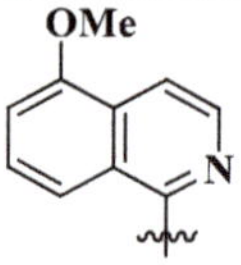		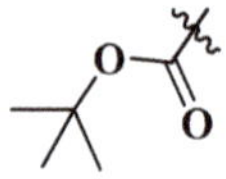	1
	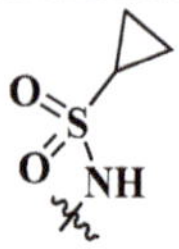	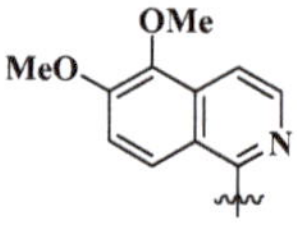		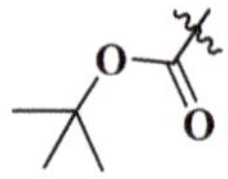	7
	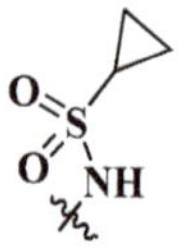	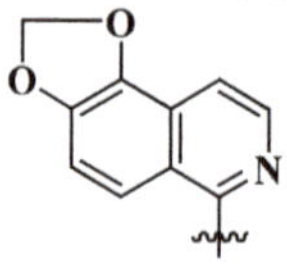		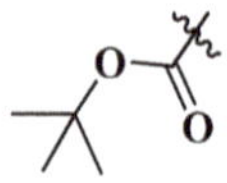	2

Mechanism of Action of Asunaprevir: Asunaprevir is highly active against HCV NS3 protease [150], which is responsible for processing the HCV polyprotein into individual viral proteins. The production of new viral proteins is prevented with the use of this drug compound, and as a result, the progression of HCV-related liver disease is reduced. It is typically used in combination with other antiviral drugs to enhance efficacy and minimize drug resistance [52].

5.2.2. Anti-HCV Agent 3-(1,2,4-oxadiazole)-quinolone

Studies have shown that 3-heterocyclic quinolones (Figure 13) can inhibit NS5B polymerase activity by binding to an allosteric site. This binding triggers a change in the protein's structure, resulting in the inhibition of RNA replication. In vitro and in vivo studies have demonstrated the significant antiviral activity of these compounds against HCV and indicated their roles as promising lead candidates for further development [151].

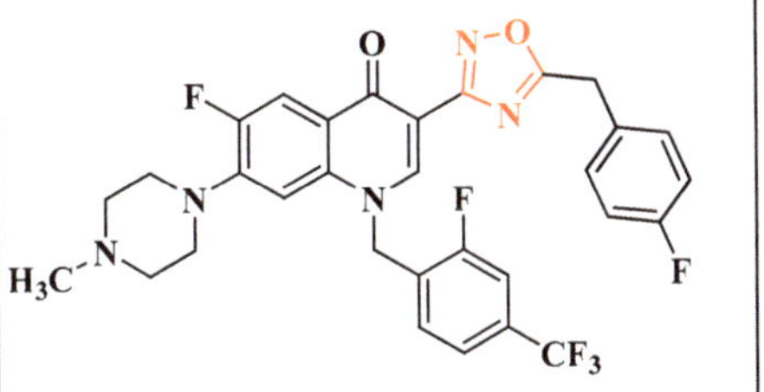

Figure 13. 3-(1,2,4-oxadiazole)-quinolone derivative (**5**) with anti-HCV activities.

Key synthetic steps for Scheme 5 [151]: (a) The nitrile moiety of quinolone is converted to the intermediate hydroxyamidine through reaction with hydroxyl amine; (b) the resulting hydroxyamidine intermediates are converted to a variety of 1,2,4-oxadiazole target compounds by reacting with appropriate carboxylic acids to obtain product **5**.

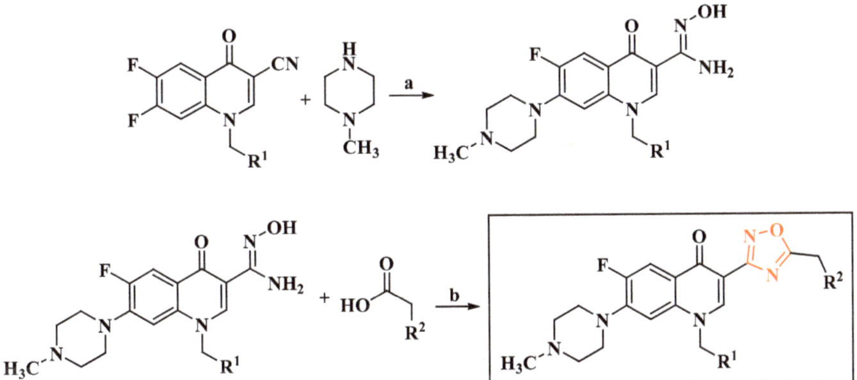

Scheme 5. Synthesis of 3-(1,2,4-oxadiazole)-quinolone derivatives.

Reagents and conditions: (a) CH_3CN, reflux (b) NH_2OH, DIEA, aq. EtOH; (c) HBTU, DIPEA, 200 °C, microwave.

Structure–Activity Relationship of 3-(1,2,4-oxadiazole)-quinolone derivatives: The activity (IC_{50}) against the NS5B polymerase enzyme (Table 6) was determined by scintillation proximity assays (SPA) [152]. For a high-throughput screening (HTS) campaign to identify inhibitors of NS5B polymerase, a scintillation proximity assay (SPA) format was employed. Scintillation proximity assays (SPAs) are an efficient technique that can be used to detect enzymes, receptors, radioimmunoassays and molecular interac-tions. This assay format allowed for the screening of compounds that effectively inhibited the enzymatic function of NS5B polymerase, specifically targeting the wild-type (genotype 1b) enzyme. The IC_{50} values of the compounds were assessed against NS5B polymerase, and their efficacy was determined in using a cell-based viral replication surrogate assay called the replicon system [152]. $\mathbf{R^1}$ and $\mathbf{R^2}$ mainly stabilize the compounds in the hydrophobic pockets. The inhibition activity of these two functionalities present in the quinolone moiety is shown in Table 6 [153]. It is clearly found that the presence of the –F or –CF$_3$ group in either $\mathbf{R^1}$ or $\mathbf{R^2}$ is very much responsible for modulating the corresponding activity [153].

Table 6. SAR analysis of 3-(1,2,4-oxadiazole)-quinolone derivatives.

Compound	Ubstitution		Activity
	$-\mathbf{R^1}$	$-\mathbf{R^2}$	IC_{50} (µM)
	4-Cl-C$_6$H$_4$	4-F$_4$-C$_6$H$_4$	0.019
	4-Cl-C$_6$H$_4$	2,4-DiF-C$_6$H$_3$	0.075
	4-CF$_3$-C$_6$H$_4$	C$_6$H$_5$	0.024
	2-F,4-CF$_3$-C$_6$H$_3$	C$_6$H$_5$	0.014
	2-F,4-CF$_3$-C$_6$H$_3$	4-F$_4$-C$_6$H$_4$	0.015

Mechanism of Action of 3-(1,2,4-oxadiazole)-quinolone derivatives: 3-(1,2,4-oxadiazole)-quinolone derivatives show promising inhibitory activity against HCV NS5B polymerase and NS3 protease. Cyclophilin, a protein present in the host cell of HCV, also is affected by the interaction of such heterocyclic compounds; consequently, the replication process of this virus becomes affected. By targeting multiple stages of the HCV life-cycle, these compounds can reduce the amount of virus and slow or stop the progression of HCV-related liver diseases [154].

5.2.3. Anti-HCV Agent Grazoprevir

Grazoprevir (Figure 14) is a potent inhibitor of RNA synthesis in HCV (due to the action of two different DAAs as NS5A and NS3/4A inhibitors), representing high genetic barriers to resistance. The mechanism of action and pharmacodynamic properties, as well as the pharmacokinetics, clinical uses, safety and efficacy of elbasvir/grazoprevir in managing a large variety of conditions, including cases in the presence of cirrhosis, co-infection with HIV and patients having inherited blood disorders, were nicely reviewed by Kassas et al. [155]. Sofosbuvir [156], ledipasvir [157] and telaprevir [158] are the reported anti-HCV drugs with similar synthetic procedures.

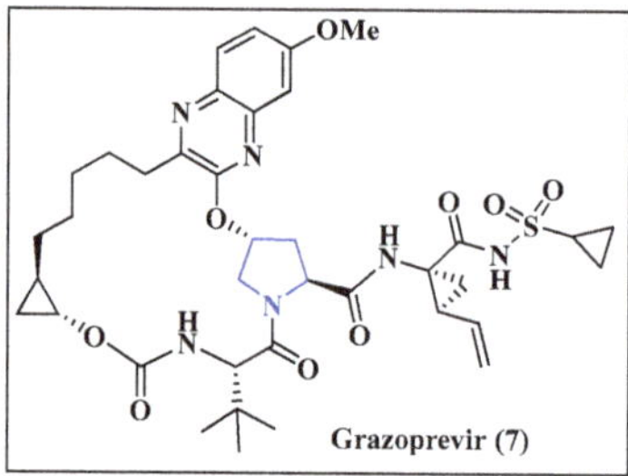

Figure 14. Chemical structure of grazoprevir (7).

Key synthetic steps for Scheme 6 [159]: Grazoprevir was synthesized starting from the (a) cross coupling strategy of 2,3-dichloro-6-methoxyquinoxaline with substituted proline derivative to obtain the corresponding five membered heterocyclic core followed by (b) esterification of the (1*R*,2*R*)-2-(pent-4-yn-1-yl)cyclopropan-1-ol with (*S*)-2-amino-3,3-dimethylbutanoic acid and (c) metal catalyzed coupling reactions of the fragments, followed by (d) cyclization through intramolecular peptide coupling (e). The final product **7** was obtained by the peptide coupling with the allylic sulfonamide, as shown in Scheme 6.

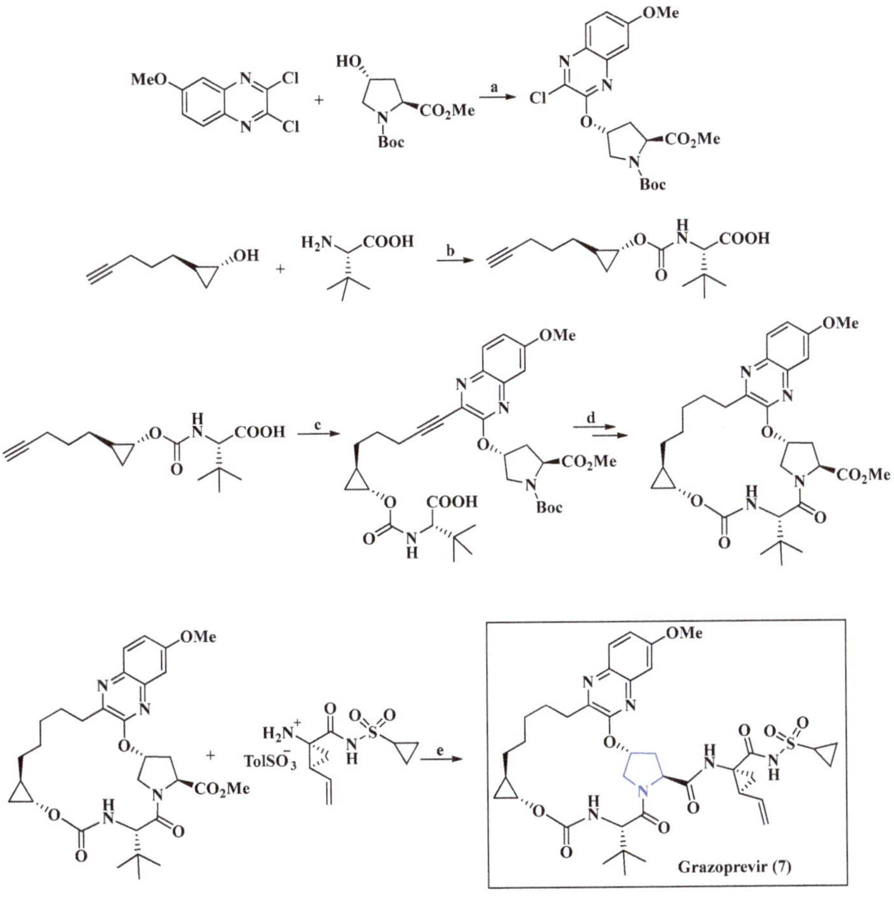

Scheme 6. Synthesis of grazoprevir.

Reagents and conditions: (a) DBU (1.05 equiv.), DMAc, 50 °C; (b) CDI, Hunig's base, 95 °C, 2.5 h; (c) Pb(OAc)$_2$, P(*t*-Bu)$_3$BF$_4$, K$_2$CO$_3$, CPME/MeCN; (d) (1) Pd/C, H$_2$, IPAc/MeOH; (2) (i) PhSO$_3$H, (ii) HATU, NEt$_3$, MeCN; (e) DEC, pyridine, MeCN.

Structure–Activity Relationship of Grazoprevir: A group of linear HCV NS3/4A protease inhibitors was created by removing the macrocyclic linker found in grazoprevir. This allows for the exploration of diverse quinoxalines while conferring conformational flexibility. Inhibitors with small substituents at the 3-position (**R^1**) of quinoxaline were found to be effective in maintaining potency. The 3-chloroquinoxaline demonstrated outstanding potency against wild-type HCV NS3/4A protease. Replacing the cyclopropyl-sulfonamide with a more hydrophobic 1-methyl cyclopropyl-sulfonamide group (**R^2**) generally enhances the potency of the resulting analogues. Similarly, substituting the tert-butyl group (**R^3**)

with a bulkier cyclopentyl moiety led to the development of compounds with improved potency (Table 7) [160]. The enzyme inhibition constants (K_i) were determined for the wild-type genotype 1a NS3/4A protease, as well as the resistant variants R155K and D168A. Additionally, a subset of compounds underwent testing to determine their cellular antiviral potencies (EC_{50}) using replicon-based antiviral assays. These assays, which assessed the efficacy of the compounds, were not only run against the wild-type HCV strain but also against the drug-resistant variants R155K, A156T, D168A, and D168V [161].

Table 7. SAR Analysis of grazoprevir derivatives.

Compound	Substitution			Activity	
	$-R^1$	$-R^2$	$-R^3$	EC_{50} (nM)	Inhibitor Constant (K_i) (nM)
	-CH₂CH₃	-H	*tert*-butyl	24	19 ± 2.7
	Cl	-H	*tert*-butyl	6.6	7.8 ± 1.1
	Cl	-CH₃	*tert*-butyl	6.3	6.1 ± 1.1
	-CH₃	-H	cyclopentyl	10	9.2 ± 0.9
	-CH₃	-CH₃	cyclopentyl	4.5	7.1 ± 1.1
	Cl	-CH₃	cyclopentyl	3.1	3.9 ± 0.7

Mechanism of Action of Grazoprevir: Grazoprevir is a potent and selective inhibitor of the NS3/4A protease enzyme in the hepatitis C virus (HCV). The NS3/4A protease enzyme plays a critical role in HCV replication by cleaving the HCV polyprotein into the individual functional proteins necessary for the virus to replicate and propagate. Grazoprevir's mechanism of action has been extensively studied and documented. Grazoprevir effectively inhibited the NS3/4A protease enzyme by binding to the enzyme's active site, thereby preventing the cleavage of the HCV polyprotein and inhibiting HCV replication [162]. There are other drugs available on the market for the treatment of HCV, such as boceprevir, sofosbuvir, etc. Their mechanisms of action, ways of use and side effects are given in Table 8.

Table 8. Anti-HCV drugs.

Sl. No.	Drug Name	Drug Target	Mechanism of Action	Ways of Use	Side Effect	Brand Name
1.	Asunaprevir [147]	Genome polyprotein (Hepatitis C virus genotype 1b)	Inhibits the proteolytic activity of HCV NS3/4A protease	Oral	Flu-like symptoms, skin rash, irritability, headache	Sunvepra
2.	Boceprevir [163]	NS3/4A protein (Hepatitis C Virus)	Inhibits the proteolytic activity of HCV NS3/4A protease	Oral	Vomiting, dry skin, fever, sore throat	Victrelis

Table 8. *Cont.*

Sl. No.	Drug Name	Drug Target	Mechanism of Action	Ways of Use	Side Effect	Brand Name
3.	Grazoprevir [164]	NS3/4A protein (Hepatitis C Virus)	Inhibits the proteolytic activity of HCV NS3/4A protease	Oral	Headache, nausea, trouble breathing	Zepatier
4.	Sofosbuvir [165]	RNA-dependent RNA-polymerase (Hepatitis C Virus)	Inhibits RNA-dependent RNA polymerases of HCV NS5B (non-structural protein 5B)	Oral	Fatigue, headache, decreased appetite	Vosevi
5.	Daclatasvir [166]	Nonstructural protein 5A (Hepatitis C Virus)	Disrupts the NS5A proteins that have undergone hyper-phosphorylation; interferes with the functioning of newly formed HCV replication complexes.	Oral	Headache, feeling tired, nausea	Daklinza

Some other synthesized compounds that show activity against HCV are given in Table 9.

Table 9. Synthesized anti-HCV compounds.

Sl. No.	Antiviral Agent	Drug Target	Activity
1.		Hepatitis C Virus (HCV) [167]	The fatty acid synthase inhibitors showed good antiviral activity in a cell-based HCV replicon assay and an acceptable selectivity index. Their observed cell permeability in an MDCK permeability assay supports these findings. The antiviral activities align with the biochemical inhibition (IC_{50} values > 30 µM) of Hfasn [168].
2.		HCV [169]	Imidazo[2,1-b]thiazole targets HCV NS4B, specifically the second amphipathic α helix (4BAH2). The effectiveness of the compound against Huh-7 cells was assessed, screening an EC_{50} value of 18 nM [170].
3.		HCV [171]	The efficacy of the compound against Huh-7 cells shows promising activity, which reflects in the EC_{50} value (16 nM).

5.3. Anti-HBV Agent

5.3.1. Anti-HBV Agent Lamivudine

Lamivudine (Figure 15) is a nucleoside reverse transcriptase inhibitor that inhibits the reverse transcriptase of the human hepatitis B virus (HBV). It is a safe medicine with minimal side effects and can be prescribed for pregnant women and children over five years of age [172].

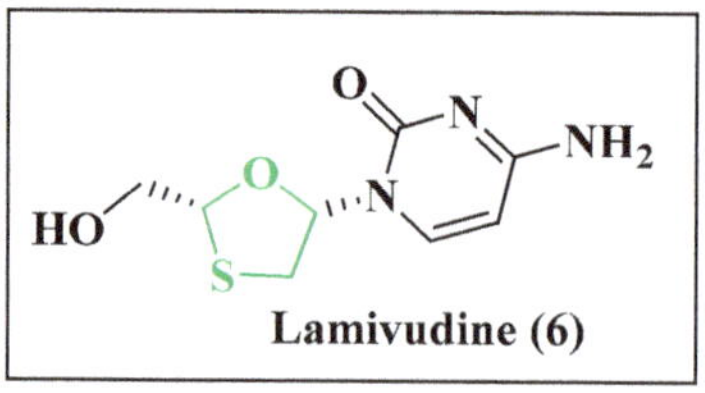

Figure 15. Chemical structure of lamivudine (**6**).

Key synthetic steps for Scheme 7 [173,174]: Formation of mixture of diastereomer at 0 °C (b) separation of diastereomers is done by recrystallization, (c) diastereo pure compound is treated with methanolic K_2CO_3 to obtain product **6**.

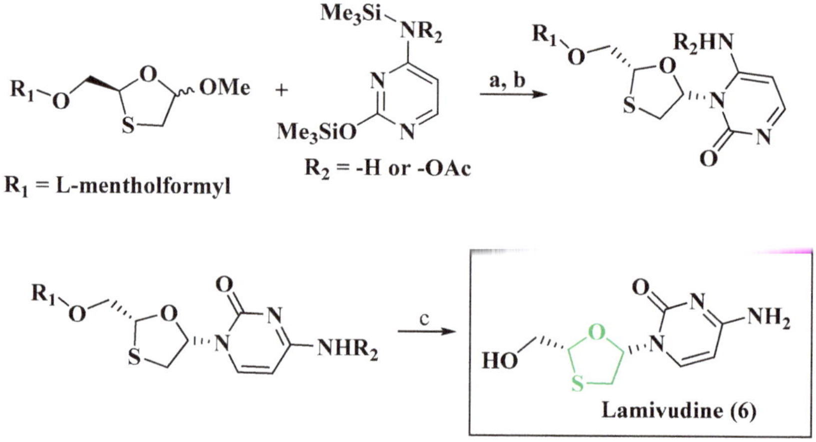

Scheme 7. Synthesis of anti-HBV agent lamivudine.

Reagents and conditions: (a) Reactants are mixed and cooled to 0 °C in MeOH; (b) separation of diastereomers; (c) MeOH, K_2CO_3.

Mechanism of Action of Lamivudine: Lamivudine is a nucleoside analogue that is used in the treatment of hepatitis B virus (HBV) infection. The mechanism of action of lamivudine is based on its ability to inhibit HBV reverse transcriptase, which is a critical enzyme for viral replication. Once inside the infected cell, lamivudine is phosphorylated by cellular enzymes into its active triphosphate form. This active form of lamivudine competes with the natural nucleotide building blocks for incorporation into the growing viral DNA chain. However, lamivudine lacks the 3′-OH group required for further chain extension, thereby resulting in the termination of viral DNA synthesis [175].

5.3.2. Anti-HBV Agent Entecavir

Entecavir (Figure 16) is a guanosine nucleoside analogue active against hepatitis B (HBV). It is highly efficient in preventing all stages of replication. Compared to the other Hepatitis B drugs, lamivudine and entecavir are more effective; the corresponding triphosphate binds with HBVpol with amino acid residues ARG A: 23, LYS A: 14, ASN A: 18 and ALA A: 68 and effectively inhibits its activity (Figure 17) [176,177].

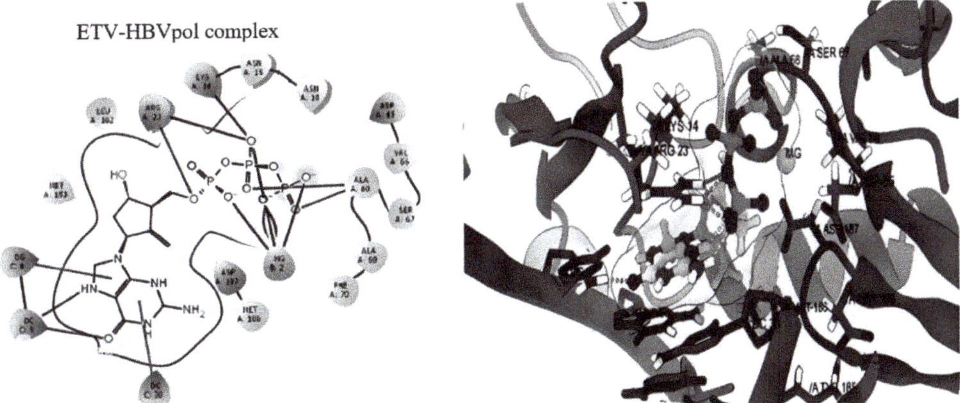

Figure 16. Chemical structure of entecavir (**28**).

Figure 17. Entecavir triphosphate docking with HBVpol.

Key synthetic steps for Scheme 8 [178]: (a) Protection of aliphatic alcohol; (b) activation of the terminal alkyne; (c,d) synthesis of the epoxide followed by (e) intramolecular cyclization; (f,g) protection and deprotection of the alcohols; (h) Mitsunobu reaction with 2-amino-6-chloropurine; followed by (i) acid treatment; and (j) saponification to obtain the Entecavir **28**.

Reagents and conditions: (a) TBSCI (1.1 equiv.), imidazole, THF, rt; (b) K_2CO_3 cat., MeOH; (c) m-CPBA, CH_2Cl_2; (d) Ac_2O, NEt_3, DMAP cat., CH_2Cl_2; (e) Cp_2TiCl_2 20 mol%, $IrCl(CO)(PPh_3)_2$ 10 mol%, Mn (2 equiv.), collidine, TMSCl, H_2 (4 bar), THF; (f) p-O_2NBzCl, NEt_3, CH_2Cl_2; (g) 5% (+)-CSA, MeOH; (h) 2-amino-6-chloropurine, DIAD, PPh_3, THF, $-10\,°C$; (i) HCOOH, $50\,°C$; (j) MeONa, MeOH.

Structure–Activity Relationship of Entecavir: An extensive investigation of the structure–activity relationship (SAR) of entecavir and its analogues is shown in Table 10. It was discovered that these compounds are the most potent inhibitors of HBV replication, with the ability to effectively inhibit lamivudine-resistant HBV. These compounds are carbocyclic guanosine nucleoside analogues (R^1) and are highly effective when tested against HBV in HepG2.2.15 cells [179]. The plasma half-life of entecavir in rats and dogs is 4–9 h. It is metabolized by HepG2 cells to the corresponding mono-, di-, and triphosphates.

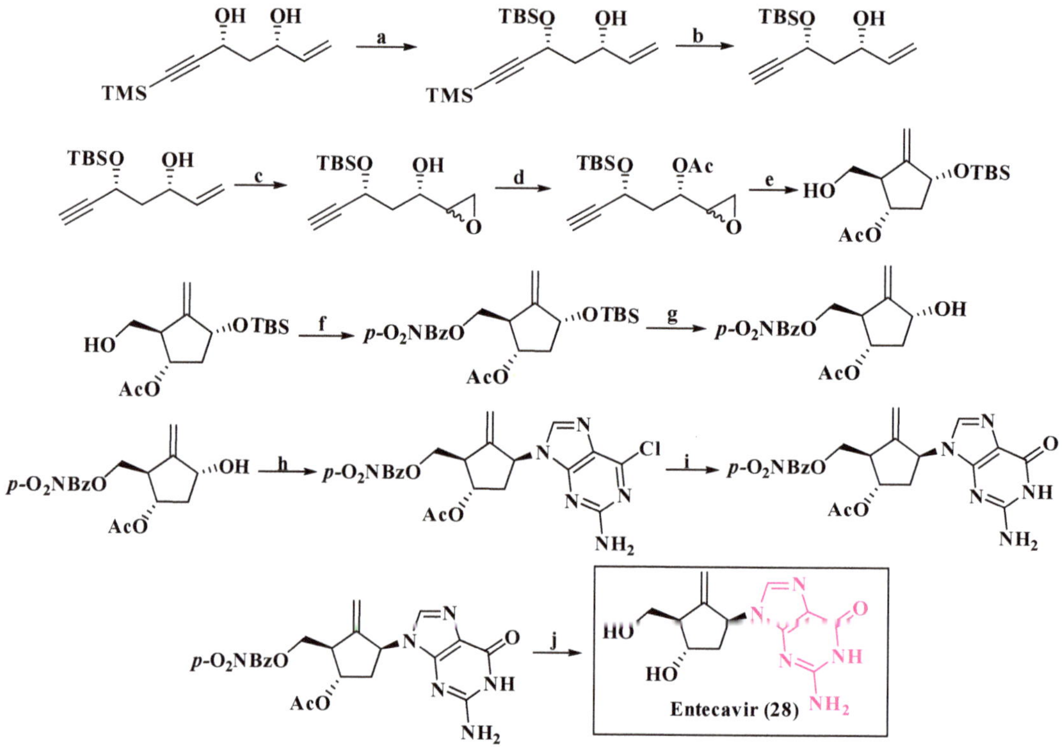

Scheme 8. Synthesis of entecavir.

Table 10. SAR analysis of entecavir derivatives.

Compound	Substitution		Activity
	$-R^1$	$-R^2$	IC$_{50}$ (µM)
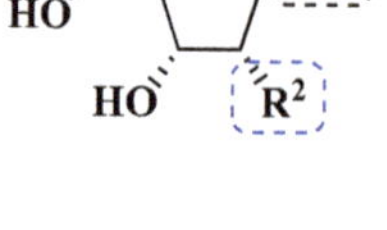	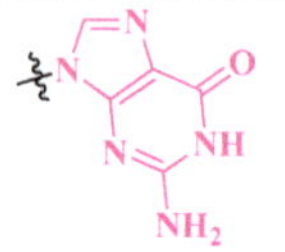 (guanine)	$-H$	0.03
	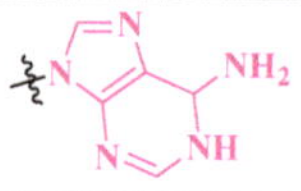 (6-amino purine)	$-H$	0.128
	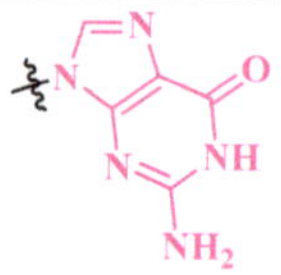 (guanine)	$-OH$	>100

Mechanism of Action of Entecavir: Entecavir is a nucleoside analogue that inhibits hepatitis B virus (HBV) DNA replication by interfering with the activity of viral polymerase, an enzyme essential for the virus to replicate its genetic material. In HBV-infected cells, it is phosphorylated into its active form, entecavir triphosphate, which competes with the natural substrate, deoxyguanosine triphosphate, for incorporation into the elongating viral DNA chain. The incorporation of entecavir triphosphate into the viral DNA chain leads to chain termination, preventing further extension of the viral DNA and ultimately inhibiting HBV replication. Entecavir's mechanism of action has been extensively studied and has been shown to be highly effective in suppressing HBV replication and reducing liver damage. Due to its high potency and low risk of developing viral resistance, entecavir has become one of the preferred first-line treatments for chronic HBV infection [180,181].

5.3.3. Anti-HBV Agent Dehydro-Andrographolide and Andrographolide Derivatives

Dehydro-andrographolide and andrographolide compounds have demonstrated the ability to inhibit the replication of HBV DNA, with IC_{50} values of 22.58 and 54.07 μM, respectively [172].

Key synthetic steps for Scheme 9 [172]: (a) Compounds are obtained with the help of esterification reaction with acids in the presence of 4-dimethylaminopyridine (DMAP) and N',N'-dicyclohexylcarbodiimide (DCC) in anhydrous dichloromethane to obtain the product **35**.

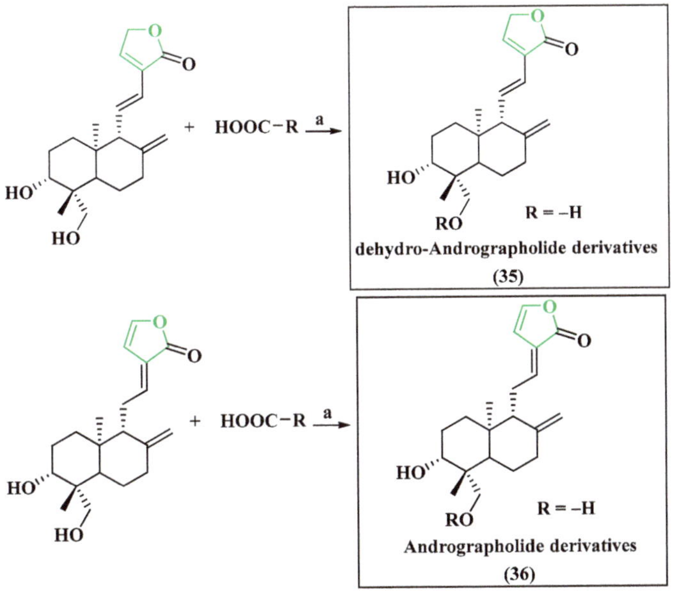

Scheme 9. Synthesis of dehydro-andrographolide (**35**) and andrographolide (**36**) derivatives.

Reagents and conditions: (a) corresponding acids, DMAP, DCC, CH_2Cl_2, rt.

Structure–Activity Relationship of dehydro-Andrographolide: The SARs of the derivatives indicate that having a free hydroxyl group at C-2 can result in enhanced anti-HBV properties. Additionally, maintaining the double bond between C-8 and C-17, as well as the conjugated double bonds between C-11 and C-14, or C-12 and C-15, is crucial for preserving anti-HBV activity and decreasing cytotoxicity. To improve the anti-HBV activity, it is useful to incorporate the 3-methoxycinnamoyl, nicotinoyl, 2-furoyl, or 2-thenoyl groups shown in Table 11 [105]. The anti-HBV activity of the compounds derived from dehydro-andrographolide and andrographolide were investigated. Specifically, their ability to inhibit the secretion of HBsAg and HBeAg, as well as HBV DNA replication, was evaluated using HepG 2.2.15 cells in vitro. Tenofovir, a known antiviral agent, was used as the positive control in the study [182].

Table 11. SAR analysis of dehydro-andrographolide and andrographolide derivatives.

Compound	Substitution −R	Activity IC$_{50}$ (μM)
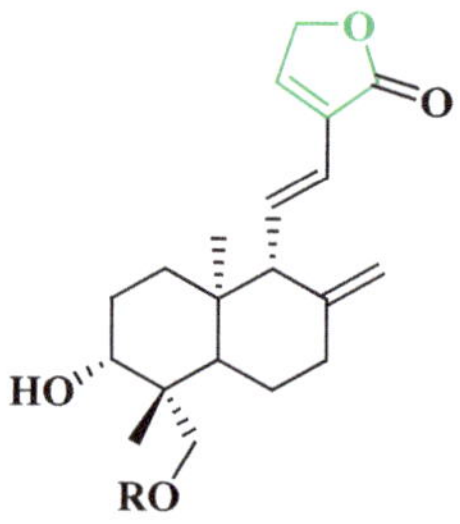	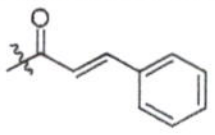	18
	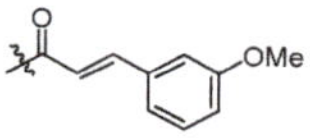	1970
	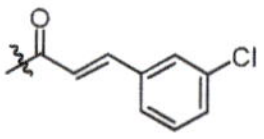	207
	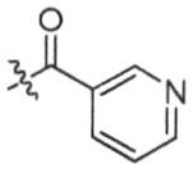	40
	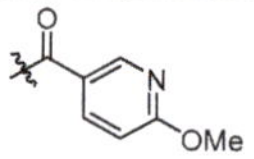	2313
	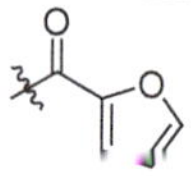	19.2
	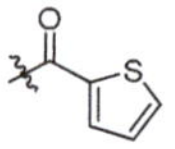	162
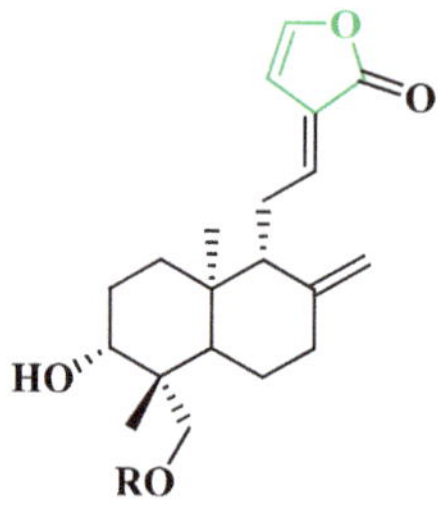	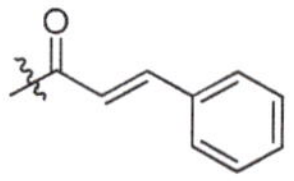	210
	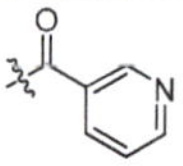	1137
	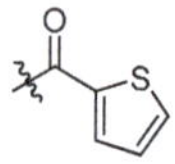	121
	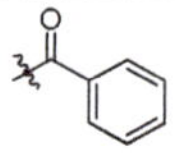	880
	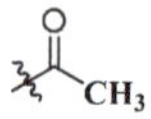	280
	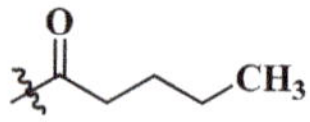	565
	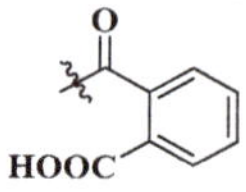	1317

Mechanism of Action of dehydro-Andrographolide: Dehydro-andrographolide inhibits HBV replication by blocking the binding of the HBV core protein to viral RNA. It also activates the host immune response, which can help control HBV replication and clear infected cells [183]. Overall, dehydro-andrographolide has shown promising anti-HBV activity through its ability to inhibit viral replication and enhance the host immune response. However, further studies are needed to fully understand its mechanisms of action and potential clinical applications [184]. The details of other anti-HBV drugs, along with their mechanisms of action, are mentioned in Table 12.

Table 12. Anti-HBV drugs.

Sl. No.	Drug Name	Drug Target	Mechanism of Action	Ways of Use	Side Effect	Brand Name
1.	Entecavir [185]	DNA polymerase	Inhibits the activity of HBV DNA polymerase	Oral	Headache, dizziness, nausea	Baraclude
2.	Telbivudine [186]	DNA polymerase	Inhibits the activity of HBV DNA polymerase	Oral	Diarrhea, cough, headache, dizziness	Tyzeka
3.	Lamivudine [187]	Protein P (HBV-F)	Inhibits reverse transcriptase	Oral	Nausea, diarrhea, headaches	Lamivudine

Some other synthesized compounds with activity against HBV are tabulated below (Table 13).

Capsid assembly modulators (CpAMs) belong to a novel category of antiviral compounds that specifically target the core protein of the hepatitis B virus (HBV) to interfere with the assembly process. HepG2.2.15 cells are a type of human hepatoblastoma cells that have been genetically modified to stably express the hepatitis B virus (HBV). The compounds under investigation inhibit HBV replication by interfering with the assembly of the HBV capsid protein [188].

Table 13. Synthesized anti-HBV compounds.

Sl. No.	Antiviral Agent	Drug Target	Activity
1.	[chemical structure]	HBV [189]	Capsid assembly modulators (CpAMs) are antiviral compounds that target the core protein of the hepatitis B virus (HBV) to disrupt assembly. In HepG2.2.15 cells, which express HBV, these compounds inhibit HBV replication by interfering with capsid protein assembly with EC_{50} = 511 nM.
2.	[chemical structure]	HBV [190]	The anti-HCV activities were tested in the Huh-Luc/neo cell line and cytotoxicity of the test compound was determined on both MT-2 cell lines with EC_{50} = 10 μM.

5.4. Anti-RSV Agent

Anti-RSV Agent Ribavirin

The antiviral property of this drug (Figure 18) was studied in 1972 [191]. This drug is presently used for the treatment of RSV. DeVincenzo et al. have studied the inhibitory activities of ribavirin against the RSV F-protein [192].

Key synthetic step for the Scheme 10 [193]: (a) Nucleophilic substitution in the presence of bio-catalyst, (b) amide formation in the presence of NH_3 to obtain product **10**.

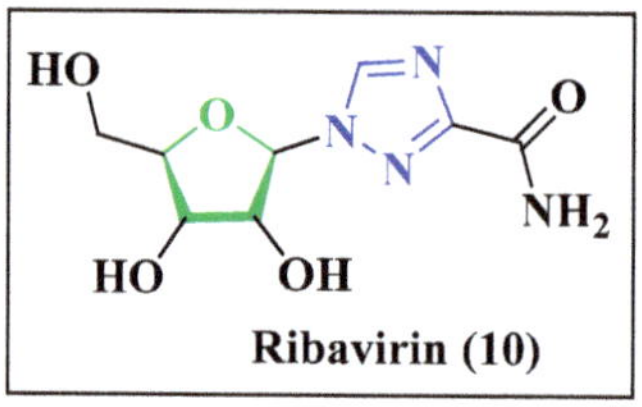

Figure 18. Chemical structure of ribavirin (**10**).

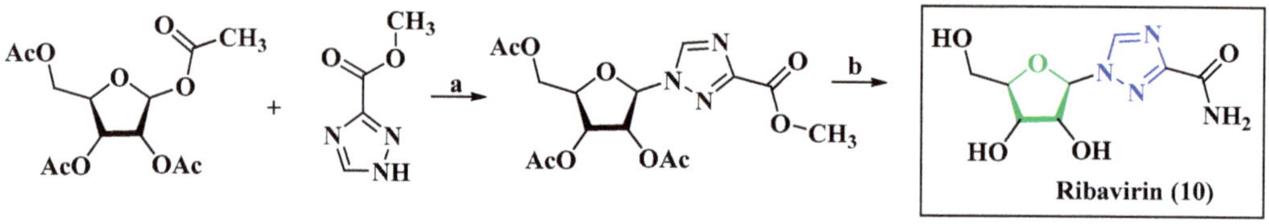

Scheme 10. Synthesis of ribavirin.

Reagents and conditions: (a) Purine nucleoside phosphorylase, buffer solution of pH = 7; (b) NH$_3$, MeOH.

Mechanism of Action of Ribavirin: Ribavirin is a broad-spectrum antiviral agent that has activity against a range of RNA and DNA viruses, including respiratory syncytial virus (RSV). The mechanism of action of ribavirin is not completely understood, but it is believed to involve several different mechanisms [194]. One proposed mechanism is that ribavirin interferes with the synthesis of viral RNA by inhibiting the activity of the viral RNA-dependent RNA polymerase. Another proposed mechanism is that ribavirin induces mutations in the viral genome, leading to non-functional or less-virulent viral particles. Additionally, ribavirin has been shown to stimulate the host's immune response, which may contribute to its antiviral effects [195–197]. RSV-IGIV and palivizumab are the available drugs on the market for the treatment of RSV. Their mechanisms of action, ways of use and side effects are given in Table 14.

Table 14. Anti-RSV drugs.

Sl. No.	Drug Name	Drug Target	Mechanism of Action	Ways of Use	Side Effect	Brand Name
1.	Ribavirin [198]	RNA-directed RNA polymerase L (HPIV-2)	Targets viral RNA polymerase to inhibit mRNA synthesis	Oral or inhaled	Hemolytic anemia, asthenia, rigors, fevers	Rebetol
2.	RSV-IGIV [199]	RSV G protein	Prevents the binding of RSV surface glycoproteins F and G	-	-	RespiGam
3.	Palivizumab [200]	Fusion glycoprotein F0 (Human respiratory syncytial virus B)	Prevents the binding of RSV surface glycoprotein F	Injection	Sore throat, runny nose, vomiting	Synagis

Some other synthesized compounds that show activity against RSV are given in Table 15.

Table 15. Synthesized anti-RSV compounds.

Sl. No.	Antiviral Agent	Drug Target	Activity
1.	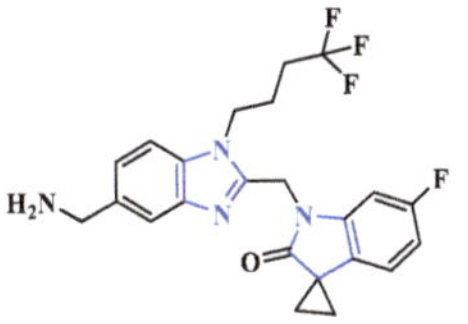	Respiratory syncytial virus [201]	The anti-RSV property of the compound tested with Hep2 cells with IC$_{50}$ = 1.2 nM. The cell viability was measured using MTT reagents and cell cytotoxicity was evaluated through parallel assessments with plaque reduction assays.
2.	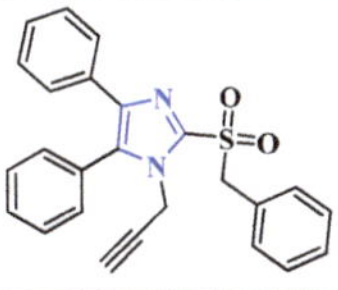	Respiratory syncytial virus [202]	The screening of the compounds against nonattenuated respiratory syncytial virus was done by high-throughput protocol with EC$_{50}$ = 0.36–0.55 μM.

5.5. Anti-HCMV Agent

5.5.1. Anti-HCMV Agent Iso-Valganciclovir Hydrochloride

Iso-Valganciclovir hydrochloride (Figure 19) is used for the treatment of cytomegalovirus (CMV). It is a type of nucleoside analogue and is the cutting-edge drug candidate against CMV [203].

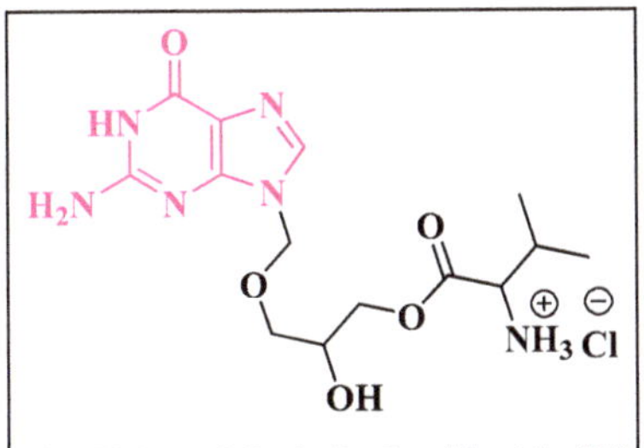

Figure 19. Chemical structure of iso-Valganciclovir hydrochloride (**21**).

Key synthetic step for Scheme 11 [204]: (a) Addition of reaction of 3-(chloromethoxy)prop-1-ene to *o*-benzyl guanine in the presence of a base, followed by oxidation; (b) addition reaction with *S*-2-azido-3-methylbutanoic acid and further reduction gives the final product **21**, as shown in Scheme 11.

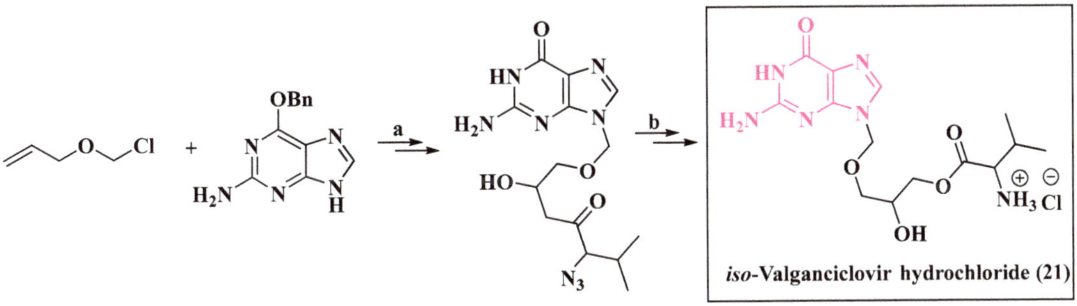

Scheme 11. Synthesis of *iso*-valganciclovir hydrochloride.

Reagents and conditions: (a) (i) NaH, DMF; (ii) KMnO$_4$, acetone; (ii) 10% Pd/C, MeOH; (iv) (*S*)-2-azido-3-methylbutanoic acid, DCC, DMSO; (b) 10% Pd/C, MeOH.

Mechanism of Action of *Iso*-Valganciclovir Hydrochloride: *Iso*-Valganciclovir hydrochloride is a prodrug of the antiviral agent ganciclovir, which is converted to its active form by hydrolysis of the valine ester in the liver and blood. The active form of ganciclovir works by inhibiting the viral DNA polymerase, which is essential for the replication of HCMV. By inhibiting the viral DNA polymerase, ganciclovir prevents the formation of new viral DNA chains, which ultimately inhibits HCMV replication [205].

5.5.2. Anti-HCMV Agent Ganciclovir

Ganciclovir (Figure 20) is a marketed drug for the treatment of HCMV; it acts as a DNA polymerase to inhibit synthesis of viral DNA [206].

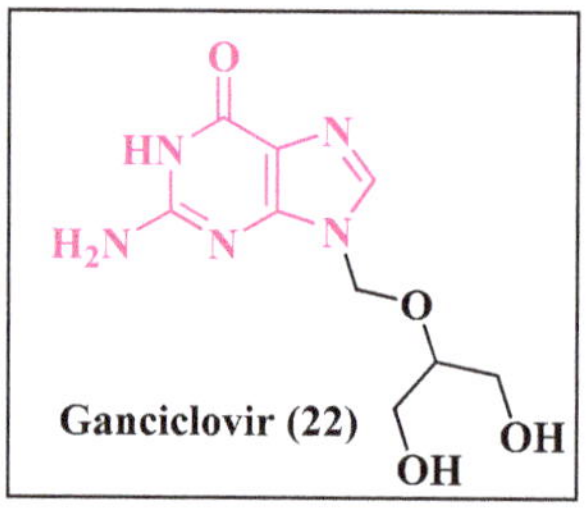

Figure 20. Chemical structure of ganciclovir (**22**).

Key synthetic steps for Scheme 12 [207]: (a) *N*-Acylation of the guanine then (d) reacts with 2-(acetoxymethoxy)propane-1,3-diyl diacetate to form N-(9-(((1,3-dihydroxypropan-2-yl)oxy)methyl)-6-oxo-6,9-dihydro-1H-purin-2-yl)acetamide (prepared from 4-(chloromethyl)-1,3-dioxolane) (e) by the deprotection of the amine and alcohol group; the final product **22** was thereby obtained.

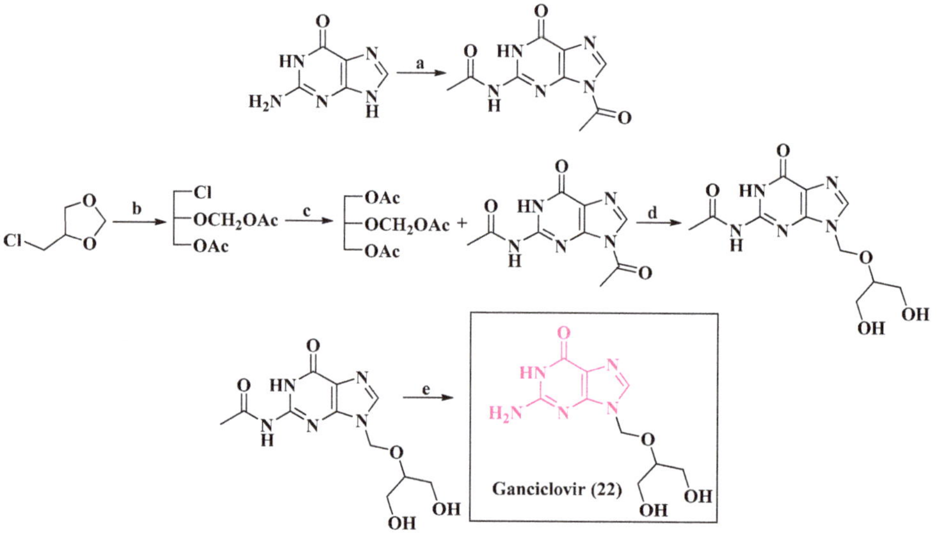

Scheme 12. Synthesis of Ganciclovir.

Reagents and conditions: (a) Ac$_2$O/HOAc, 140 °C; (b) Ac$_2$O/HOAc/ZnCl$_2$, r.t.; (c) KOAc/DMF, 150 °C; (d) EtSO$_3$H, 165–170 °C; (e) 40% aq. MeNH$_2$, 75 °C.

Mechanism of Action of Ganciclovir: Ganciclovir is an antiviral drug used to treat HCMV infections. It stops viral DNA synthesis by acting as a chain terminator, which inhibits the elongation of the viral DNA strand. Ganciclovir triphosphate, its active form, is similar to guanosine and is selectively toxic to infected cells as it is preferentially incorporated into viral DNA, reducing viral replication and controlling infections [206].

5.5.3. Anti-HCMV Agent 1,2,4-Triazol-Quinoxalin Derivative

Another potential anti-HCMV agent is represented by quinoxaline derivatives, which have been found in recent research studies. Quinoxaline is a heterocyclic compound containing a benzene ring fused to a pyrazine ring, and its derivatives have diverse biological activities, including antiviral properties. These compounds have exhibited greater antiviral activity against HCMV compared to the standard drug ganciclovir [208]. The triazole and quinoxaline moieties in 1,2,4-triazoloquinoxaline (Figure 21) have been reported to exhibit antiviral activity against HCMV. The triazole moiety is a five-membered heterocyclic ring containing three nitrogen atoms, which has been reported to possess antiviral activity. The quinoxaline moiety is a bicyclic aromatic ring system that has also been reported to exhibit antiviral activity. Studies have shown that 1,2,4-triazoloquinoxaline derivatives can inhibit HCMV replication by targeting the viral DNA polymerase, which is a key enzyme involved in viral replication. These compounds have also been reported to have low cytotoxicity toward human cells, making them potentially useful as antiviral agents [209].

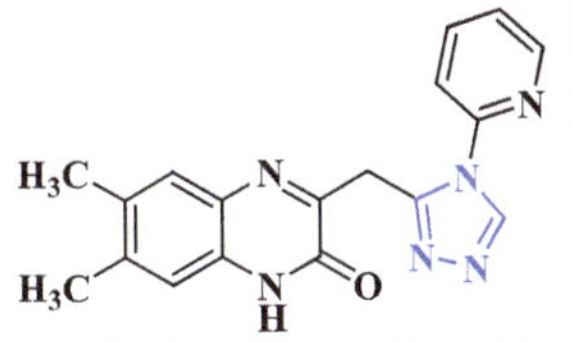

Figure 21. 1,2,4-triazol-quinoxalin derivative (**19**), a new anti-human-cytomegalovirus (HCMV) agent.

Key synthetic steps for Scheme 13 [210]: (a) The compound 2-(6,7-dimethyl-3-oxo-3,4-dihydroquinoxalin-2-yl)acetohydrazide was reacted with triethylorthoformate in ethanol to afford ethyl [(6,7-dimethyl-3-oxo-3,4- dihydroquinoxalin-2-yl)acetyl]hydrazonoformate; further, (b) treatment of the hydrazonoformate with 2-aminopyridine in acetic acid reflux afforded 6,7-dimethyl-3-{[4-(pyridin-2-yl)- 4H-1,2,4-triazol-3-yl]methyl}quinoxalin-2(1H)-one to obtain the product **19**.

Reagents and conditions: (a) C_2H_5OH, rt; (b) CH_3COOH, reflux.

Mechanism of Action of 1,2,4-triazol-quinoxalin Derivatives: 1,2,4-triazol-quinoxalin derivatives have potential as antiviral agents against HCMV, but their exact mechanism of action is not fully understood. They may inhibit viral DNA replication and interfere with viral gene expression or virion assembly. Additionally, they may have immunomodulatory effects that enhance antiviral activity or inhibit immune evasion strategies against HCMV [211]. Valganciclovir is another available drug for the treatment of HCMV. The mechanism of action, ways of use and side effects of valganciclovir and ganciclovir are given in Table 16.

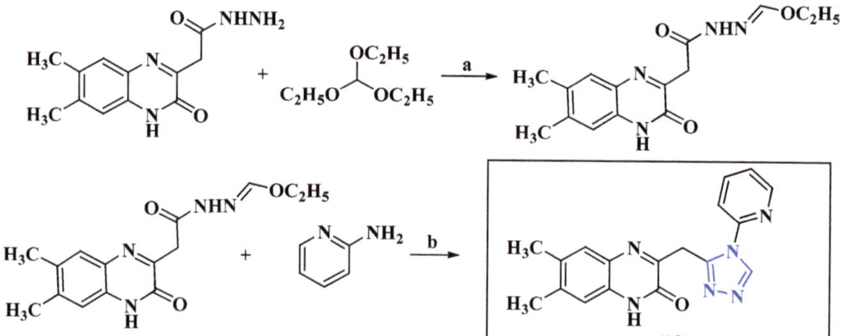

Scheme 13. Synthesis of 1,2,4-triazol-quinoxalin derivative.

Table 16. Anti-HCMV drugs.

Sl. No.	Drug Name	Drug Target	Mechanism of Action	Ways of Use	Side Effects	Brand Name
1.	Valganciclovir [212]	DNA polymerase	Inhibits the activity of viral DNA polymerase	Oral	Diarrhea, upset stomach, dizziness	Valcyte
2.	Ganciclovir [213]	DNA polymerase catalytic subunit	DNA polymerase to inhibit viral DNA synthesis	Oral	Diarrhea, loss of appetite, increased sweating	Cytovene

Some other synthesized compounds that show activity against HCMV are given in the Table 17.

Table 17. Synthesized anti-HCMV compounds.

Sl. No.	Antiviral Agent	Drug Target	Activity
1.	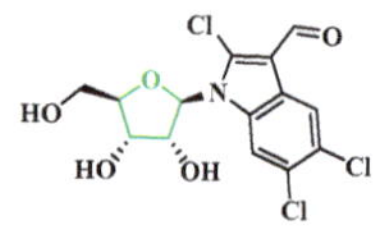	HCMV [214]	Synthesized trichlorinated indole nucleosides were tested for activity against human cytomegalovirus (HCMV). Cytotoxicity was assessed using two methods: microscopic inspection of uninfected HFF cells [215] and crystal violet staining with spectrophotometric quantitation in KB cells [216]. The IC$_{50}$ was found to be 0.23 µM.
2.	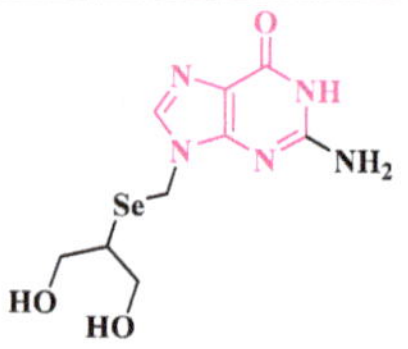	HCMV [217]	The compound was assayed for antiviral activity against HCMV (Davis, VR-807) cells in a cytotoxic assay. The EC$_{50}$ value of 53.1 µM showed only moderate antiviral activity against HCMV.

5.6. Anti-HSV Agent

Lycogarubins have been reported as the first naturally occurring dimethyl pyrrole-dicarboxylate attached to two indole moieties [218]. These compounds were isolated from the fruit bodies of the slime molds *Arcyria denudate*, and are closely related to Arcyriarubins and Arycyriaflavins. Three novel dimethyl pyrrole dicarboxylates named Lycogarubins A–C were isolated by Haahimoto et al. from the *Myxomycetes Lycogala epidendrum*, among which Lycogarubin C showed the effective potency against HSV-I [219]. Idoxuridine, tri-fluridine and brivudine are marketed anti-HSV drugs used as ointments for the treatment of eye infections due to HSV. They act by inhibiting DNA polymerase of HSV and interrupting viral DNA synthesis. The mechanisms of action, ways of use, and side effects of idoxuridine, trifluridine and brivudine are given in Table 18.

Table 18. Anti-HSV drugs.

Sl. No.	Drug Name	Drug Target	Mechanism of Action	Ways of Use	Side Effect	Brand Name
1.	Idoxuridine [95]	DNA polymerase	HSV DNA polymerase to inhibit viral DNA synthesis	Used as an ointment	Eye irritation or pain, swelling of the eye	Dendrid
2.	Trifluridine [220]	DNA polymerase	Inhibits HSV DNA replication	Eye drop	Eye pain, mild burning of eyes	Viroptic
3.	Brivudine [221]	DNA polymerase	Inhibits HSV DNA replication	Oral	No side effect	Zostex

Mechanism of Action of Anti-HSV Drugs: Anti-HSV drugs target the herpes simplex virus (HSV) and work by inhibiting viral replication and/or reducing the severity and duration of HSV symptoms. There are three main classes of anti-HSV drugs:

Nucleoside analogues: These drugs mimic the structure of the nucleotides that the virus needs to replicate its DNA. When the virus interacts with the DNA part of the nucleoside, it disrupts the replication process, preventing the virus from making new copies of it. Examples of nucleoside derivatives used to treat HSV include acyclovir, valacyclovir, and famciclovir [58].

Non-nucleoside inhibitors: These drugs target specific viral enzymes that are essential for viral replication. They work by binding to the enzyme and blocking its activity, thereby preventing the virus from replicating. Examples of non-nucleoside inhibitors used to treat HSV include foscarnet and cidofovir [222].

Interferons: These drugs are proteins that the body naturally produces in response to viral infections. They provoke activity by stimulating the immune system to produce antiviral proteins that can inhibit viral replication. Examples of interferons used to treat HSV include interferon alpha and interferon beta [223]. Here, it is important to note that while these drugs can help reduce the severity and duration of HSV symptoms, they do not cure the infection. The virus remains in the body and can reactivate, causing recurrent outbreaks.

5.7. Anti-Ebola Agent

Anti-Ebola agents are drugs that target the Ebola virus by preventing its replication or entry into human cells. Examples include ZMapp, a combination of three monoclonal antibodies, and remdesivir (Figure 22) [224,225], which is used as a broad-spectrum antiviral drug. The other treatments developed include RNA-based therapies and gene therapies. Such advanced treatments offer genuine hope for a better future, in which EVD will not be considered to be a major concern of public health.

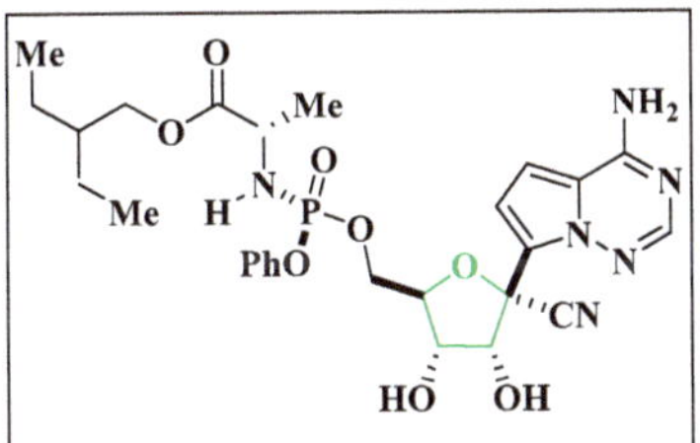

Figure 22. Anti Ebola agent remdesivir (**15**).

Key synthetic steps for Scheme 14 [226]: (a) The iodopyrazole was dissolved in THF and cooled to 0 °C, TMSCl was added, and after 1 h, phenylmagnesium chloride was added. The reaction mixture was cooled to −20 °C and iso-propylmagnesium chloride was added slowly to (b) a pre-cooled (−40 °C.) solution of (3R,4R,5R)-2-(4-aminopyrrolo[2,1-f][1.2.4]triazin-7-yl)-3,4-bis(benzyloxy)-5-((benzyloxy)methyl)tetrahydrofuran-2-ol in DCM trifluoroacetic acid was added, followed by a pre-cooled (−30 °C.) solution of TMSOTf and TMSCN in DCM at rt; (c) the tribenzyl cyano nucleoside was dissolved in anhydrous CH_2Cl_2 and cooled to about −20 °C. A solution of BCl_3, the reaction mixture, was stirred for 1 h at about −20 °C. MeOH was added dropwise (d) to a mixture of (2R,3R,4S.5R)-2-(4-aminopyrrolo[2,1-f][1.2.4] triazin-7-yl)-3,4-dihydroxy-5-(hydroxymethyl)tetrahydrofuran-2-carbonitrile, 2,2-dimethoxypropane and acetone at ambient temperature, to which sulfuric acid was added. The mixture was warmed to about 45 °C and (e) N,N-dimethylacetamide was added to a mixture of (2R,3R,4S.5R)-2-(4-aminopyrrolo[2,1-f][1.2.4]triazin-7-yl)-3,4-dihydroxy-5-(hydroxymethyl)tetrahydrofuran-2-carbonitrile, (S)-2-ethylbutyl2-(((S)-(4-nitrophenoxy) (phenoxy)phosphoryl)amino)propanoate and $MgCl_2$. Then the resulting reaction mixture was warmed at 30 °C with constant stirring and N,N-diisopropylethylamine was added slowly, (f) the deprotection of the alcohols was performed by conc. HCl to obtain the product 15.

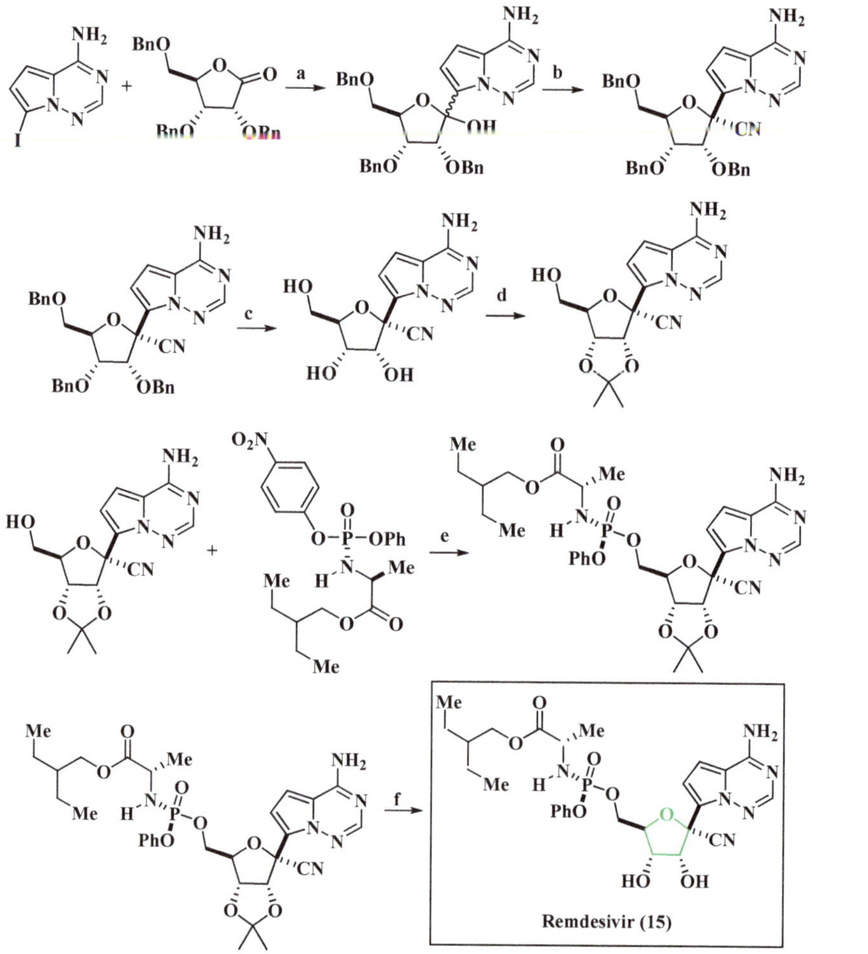

Scheme 14. Synthesis of remdesivir.

Reagents and conditions: (a) TMSCl, PhMgCl, ^{i}PrMgCl·LiCl, THF, $-20\,°C$; (b) TMSCN, TMSOTf, TfOH, CH_2Cl_2, $-78\,°C$; (c) (1) BCl_3, CH_2Cl_2, $-40\,°C$; (2) Et_3N, MeOH, $-78\,°C$–rt; (d) 2,2-DMP, H_2SO_4, Me_2CO, rt; $45\,°C$; (e) $MgCl_2$, DIPEA, MeCN, $50\,°C$; (f) 12 N HCl, THF (1:5), rt.

Mechanism of Action of Remdesivir: Remdesivir interferes with the Ebola virus's replication by inhibiting its RNA-dependent RNA polymerase enzyme. It acts as a chain terminator, preventing the virus from replicating further and causing harm to the host [97]. The mechanism of action and ways of use are given in Table 19.

Table 19. Anti-EBOV drugs.

Sl. No.	Drug Name	Drug Target	Mechanism of Action	Ways of Use	Side Effect	Brand Name
1.	BCX4430 [227]	RNA-directed RNA polymerase L	Acts as a nonobligate RNA chain terminator upon incorporation into viral RNA	Intramuscular or oral	-	BioCryst
2.	Neplanocin A [228]	S-adenosyl-l-homocysteine (SAH) hydrolase inhibitor	-	-	-	-
3.	Lectins [229]	-	-	-	-	-
4.	Remdesivir [230]	RNA-directed RNA polymerase L	Inhibits the viral RNA polymerase enzyme, which is essential for the replication of the virus	Intravenously	Nausea, vomiting, diarrhea, and elevated liver enzymes	Veklury
5.	TKM-130803 [231]	mRNA of EBOV	Degrades the viral RNA of the Ebola virus	Intravenous infusion	Immune reactions, off-target effects, and toxicity due to high doses	-

Synthesized compounds that show activity against the Ebola virus are given in Table 20.

Table 20. Synthesized anti-Ebola compounds.

Sl. No.	Antiviral Agent	Drug Target	Activity
1.		Ebola virus [232]	3,5-Disubstitutedpyrrolo[2,3 b]pyridines demonstrate potent activity against DENV-infected primary dendritic cells and exhibit anti-EBOV activity. A LanthaScreen binding assay was conducted to assess their effectiveness against the Ebola virus ($EC_{50} = 0.59\ \mu M$).
2.		Ebola virus [233]	Isoxazole analogues were tested as inhibitors of Ebola GP-mediated cell entry, with a promising IC_{50} value in the range of 30 μM.
3.		Ebola virus [200]	-

5.8. Anti-SARS-COV-2 Agent

SARS-CoV-2 is a beta-coronavirus in the B lineage that is closely related to the SARS-CoV virus [234]. The major structural genes include N, S, SM and M, while an additional glycoprotein HE occurs in HCoV-OC43 and HKU1 beta-coronaviruses. SARS-CoV-2 shares 96% of its genome with a bat coronavirus. There are several types of anti-SARS-CoV-2 medications, each with its own mechanism of action. The examples are mentioned below:

Vaccines: Vaccines stimulate the immune system to produce antibodies that can neutralize the virus before it can cause an infection. There are currently several COVID-19 vaccines available, including mRNA vaccines, viral vector vaccines, and inactivated or protein subunit vaccines [235].

As of 12 January 2022, the following vaccines have been granted Emergency Use Listing:

- Comirnaty vaccine by Pfizer/BioNTech, approved 31 December 2020.
- SII/COVISHIELD and AstraZeneca/AZD1222 vaccines, approved 16 February 2021.
- Janssen/Ad26.COV 2.S vaccine developed by Johnson & Johnson, approved 12 March 2021.
- Moderna COVID-19 vaccine (mRNA 1273), approved 30 April 2021.
- Sinopharm COVID-19 vaccine, approved 7 May 2021.
- Sinovac-CoronaVac vaccine, approved 1 June 2021.
- Bharat Biotech BBV152 COVAXIN vaccine, approved 3 November 2021.
- Covovax (NVX-CoV2373) vaccine, approved 17 December 2021.
- Nuvaxovid (NVX-CoV2373) vaccine, approved 20 December 2021.

Monoclonal antibodies: Monoclonal antibodies are laboratory-made proteins that mimic the immune system's ability to fight off harmful pathogens. They can neutralize the virus by binding to specific proteins on its surface and preventing it from entering host cells [236].

Antiviral drugs (Table 21): Antiviral drugs can inhibit viral replication by targeting specific viral proteins or enzymes. For example, remdesivir is an antiviral drug that inhibits the viral RNA polymerase enzyme which is essential for the replication of the virus [237].

Immune-modulators: Immune modulators help modulate the immune response to the virus. For example, dexamethasone is a corticosteroid drug that reduces inflammation and has been shown to improve outcomes in severe COVID-19 cases [238].

Table 21. Anti-SARS-CoV-2 drugs.

Sl. No.	Drug Name	Drug Target	Mechanism of Action	Ways of Use	Side Effect	Brand Name
1.	Nirmatrelvir	SARS-CoV-2 [239]	Nirmatrelvir inhibits cysteine residue in the 3C-like protease (3CL^{PRO}) of SARS-CoV-2	Oral	There is no such side effect observed	Paxlovid
2.	Baricitinib	COVID-19 [240]	Baricitinib inhibits the activity of JAK proteins and modulates the signaling pathway of various interleukins, interferons	Oral	There is no such side effect observed	Olumiant

Some other synthesized compounds that show activity against SARS-COV-2 are given in Table 22.

Table 22. Synthesized anti-SARS-COV-2 compounds.

Sl. No.	Antiviral Agent	Drug Target	Activity
1.		SARS-CoV-2 [241]	Indomethacin exhibits potent antiviral activity against SARS coronavirus by selectively inhibiting viral RNA synthesis, independent of its COX inhibition and anti-inflammatory properties [242], with a promising EC_{50} value of 50 µM.
2.		SARS-CoV-2 [243]	$3'$-Deoxy-$3'$,$4'$-didehydro-cytidine triphosphate (ddhCTP) is a novel antiviral molecule that specifically targets non-native RNA polymerases. Its production does not affect cell viability or growth rate.
3.		SARS-CoV-2 [244]	Nitazoxanide (NTZ), originally an antiparasitic agent, shows potent activity against various RNA and DNA viruses, including SARS-CoV-2 (EC_{50} = 3.16 µM). Further studies are needed to understand its mode of action and specificity [245].
4.		SARS-CoV-2 [246]	This compound targets the SARS-CoV-2 methyltransferases MTase and Nsp14. Which demonstrates superior anti-SARS-CoV-2 inhibition (EC_{50} = 0.72 µM), as evidenced by the compound's high antiviral activity and low cytotoxicity.

5.9. Anti-HPV Agent

HPV can infect both men and women, and it is predicted that most sexually active adults will become infected at some point in their lives. It can cause genital warts and certain types of cancer, including cervical, anal and oropharyngeal cancer [247]. There is no cure for HPV; however, there are various treatment options available (as below) to minimize the symptoms in a controlled way [248].

Imiquimod (Aldara): This topical cream stimulates the immune system to fight the virus and is used to treat external genital warts and certain pre-cancerous skin lesions caused by HPV.

Podofilox (Condylox): This topical solution works by destroying the skin cells infected with HPV and is used to treat external genital warts.

Trichloroacetic acid (TCA): This topical solution is used to treat genital warts and certain pre-cancerous skin lesions caused by HPV.

Cidofovir (Vistide): This antiviral drug is used to treat severe cases of HPV infections, including those that have spread to other parts of the body.

Gardasil and Cervarix: These are vaccines that protect against several strains of HPV, including those that are known to cause most cases of cervical cancer.

A synthesized compound that shows activity as an anti-HPV agent is given in Table 23.

Table 23. Synthesized anti-HPV compound.

Sl. No.	Antiviral Agent	Drug Target	Activity
1.		Human papillomavirus (HPV-18) [249]	The compound octadecyloxyethyl benzyl 9-[(2-phosphonomethoxy)ethyl]guanine effectively inhibited the amplification of HPV-11 plasmid DNA in transfected cells (EC_{50} = 0.10 µM). Cell viability was assessed using CellTiter-Glo reagent (Promega) and measured with a luminometer.

5.10. Anti-Rabies Agent

There are two main ways to prevent and treat rabies virus infection: vaccination and post-exposure prophylaxis (PEP) with immunoglobulin and vaccines. In the case of suspected rabies virus exposure, PEP is recommended to prevent the virus from causing an infection. PEP typically involves the administration of both rabies immunoglobulin (RIG), which contains antibodies against the virus, and a series of rabies vaccine injections [250].

5.11. Anti-Zika Agent

There is currently no specific antiviral treatment for Zika virus infection, and treatment is generally supportive. For example, drugs that are used to treat other Flaviviruses, such as dengue and yellow fever, are being tested in clinical trials to see if they can also be effective against the Zika virus [251,252]. In addition, multiple vaccine candidates, such as DNA vaccines and RNA vaccines, are in various stages of development. These vaccines are being tested in preclinical and clinical trials to determine their safety and effectiveness in preventing Zika virus infection [253].

Synthesized compounds that show activity against the Zika virus are given in Table 24.

Several compounds have shown a broad inhibition activity against Flavivirus proteases and have been extensively studied.

Table 24. Synthesized anti-Zika compounds.

Sl. No.	Antiviral Agent	Drug Target	Activity
1.		Zika virus [254]	These compounds have demonstrated significant efficacy in both cellular and in vivo studies, particularly in U87 glioma cells infected with the ZIKV FLR strain [255], with IC_{50} values ranging from 200–790 nM. Their mode of action involves binding to an allosteric pocket of NS3, which is a viable target within the Flavivirus protease. This binding occurs in contrast to the shallow active site, which typically recognizes polar and positively charged residues in the substrate, such as arginine (Arg) or lysine (Lys) [256].
2.		Zika virus [257]	Doramectin has been investigated as a potential broad-spectrum antiviral agent against ZIKV. It has shown strong inhibitory effects against ZIKV infection in SNB19 cells. It shows an EC_{50} value of less than 3 μM.
3.		Zika virus [258]	Another substituted Glycosylated diphyllin has shown strong inhibitory effects against ZIKV infection in various cell lines, including CHME3 cells (human microglia cells), with IC_{50} values ranging from 10–70 Nm.
4.		Zika virus [259]	Cyanohydrazones have also exhibited noteworthy antiviral activity against ZIKV, with an IC_{90} value of 4.2 ± 0.2. These compounds target the E-mediated membrane fusion process, effectively inhibiting ZIKV infection [260].

5.12. Anti-Polio Agent

Remediation of polio involves immunization through the administration of the oral polio vaccine (OPV) or the inactivated polio vaccine (IPV). OPV is the preferred vaccine for most countries, as it is easy to administer and can also provide herd immunity by interrupting the transmission of the virus from person to person [261].

5.13. Anti-West Nile Agent

There is currently no specific treatment or vaccine for West Nile virus, but several vaccines are being developed and tested in clinical trials. Nonetheless, several promising vaccine candidates are currently being studied, and ongoing research in this area is very promising and provides hope for the future [262].

5.14. Anti-Chickenpox Agent

Remediation for chickenpox includes management of the symptoms, such as the use of antihistamines to alleviate itching and pain relievers to reduce fever. There are two major ways to prevent and treat chickenpox: through vaccination and antiviral medications. The chickenpox vaccine is a live, attenuated vaccine that contains a weakened form of the varicella–zoster virus [263]. Antiviral medications such as acyclovir, valacyclovir, and famciclovir can also be used to treat chickenpox. These drugs work by inhibiting the replication of the virus and are typically used in individuals who are at high risk of complications, such as pregnant women, immunocompromised individuals, and those with severe symptoms [264].

5.15. Anti-Influenza Agent

The most commonly used drugs for the treatment of the flu disease are neuraminidase inhibitors, which generally work by blocking the intracellular spread of the virus in the body. The two main neuraminidase inhibitors used for this purpose are oseltamivir (Tamiflu) [265] and zanamivir (Relenza) [266]. These drugs are effective in reducing the duration and severity of flu symptoms, as well as preventing complications.

Synthesized compounds with prominent activity against the influenza virus are mentioned in Table 25.

Table 25. Synthesized anti-influenza compounds.

Sl. No.	Antiviral Agent	Drug Target	Activity
1.	*(chemical structure: sialosyl sulfonate)*	Influenza virus [267]	Sialosyl sulfonate is a potent inhibitor of influenza virus replication. Sialosyl sulfonates have shown their ability to block influenza A virus H3N2 (Perth/16/2009) infection of MDCK cells in vitro, using an in-situ ELISA method with $IC_{50} > 1000$ μM.
2.	*(chemical structure)*	[267]	The above-mentioned method (Table 25, Sl. No. 1) was used to determine the activity of the compound. The IC_{50} value was found to be 0.7 μM.
3.	*(chemical structure)*	[267]	The above-mentioned method (Table 25, Sl. No. 1) was used to determine the activity of the compound. It showed an $IC_{50} > 0.02$ μM.

Table 25. *Cont.*

Sl. No.	Antiviral Agent	Drug Target	Activity
4.		[267]	The above-mentioned method (Table 25, Sl. No. 1) was used to determine the activity of the compound. The IC_{50} value for this compound was found to be 550 µM.
5.		Influenza virus [268]	Through intervention in these host–virus interactions, the substituted furan-succinimide derivative can successfully impede viral replication and restrict the advancement of the infection, with the EC_{50} value determined to be 50 pM [269].
6.		Influenza virus [270]	In human cells (HeLa), the compound was evaluated for its antiviral activity (EC_{90} = 11.4–15.9 µM), more precisely the polymerase activity of influenza A virus PR8. To assess this, a negative-sense EGFP gene, flanked with 5′ and 3′ UTRs derived from the NS segment, was cloned under the control of the human RNA polymerase I promoter.
7.		Influenza virus [271]	The compound inhibits human dihydroorotate dehydrogenase (DHODH) and viral replication of WSN-Influenza, with an EC_{50} of 41 nM.

5.16. Anti-Yellow Fever Agent

Currently, there is no specific antiviral drug available to treat yellow fever. Treatment is primarily supportive, with measures such as fluid replacement, pain relief and management of other symptoms [272].

6. Conclusions

In this review article, we have precisely discussed the antiviral activities of structurally diverse oxa- and aza-cycles with respect to different diseases. We have highlighted the role of representative small molecules, from natural products to synthetic compounds with heterocyclic subunits, and demonstrated their antiviral features. Taking into consideration of the severity of the viral infections, it is undoubtedly necessary to have a complete data set, along with the structure–activity relationship (SAR) of various drug candidates against the infectious viruses. In this regard, this review article could definitely play a crucial role in the discovery of antiviral drugs.

7. Scope, Limitation and the Presentation of the Future Trend of Antiviral Drugs

Antiviral drugs have revolutionized the treatment and control of infections caused by viral diseases. They target specific viral mechanisms, by such means as inhibiting replication, preventing viral entry into cells and blocking the activity of the viral enzyme. These drugs have significantly improved patient outcomes and reduced the spread of contagions. They are capable enough to tackle infections like HIV, hepatitis, influenza, herpes, and more. In spite of the suitability of such drugs in terms of the proper treatment and control of viral infections, they do have limitations. Viruses can develop resistance to certain drugs, necessitating the development of new classes of drugs or combination therapies. Moreover, antiviral drugs may cause side effects and interact with other medications, a situation which requires careful management. The challenges are even greater when facing the newly emerged viruses, and the lack of specific and effective treatment options becomes apparent. The development of drugs targeting such viruses is very complicated and involves extensive research and development, advanced computational support and dedicated clinical trials. Additionally, viral mutations and the potential for drug resistance further impede the effectiveness of existing therapies against newly arrived viruses.

Furthermore, emerging technologies like CRISPR/Cas9-based gene editing hold the potential for targeted viral genome disruption, offering innovative approaches to combat viral infections. The ongoing research and development of antiviral drugs aim to address current limitations, including drug resistance, side effects and access issues, while providing more effective, safer and accessible treatments for viral infections in the future.

Author Contributions: Investigation, S.S. and E.V.N.; resources, S.S. and E.V.N.; writing—original draft preparation, K.D. and S.H.; writing—review and editing, S.S.; E.V.N. and G.V.Z.; literature Search, S.M. and K.D.; structure Draw, S.M. and K.D.; referencing, S.H.; editing, S.H. All authors have read and agreed to the published version of the manuscript.

Funding: Science and Engineering Research Board; ECR/2017/000966.

Acknowledgments: S.H. is thankful to SERB, India (grant no. ECR/2017/000966) and DST, India (grant no. INT/RUS/RFBR/P-293/G) for the financial support. S. Santra and G.V. Zyryanov are grateful to the Ministry of Science and Higher Education of the Russian Federation (Agreement # 075-15-2022-1118 dated 29.06.2022) for funding. S.M. thanks VNIT Nagpur, India, for the research fellowship. K.D. is grateful to DST, India for providing a research fellowship. We all are thankful to the department of Chemistry VNIT Nagpur for providing infrastructure and a research facility.

Conflicts of Interest: The authors declare no conflict of interest.

Abbreviations

Abbreviation	Full Name
DNA	Deoxyribonucleic Acid
RNA	Ribonucleic Acid
AIDS	Acquired Immune Deficiency Syndrome
HIV	Human Immunodeficiency Virus
HCV	Hepatitis C Virus
HBV	Hepatitis B Virus
RSV	Respiratory Syncytial Virus
HCMV	Human Cytomegalovirus
HSV	Herpes Simplex Virus
EBOV	Ebola Virus
SARS-CoV-2	Severe Acute Respiratory Syndrome CoV-2
PNBA	Para Nitro Benzoic Acid
GP120-CCR5	Beta chemokine receptors
dATP	Deoxyadenosine triphosphate
mRNA	Messenger Ribonucleic Acid
eIF4A	Eukaryotic initiation factor 4A
ToS	Toluenesulfonyl
OiPr	Isopropoxide
OSBT	O-(Tert-Butyldimethylsilyl)hydroxylamine
Boc	tert-butoxycarbonyl
PMP	Polymethylpentene
OBn	Benzyl group
mCPBA	meta-chloroperoxybenzoic acid
DMSO	Dimethyldioxirane
TFAA	Trifluoroacetic anhydride
TFA	Trifluoroacetic acid
DSC	N,N′-Disuccinimidyl carbonate
DEC	Diethylcarbamazine

DMAP	4-Dimethylaminopyridine
LDA	Lithium diisopropylamide
DDQ	2,3-Dichloro-5,6-Dicyanobenzoquinone
THF	Tetrahydrofuran
NBS	N-Bromosuccinimide
DPPA	Diphenylphosphoryl azide
HATU	Hexafluorophosphate Azabenzotriazole Tetramethyl Uronium
DCM	Dichloromethane
DMAc	N,N-Dimethylacetamide
DBU	1,8-Diazabicyclo(5.4.0)undec-7-ene
CDI	Carbonyldiimidazole
CPME	Cyclopentyl methyl ether
DIAD	Diisopropyl azodicarboxylate
IPAc	Isopropyl acetate
DMF	Dimethylformamide
DCC	N,N′-Dicyclohexylcarbodiimide
DEAD	Diethyl azodicarboxylate
TBAF	Tetra-n-butylammonium fluoride
4Å MS	4Å Molecular Sieve

References

1. Parvez, M.K.; Parveen, S. Evolution and Emergence of Pathogenic Viruses: Past, Present, and Future. *Intervirology* **2017**, *60*, 1–7. [CrossRef]
2. Gaba, A.; Ayalew, L.E.; Tikoo, S.K. Animal Adenoviruses. In *Recent Advances in Animal Virology*; Malik, Y.S., Singh, R.K., Yadav, M.P., Eds., Springer: Singapore, 2019; ISBN 978-981-13-9072-2. [CrossRef]
3. Zhu, Z.; Lian, X.; Su, X.; Wu, W.; Marraro, G.A.; Zeng, Y. From SARS and MERS to COVID-19: A brief summary and comparison of severe acute respiratory infections caused by threeighly pathogenic human coronaviruses. *Respir. Res.* **2020**, *21*, 224. [CrossRef] [PubMed]
4. Wirth, T. *Biodiversity and Evolution*; Elsevier: Amsterdam, The Netherlands, 2018; Volume 3, pp. 123–137. [CrossRef]
5. Morens, D.M.; Fauci, A.S. Emerging Pandemic Diseases: How We Got to COVID-19. *Cell* **2020**, *182*, 1077–1092. [CrossRef] [PubMed]
6. Pagidipati, N.J.; Gaziano, T.A. Estimating Deaths From Cardiovascular Disease: A Review of Global Methodologies of Mortality Measurement. *Circulation* **2013**, *127*, 749–756. [CrossRef] [PubMed]
7. Joint United Nations Programme on HIV/AIDS (UNAIDS). *2006 Report on the Global AIDS Epidemic*; UNAIDS: Geneva, Switzerland, 2006.
8. Mathers, C.D.; Boerma, T.; Ma Fat, D. Global and regional causes of death. *Br. Med. Bull.* **2009**, *92*, 7–32. [CrossRef] [PubMed]
9. World Health Organization. *Division of Emerging and Other Communicable Diseases Surveillance and Control. World Survey of Rabies: No. 30: For the Year 1994*; World Health Organization: Geneva, Switzerland, 1996.
10. Piot, P.; Muyembe, J.J.; Edmunds, W.J. Ebola in west Africa: From disease outbreak to humanitarian crisis. *Lancet Infect. Dis.* **2014**, *14*, 1034–1035. [CrossRef]
11. Gostin, L.O.; Lucey, D.; Phelan, A. The Ebola Epidemic A Global Health Emergency. *JAMA J. Am. Med. Assoc.* **2014**, *312*, 1095–1096. [CrossRef]
12. Meyers, L.; Frawley, T.; Goss, S.; Kang, C. Ebola Virus Outbreak 2014: Clinical Review for Emergency Physicians. *Ann. Emerg. Med.* **2015**, *65*, 101–108. [CrossRef]
13. Gogineni, V.; Schinazi, R.F.; Hamann, M.T. Role of Marine Natural Products in the Genesis of Antiviral Agents. *Chem. Rev.* **2015**, *115*, 9655–9706. [CrossRef]
14. More, A.F.; Loveluck, C.P.; Clifford, H.; Handley, M.J.; Korotkikh, E.V.; Kurbatov, A.V.; McCormick, M.; Mayewski, P.A. The health benefit of physical exercise on COVID-19 pandemic: Evidence from mainland China. *GeoHealth* **2020**, *4*, 277–284. [CrossRef]
15. Mammas, I.; Spandidos, D.A. [Comment] COVID-19 threat and the 1918 Spanish flu outbreak: The following day. *Exp. Ther. Med.* **2019**, *20*, 292. [CrossRef]
16. Taubenberger, J.K.; Morens, D.M. 1918 Influenza: The mother of all pandemics. *Emerg. Infect. Dis.* **2006**, *12*, 15–22. [CrossRef] [PubMed]
17. Trilla, A.; Trilla, G.; Daer, C. The 1918 "Spanish Flu" in Spain. *Clin. Infect. Dis.* **2008**, *47*, 668–673. [CrossRef] [PubMed]
18. Li, L.; Huang, T.; Wang, Y.; Wang, Z.; Liang, Y.; Huang, T.; Zhang, H.; Sun, W.; Wang, Y.J. COVID-19 patients' clinical characteristics, discharge rate, and fatality rate of meta-analysis. *Med. Virol.* **2020**, *92*, 577–583. [CrossRef] [PubMed]
19. Dagan, N.; Barda, N.; Kepten, E.; Miron, O.; Perchik, S.; Katz, M.A.; Hernán, M.A.; Lipsitch, M.; Reis, B.; Balicer, R.D. BNT162b2 mRNA COVID-19 Vaccine in a Nationwide Mass Vaccination Setting. *N. Engl. J. Med.* **2021**, *384*, 1412–1423. [CrossRef] [PubMed]
20. Haynes, B.F.; Corey, L.; Fernandes, P.; Gilbert, P.B.; Hotez, P.J.; Rao, S.; Santos, M.R.; Schuitemaker, H.; Watson, M.; Arvin, A. Prospects for a safe COVID-19 vaccine. *Sci. Transl. Med.* **2020**, *12*, eabe0948. [CrossRef] [PubMed]

21. Corey, L.; Mascola, J.R.; Fauci, A.S.; Collins, F.S. A strategic approach to COVID-19 vaccine R&D. *Science* **2020**, *368*, 948–950. [CrossRef]

22. Kim, J.H.; Marks, F.; Clemens, J.D. Looking beyond COVID-19 vaccine phase 3 trials. *Nat. Med.* **2021**, *27*, 205–211. [CrossRef]

23. Cao, Y.; Deng, Q.; Dai, S. Remdesivir for severe acute respiratory syndrome coronavirus 2 causing COVID-19: An evaluation of the evidence. *Travel Med. Infect. Dis.* **2020**, *35*, 101647–101677. [CrossRef]

24. World Health Organization. Newsroom: Coronavirus Disease (COVID-19): Vaccines. Available online: https://www.who.int/news-room/q-a-detail/coronavirus-disease-(COVID-19)-vaccines (accessed on 15 August 2021).

25. Onnuswamy, M.N.; Gromiha, M.M.; Sony, S.M.M.; Saraboji, K. *QSAR and Molecular Modeling Studies in Heterocyclic Drugs I;* Springer: Berlin/Heidelberg, Germany, 2006; Volume 3, pp. 81–147. [CrossRef]

26. Sliwoski, G.R.; Meiler, J.; Lowe, E.W. Computational methods in drug discovery. *Comput. Methods Drug Discov.* **2014**, *66*, 334–395. [CrossRef]

27. Jambhekar, S.S.; Breen, P.J. Drug dissolution: Significance of physicochemical properties and physiological conditions. *Drug Discov. Today* **2013**, *18*, 1173–1184. [CrossRef] [PubMed]

28. Syam, Y.; Kamel, M. Structure and physicochemical properties in relation to drug action. *Egypt. Pharm. J.* **2013**, *12*, 95–108. [CrossRef]

29. Martin-Benlloch, X.; Haid, S.; Novodomska, A.; Rominger, F.; Pietschmann, T.; Davioud-Charvet, E.; Elhabiri, M. Physicochemical properties govern the activity of potent antiviral flavones. *ACS Omega* **2019**, *4*, 4871–4887. [CrossRef]

30. Klimenko, K.; Marcou, G.; Horvath, D.; Varnek, A.J. Chemical Space Mapping and Structure–Activity Analysis of the ChEMBL Antiviral Compound Set. *Chem. Inf. Model.* **2016**, *56*, 1438–1454. [CrossRef]

31. Prashantha Kumar, B.R.; Soni, M.; Bharvi Bhikhalal, U.; Kakkot, I.R.; Jagadeesh, M.; Bommu, P.; Nanjan, M.J. Analysis of physicochemical properties for drugs from nature. *Nat. Med. Chem. Res.* **2010**, *19*, 984–992. [CrossRef]

32. Evstigneev, M.P. Physicochemical mechanisms of synergistic biological action of combinations of aromatic heterocyclic compounds. *Org. Chem. Int.* **2013**, *2013*, 278143. [CrossRef]

33. Kerru, N.; Gummidi, L.; Maddila, S.; Gangu, K.K.; Jonnalagadda, S.B. A review on recent advances in nitrogen-containing molecules and their biological applications. *Molecules* **2020**, *25*, 1909. [CrossRef]

34. Gomtsyan, AHeterocycles in drugs and drug discovery. *Chem. Heterocycl. Compd.* **2012**, *48*, 7–10. [CrossRef]

35. De Vivo, M.; Masetti, M.; Bottegoni, G.; Cavalli, A. Role of molecular dynamics and related methods in drug discovery. *J. Med. Chem.* **2016**, *59*, 4035–4061. [CrossRef]

36. Pastorino, B.; Nougairède, A.; Wurtz, N.; Gould, E.; de Lamballerie, X. Role of host cell factors in flavivirus infection: Implications for pathogenesis and development of antiviral drugs. *Antivir. Res.* **2010**, *87*, 281–294. [CrossRef]

37. Monto, A.S. Vaccines and antiviral drugs in pandemic preparedness. *Emerg. Infect. Dis.* **2006**, *12*, 55–60. [CrossRef]

38. Schmid, M.; Speiseder, T.; Dobner, T.; Gonzalez, R.A. DNA virus replication compartments. *J. Virol.* **2014**, *88*, 1404–1420. [CrossRef]

39. Schwartz, M.; Chen, J.; Janda, M.; Sullivan, M.; Den Boon, J.; Ahlquist, P. A positive-strand RNA virus replication complex parallels form and function of retrovirus capsids. *Mol. Cell* **2002**, *9*, 505–514. [CrossRef]

40. Vahlne, A. A historical reflection on the discovery of human retroviruses. *Retrovirology* **2009**, *6*, 40. [CrossRef]

41. Levin, J.G.; Mitra, M.; Mascarenhas, A.; Musier-Forsyth, K. Role of HIV-1 nucleocapsid protein in HIV-1 reverse transcription. *RNA Biol.* **2010**, *7*, 754–774. [CrossRef]

42. Gallo, R.C.; Montagnier, L. The discovery of HIV as the cause of AIDS. *N. Engl. J. Med.* **2003**, *349*, 2283–2285. [CrossRef]

43. World Health Organization. Newsroom: HIV and AIDS. Available online: https://www.who.int/news-room/fact-sheets/detail/hiv-aids (accessed on 4 May 2023).

44. Ghosh, A.K.; Fyvie, W.S.; Brindisi, M.; Steffey, M.; Agniswamy, J.; Wang, Y.F.; Aoki, M.; Amano, M.; Weber, I.T.; Mitsuya, H. Design, Synthesis, Biological Evaluation, and X-ray Studies of HIV-1 Protease Inhibitors with Modified P2' Ligands of Darunavir. *ChemMedChem* **2017**, *12*, 1942–1952. [CrossRef]

45. Du, L.; Jia, J.; Ge, P.; Jin, Y. Self-assemblies of 5'-cholesteryl-ethyl-phosphoryl zidovudine. *Colloids Surf. B.* **2016**, *148*, 385–391. [CrossRef]

46. Belk, D.; Belk, P. *The Great American Healthcare Scam: How Kickbacks, Collusion and Propaganda Have Exploded Healthcare Costs in the United States Paperback;* David Belk: Greensboro, NC, USA, 2020; ISBN 13: 9781734709018.

47. Vasilyeva, S.V.; Shtil, A.A.; Petrova, A.S.; Balakhnin, S.M.; Achigecheva, P.Y.; Stetsenko, D.A.; Silnikov, V.N. Conjugates of phosphorylated zalcitabine and lamivudine with SiO$_2$ nanoparticles: Synthesis by CuAAC click chemistry and preliminary assessment of anti-HIV and antiproliferative activity. *Bioorg. Med. Chem.* **2017**, *25*, 1696–1702. [CrossRef]

48. Paton, N.I.; Musaazi, J.; Kityo, C.; Walimbwa, S.; Hoppe, A.; Balyegisawa, A.; Kaimal, A.; Mirembe, G.; Tukamushabe, P.; Ategeka, G. Dolutegravir or darunavir in combination with zidovudine or tenofovir to treat HIV. *N. Engl. J. Med.* **2021**, *385*, 330–341. [CrossRef]

49. Mayer, K.H.; Molina, J.-M.; Thompson, M.A.; Anderson, P.L.; Mounzer, K.C.; De Wet, J.J.; DeJesus, E.; Jessen, H.; Grant, R.M.; Ruane, P.J. Emtricitabine and tenofovir alafenamide vs emtricitabine and tenofovir disoproxil fumarate for HIV pre-exposure prophylaxis (DISCOVER): Primary results from a randomised, double-blind, multicentre, active-controlled, phase 3, non-inferiority trial. *Lancet* **2020**, *396*, 239–254. [CrossRef]

50. Alter, M.J. Epidemiology of hepatitis C virus infection. *World J. Gastroenterol.* **2007**, *13*, 2436–2441. [CrossRef]

51. Massengill, M.T.; Park, J.C.; McAnany, J.J.; Hyde, R.A. Occult retinopathy following treatment of Hepatitis C with glecaprevir/pibrentasvir (Mavyret). *Doc. Ophthalmol.* **2023**, *146*, 191–197. [CrossRef]

52. Gentile, I.; Buonomo, A.R.; Zappulo, E.; Minei, G.; Morisco, F.; Borrelli, F.; Coppola, N.; Borgia, G. Asunaprevir, a protease inhibitor for the treatment of hepatitis C infection. *Ther. Clin. Risk Manag.* **2014**, *10*, 493–504. [CrossRef]

53. Chang, M.H.; Gordon, L.A.; Fung, H.B. Infection of common marmosets with GB virus B chimeric virus encoding the major nonstructural proteins NS2 to NS4A of hepatitis C virus. *Clin. Ther.* **2012**, *34*, 2021–2038. [CrossRef]

54. Kiang, T.K.L. Clinical pharmacokinetics and drug–drug interactions of elbasvir/grazoprevir. *Eur. J. Drug Metab. Pharmacokinet.* **2018**, *43*, 509–531. [CrossRef]

55. Blumberg, B.S. Hepatitis B virus, the vaccine, and the control of primary cancer of the liver. *Proc. Natl. Acad. Sci. USA* **1997**, *94*, 7121–7125. [CrossRef]

56. Liu, J.; Li, T.; Zhang, L.; Xu, A. The role of hepatitis B surface antigen in nucleos (t) ide analogues cessation among asian patients with chronic hepatitis B: A systematic review. *Hepatology* **2019**, *70*, 1045–1055. [CrossRef]

57. Sulkowski, M.S.; Agarwal, K.; Ma, X.; Nguyen, T.T.; Schiff, E.R.; Hann, H.W.L.; Dieterich, D.T.; Nahass, R.G.; Park, J.S.; Chan, S.J. Safety and efficacy of vebicorvir administered with entecavir in treatment-naïve patients with chronic hepatitis B virus infection. *Hepatol.* **2022**, *77*, 1265–1275. [CrossRef]

58. Zenchenko, A.A.; Drenichev, M.S.; Il'icheva, I.A.; Mikhailov, S.N. Antiviral and antimicrobial nucleoside derivatives: Structural features and mechanisms of action. *Mol. Biol.* **2021**, *55*, 786–812. [CrossRef]

59. Menéndez-Arias, L.; Álvarez, M.; Pacheco, B. Nucleoside/nucleotide analog inhibitors of hepatitis B virus polymerase: Mechanism of action and resistance. *Curr. Opin. Virol.* **2014**, *8*, 1–9. [CrossRef] [PubMed]

60. Acosta, P.L.; Caballero, M.T.; Polack, F.P. Brief history and characterization of enhanced respiratory syncytial virus disease. *Clin. Vaccine Immunol.* **2016**, *23*, 189–195. [CrossRef] [PubMed]

61. Li, Y.; Wang, X.; Blau, D.M.; Caballero, M.T.; Feikin, D.R.; Gill, C.J.; Madhi, S.A.; Omer, S.B.; Simões, E.A.F.; Campbell, H. Brief history and characterization of enhanced respiratory syncytial virus disease. *Lancet* **2022**, *399*, 2047–2064. [CrossRef] [PubMed]

62. Shi, T.; Vennard, S.; Mahdy, S.; Nair, H.J. Risk factors for RSV associated acute lower respiratory infection poor outcome and mortality in young children: A systematic review and meta-analysis. *Infect. Dis.* **2022**, *226*, S10–S16. [CrossRef]

63. Nair, H.; Nokes, D.J.; Gessner, B.D.; Dherani, M., Madhi, S.A., Singleton, R.J., O'Brien, K.L., Roca, A., Wright, P.F.; Bruce, N.; et al. Global burden of acute lower respiratory infections due to respiratory syncytial virus in young children: A systematic review and meta-analysis. *Lancet* **2010**, *375*, 1545–1555. [CrossRef] [PubMed]

64. Wang, Y.Q.; Zhao, X.Y. Human cytomegalovirus primary infection and reactivation: Insights from virion-carried molecules. *Front. Microbiol.* **2020**, *11*, 1511. [CrossRef]

65. Stoelben, S.; Arns, W.; Renders, L.; Hummel, J.; Mühlfeld, A.; Stangl, M.; Fischereder, M.; Gwinner, W.; Suwelack, B.; Witzke, O.; et al. Preemptive treatment of Cytomegalovirus infection in kidney transplant recipients with letermovir: Results of a Phase 2a study. *Transpl. Int.* **2014**, *27*, 77–86. [CrossRef]

66. Biron, K.K. Antiviral drugs for cytomegalovirus diseases. *Antivir. Res.* **2006**, *71*, 154–163. [CrossRef]

67. Corey, L.; Sper, P.G. Infections with herpes simplex viruses. *N. Engl. J. Med.* **1986**, *314*, 686–691. [CrossRef]

68. Gottlieb, S.L.; Giersing, B.K.; Hickling, J.; Jones, R.; Deal, C.; Kaslow, D.C. Meeting report: Initial World Health Organization consultation on herpes simplex virus (HSV) vaccine preferred product characteristics, March 2017. *Vaccine* **2019**, *37*, 7408–7418. [CrossRef]

69. Feldmann, H.; Jones, S.; Klenk, H.-D.; Schnittler, H.-J. Ebola virus: From discovery to vaccine. *Nat. Rev. Immunol.* **2003**, *3*, 677–685. [CrossRef] [PubMed]

70. Mirza, M.U.; Vanmeert, M.; Ali, A.; Iman, K.; Froeyen, M.; Idrees, M. Perspectives towards antiviral drug discovery against Ebola virus. *J. Med. Virol.* **2019**, *91*, 2029–2048. [CrossRef] [PubMed]

71. Grellet, E.; Goulet, A.; Imbert, I. Replication of the Coronavirus Genome: A Paradox among Positive-Strand RNA Viruses. *J. Biol. Chem.* **2022**, *5*, 101923–101938. [CrossRef]

72. Tyrrell, D.A.J.; Bynoe, M.L. Cultivation of viruses from a high proportion of patients with colds. *Lancet* **1966**, *287*, 76–77. [CrossRef] [PubMed]

73. Kahn, J.S.; McIntosh, K. History and Recent Advances in Coronavirus Discovery. *Pediatr. Infect. Dis. J.* **2005**, *24*, S223–S227. [CrossRef] [PubMed]

74. Hu, B.; Guo, H.; Zhou, P.; Shi, Z.L. Characteristics of SARS-CoV-2 and COVID-19. *Nat. Rev. Microbiol.* **2021**, *19*, 141–154. [CrossRef]

75. Chang, F.; Syrjänen, S.; Kellokoski, J.; Syrjänen, K. Human Papillomavirus (HPV) Infections and Their Associations with Oral Disease. *J. Oral Pathol. Med.* **1991**, *20*, 305–317. [CrossRef]

76. World Health Organization. Newsroom: Fact Sheets. Cervical Cancer. Available online: https://www.who.int/news-room/fact-sheets/detail/cervical-cancer (accessed on 1 May 2023).

77. Fekadu, M.; Shaddock, J.H.; Chandler, F.W.; Baer, G.M. Rabies Virus in the Tonsils of a Carrier Dog. *Arch. Virol.* **1983**, *78*, 37–47. [CrossRef]

78. Dick, G.W.A.; Kitchen, S.F.; Haddow, A.J. zika virus (i). Isolations and serological specificity. *Trans. R. Soc. Trop. Med. Hyg.* **1952**, *46*, 509–520. [CrossRef]

79. Posen, H.J.; Keystone, J.S.; Gubbay, J.B.; Morris, S.K. Epidemiology of Zika Virus, 1947–2007. *BMJ Glob. Health* **2016**, *1*, e000087. [CrossRef]

80. Chen, H.L.; Tang, R.B.J. Why Zika virus infection has become a public health concern? *Chin. Med. Assoc.* **2016**, *79*, 174–178. [CrossRef] [PubMed]

81. Gorshkov, K.; Shiryaev, S.A.; Fertel, S.; Lin, Y.W.; Huang, C.T.; Pinto, A.; Farhy, C.; Strongin, A.Y.; Zheng, W.; Terskikh, A.V. Transdermal permeation of bacteriophage particles by choline oleate: Potential for treatment of soft-tissue infections. *Front. Microbiol.* **2019**, *9*, 3252–3296. [CrossRef] [PubMed]

82. Skern, T. 100 years poliovirus: From discovery to eradication. A meeting report. *Arch. Virol.* **2010**, *155*, 1371–1381. [CrossRef] [PubMed]

83. Melnick, J.L. Current Status of Poliovirus Infections. *Clin. Microbiol. Rev.* **1996**, *9*, 293–300. [CrossRef] [PubMed]

84. Dowdle, W.R.; Birmingham, M.E. The Biologic Principles of Poliovirus Eradication. *J. Infect. Dis.* **1997**, *175*, S286–S292. [CrossRef]

85. World Health Organization. Health Topics: Poliomyelitis. Available online: https://www.who.int/health-topics/poliomyelitis (accessed on 1 May 2023).

86. Smithburn, K.C.; Hughes, T.P.; Burke, A.W.; Paul, J.H. A neurotropic virus isolated from the blood of a native of Uganda. *Am. J. Trop. Med.* **1940**, *20*, 471–472. [CrossRef]

87. World Health Organization. Newsroom: Fact Sheets. West Nile Virus. Available online: https://www.who.int/news-room/fact-sheets/detail/west-nile-virus (accessed on 1 May 2023).

88. Takahashi, M. Chickenpox Virus. *Adv. Virus Res.* **1983**, *28*, 285–356. [CrossRef]

89. Hutchinson, E.C. Influenza Virus. *Trends Microbiol.* **2018**, *26*, 809–810. [CrossRef]

90. World Health Organization. Newsroom: Fact Sheets. Detail. Influenza (Seasonal). Available online: https://www.who.int/news-room/fact-sheets/detail/influenza-(seasonal) (accessed on 1 May 2023).

91. Monath, T.P.; Vasconcelos, P.F.C. Yellow Fever. *J. Clin. Virol.* **2015**, *64*, 160–173. [CrossRef]

92. World Health Organization. Health Topics: Yellow-Fever. Available online: https://www.who.int/health-topics/yellow-fever (accessed on 1 May 2023).

93. Prusoff, W.H. Synthesis and biological activities of iododeoxyuridine, an analog of thymidine. *BBA—Biochim. Biophys. Acta* **1959**, *32*, 295–296. [CrossRef]

94. Cheng, Y.; Prusoff, W.H. Relationship between the inhibition constant (&) and the concentration of inhibitor which causes 50 per cent inhibition (iso) of an enzymatic reaction. *Biochem. Pharmacol.* **1973**, *22*, 3099–3108. [CrossRef] [PubMed]

95. Kaufman, H.E.; Martola, E.-L.; Dohlman, C. Use of 5-Iodo-2'-Deoxyuridine(IDU) in Treatment of Herpes Simplex Keratitis. *Arch. Ophthalmol.* **1962**, *68*, 235–239. [CrossRef] [PubMed]

96. De Clercq, E.; Li, G. Approved Antiviral Drugs over the Past 50 Years. *Clin. Microbiol. Rev.* **2016**, *29*, 695–747. [CrossRef] [PubMed]

97. Tchesnokov, E.P.; Feng, J.Y.; Porter, D.P.; Götte, M. Mechanism of Inhibition of Ebola Virus RNA-Dependent RNA Polymerase by Remdesivir. *Viruses* **2019**, *11*, 326. [CrossRef]

98. Bergmann, W.; Feeney, R.J.J. The isolation of a new thymine pentoside from sponges. *J. Am. Chem. Soc.* **1950**, *72*, 2809–2810. [CrossRef]

99. Wahyuni, T.S.; Widyawaruyanti, A.; Lusida, M.I.; Fuad, A.; Soetjipto; Fuchino, H.; Kawahara, N.; Hayashi, Y.; Aoki, C.; Hotta, H. Inhibition of hepatitis C virus replication by chalepin and pseudane IX isolated from Ruta angustifolia leaves. *Fitoterapia* **2014**, *99*, 276–283. [CrossRef]

100. Andersen, R.J.; Faulkner, D.J.; He, C.H.; Van Duyne, G.D.; Clardy, J.J. Metabolites of the Marine Prosobranch Mollusc *Lamellaria* sp. *Am. Chem. Soc.* **1985**, *107*, 5492–5495. [CrossRef]

101. Hashimoto, T.; Akiyo, Y.; Akazawa, K.; Takaoka, S.; Tori, M.; Asakawa, Y. Three novel dimethyl pyrroledicarboxylate, lycogarubins A C, from the myxomycetes lycogala epidendrum. *Tetrahedron Lett.* **1994**, *35*, 2559–2560. [CrossRef]

102. Biedenkopf, N.; Lange-Grünweller, K.; Schulte, F.W.; Weiber, A.; Muller, C.; Becker, D.; Becker, S.; Hartmann, R.K.; Grünweller, A. The natural compound silvestrol is a potent inhibitor of Ebola virus replication. *Antivir. Res.* **2017**, *137*, 76–81. [CrossRef]

103. Cui, H.; Xu, B.; Wu, T.; Xu, J.; Yuan, Y.; Gu, Q.J. Potential Antiviral Lignans from the Roots of Saururus chinensis with Activity against Epstein−Barr Virus Lytic Replication. *Nat. Prod.* **2014**, *77*, 100–110. [CrossRef]

104. Perez, R.M. Antiviral Activity of Compounds Isolated From Plants. *Pharm. Biol.* **2003**, *41*, 107–157. [CrossRef]

105. Chen, H.; Ma, Y.B.; Huang, X.Y.; Geng, C.A.; Zhao, Y.; Wang, L.J.; Guo, R.H.; Liang, W.J.; Zhang, X.M.; Chen, J.J. Synthesis, structure–activity relationships and biological evaluation of dehydroandrographolide and andrographolide derivatives as novel anti-hepatitis B virus agents. *Bioorg. Med. Chem. Lett.* **2014**, *24*, 2353–2359. [CrossRef] [PubMed]

106. Geng, C.A.; Chen, J.J. The Progress of Anti-HBV Constituents from Medicinal Plants in China. *Nat. Prod. Bioprospect.* **2018**, *8*, 227–244. [CrossRef]

107. Zhou, N.J.; Geng, C.A.; Huang, X.Y.; Ma, Y.-B.; Zhang, X.M.; Wang, J.L.; Chen, J.J. Anti-Hepatitis B Virus Active Constituents from Swertia Chirayita. *Fitoterapia* **2015**, *100*, 27–34. [CrossRef]

108. Baumann, M.; Baxendale, I.R.; Ley, S.V.; Nikbin, N. An overview of the key routes to the best selling 5-membered ring heterocyclic pharmaceuticals. *Beilstein J. Org. Chem.* **2011**, *7*, 442–495. [CrossRef]

109. Vitaku, E.; Smith, D.T.; Njardarson, J.T. Analysis of the Structural Diversity, Substitution Patterns, and Frequency of Nitrogen Heterocycles among U.S. FDA Approved Pharmaceuticals. *J. Med. Chem.* **2014**, *57*, 10257–10274. [CrossRef] [PubMed]

110. Delost, M.D.; Smith, D.T.; Anderson, B.J.; Njardarson, J.T. From Oxiranes to Oligomers: Architectures of U.S. FDA Approved Pharmaceuticals Containing Oxygen Heterocycles. *J. Med. Chem.* **2018**, *61*, 10996–11020. [CrossRef] [PubMed]
111. De Clercq, E. Fifty Years in Search of Selective Antiviral Drugs. *J. Med. Chem.* **2019**, *62*, 7322–7339. [CrossRef]
112. McKeage, K.; Perry, C.M.; Keam, S.J. Darunavir: A review of its use in the management of HIV infection in adults. *Drugs* **2009**, *69*, 477–503. [CrossRef]
113. Kevin, M.; BelykHenry, G.; MorrisonAmar, J.; MahajanDaniel, J. KumkeHsien-Hsin TungLawrence WaiVanessa Pruzinsky. Potassium Salt of an HIV Integrase Inhibitor. U.S. Patent No.: US 7,754,731 B2, 8 July 2006.
114. Ray, P.C.; Tummanapalli, J.M.C.; Gorantla, S.R. Process for the Largescale Production of Stavudine. U.S. Patent No.: US 8,026,356 B2, 31 May 2007.
115. Leonis, G.; Czyżnikowska, Ż.; Megariotis, G.; Reis, H.; Papadopoulos, M.G.J. Computational Studies of Darunavir into HIV-1 Protease and DMPC Bilayer: Necessary Conditions for Effective Binding and the Role of the Flaps. *Chem. Inf. Model.* **2012**, *52*, 1542–1558. [CrossRef]
116. Vellanki, S.R.P.; Sahu, A.; Katukuri, A.K.; Vanama, V.; Kothari, S.; Ponnekanti, V.S.; Datta, D. Process for the Preparation of Darunavir. U.S. Patent No. US8703980B2, 17 September 2019.
117. Ghosh, A.K.; Martyr, C.D. Darunavir (Prezista): A HIV-1 Protease Inhibitor for Treatment of Multidrug-Resistant HIV. In *Modern Drug Synthesis*; John Wiley & Sons, Inc.: Hoboken, NJ, USA, 2010; pp. 29–44. [CrossRef]
118. Koh, Y.; Matsumi, S.; Das, D.; Amano, M.; Davis, D.A.; Li, J.F.; Leschenko, S.; Baldridge, A.; Shioda, T.; Yarchoan, R.; et al. Potent Inhibition of HIV-1 Replication by Novel Non-peptidyl Small Molecule Inhibitors of Protease Dimerization. *J. Biol Chem.* **2007**, *282*, 28709–28720. [CrossRef] [PubMed]
119. Fujimoto, H.; Higuchi, M.; Watanabe, H.; Koh, Y.; Ghosh, A.K.; Mitsuya, H.; Tanoue, N.; Hamada, A.; Saito, H. P-Glycoprotein Mediates Efflux Transport of Darunavir in Human Intestinal Caco-2 and ABCB1 Gene-Transfected Renal LLC-PK1 Cell Lines. *Biol. Pharm. Bull.* **2009**, *32*, 1588–1593. [CrossRef] [PubMed]
120. Koh, Y.; Nakata, H.; Maeda, K.; Ogata, H.; Bilcer, G.; Devasamudram, T.; Kincaid, J.F.; Boross, P.; Wang, Y.F.; Tie, Y.; et al. Novel bis-Tetrahydrofuranylurethane-Containing Nonpeptidic Protease Inhibitor (PI) UIC-94017 (TMC114) with Potent Activity against Multi-PI-Resistant Human Immunodeficiency Virus In Vitro. *Antimicrob. Agents Chemother.* **2003**, *47*, 3123–3129. [CrossRef]
121. Ghosh, A.K.; Sridhar, P.R.; Leshchenko, S.; Hussain, A.K.; Li, J.; Kovalevsky, A.Y.; Walters, D.E.; Wedekind, J.E.; Grum Tokars, V.; Das, D.; et al. Structure-Based Design of Novel HIV-1 Protease Inhibitors To Combat Drug Resistance. *J. Med. Chem.* **2006**, *49*, 5252–5261. [CrossRef]
122. Davis, D.A.; Soule, E.E.; Davidoff, K.S.; Daniels, S.I.; Naiman, N.E.; Yarchoan, R. Activity of Human Immunodeficiency Virus Type 1 Protease Inhibitors against the Initial Autocleavage in Gag-Pol Polyprotein Processing. *Antimicrob. Agents Chemother.* **2012**, *56*, 3620–3628. [CrossRef]
123. Modh, R.P.; De Clercq, E.; Pannecouque, C.; Chikhalia, K.H.J. Design, synthesis, antimicrobial activity and anti-HIV activity evaluation of novel hybrid quinazoline–triazine derivatives. *Enzym. Inhib. Med. Chem.* **2014**, *29*, 100–108. [CrossRef] [PubMed]
124. Xu, Z.; Zhao, S.J.; Lv, Z.S.; Gao, F.; Wang, Y.; Zhang, F.; Bai, L.; Deng, J.L. Fluoroquinolone-isatin hybrids and their biological activities. *Eur. J. Med. Chem.* **2019**, *162*, 396–406. [CrossRef]
125. Banerjee, D.; Yogeeswari, P.; Bhat, P.; Thomas, A.; Srividya, M.; Sriram, D. Novel isatinyl thiosemicarbazones derivatives as potential molecule to combat HIV-TB co-infection. *Eur. J. Med. Chem.* **2011**, *46*, 106–121. [CrossRef] [PubMed]
126. Pandeya, S.N.; Sriram, D.; Nath, G.; DeClercq, E. Synthesis, antibacterial, antifungal and anti-HIV activities of Schiff and Mannich bases derived from isatin derivatives and N-[4-(49-chlorophenyl)thiazol-2-yl] thiosemicarbazide. *Eur. J. Pharm. Sci.* **1999**, *9*, 25–31. [CrossRef]
127. Bal, T.R.; Anand, B.; Yogeeswari, P.; Sriram, D. Synthesis and evaluation of anti-HIV activity of isatin b-thiosemicarbazone derivatives. *Bioorg. Med. Chem. Lett.* **2005**, *15*, 4451–4455. [CrossRef] [PubMed]
128. Shen, C.-H.; Wang, Y.-F.; Kovalevsky, A.Y.; Harrison, R.W.; Weber, I.T. Amprenavir complexes with HIV-1 protease and its drug-resistant mutants altering hydrophobic clusters. *FEBS J.* **2010**, *277*, 3699–3714. [CrossRef] [PubMed]
129. Weber, I.T.; Waltman, M.J.; Mustyakimov, M.; Blakeley, M.P.; Keen, D.A.; Ghosh, A.K.; Langan, P.; Kovalevsky, A.Y. Joint X-ray/Neutron Crystallographic Study of HIV-1 Protease with Clinical Inhibitor Amprenavir: Insights for Drug Design. *J. Med. Chem.* **2013**, *56*, 5631–5635. [CrossRef] [PubMed]
130. Gadakh, S.K.; Santhosh Reddy, R.; Sudalai, A. Enantioselective synthesis of HIV protease inhibitor amprenavir via Co-catalyzed HKR of 2-(1-azido-2-phenylethyl)oxirane. *Tetrahedron Asymmetry* **2012**, *23*, 898–903. [CrossRef]
131. Fan, L.L.; Liu, W.Q.; Xu, H.; Yang, L.M.; Lv, M.; Zheng, Y.T. Anti Human Immunodeficiency Virus-1 (HIV-1) Agents 3. Synthesis and in Vitro Anti-HIV-1 Activity of Some N-Arylsulfonylindoles. *Chem. Pharm. Bull.* **2009**, *57*, 797–800. [CrossRef] [PubMed]
132. Ali, A.; Reddy, G.S.K.K.; Cao, H.; Anjum, S.G.; Nalam, M.N.L.; Schiffer, C.A.; Rana, T.M. Discovery of HIV-1 Protease Inhibitors with Picomolar Affinities Incorporating N-Aryl-oxazolidinone-5-carboxamides as Novel P2 Ligands. *J. Med. Chem.* **2006**, *49*, 7342–7356. [CrossRef] [PubMed]
133. Yan, J.; Huang, N.; Li, S.; Yang, L.M.; Xing, W.; Zheng, Y.T.; Hu, Y. Synthesis and biological evaluation of novel amprenavir-based P1-substituted bi-aryl derivatives as ultra-potent HIV-1 protease inhibitors. *Bioorg. Med. Chem. Lett.* **2012**, *22*, 1976–1979. [CrossRef]

134. Tie, Y.; Boross, P.I.; Wang, Y.F.; Gaddis, L.; Hussain, A.K.; Leshchenko, S.; Ghosh, A.K.; Louis, J.M.; Harrison, R.W.; Weber, I.T.J. High Resolution Crystal Structures of HIV-1 Protease with a Potent Non-peptide Inhibitor (UIC-94017) Active Against Multi-drug-resistant Clinical Strains. *Mol. Biol.* **2004**, *338*, 341–352. [CrossRef]

135. Shahabadi, N.; Abbasi, A.R.; Moshtkob, A.; Shiri, F.J. DNA-binding studies of a new Cu(II) complex containing reverse transcriptase inhibitor and anti-HIV drug zalcitabine. *Coord. Chem.* **2019**, *72*, 1957–1972. [CrossRef]

136. Brower, E.T.; Bacha, U.M.; Kawasaki, Y.; Freire, E. Title of article. *Chem. Biol. Drug Des.* **2008**, *71*, 298–305. [CrossRef]

137. Gulick, R.M.; Lalezari, J.; Goodrich, J.; Clumeck, N.; DeJesus, E.; Horban, A.; Nadler, J.; Clotet, B.; Karlsson, A.; Wohlfeiler, M.; et al. Maraviroc for Previously Treated Patients with R5 HIV-1 Infection. *N. Engl. J. Med.* **2008**, *359*, 1429–1441. [CrossRef] [PubMed]

138. De Clercq, E. Where rilpivirine meets with tenofovir, the start of a new anti-HIV drug combination era. *Biochem. Pharmacol.* **2012**, *84*, 241–248. [CrossRef] [PubMed]

139. Orkin, C.; Llibre, J.M.; Gallien, S.; Antinori, A.; Behrens, G.M.N.; Carr, A. Nucleoside reverse transcriptase inhibitor-reducing strategies in HIV treatment: Assessing the evidence. *HIV Med.* **2018**, *19*, 18–32. [CrossRef] [PubMed]

140. Clement, M.E.; Kofron, R.; Landovitz, R.J. Long-acting injectable cabotegravir for the prevention of HIV infection. *Curr. Opin. HIV AIDS* **2020**, *15*, 19. [CrossRef] [PubMed]

141. Molina, J.-M.; Segal-Maurer, S.; Stellbrink, H.J.; Castagna, A.; Berhe, M.; Richmond, G.J.; Ruane, P.J.; Sinclair, G.I.; Siripassorn, K.; Wang, H.J.; et al. Efficacy and Safety of Long-Acting Subcutaneous Lenacapavir in Phase 2/3 in Heavily Treatment-Experienced People with HIV: Week 26 Results (Capella Study). Available online: https://theprogramme.ias2021.org/Abstract/Abstract/2605 (accessed on 29 November 2021).

142. Curreli, F.; Choudhury, S.; Pyatkin, I.; Zagorodnikov, V.P.; Bulay, A.K.; Altieri, A.; Kwon, Y.D.; Kwong, P.D.; Debnath, A.K.J. Design, Synthesis, and Antiviral Activity of Entry Inhibitors That Target the CD4-Binding Site of HIV-1. *Med. Chem.* **2012**, *55*, 4764–4775. [CrossRef] [PubMed]

143. Lobatón, E.; Rodríguez-Barrios, F.; Gago, F.; Pérez-Pérez, M.-J.; De Clercq, E.; Balzarini, J.; Camarasa, M.-J.; Velázquez, S. Synthesis of 3″-Substituted TSAO Derivatives with Anti-HIV-1 and Anti-HIV-2 Activity through an Efficient Palladium-Catalyzed Cross-Coupling Approach. *J. Med. Chem.* **2002**, *45*, 3934–3945. [CrossRef]

144. Kageyama, M.; Nagasawa, T.; Yoshida, M.; Ohrui, H.; Kuwahara, S. Enantioselective Total Synthesis of the Potent Anti-HIV Nucleoside EFdA. *Org. Lett.* **2011**, *13*, 5264–5266. [CrossRef]

145. Wu, T.; Froeyen, M.; Kempeneers, V.; Pannecouque, C.; Wang, J.; Busson, R.; De Clercq, E.; Herdewijn, P.J. Deoxythreosyl Phosphonate Nucleosides as Selective Anti-HIV Agents. *Am. Chem. Soc.* **2005**, *127*, 5056–5065. [CrossRef]

146. Ali, A.; Ghosh, A.; Nathans, R.S.; Sharova, N.; O'Brien, S.; Cao, H.; Stevenson, M.; Rana, T.M. Identification of Flavopiridol Analogues that Selectively Inhibit Positive Transcription Elongation Factor (P-TEFb) and Block HIV-1 Replication. *ChemBioChem* **2009**, *10*, 2072–2080. [CrossRef]

147. Suzuki, Y.; Ikeda, K.; Suzuki, F.; Toyota, J.; Karino, Y.; Chayama, K.; Kawakami, Y.; Ishikawa, H.; Watanabe, H.; Hu, W.; et al. Dual oral therapy with daclatasvir and asunaprevir for patients with HCV genotype 1b infection and limited treatment options. *J. Hepatol.* **2013**, *58*, 655–662. [CrossRef]

148. Soumana, D.I.; Ali, A.; Schiffer, C.A. Structural Analysis of Asunaprevir Resistance in HCV NS3/4A Protease. *ACS Chem. Biol.* **2014**, *9*, 2485–2490. [CrossRef]

149. Scola, P.M.; Sun, L.Q.; Wang, A.X.; Chen, J.; Sin, N.; Venables, B.L.; Sit, S.Y.; Chen, Y.; Cocuzza, A.; Bilder, D.M.; et al. The Discovery of Asunaprevir (BMS-650032), An Orally Efficacious NS3 Protease Inhibitor for the Treatment of Hepatitis C Virus Infection. *J. Med. Chem.* **2014**, *57*, 1730–1752. [CrossRef]

150. Lok, A.S.; Gardiner, D.F.; Lawitz, E.; Martorell, C.; Everson, G.T.; Ghalib, R.; Reindollar, R.; Rustgi, V.; McPhee, F.; Wind-Rotolo, M. Preliminary Study of Two Antiviral Agents for Hepatitis C Genotype 1. *N. Engl. J. Med.* **2012**, *366*, 216–224. [CrossRef]

151. Kumar, D.V.; Rai, R.; Brameld, K.A.; Riggs, J.; Somoza, J.R.; Rajagopalan, R.; Janc, J.W.; Xia, Y.M.; Ton, T.L.; Hu, H.; et al. 3-Heterocyclyl quinolone inhibitors of the HCV NS5B polymerase. *Bioorg. Med. Chem. Lett.* **2012**, *22*, 300–304. [CrossRef]

152. McKercher, G.; Beaulieu, P.L.; Lamarre, D.; LaPlante, S.; Lefebvre, S.; Pellerin, C.; Thauvette, L.; Kukolj, G. Specific inhibitors of HCV polymerase identified using an NS5B with lower affinity for template/primer substrate. *Nucleic Acids Res.* **2004**, *32*, 422–431. [CrossRef]

153. Kumar, D.V.; Rai, R.; Brameld, K.A.; Somoza, J.R.; Rajagopalan, R.; Janc, J.W.; Xia, Y.M.; Ton, T.L.; Shaghafi, M.B.; Hu, H. Quinolones as HCV NS5B polymerase inhibitors. *Bioorg. Med. Chem. Lett.* **2011**, *21*, 82–87. [CrossRef]

154. Han, J.; Lee, M.K.; Jang, Y.; Cho, W.J.; Kim, M. Repurposing of cyclophilin A inhibitors as broad-spectrum antiviral agents. *Drug Discov. Today.* **2022**, *27*, 1895–1912. [CrossRef]

155. El Kassas, M.; Elbaz, T.; Abd El Latif, Y.; Esmat, G. Elbasvir and grazoprevir for chronic hepatitis C genotypes 1 and 4. *Expert Rev. Clin. Pharmacol.* **2016**, *9*, 1413–1421. [CrossRef]

156. Sofia, M.J.; Bao, D.; Chang, W.; Du, J.; Nagarathnam, D.; Rachakonda, S.; Reddy, P.G.; Ross, B.S.; Wang, P.; Zhang, H.R.; et al. Discovery of a β-D-20 -Deoxy-20 -r-fluoro-20 -β-C-methyluridine Nucleotide Prodrug (PSI-7977) for the Treatment of Hepatitis C Virus. *J. Med. Chem.* **2010**, *53*, 7202–7218. [CrossRef]

157. Link, J.O.; Taylor, J.G.; Xu, L.; Mitchell, M.; Guo, H.; Liu, H.; Kato, D.; Kirschberg, T.; Sun, J.; Squires, N.; et al. Discovery of Ledipasvir (GS-5885): A Potent, Once-Daily Oral NS5A Inhibitor for the Treatment of Hepatitis C Virus Infection. *J. Med. Chem.* **2014**, *57*, 2033–2046. [CrossRef]

158. Znabet, A.; Polak, M.M.; Janssen, E.; De Kanter, F.J.J.; Turner, N.J.; Orru, R.V.A.; Ruijter, E. A highly efficient synthesis of telaprevir by strategic use of biocatalysis and multicomponent reactions. *Chem. Commun.* **2010**, *46*, 7918–7920. [CrossRef]
159. Xu, F.; Kim, J.; Waldman, J.; Wang, T.; Devine, P. Synthesis of Grazoprevir, a Potent NS3/4a Protease Inhibitor for the Treatment of Hepatitis C Virus. *Org. Lett.* **2018**, *20*, 7261–7265. [CrossRef]
160. Rusere, L.N.; Matthew, A.N.; Lockbaum, G.J.; Jahangir, M.; Newton, A.; Petropoulos, C.J.; Huang, W.; Kurt Yilmaz, N.; Schiffer, C.A.; Ali, A. Quinoxaline-Based Linear HCV NS3/4A Protease Inhibitors Exhibit Potent Activity Against Drug Resistant Variants. *ACS Med. Chem. Lett.* **2018**, *9*, 691–696. [CrossRef] [PubMed]
161. Matthew, A.N.; Zephyr, J.; Hill, C.J.; Jahangir, M.; Newton, A.; Petropoulos, C.J.; Huang, W.; Kurt-Yilmaz, N.; Schiffer, C.A.; Ali, A. Hepatitis C Virus NS3/4A Protease Inhibitors Incorporating Flexible P2 Quinoxalines Target Drug Resistant Viral Variants. *J. Med. Chem.* **2017**, *60*, 5699–5716. [CrossRef] [PubMed]
162. Roth, D.; Nelson, D.R.; Bruchfeld, A.; Liapakis, A.; Silva, M.; Monsour, H.; Martin, P.; Pol, S.; Londoño, M.C.; Hassanein, T. Grazoprevir plus elbasvir in treatment-naive and treatment-experienced patients with hepatitis C virus genotype 1 infection and stage 4–5 chronic kidney disease (the C-SURFER study): A combination phase 3 study. *Lancet* **2015**, *386*, 1537–1545. [CrossRef] [PubMed]
163. De Clercq, E. Current race in the development of DAAs (direct-acting antivirals) against HCV. *Biochem. Pharmacol.* **2014**, *89*, 441–452. [CrossRef]
164. DeCarolis, D.D.; Chen, Y.C.; Westanmo, A.D.; Conley, C.; Gravely, A.A.; Khan, F.B. Decreased warfarin sensitivity among patients treated with elbasvir and grazoprevir for hepatitis C infection. *Am. J. Health Pharm.* **2019**, *76*, 1273–1280. [CrossRef]
165. Keating, G.M.; Vaidya, A. Sofosbuvir: First Global Approval. *Drugs* **2014**, *74*, 273–282. [CrossRef]
166. Lee, C. Daclatasvir: Potential role in hepatitis C. *Drug Des. Dev. Ther.* **2013**, *7*, 1223–1233. [CrossRef]
167. Shannon, A.; Fattorini, V.; Sama, B.; Selisko, B.; Feracci, M.; Falcou, C.; Gauffre, P.; El Kazzi, P.; Delpal, A.; Decroly, E.; et al. A dual mechanism of action of AT-527 against SARS-CoV-2 polymerase. *Nat. Commun.* **2022**, *13*, 621–630. [CrossRef]
168. Abdel-Magid, A.F. Fatty Acid Synthase Inhibitors as Possible Treatment for Cancer. *ACS Med. Chem. Lett.* **2012**, *3*, 612–613. [CrossRef] [PubMed]
169. Oslob, J.D.; Johnson, R.J.; Cai, H.; Feng, S.Q.; Hu, L.; Kosaka, Y.; Lai, J.; Sivaraja, M.; Tep, S.; Yang, H.; et al. Imidazopyridine-Based Fatty Acid Synthase Inhibitors That Show Anti-HCV Activity and in Vivo Target Modulation. *ACS Med. Chem. Lett.* **2013**, *4*, 113–117. [CrossRef] [PubMed]
170. Lohmann, V.; Korner, F.; Koch, J.; Herian, U.; Theilmann, L.; Bartenschlager, R. Replication of Subgenomic Hepatitis C Virus RNAs in a Hepatoma Cell Line. *Science* **1999**, *285*, 110–113. [CrossRef] [PubMed]
171. Wang, N.Y.; Xu, Y.; Zuo, W.Q.; Xiao, K.J.; Liu, L.; Zeng, X.X.; You, X.Y.; Zhang, L.D.; Gao, C.; Liu, Z.H.; et al. Discovery of imidazo[2,1-b]thiazole HCV NS4B inhibitors exhibiting synergistic effect with other direct-acting antiviral agents. *J. Med. Chem.* **2015**, *58*, 2764–2778. [CrossRef] [PubMed]
172. Matthews, G.V.; Seaberg, E.; Dore, G.J.; Bowden, S.; Lewin, S.R.; Sasadeusz, J.; Marks, P.; Goodman, Z.; Philp, F.H.; Tang, Y. Combination HBV therapy is linked to greater HBV DNA suppression in a cohort of lamivudineexperienced HIV/HBV coinfected individuals. *AIDS* **2009**, *23*, 1707–1715. [CrossRef] [PubMed]
173. Li, J.; Lv, F. Process for Stereoselective Synthesis of Lamvudine. U.S. Patent No.: US 8,304,540 B2, 18 May 2007.
174. Caso, M.F.; Dalonzo, D.; Derrico, S.; Palumbo, G.; Guaragna, A. Highly Stereoselective Synthesis of Lamivudine (3TC) and Emtricitabine (FTC) by a Novel N-Glycosidation Procedure. *Org. Lett.* **2015**, *17*, 2626–2629. [CrossRef]
175. Lok, A.S.F.; McMahon, B.J. Chronic Hepatitis B: Update 2009. *Hepatology* **2009**, *50*, 661–662. [CrossRef]
176. Yasutake, Y.; Hattori, S.; Hayashi, H.; Matsuda, K.; Tamura, N.; Kohgo, S.; Maeda, K.; Mitsuya, H. HIV-1 with HBV-associated Q151M substitution in RT becomes highly susceptible to entecavir: Structural insights into HBV-RT inhibition by entecavir. *Sci. Rep.* **2018**, *8*, 1624–1636. [CrossRef]
177. Parvez, M.K.; Al-Dosari, M.S.; Abdelwahid, M.A.S.; Alqahtani, A.S.; Alanzi, A.R. Novel anti-hepatitis B virus-active catechin and epicatechin from Rhus tripartita. *Exp. Ther. Med.* **2022**, *23*, 398. [CrossRef]
178. Velasco, J.; Ariza, X.; Badía, L.; Bartra, M.; Berenguer, R.; Farràs, J.; Gallardo, J.; Garcia, J.; Gasanz, Y.J. Total Synthesis of Entecavir. *Org. Chem.* **2013**, *78*, 5482–5491. [CrossRef]
179. Li, J.J. *Innovative Drug Synthesis*; John Wiley & Sons, Inc.: Hoboken, NJ, USA, 2015; pp. 1–14. [CrossRef]
180. Shaw, T.; Locarnini, S. Entecavir for the treatment of chronic hepatitis B. *Expert Rev. Anti. Infect. Ther.* **2004**, *2*, 853–871. [CrossRef] [PubMed]
181. Matthews, S.J. Entecavir for the Treatment of Chronic Hepatitis B Virus Infection. *Clin. Ther.* **2006**, *28*, 184–203. [CrossRef] [PubMed]
182. Dai, G.F.; Xu, H.W.; Wang, J.F.; Liu, F.W.; Liu, H.M. Studies on the novel a-glucosidase inhibitory activity and structure–activity relationships for andrographolide analogues. *Bioorg. Med. Chem. Lett.* **2006**, *16*, 2710–2713. [CrossRef] [PubMed]
183. Wu, Y.H. Naturally derived anti-hepatitis B virus agents and their mechanism of action. *World J. Gastroenterol.* **2016**, *22*, 188–204. [CrossRef] [PubMed]
184. Sadiea, R.Z.; Sultana, S.; Chaki, B.M.; Islam, T.; Dash, S.; Akter, S.; Islam, M.S.; Kazi, T.; Nagata, A.; Spagnuolo, R. Phytomedicines to Target Hepatitis B Virus DNA Replication: Current Limitations and Future Approaches. *Int. J. Mol. Sci.* **2022**, *23*, 1617. [CrossRef]

185. Poordad, F.; Sievert, W.; Mollison, L.; Bennett, M.; Tse, E.; Bräu, N.; Levin, J.; Sepe, T.; Lee, S.S.; Angus, P.; et al. Fixed-Dose Combination Therapy With Daclatasvir, Asunaprevir, and Beclabuvir for Noncirrhotic Patients With HCV Genotype 1 Infection. *JAMA* **2015**, *313*, 1728–1735. [CrossRef]
186. Lai, C.L.; Gane, E.; Liaw, Y.F.; Hsu, C.W.; Thongsawat, S.; Wang, Y.; Chen, Y.; Heathcote, E.J.; Rasenack, J.; Bzowej, N.; et al. Telbivudine versus Lamivudine in Patients with Chronic Hepatitis B. *N. Engl. J. Med.* **2007**, *357*, 2576–2588. [CrossRef]
187. Lai, C.L.; Shouval, D.; Lok, A.S.; Chang, T.T.; Cheinquer, H.; Goodman, Z.; DeHertogh, D.; Wilber, R.; Zink, R.C.; Cross, A.; et al. Entecavir versus Lamivudine for Patients with HBeAg-Negative Chronic Hepatitis B. *N. Engl. J. Med.* **2006**, *354*, 1011–1020. [CrossRef]
188. Testoni, B.; Durantel, D.; Zoulim, F. Novel targets for hepatitis B virus therapy. *Liver Int.* **2017**, *37*, 33–39. [CrossRef]
189. Li, X.; Zhang, Z.; Chen, Y.; Wang, B.; Yang, G.; Xu, X.; Yechao, B.; Bai, D.; Feng, B.; Mao, Y.; et al. Discovery of SHR5133, a Highly Potent and Novel HBV Capsid Assembly Modulator. *ACS Med. Chem. Lett.* **2022**, *13*, 507–512. [CrossRef]
190. Yeo, H.; Li, Y.; Fu, L.; Zhu, J.L.; Gullen, E.A.; Dutschman, G.E.; Lee, Y.; Chung, R.; Huang, E.-S.; Austin, D.J.; et al. Synthesis and Antiviral Activity of Helioxanthin Analogues. *J. Med. Chem.* **2005**, *48*, 534–546. [CrossRef] [PubMed]
191. Crotty, S.; Maag, D.; Arnold, J.J.; Zhong, W.; Lau, J.Y.N.; Hong, Z.; Andino, R.; Cameron, C.E. The broad-spectrum antiviral ribonucleoside ribavirin is an RNA virus mutagen. *Nat. Med.* **2000**, *6*, 1375–1379. [CrossRef] [PubMed]
192. Kim, Y.-I.; Pareek, R.; Murphy, R.; Harrison, L.; Farrell, E.; Cook, R.; DeVincenzo, J. The antiviral effects of RSV fusion inhibitor, MDT-637, on clinical isolates, versus its achievable concentrations in the human respiratory tract and comparison to ribavirin. *Influenza Other Respi. Viruses* **2017**, *11*, 525–530. [CrossRef]
193. Sakharov, V.; Baykov, S.; Konstantinova, I.; Esipov, R.; Dorogov, M. An Efficient Chemoenzymatic Process for Preparation of Ribavirin. *Int. J. Chem. Eng.* **2015**, *2015*, 734851. [CrossRef]
194. Feld, J.J.; Hoofnagle, J.H. Mechanism of action of interferon and ribavirin in treatment of hepatitis C. *Nature* **2005**, *436*, 967–972. [CrossRef]
195. Yang, Y.; Rijnbrand, R.; McKnight, K.L.; Wimmer, E.; Paul, A.; Martin, A.; Lemon, S.M. Sequence Requirements for Viral RNA Replication and VPg Uridylylation Directed by the Internal cis-Acting Replication Element (cre) of Human Rhinovirus Type 14. *J. Virol.* **2002**, *76*, 7485–7494. [CrossRef] [PubMed]
196. Nyström, K.; Waldenström, J.; Tang, K.W.; Lagging, M. Ribavirin: Pharmacology, Multiple Modes of Action and Possible Future Perspectives. Sequence Requirements for Viral RNA Replication and VPg Uridylylation Directed by the Internal cis-Acting Replication Element (cre) of Human Rhinovirus Type 14. *Future Virol.* **2019**, *14*, 153–160. [CrossRef]
197. Cameron, C.E.; Castro, C. The mechanism of action of ribavirin: Lethal mutagenesis of RNA virus genomes mediated by the viral RNA-dependent RNA polymerase. *Curr. Opin. Infect. Dis.* **2001**, *14*, 757–764. [CrossRef]
198. Thomas, E.; Ghany, M.G.; Liang, T.J. The application and mechanism of action of ribavirin in therapy of hepatitis C. *Antivir. Chem. Chemother.* **2012**, *23*, 1–12. [CrossRef]
199. Robinson, R.F.; Nahata, M.C. Respiratory syncytial virus (RSV) immune globulin and palivizumab for prevention of RSV infection. *Am. J. Health Pharm.* **2000**, *57*, 259–264. [CrossRef]
200. Wright, M.; Piedimonte, G. Respiratory Syncytial Virus Prevention and Therapy: Past, Present, and Future. *Pediatr. Pulmonol.* **2011**, *46*, 324–347. [CrossRef] [PubMed]
201. Cockerill, G.S.; Angell, R.M.; Bedernjak, A.; Chuckowree, I.; Fraser, I.; Gascon-Simorte, J.; Gilman, M.S.A.; Good, J.A.D.; Harland, R.; Johnson, S.M.; et al. Discovery of Sisunatovir (RV521), an Inhibitor of Respiratory Syncytial Virus Fusion. *J. Med. Chem.* **2021**, *64*, 3658–3676. [CrossRef] [PubMed]
202. Yoon, J.J.; Chawla, D.; Paal, T.; Ndungu, M.; Du, Y.H.; Kurtkaya, S.; Sun, A.M.; Snyder, J.P.; Plemper, R.K.J. High-Throughput Screening–Based Identification of Paramyxovirus Inhibitors. *Biomol. Screen.* **2008**, *13*, 591–608. [CrossRef]
203. Yust, I.; Fox, Z.; Burke, M.; Johnson, A.; Turner, D.; Mocroft, A.; Katlama, C.; Ledergerber, B.; Reiss, P.; Kirk, O. Retinal and extraocular cytomegalovirus end-organ disease in HIV-infected patients in Europe: A EuroSIDA study, 1994–2001. *Eur. J. Clin. Microbiol. Infect. Dis.* **2004**, *23*, 550–559. [CrossRef]
204. Babu, K.S.; Rao, M.R.; Goverdhan, G.; Srinivas, P.; Reddy, P.P.; Venkateswarlu, G.; Anand, R.V. Synthesis of Valganciclovir Hydrochloride Congeners. *Synth. Commun.* **2013**, *43*, 1751–1758. [CrossRef]
205. Wiltshire, H.; Paya, C.V.; Pescovitz, M.D.; Humar, A.; Dominguez, E.; Washburn, K.; Blumberg, E.; Alexander, B.; Freeman, R.; Heaton, N. Pharmacodynamics of Oral Ganciclovir and Valganciclovir in Solid Organ Transplant Recipients. *Transplantation* **2005**, *79*, 1477–1483. [CrossRef] [PubMed]
206. Matthews, T.; Boehme, R. Antiviral Activity and Mechanism of Action of Ganciclovir. *Rev. Infect. Dis.* **1988**, *10*, 490–494. [CrossRef]
207. Gao, H.; Mitra, A.K. Synthesis of Acyclovir, Ganciclovir and Their Prodrugs: A Review. *Synthesis* **2000**, *2000*, 329–351. [CrossRef]
208. Montana, M.; Montero, V.; Khoumeri, O.; Vanelle, P. Quinoxaline Derivatives as Antiviral Agents: A Systematic Review. *Molecules* **2020**, *25*, 2784. [CrossRef]
209. Wang, Y.; Mukhopadhyay, R.; Roy, S.; Kapoor, A.; Su, Y.-P.; Charman, S.A.; Chen, G.; Wu, J.; Wang, X.; Vennerstrom, J.L. Inhibition of Cytomegalovirus Replication with Extended-HalfLife Synthetic Ozonides. *Antimicrob. Agents Chemother.* **2019**, *63*, e01735-18. [CrossRef]
210. El-Zahab, H.S.A. Synthesis, Characterization, and Biological Evaluation of Some Novel Quinoxaline Derivatives as Antiviral Agents. *Arch. Pharm. Chem. Life Sci.* **2017**, *350*, 1700028. [CrossRef] [PubMed]

211. El-Sebaey, S.A. Recent Advances in 1,2,4-Triazole Scaffolds as Antiviral Agents. *ChemistrySelect* **2020**, *5*, 11654–11680. [CrossRef]
212. Kimberlin, D.W.; Jester, P.M.; Sánchez, P.J.; Ahmed, A.; Arav-Boger, R.; Michaels, M.G.; Ashouri, N.; Englund, J.A.; Estrada, B.; Jacobs, R.F.; et al. Valganciclovir for Symptomatic Congenital Cytomegalovirus Disease. *N. Engl. J. Med.* **2015**, *372*, 933–943. [CrossRef] [PubMed]
213. Martin, D.F.; Kuppermann, B.D.; Wolitz, R.A.; Palestine, A.G.; Li, H.; Robinson, C.A. Oral ganciclovir for patients with cytomegalovirus retinitis treated with a ganciclovir implant. *N. Engl. J. Med.* **1999**, *340*, 1063–1070. [CrossRef] [PubMed]
214. Williams, J.D.; Chen, J.J.; Drach, J.C.; Townsend, L.B. Design, Synthesis, and Antiviral Activity of Certain 3-Substituted 2,5,6-Trichloroindole Nucleosides. *J. Med. Chem.* **2004**, *47*, 5753–5765. [CrossRef] [PubMed]
215. Turk, S.R.; Shipman, C., Jr.; Nassiri, M.R.; Genzlinger, G.; Krawczyk, S.H.; Townsend, L.B.; Drach, J.C. Pyrrolo[2,3-d]Pyrimidine Nucleosides as Inhibitors of Human Cytomegalovirus. *Antimicrob. Agents Chemother.* **1987**, *31*, 544–550. [CrossRef]
216. Prichard, M.N.; Prichard, L.E.; Baguley, W.A.; Nassiri, M.R.; Shipman, C., Jr. Three-Dimensional Analysis of the Synergistic Cytotoxicity of Ganciclovir and Zidovudine. *Antivir. Res.* **1991**, *35*, 1060–1065. [CrossRef]
217. Sahu, P.K.; Umme, T.; Yu, J.; Nayak, A.; Kim, G.; Noh, M.; Lee, J.Y.; Kim, D.-D.; Jeong, L.S. Seleno-acyclovir and –ganciclovir: A Discovery of a New Template for Antiviral Agents. *J. Med. Chem.* **2015**, *58*, 8734–8738. [CrossRef]
218. Zhou, N.; Xie, T.; Liu, L.; Xie, Z.J. Cu/Mn Co-oxidized Cyclization for the Synthesis of Highly Substituted Pyrrole Derivatives from Amino Acid Esters: A Strategy for the Biomimetic Syntheses of Lycogarubin C and Chromopyrrolic Acid. *Org. Chem.* **2014**, *79*, 6061–6068. [CrossRef]
219. Lin, Z.Q.; Li, C.D.; Zhou, Z.C.; Xue, S.; Gao, J.R.; Ye, Q.; Li, Y.J. Copper(II)-Promoted Oxidation/[3+2]Cycloaddition/Aromatization Cascade: Efficient Synthesis of Tetrasubstituted NH-Pyrrole from Chalcones and Iminodiacetates. *Synlett* **2019**, *30*, 1442–1446. [CrossRef]
220. Kaufman, H.E.; Heidelberger, C. Therapeutic Antiviral Action of 5-Trifluoromethyl-2′-deoxyuridine in Herpes Simplex Keratitis. *Science* **1964**, *145*, 585–586. [CrossRef]
221. De Clercq, E. Discovery and development of BVDU (brivudin) as a therapeutic for the treatment of herpes zoster. *Biochem. Pharmacol.* **2004**, *68*, 2301–2315. [CrossRef]
222. McClain, L.; Zhi, Y.; Cheng, H.; Ghosh, A.; Piazza, P.; Yee, M.B.; Kumar, S.; Milosevic, J.; Bloom, D.C.; Arav-Boger, R. Broad-spectrum non-nucleoside inhibitors of human herpesviruses. *Antivir. Res.* **2015**, *121*, 16–20. [CrossRef]
223. Taylor, J.L.; Punda-Polic, V.; O'brien, W.J. Combined anti-herpes virus activity of nucleoside analogs and interferon. *Curr. Eye Res.* **1991**, *10*, 205–211. [CrossRef]
224. Warren, T.K.; Jordan, R.; Lo, M.K.; Ray, A.S.; Mackman, R.L.; Soloveva, V.; Siegel, D.; Perron, M.; Bannister, R.; Hui, H.C.; et al. Therapeutic efficacy of the small molecule GS-5734 against Ebola virus in rhesus monkeys. *Nature* **2016**, *531*, 381–385. [CrossRef] [PubMed]
225. Pardo, J.; Shukla, A.M.; Chamarthi, G.; Gupte, A. The journey of remdesivir: From Ebola to COVID-19. *Drugs Context* **2020**, *9*, 1–9. [CrossRef] [PubMed]
226. Chun, B.K.; Clarke, M.O.H.; Doerffler, E.; Hui, H.C.; Jordan, R.; Mackman, R.L.; Parrish, J.P.; Ray, A.S.; Siegel, D.; Gilead Sciences, Inc. Methods for Treating Filoviridae Virus Infections. Patent WO 2016069826 A1, 5 July 2016.
227. Taylor, R.; Kotian, P.; Warren, T.; Panchal, R.; Bavari, S.; Julander, J.; Dobo, S.; Rose, A.; El-Kattan, Y.; Taubenheim, B.; et al. BCX4430 – A broad-spectrum antiviral adenosine nucleoside analog under development for the treatment of Ebola virus disease. *Infect. Public Health* **2016**, *9*, 220–226. [CrossRef]
228. De Clercq, E. Ebola virus (EBOV) infection: Therapeutic strategies. *Biochem. Pharmacol.* **2015**, *93*, 1–10. [CrossRef]
229. Picazo, E.; Giordanetto, F. Small molecule inhibitors of ebola virus infection. *Drug Discov. Today* **2015**, *20*, 277–286. [CrossRef] [PubMed]
230. Shannon, A.; Canard, B. Development of a robust and convenient dual-reporter high-throughput screening assay for SARS-CoV-2 antiviral drug discovery. *Antivir. Res.* **2023**, *210*, 105501–105516. [CrossRef]
231. Scott, J.T.; Sharma, R.; Meredith, L.W.; Dunning, J.; Moore, C.E.; Sahr, F.; Ward, S.; Goodfellow, I.; Horby, P. Pharmacokinetics of TKM-130803 in Sierra Leonean patients with Ebola virus disease: Plasma concentrations exceed target levels, with drug accumulation in the most severe patients. *EBioMedicine* **2020**, *52*, 102601–102611. [CrossRef] [PubMed]
232. Verdonck, S.; Pu, S.Y.; Sorrell, F.J.; Elkins, J.M.; Froeyen, M.; Gao, L.J.; Prugar, L.I.; Dorosky, D.E.; Brannan, J.M.; Barouch-Bentov, R.; et al. Synthesis and Structure−Activity Relationships of 3,5-Disubstitutedpyrrolo[2,3-b]pyridines as Inhibitors of Adaptor-Associated Kinase 1 with Antiviral Activity. *J. Med. Chem.* **2019**, *62*, 5810–5831. [CrossRef] [PubMed]
233. Janeba, Z. Development of Small-Molecule Antivirals for Ebola. *Med. Res. Rev.* **2015**, *35*, 1175–1194. [CrossRef] [PubMed]
234. Zhou, P.; Yang, X.L.; Wang, X.G.; Hu, B.; Zhang, L.; Zhang, W.; Si, H.R.; Zhu, Y.; Li, B.; Huang, C.L.; et al. A pneumonia outbreak associated with a new coronavirus of probable bat origin. *Nature* **2020**, *579*, 270–273. [CrossRef]
235. Mascellino, M.T.; Di Timoteo, F.; De Angelis, M.; Oliva, A. Overview of the Main Anti-SARS-CoV-2 Vaccines: Mechanism of Action, Efficacy and Safety. *Infect. Drug Resist.* **2021**, *14*, 3459–3476. [CrossRef]
236. Quiros-Roldan, E.; Amadasi, S.; Zanella, I.; Degli Antoni, M.; Storti, S.; Tiecco, G.; Castelli, F. Monoclonal Antibodies against SARS-CoV-2: Current Scenario and Future Perspectives. *Pharmaceuticals* **2021**, *14*, 1272. [CrossRef]
237. Wang, X.; Sacramento, C.Q.; Jockusch, S.; Chaves, O.A.; Tao, C.; Fintelman-Rodrigues, N.; Chien, M.; Temerozo, J.R.; Li, X.; Kumar, S.; et al. Combination of antiviral drugs inhibits SARS-CoV-2 polymerase and exonuclease and demonstrates COVID-19 therapeutic potential in viral cell culture. *Commun. Biol.* **2022**, *5*, 154–168. [CrossRef]

238. Dube, T.; Ghosh, A.; Mishra, J.; Kompella, U.B.; Panda, J.J. Repurposed Drugs, Molecular Vaccines, Immune-Modulators, and Nanotherapeutics to Treat and Prevent COVID-19 Associated with SARS-CoV-2, a Deadly Nanovector. *Adv. Ther.* **2021**, *4*, 2000172–2000202. [CrossRef]
239. Marzolini, C.; Kuritzkes, D.R.; Marra, F.; Boyle, A.; Gibbons, S.; Flexner, C.; Pozniak, A.; Boffito, M.; Waters, L.; Burger, D. Recommendations for the Management of Drug–Drug Interactions Between the COVID-19 Antiviral Nirmatrelvir/Ritonavir (Paxlovid) and Comedications. *Clin. Pharmacol. Ther.* **2022**, *112*, 1191–1200. [CrossRef]
240. Rubin, R. Baricitinib Is First Approved COVID-19 Immunomodulatory Treatment. *JAMA* **2022**, *327*, 2281. [CrossRef]
241. Amici, C.; Di Caro, A.; Ciucci, A.; Chiappa, L.; Castilletti, C.; Martella, V.; Decaro, N.; Buonavoglia, C.; Capobianchi, M.R.; Santoro, M.G. Indomethacin has a potent antiviral activity against SARS coronavirus. *Antivir. Ther.* **2006**, *11*, 1021–1030. [CrossRef] [PubMed]
242. Tegeder, I.; Pfeilschifter, J.; Geisslinger, G. Cyclooxygenase-independent actions of cyclooxygenase inhibitors. *FASEB J.* **2001**, *15*, 2057–2072. [CrossRef] [PubMed]
243. Wood, J.M.; Evans, G.B.; Grove, T.L.; Almo, S.C.; Cameron, S.A.; Furneaux, R.H.; Harris, L.D.J. Chemical Synthesis of the Antiviral Nucleotide Analogue ddhCTP. *Org. Chem.* **2021**, *86*, 8843–8850. [CrossRef] [PubMed]
244. Stachulski, A.V.; Taujanskas, J.; Pate, S.L.; Rajoli, R.K.R.; Aljayyoussi, G.; Pennington, S.H.; Ward, S.A.; Hong, W.D.; Biagini, G.A.; Owen, A.; et al. Therapeutic Potential of Nitazoxanide: An Appropriate Choice for Repurposing versus SARS-CoV-2? *ACS Infect. Dis.* **2021**, *7*, 1317–1331. [CrossRef]
245. Gizzi, A.S.; Grove, T.L.; Arnold, J.J.; Jose, J.; Jangra, R.K.; Garforth, S.J.; Du, Q.; Cahill, S.M.; Dulyaninova, N.G.; Love, J.D.; et al. A naturally occurring antiviral ribonucleotide encoded by the human genome. *Nature* **2018**, *558*, 610–614. [CrossRef]
246. Jung, E.; Soto-Acosta, R.; Xie, J.; Wilson, D.J.; Dreis, C.D.; Majima, R.; Edwards, T.C.; Geraghty, R.J.; Chen, L. Bisubstrate Inhibitors of Severe Acute Respiratory Syndrome Coronavirus-2 Nsp14 Methyltransferase. *ACS Med. Chem. Lett.* **2022**, *13*, 1477–1484. [CrossRef]
247. Braaten, K.P.; Laufer, M.R. Human Papillomavirus (HPV), HPV-Related Disease, and the HPV Vaccine. *Rev. Obstet. Gynecol.* **2008**, *1*, 2–10. [PubMed]
248. Rivera, A.; Tyring, S.K. Therapy of cutaneous human Papillomavirus infections. *Dermatol. Ther.* **2004**, *17*, 441–448. [CrossRef]
249. Beadle, J.R.; Valiaeva, N.; Yang, G.; Yu, J.H.; Broker, T.R.; Aldern, K.A.; Harden, E.A.; Keith, K.A.; Prichard, M.N.; Hartman, T.; et al. Synthesis and Antiviral Evaluation of Octadecyloxyethyl Benzyl 9-[(2- Phosphonomethoxy)ethyl]guanine (ODE-Bn-PMEG), a Potent Inhibitor of Transient HPV DNA Amplification. *J. Med. Chem.* **2016**, *59*, 10470–10478. [CrossRef]
250. Briggs, D.J. The role of vaccination in rabies prevention. *Curr. Opin. Virol.* **2012**, *2*, 309–314. [CrossRef]
251. Da Silva, S.; Oliveira Silva Martins, D.; Jardim, A.C. A Review of the Ongoing Research on Zika Virus Treatment. *Viruses* **2018**, *10*, 255. [CrossRef] [PubMed]
252. Barrows, N.J.; Campos, R.K.; Powell, S.T.; Prasanth, K.R.; Schott-Lerner, G.; Soto-Acosta, R.; Galarza-Muñoz, G.; McGrath, E.L.; Urrabaz-Garza, R.; Gao, J. A Screen of FDA-Approved Drugs for Inhibitors of Zika Virus Infection. *Cell Host Microbe* **2016**, *20*, 259–270. [CrossRef]
253. Zhou, K.; Li, C.; Shi, W.; Hu, X.; Nandakumar, K.S.; Jiang, S.; Zhang, N. Current Progress in the Development of Zika Virus Vaccines. *Vaccines* **2021**, *9*, 1004. [CrossRef] [PubMed]
254. Yao, Y.; Huo, T.; Lin, Y.L.; Nie, S.; Wu, F.; Hua, Y.; Wu, J.; Kneubehl, A.R.; Vogt, M.B.; Rico-Hesse, R.; et al. Discovery, X-ray Crystallography and Antiviral Activity of Allosteric Inhibitors of Flavivirus NS2B-NS3 Protease. *Am. Chem. Soc.* **2019**, *141*, 6832–6836. [CrossRef] [PubMed]
255. Tricarico, P.M.; Caracciolo, I.; Crovella, S.; D'Agaro, P. Zika virus induces inflammasome activation in the glial cell line U87-MG. *Biochem. Biophys. Res. Commun.* **2017**, *492*, 597–602. [CrossRef] [PubMed]
256. Lahon, A.; Arya, R.P.; Kneubehl, A.R.; Vogt, M.B.; Dailey Garnes, N.J.M.; Rico-Hesse, R. Characterization of a Zika Virus Isolate from Colombia. *PLoS Negl. Trop. Dis.* **2016**, *10*, e0005019. [CrossRef]
257. Zhu, Y.; Liang, M.; Yu, J.; Zhang, B.; Zhu, G.; Huang, Y.; He, Z.; Yuan, J. Repurposing of Doramectin as a New Anti-Zika Virus Agent. *Viruses* **2023**, *15*, 1068. [CrossRef]
258. Martinez-Lopez, A.; Persaud, M.; Chavez, M.P.; Zhang, H.; Rong, L.; Liu, S.; Wang, T.T.; Sarafianos, S.G.; Diaz-Griffero, F. Glycosylated diphyllin as a broad-spectrum antiviral agent against Zika virus. *EBioMedicine* **2019**, *47*, 269–283. [CrossRef]
259. Li, P.C.; Jang, J.; Hsia, C.Y.; Groomes, P.V.; Lian, W.; de Wispelaere, M.; Pitts, J.D.; Wang, J.; Kwiatkowski, N.; Gray, N.S.; et al. Small Molecules Targeting the Flavivirus E Protein with Broad-Spectrum Activity and Antiviral Efficacy in Vivo. *ACS Infect. Dis.* **2019**, *5*, 460–472. [CrossRef]
260. Lian, W.; Jang, J.; Potisopon, S.; Li, P.C.; Rahmeh, A.; Wang, J.; Kwiatkowski, N.P.; Gray, N.S.; Yang, P.L. Discovery of Immunologically Inspired Small Molecules That Target the Viral Envelope Protein. *ACS Infect. Dis.* **2018**, *4*, 1395–1406. [CrossRef]
261. Murdin, A.D.; Barreto, L.; Plotkin, S. Inactivated poliovirus vaccine: Past and present experience. *Vaccine* **1996**, *14*, 735–746. [CrossRef]
262. Amanna, I.J.; Slifka, M.K. Current trends in West Nile virus vaccine development. *Expert Rev. Vaccines* **2014**, *13*, 589–608. [CrossRef]
263. Harder, T.; Siedler, A. Systematic Review and Meta-analysis of Chickenpox Vaccination and Risk of Herpes Zoster: A Quantitative View on the "Exogenous Boosting Hypothesis". *Clin. Infect. Dis.* **2019**, *69*, 1329–1338. [CrossRef] [PubMed]

264. Davidson, R.N.; Lynn, W.; Savage, P.; Wansbrough-Jones, M.H. Chickenpox pneumonia: Experience with antiviral treatment. *Thorax* **1988**, *43*, 627–630. [CrossRef] [PubMed]
265. Ward, P.; Small, I.; Smith, J.; Suter, P.; Dutkowski, R.J. Oseltamivir (Tamiflu®) and its potential for use in the event of an influenza pandemic. *Antimicrob. Chemother.* **2005**, *55*, 5–21. [CrossRef] [PubMed]
266. Lin, L.Z.; Fang, J.M. Total Synthesis of Anti-Influenza Agents Zanamivir and Zanaphosphor via Asymmetric Aza-Henry Reaction. *Org. Lett.* **2016**, *18*, 4400–4403. [CrossRef] [PubMed]
267. Shie, J.J.; Fang, J.M.J. Development of effective anti-influenza drugs: Congeners and conjugates – a review. *Biomed. Sci.* **2019**, *26*, 84–104. [CrossRef]
268. Rowse, M.; Qiu, S.; Tsao, J.; Yamauchi, Y.; Wang, G.; Luo, M. Reduction of Influenza Virus Envelope's Fusogenicity by Viral Fusion Inhibitors. *ACS Infect. Dis.* **2016**, *2*, 47–53. [CrossRef]
269. DeFilippis, V.; Früh, K. Host cell targets for antiviral therapy: An update. *Future Virol.* **2006**, *1*, 509–518. [CrossRef]
270. Cha, H.M.; Kim, U.I.; Ahn, S.B.; Lee, M.K.; Lee, H.; Bang, H.; Jang, Y.; Kim, S.S.; Bae, M.A.; Kim, K.; et al. Evaluation of Antiviral Activity of Gemcitabine Derivatives against Influenza Virus and Severe Acute Respiratory Syndrome Coronavirus. *ACS Infect. Dis.* **2023**, *9*, 1033–1045. [CrossRef]
271. Das, P.; Deng, X.; Zhang, L.; Roth, M.G.; Fontoura, B.M.A.; Phillips, M.A.; De Brabander, J.K. SAR-Based Optimization of a 4-Quinoline Carboxylic Acid Analogue with Potent Antiviral Activity. *ACS Med. Chem. Lett.* **2013**, *4*, 517–521. [CrossRef] [PubMed]
272. Monath, T.P. SAR-Based Optimization of a 4-Quinoline Carboxylic Acid Analogue with Potent Antiviral Activity. *Lancet Infect. Dis.* **2001**, *1*, 11–20. [CrossRef] [PubMed]

MDPI
St. Alban-Anlage 66
4052 Basel
Switzerland
www.mdpi.com

Viruses Editorial Office
E-mail: viruses@mdpi.com
www.mdpi.com/journal/viruses